Novel food packaging techniques

Related titles from Woodhead's food science, technology and nutrition list:

Food preservation techniques (ISBN: 1 85573 530 X)

Extending the shelf-life of foods whilst maintaining safety and quality is a critical issue for the food industry. As a result there have been major developments in food preservation techniques which are summarised in this authoritative collection. The first part of the book examines the key issue of maintaining safety as preservation methods become more varied and complex. The rest of the book looks both at individual technologies and how they are combined to achieve the right balance of safety, quality and shelf-life for particular products.

Natural antimicrobials for the minimal processing of foods (ISBN: 1 85573 669 1)

Consumers demand products with fewer synthetic additives but increased safety and shelf-life. These demands have increased the importance of natural antimicrobials which prevent the growth of pathogenic and spoilage micro-organisms. Edited by a leading expert in the field, this important collection reviews the range of key antimicrobials such as nisin and chitosan, applications in such areas as postharvest storage of fruits and vegetables, and ways of combining antimicrobials with other preservation techniques to enhance the safety and quality of foods.

Minimal processing technologies in the food industry (ISBN: 1 85573 547 4)

The emergence of 'minimal' processing techniques, which have limited impact on a food's nutritional and sensory properties, has been a major new development in the food industry. This book provides an authoritative review of the range of minimal techniques currently available, their applications and safety and quality issues.

Selected food science and technology titles are also available in electronic form. Details of these books and a complete list of Woodhead's food science, technology and nutrition titles can be obtained by:

- visiting our web site at www.woodhead-publishing.com
- contacting Customer Services (email: sales@woodhead-publishing.com; fax: +44 (0) 1223 893694; tel.: +44 (0) 1223 891358 ext. 30; address: Woodhead Publishing Limited, Abington Hall, Abington, Cambridge CB1 6AH, England)

If you would like to receive information on forthcoming titles in this area, please send your address details to: Francis Dodds (address, tel. and fax as above; e-mail: francisd@woodhead-publishing.com). Please confirm which subject areas you are interested in.

Novel food packaging techniques

Edited by
Raija Ahvenainen

CRC Press
Boca Raton Boston New York Washington, DC

WOODHEAD PUBLISHING LIMITED
Cambridge England

Published by Woodhead Publishing Limited
Abington Hall, Abington
Cambridge CB1 6AH
England
www.woodhead-publishing.com

Published in North America by CRC Press LLC
2000 Corporate Blvd, NW
Boca Raton FL 33431
USA

First published 2003, Woodhead Publishing Limited and CRC Press LLC

Reprinted 2005

British Library Cataloguing in Publication Data
A catalogue record for this book is available from the British Library.

Library of Congress Cataloging in Publication Data
A catalog record for this book is available from the Library of Congress.

Woodhead Publishing Limited ISBN 1 85573 675 6 (book); 1 85573 702 7 (e-book)
CRC Press LLC ISBN 0-8493-1789-4

Cover design by The ColourStudio
Project managed by Macfarlane Production Services, Markyate, Hertfordshire
(e-mail: macfarl@aol.com)
Typeset by MHL Typesetting Limited, Coventry, Warwickshire
Printed and bound by Replika Press Pvt. Ltd., India

Contents

Contributor contact details xiii

1 Introduction 1
R. Ahvenainen, VTT Biotechnology, Finland

Part I Types and roles of active and intelligent packaging 3

2 Active and intelligent packaging: an introduction 5
R. Ahvenainen, VTT Biotechnology, Finland
2.1 Introduction: the role of packaging in the food chain 5
2.2 Active packaging techniques 6
2.3 Intelligent packaging techniques 11
2.4 Current use of novel packaging techniques 12
2.5 Current research 13
2.6 The legislative context 13
2.7 Consumers and novel packaging 15
2.8 Future trends 16
2.9 Sources of further information and advice 18
2.10 References 19

3 Oxygen, ethylene and other scavengers 22
L. Vermeiren, L. Heirlings, F. Devlieghere and J. Debevere, Ghent University, Belgium ....
3.1 Introduction 22
3.2 Oxygen scavenging technology 22
3.3 Selecting the right type of oxygen scavenger 25

3.4 Ethylene scavenging technology 34
3.5 Carbon dioxide and other scavengers 41
3.6 Future trends 44
3.7 References 45

4 Antimicrobial food packaging 50
J.H. Han, The University of Manitoba, Canada
4.1 Introduction 50
4.2 Antimicrobial agents 52
4.3 Constructing an antimicrobial packaging system 58
4.4 Factors affecting the effectiveness of antimicrobial packaging 60
4.5 Conclusion 65
4.6 References 65

5 Non-migratory bioactive polymers (NMBP) in food packaging .. 71
M. D. Steven and J. H. Hotchkiss, Cornell University, USA
5.1 Introduction 71
5.2 Advantages of NMBP 72
5.3 Current limitations 77
5.4 Inherently bioactive synthetic polymers: types and applications 79
5.5 Polymers with immobilised bioactive compounds 84
5.6 Applications of polymers with immobilised bioactive compounds 90
5.7 Future trends 95
5.8 References 95

6 Time-temperature indicators (TTIs) 103
P. S. Taoukis, National Technical University of Athens, Greece and T. P. Labuza, University of Minnesota, USA
6.1 Introduction 103
6.2 Defining and classifying TTIs 104
6.3 Requirements for TTIs 106
6.4 The development of TTIs 106
6.5 Current TTI systems 108
6.6 Maximising the effectiveness of TTIs 111
6.7 Using TTIs to monitor shelf-life during distribution 112
6.8 Using TTIs to optimise distribution and stock rotation 116
6.9 Future trends 121
6.10 References 122

7 The use of freshness indicators in packaging 127
M. Smolander, VTT Biotechnology, Finland
7.1 Introduction 127

7.2 Compounds indicating the quality of packaged food products 128
7.3 Freshness indicators 132
7.4 Pathogen indicators 136
7.5 Other methods for spoilage detection 137
7.6 Future trends 138
7.7 References 139

8 Packaging-flavour interactions 144
J. P. H. Linssen, R. W. G. van Willige and M. Dekker, Wageningen University, The Netherlands
8.1 Introduction 144
8.2 Factors affecting flavour absorption 145
8.3 The role of the food matrix 149
8.4 The role of differing packaging materials 153
8.5 Flavour modification and sensory quality 156
8.6 Case study: packaging and lipid oxidation 159
8.7 Modelling flavour absorption 160
8.8 Packaging–flavour interactions and active packaging 164
8.9 References 166

9 Moisture regulation 172
T. Powers and W. J. Calvo, Multisorb Technologies, USA
9.1 Introduction 172
9.2 Silica gel 173
9.3 Clay 174
9.4 Molecular sieve 176
9.5 Humectant salts 178
9.6 Irreversible adsorption 179
9.7 Planning a moisture defense 180
9.8 Future trends 185

Part II Developments in modified atmosphere packaging (MAP) . 187

10 Novel MAP applications for fresh-prepared produce 189
B. P. F. Day, Food Science Australia
10.1 Introduction 189
10.2 Novel MAP gases 191
10.3 Testing novel MAP applications 193
10.4 Applying high O_2 MAP 196
10.5 Future trends 202
10.6 References 204
10.7 Acknowledgements 207

11 MAP, product safety and nutritional quality 208
F. Devlieghere and J. Debevere, Ghent University, Belgium and M. Gil CEBAS-CSIC, Spain
11.1 Introduction 208
11.2 Carbon dioxide as an antimicrobial gas 208
11.3 The microbial safety of MAP: *Clostridium botulinum* and *Listeria monocytogenes* 210
11.4 The microbial safety of MAP: *Yersinia enterocolitica* and *Aeromonas spp* 212
11.5 The effect of MAP on the nutritional quality of non-respiring food products 215
11.6 The effect of MAP on the nutritional quality of fresh fruits and vegetables: vitamin C and carotenoids 215
11.7 The effect of MAP on the nutritional quality of fresh fruits and vegetables: phenolic compounds and glucosinolates 219
11.8 References 222

12 Reducing pathogen risks in MAP-prepared produce 231
D. O'Beirne and G. A. Francis, University of Limerick, Ireland
12.1 Introduction 231
12.2 Measuring pathogen risks 232
12.3 Factors affecting pathogen survival 242
12.4 Improving MAP to reduce pathogen risks 251
12.5 Future trends 254
12.6 Sources of further information and advice 256
12.7 References 257

13 Detecting leaks in modified atmosphere packaging 276
E. Hurme, VTT Biotechnology, Finland
13.1 Introduction 276
13.2 Leakage, product safety and quality 276
13.3 Package leak detection during processing 277
13.4 Package leak indicators during distribution 279
13.5 Future trends 282
13.6 References 283

14 Combining MAP with other preservation techniques 287
J. T. Rosnes, M. Sivertsvik and T. Skåra, NORCONSERV, Norway
14.1 Introduction 287
14.2 Combining MAP with other preservative techniques 288
14.3 Heat treatment and irradiation 293
14.4 Preservatives 297
14.5 Other techniques 299
14.6 Consumer attitudes 301

14.7 Future trends ... 302
14.8 Sources of further information and advice ... 303
14.9 References ... 304

15 Integrating MAP with new germicidal techniques ... 312
J. Lucas, University of Liverpool, UK
15.1 Introduction ... 312
15.2 Ultra violet radiation ... 315
15.3 Ozone ... 321
15.4 Integration with MAP ... 325
15.5 Future trends ... 332
15.6 References ... 336

16 Improving MAP through conceptual models ... 337
M. L. A. T. M. Hertog, Katholieke Universiteit Leuven, Belgium and N. H. Banks, Zespri Innovation Ltd, New Zealand
16.1 Introduction ... 337
16.2 Conceptual models ... 338
16.3 Mathematical models ... 346
16.4 Dedicated MAP models ... 351
16.5 Applying models to improve MAP ... 352
16.6 The risks and benefits of applying models ... 354
16.7 Future trends ... 356
16.8 Sources of further information and advice ... 356
16.9 References ... 357

Part III Novel packaging and particular products ... 363

17 Active packaging in practice: meat ... 365
C. O. Gill, Agriculture and Agri-Food Canada
17.1 Introduction ... 365
17.2 Control of product appearance ... 366
17.3 Control of flavour, texture and other characteristics ... 368
17.4 Delaying microbial spoilage ... 369
17.5 The effects of temperature on storage life ... 371
17.6 MAP technology for meat products ... 372
17.7 Controlled atmosphere packaging for meat products ... 376
17.8 Future trends in active packaging for raw meats ... 377
17.9 References ... 379

18 Active packaging in practice: fish ... 384
M. Sivertsvik, NORCONSERV, Norway
18.1 Introduction ... 384
18.2 The microbiology of fish products ... 385

18.3 Active packaging: atmosphere modifiers 387
18.4 Active packaging: water control 390
18.5 Active packaging: antimicrobial and antioxidant applications 391
18.6 Active packaging: edible coatings and films 392
18.7 Active packaging: taint removal 393
18.8 Intelligent packaging applications 394
18.9 Future trends 395
18.10 References 396

19 Active packaging and colour control: the case of meat 401
M. Jakobsen and G. Bertelsen, The Royal Veterinary and Agricultural University, Denmark
19.1 Introduction 401
19.2 Packaging and storage factors affecting colour stability 402
19.3 Modelling the impact of MAP 403
19.4 Pre- and post-slaughter factors 410
19.5 Future trends 412
19.6 References 414

20 Active packaging and colour control: the case of fruit and vegetables 416
F. Artes Calero, Technical University of Cartagena, Spain and P. A. Gomez, National Institute for Agricultural Technology, Argentina
20.1 Introduction 416
20.2 Colour changes and stability in fruit and vegetables 417
20.3 Colour measurement 418
20.4 Processes of colour change 419
20.5 Colour stability and MAP 424
20.6 Combining low oxygen, high carbon dioxide and other gases 429
20.7 Future trends 432
20.8 References 432

Part IV General issues 439

21 Optimizing packaging 441
T. Lyijynen, E. Hurme and R. Ahvenainen, VTT Biotechnology, Finland
21.1 Introduction 441
21.2 Issues in optimizing packaging 442
21.3 The VTT Precision Packaging Concept 444
21.4 Examples of food packaging optimization 449
21.5 Conclusion: improving decision-making 458

22 Legislative issues relating to active and intelligent packaging .. 459
N. de Kruijf and R. Rijk, TNO Nutrition and Food Research, The Netherlands
22.1 Introduction 459
22.2 Initiatives to amend EU legislation: European project 461
22.3 Initiatives to amend EU legislation: Nordic report 468
22.4 Current EU legislation and recommendations for change 468
22.5 Food contact materials 470
22.6 Food additives 478
22.7 Food flavouring 482
22.8 Biocides and pesticides 483
22.9 Food hygiene 485
22.10 Food labelling, weight and volume control 487
22.11 Product safety and waste 489
22.12 References 492

23 Recycling packaging materials 497
R. Franz and F. Welle, Fraunhofer Institute for Process Engineering and Packaging, Germany
23.1 Introduction 497
23.2 The recyclability of packaging plastics 498
23.3 Improving the recyclability of plastics packaging 500
23.4 Testing the safety and quality of recycled material 504
23.5 Using recycled plastics in packaging 509
23.6 Future trends 513
23.7 Sources of further information and advice 514
23.8 References 515

24 Green plastics for food packaging 519
J. J. de Vlieger, TNO Industrial Technology, The Netherlands
24.1 Introduction: the problem of plastic packaging waste 519
24.2 The range of biopolymers 520
24.3 Developing novel biodegradable materials 524
24.4 Legislative issues 526
24.5 Current applications 529
24.6 Future trends 533
24.7 References 533

25 Integrating intelligent packaging, storage and distribution 535
T. Järvi-Kääriäinen, Association of Packaging Technology and Research, Finland
25.1 Introduction: the supply chain for perishable foods 535
25.2 The role of packaging in the supply chain 538
25.3 Creating integrated packaging, storage and distribution: alarm systems and TTIs 540

25.4 Traceability: radio frequency identification 542
25.5 Future trends 545
25.6 Sources of further information and advice 547
25.7 References 548

26 Testing consumer responses to new packaging concepts 550
L. Lähteenmäki and A. Arvola, VTT Biotechnology, Finland
26.1 Introduction: new packaging techniques and the consumer .. 550
26.2 Special problems in testing responses to new packaging 551
26.3 Methods for testing consumer responses 552
26.4 Consumer attitudes towards active and intelligent packaging 555
26.5 Consumers and the future of active and intelligent packaging 559
26.6 References 562

27 MAP performance under dynamic conditions 563
M. L. A. T. M. Hertog, Katholieke Universiteit Leuven, Belgium
27.1 Introduction 563
27.2 MAP performance 564
27.3 Temperature control and risks of MAP 566
27.4 The impact of dynamic temperature conditions on MAP performance 568
27.5 Maximising MAP performance 572
27.6 Future trends 573
27.7 References 574

Index 576

Contributor contact details

Chapters 1 and 2

Dr R. Ahvenainen
Yläkartanon kuja 6A5
FIN-02360 Espoo
Finland

Tel: +358 40 840 8480
Fax: +358 9 547 4700
E-mail: raija.ahvenainen@kolumbus.fi

Chapter 3

Ir. L. Vermeiren, Ir. L. Heirlings, Dr Ir. F. Devlieghere and Professor Dr Ir. J. Debevere
Ghent University
Faculty of Agricultural and Applied Biological Sciences
Department of Food Technology and Nutrition
Laboratory of Food Microbiology and Food Preservation
Coupure Links 653
9000 Ghent
Belgium

Tel: +32 (0) 9 264 61 77
Fax: +32 (0) 9 225 55 10
E-mail: Frank.Devlieghere@ugent.be

Chapter 4

Dr J. H. Han
Department of Food Science
The University of Manitoba
Winnipeg
Manitoba R3T 2N2
Canada

Tel: +1 204 474 8368
Fax: +1 204 474 7630
E-mail: hanjh@ms.umanitoba.ca

Chapter 5

Mr M. D. Steven and Professor J. H. Hotchkiss
Department of Food Science
Cornell University
Stocking Hall
Ithaca
NY 14853
USA

Tel: +1 607 255 7912
Fax: +1 607 254 4868
E-mail: jhh3@cornell.edu
mds34@cornell.edu

Chapter 6

Dr P. S. Taoukis
School of Chemical Engineering
National Technical University of Athens
5 Iroon Polytechniou
Zografou Campus
15780 Athens
Greece

Fax: +301 7723163
E-mail: taoukis@chemeng.ntua.gr

Professor T. Labuza
Department of Food Science and Nutrition
136 ABLMS
University of Minnesota St Paul,
MN 55108
USA

Tel: +1 612 624 9701
Fax: +1 612 625 5272
E-mail: tplabuza@tc.umn.edu

Chapter 7

Dr Maria Smolander
VTT Biotechnology
P.O.Box 1500
FIN-02044 VTT
Finland

Tel: +358 9 456 5836
Fax: +358 9 455 2103
E-mail: maria.smolander@vtt.fi

Chapter 8

Dr J P H Linssen, Dr R W G van Willige and Dr M. Dekker
Wageningen University
Department of Agrotechnology and Food Sciences
Postbox 8219
6700 EV Wageningen
The Netherlands

E-mail: jozef.linssen@wur.nl

Chapter 9

Mr Thomas H. Powers and Dr William J. Calvo
Research and Development
Multisorb Technologies
325 Harlem Road
Buffalo
NY 14224
USA

E-mail: tpowers@multisorb.com
wcalvo@multisorb.com

Chapter 10

Dr B. P. F. Day
Research Section Leader – Food Packaging and Coatings
Food Science Australia
671 Sneydes Road (Private Bag 16)
Werribee
Victoria 3030
Australia

Tel: +61 (0) 3 9731 3346
Fax: +61 (0) 3 9731 3250
E-mail: brian.day@foodscience.afisc.csiro.au

Chapter 11

Dr Ir. F. Devlieghere and Professor Dr Ir. J. Debevere
Ghent University
Department of Food Technology and Nutrition
Coupure Links, 653
9000 Ghent
Belgium

Tel: +32 9 264 61 78
Fax: +32 9 225 55 10
E-mail: Frank.Devlieghere@ugent.be
Johan.Debevere@ugent.be

Dr M. I. Gil
Department of Food Science and Technology
CEBAS-CSIC
P.O. Box 4195
30080 Murcia
Spain

E-mail: migil@cebas.csic.es

Chapter 12

Professor D. O'Beirne and Dr G. A. Francis
Department of Life Sciences
University of Limerick
Limerick
Ireland

Tel: +353 61 202845
Fax: +353 61 331490
E-mail: David.OBeirne@ul.ie
Jillian.Francis@ul.ie

Chapter 13

Mr E. Hurme
Packaging Technology
VTT Biotechnology
P.O.Box 1500
FIN-02044 VTT
Finland

Tel: +358 9 456 6191
Fax: +358 9 455 2103
E-mail: eero.hurme@vtt.fi

Chapter 14

Dr J. T. Rosnes, Dr M. Sivertsvik and Dr T. Skåra
NORCONSERV
Alexander Kiellandsgt. 2
Postboks 327
4002 Stavanger
Norway

Tel: +47 51 84 46 00
Fax: +47 51 84 46 50
E-mail: jan.thomas.rosnes@norconserv.no
morten.sivertsvik@norconserv.no
torstein.skara@norconserv.no

Chapter 15

Professor J. Lucas
The University of Liverpool
Department of Electrical Engineering and Electronics
Brownlow Hill
Liverpool
L69 3GJ

Tel: +44 151 794 4533
Fax: +44 151 794 4540
E-mail: j.lucas@liv.ac.uk

Chapters 16 and 27

Dr Maarten Hertog
Laboratory of Postharvest Technology
Department of Agro-Engineering and Economics
Katholieke Universiteit Leuven
de Croylaan 42
3001 Heverlee
Belgium

Tel: +32 (0)16 322376
Fax: +32 (0)16 322955
E-mail: Maarten.Hertog@agr.kuleuven.ac.be

Dr N. H. Banks
Zespri Innovation
PO Box 4043
Mt Maunganui South
New Zealand

Fax: +64 7 575 1645
E-mail: Nigel.Banks@zespri.com

Chapter 17

Dr Colin O. Gill
Agriculture and Agri-Food Canada
Lacombe Research Centre
6000 C & E Trail
Lacombe
Alberta
T4L 1W1
Canada

Fax: +1 403 782 6120
E-mail: gillc@agr.gc.ca

Chapter 18

Dr M. Sivertsvik
NORCONSERV
Alexander Kiellandsgt. 2
Postboks 327
4002 Stavanger
Norway

Tel: +47 51 84 46 00
Fax: +47 51 84 46 50
E-mail: morten.sivertsvik@norconserv.no

Chapter 19

Dr M. Jakobsen and Associate Professor G. Bertelsen
Department of Dairy and Food Science
The Royal Veterinary and Agricultural University
Rolighedsvej 30
1958 Frederiksberg C
Denmark

Tel: +45 35283268
Tel: +45 35283212
Fax: +45 35283344
Fax: +45 35283190
E-mail: mj@kvl.dk
grb@kvl.dk

Chapter 20

Professor Dr F. Artés Calero
Department of Food Engineering
Postharvest and Refrigeration Group
Technical University of Cartagena
P° Alfonso XIII, 48
30208 Cartagena
Murcia
Spain

Tel: +34 968 32 55 10
Fax: +34 968 32 54 33
E-mail: fr.artes@upct.es

Dr P. Gómez
INTA (National Institute for Agricultural Technology
Balcarce Research Station
cc 276. 7620 Balcarce
Argentina

Tel: +54 2266 439103
Fax: +54 2266 439101
E-mail: pegomez@balcarce.inta.gov.ar

Chapter 21

Ms T. Lyijynen, Mr E. Hurme and Dr R. Ahvenainen
VTT Biotechnology
Tietotie 2
P.O. Box 1500
02044 VTT
Finland

Tel. +358 9 456 5198
Fax +358 9 455 2103
E-mail: tuija.lyijynen@vtt.fi
eero.hurme@vtt.fi

Chapter 22

Ir N. de Kruijf and Mr R. Rijk
TNO Nutrition and Food Research
Utrechtseweg 48
P.O.Box 360
3700 AJ Zeist
The Netherlands

Tel: +31 (0) 30 694 45 21
Fax: +31 (0) 30 695 48 94
E-mail: dekruijf@voeding.tno.nl

Chapter 23

Dr R. Franz and Dr F. Welle
Fraunhofer Institute for Process Engineering and Packaging
Department of Product Safety and Analysis
Giggenhauser Str. 35
D-85354 Freising
Germany

Tel. +49 (0) 8161 491-724
Fax +49 (0) 8161 491-777
E-mail: welle@ivv.fraunhofer.de
franz@ivv.fraunhofer.de

Chapter 24

Dr J.J. de Vlieger
TNO Industrial Technology
P.O. Box 6235
5600 HE Eindhoven
The Netherlands

E-mail: j.devlieger@ind.tno.nl

Chapter 25

Ms I. Terhen Järvi-Kääriäinen
Association of Packaging Technology and Research
Mannerheimintie 156
00270 Helsinki
Finland

Tel +358 9 643 497
Fax +358 9 643 498
E-mail: ptr.ry@pakkausteknologia-ptr.fi

Chapter 26

Dr Liisa Lähteenmäki
Sensory Quality and Food Choice
VTT Biotechnology
P.O.Box 1500
FIN 02044 VTT
Finland

Tel. +358 9 456 5965
Fax +358 9 455 2103
E-mail: Liisa.lahteenmaki@vtt.fi

1

Introduction

R. Ahvenainen, VTT Biotechnology, Finland

The packaging sector is an important global industry, representing about 2% of the Gross National Product (GNP) of the developed countries. The value of the packaging industry is about 345 million euros worldwide, of which Europe represents a third. Fifty per cent of this market is packaging for food. Forecasts suggest that the sector will continue to grow in size and importance.

Many cooking and preservation processes still largely depend on effective packaging, for example canning, aseptic, sous vide and baking processes. Processes such as drying and freezing would be lost without protective packaging after processing to control product exposure to the effects of oxygen, light, water vapour, bacterial and other contaminants. However, modern food packaging no longer has just a passive role in protecting and marketing the product. It increasingly has an active role in processing, preservation and in retaining the safety and quality of foods throughout the distribution chain. Indeed, packaging development has changed the preservation methods used for food products. Ten to fifteen years ago all poultry products or industrially prepared raw minced meat were sold as frozen. Nowadays, thanks to modified-atmosphere packaging based on protective gases and novel gas-impermeable packaging materials, they are mainly sold as chilled products. The modern preparation and often international distribution of fresh-cut fruit and vegetables for retail sale is also possible today because of respirable packaging films.

Nowadays packaging plays an increasingly important role in the whole food chain ‘from the field to the consumer’s table’. As an example, many fresh agricultural products such as berries and mushrooms are picked in the field or the greenhouse directly into consumer packages and plastic or fibre-based trays. The product is thus touched only once before it reaches the consumer. Another example is ready-to-eat food and snack products which are packed in microwaveable trays which allow consumers to prepare the food immediately and even serve as an eating dish.

Food packaging has developed strongly during recent years, mainly due to increased demands on product safety, shelf-life extension, cost-efficiency, environmental issues, and consumer convenience. In order to improve the performance of packaging in meeting these varied demands, innovative modified- and controlled-atmosphere packaging, and active and intelligent packaging systems are being developed, tested and optimised in laboratories around the world. All these novel packaging technologies have great commercial potential to ensure the quality and safety of food with fewer or no additives and preservatives, thus reducing food wastage, food poisoning and allergic reactions. Intelligent packaging can also monitor product quality and trace a product's history through the critical points in the food supply chain. An intelligent product quality control system thus enables more efficient production, higher product quality and a reduced number of complaints from retailers and consumers. Intelligent packaging will also give the food industry the means to carry out in-house quality control required by food regulators.

This book covers selected trends and development in food packaging technologies and materials aiming at assuring the safety and quality of foodstuffs. In today's competitive market optimal packages are a major advantage when persuading consumers to buy a certain brand. Packaging has to satisfy various requirements effectively and economically. The food manufacturer's objective is to design an optimised package which satisfies all legislative, marketing and functional requirements sufficiently, and fulfils environmental, cost and consumer demands as well as possible.

I hope that the book will be interesting to readers, and reach a wide market amongst those working in research, industry or government, i.e., all those people who should know new trends in food packaging and the possibilities they raise to improve product safety and quality. In editing this book I would like to thank all the contributors, many of whom I have known for several years. I appreciate their willingness to share their expert knowledge and working within a tight schedule. I also want to thank my colleagues at VTT Biotechnology for many years of valuable cooperation and discussion and for helping to build such a positive and creative environment.

Part I

Types and roles of active and intelligent packaging

2

Active and intelligent packaging

An introduction

R. Ahvenainen, VTT Biotechnology, Finland

2.1 Introduction: the role of packaging in the food chain

Packaging has a significant role in the food supply chain and it is an integral part both of the food processes and the whole food supply chain. Food packaging has to perform several tasks as well as fulfilling many demands and requirements. Traditionally, a food package makes distribution easier. It has protected food from environmental conditions, such as light, oxygen, moisture, microbes, mechanical stresses and dust. Other basic tasks have been to ensure adequate labelling for providing information e.g., to the customer, and a proper convenience to the consumer, e.g., easy opening, reclosable lids and a suitable dosing mechanism. Basic requirements are good marketing properties, reasonable price, technical feasibility (e.g., suitability for automatic packaging machines, sealability), suitability for food contact, low environmental stress and suitability for recycling or refilling. A package has to satisfy all these various requirements effectively and economically. Some requirements and demands are contradictory to each other, at least to some extent. For these reasons, a modern food package should be optimised and integrated effectively with the food supply chain. In this book, package optimisation is discussed in detail in Chapter 21 and integrating active packaging, storage and distribution in Chapter 25.

For a long time packaging has also had an active role in processing, preservation and in retaining quality of foods. Changes in the way food products are produced, distributed, stored and retailed, reflecting the continuing increase in consumer demand for improved safety, quality and extended shelf-life for packaged foods, are placing greater demands on the performance of food packaging. Consumers want to be assured that the packaging is fulfilling its function of protecting the quality, freshness and safety of foods. The trend to ensure the quality and safety of food without, or at least fewer, additives and preservatives means that packaging has a more significant role in the preservation of food and in ensuring the safety of food in order to avoid wastage and food poisoning and to reduce allergies.

In this chapter active and intelligent packaging are introduced.

Various terms for new packaging methods can be found in the literature, such as active, smart, interactive, clever or intelligent packaging. These terms are often more or less undefined. For this reason, twelve partners from research and industry formulated the joint definitions for active and intelligent packaging systems in a European study 'Evaluating Safety, Effectiveness, Economic-environmental Impact and Consumer Acceptance of Active and Intelligent Packaging (ACTIPAK-FAIR CT98-4170)' in the years 1999–2001. The main objective of the study was to establish and implement active and intelligent packaging systems within the relevant regulations for food packaging in Europe. The project was coordinated by Mr Nico deKruijf, TNO, the Netherlands.[1, 2]

According to the definitions of the Actipak project, active and intelligent packaging are:

- Active packaging changes the condition of the packed food to extend shelf-life or to improve safety or sensory properties, while maintaining the quality of the packaged food.
- Intelligent packaging systems monitor the condition of packaged foods to give information about the quality of the packaged food during transport and storage.

2.2 Active packaging techniques

Food condition in the definition of active packaging includes various aspects that may play a role in determining the shelf-life of packaged foods, such as physiological processes (e.g., respiration of fresh fruit and vegetables), chemical processes (e.g., lipid oxidation), physical processes (e.g., staling of bread, dehydration), microbiological aspects (e.g., spoilage by micro-organisms) and infestation (e.g., by insects). Through the application of appropriate active packaging systems these conditions can be regulated in numerous ways and, depending on the requirements of the packaged food, food deterioration can be significantly reduced.[1]

Active packaging techniques for preservation and improving quality and safety of foods can be divided into three categories; absorbers (i.e. scavengers) (Table 2.1), releasing systems (Table 2.2) and other systems (Table 2.3). Absorbing (scavenging) systems remove undesired compounds such as oxygen, carbon dioxide, ethylene, excessive water, taints and other specific compounds (Table 2.1). Releasing systems actively add or emit compounds to the packaged food or into the head-space of the package such as carbon dioxide, antioxidants

Table 2.1 Examples of sachet, label and film type absorbing (scavenging) active packaging systems for preservation and shelf-life extension of foods or improving their quality and usability for consumers. Oxygen, carbon dioxide, ethylene and humidity absorbers have the most significant commercial use, lactose and cholesterol removers are not yet in use. Adapted from[5,6,9,22]

Packaging type	Examples of working principle/ mechanism/reagents	Purpose	Examples of possible applications
Oxygen absorbers (sachets, labels, films, corks)	Ferro-compounds, ascorbic acid, metal salts, glucose oxidases, alcohol oxidase	Reduction/preventing of mould, yeast and aerobic bacteria growth Prevention of oxidation of fats, oils, vitamins, colours Prevention of damage by worms, insects and insect eggs	Cheese, meat products, ready-to-eat products, bakery products, coffee, tea, nuts, milk powder
Carbon dioxide absorbers (sachets)	Calcium hydroxide and sodium hydroxide or potassium hydroxide Calcium oxide and silica gel	Removing of carbon dioxide formed during storage in order to prevent bursting of a package	Roasted coffee Beef jerkey Dehydrated poultry products
Ethylene absorbers (sachets, films)	Aluminium oxide and potassium permanganate (sachets) Activated carbon + metal catalyst (sachet) Zeolite (films) Clay (films) Japanese oya stone (films)	Prevention of too fast ripening and softening	Fruits like apples, apricots, banana, mango, cucumber, tomatoes, avocados and vegetables like carrots, potatoes and Brussels sprouts
Humidity absorbers (drip-absorbent sheets, films, sachets)	Polyacrylates (sheets) Propylene glycol (film) Silica gel (sachet) Clays (sachet)	Control of excess moisture in packed food Reduction of water activity on the surface of food in order to prevent the growth of moulds, yeast and spoilage bacteria	Meat, fish, poultry, bakery products, cuts of fruits and vegetables

Table 2.1 *(continued)*

Packaging type	Examples of working principle/ mechanism/reagents	Purpose	Examples of possible applications
Absorbers of off flavours, amines and aldehydes (films, sachets)	Cellulose acetate film containing naringinase enzyme Ferrous salt and citric or ascorbic acid (sachet) Specially treated polymers	Reduction of bitterness in grapefruit juice Improving the flavour of fish and oil-containing food	Fruit juices Fish Oil-containing foods such as potato chips, biscuits and cereal products Beer
UV-light absorbers	Polyolefins like polyethylene and polypropylene doped the material with a UV-absorbent agent Crystallinity modification of nylon 6 UV stabiliser in polyester bottles	Restricting light-induced oxidation	Light-sensitive foods such as ham Drinks
Lactose remover	Immobilised lactase in the packaging material	Serving milk products to the people suffering lactose intolerance	Milk and other dairy products
Cholesterol remover	Immobilised cholesterol reductase in the packaging material	Improving the healthiness of milk products	Milk and other dairy products

and preservatives (Table 2.2). Other systems may have miscellaneous tasks, such as self-heating, self-cooling and preservation (Table 2.3).

Depending on the physical form of active packaging systems, absorbers and releasers can be a sachet, label or film type. Sachets are placed freely in the head-space of the package. Labels are attached into the lid of the package. Direct contact with food should be avoided because it impairs the function of the system and, on the other hand, may cause migration problems (see Chapter 22).

Films or materials having antimicrobial properties can be divided into two types.

- Those from which an active substance emits or migrates to the head-space of the package or to the surface of the food, respectively. In the first case, the system does not need to be in direct contact with the food, but in the second case it must be in contact (Table 2.2).
- Those that are effective against microbial growth without emitting or migration of the active agents into the head-space of the package or to the

Table 2.2 Examples of sachet and film type releasing active packaging systems for preservation and shelf-life extension of foodstuffs or improving their quality. So far, none of these systems are in wide commercial use. Adapted from[5,6,12,22,23]

Packaging type	Examples of working principle/ mechanism/reagents	Purpose	Examples of possible applications
Carbon dioxide emitters (sachets)	Ascorbic acid Sodium hydrogen carbonate and ascorbate	Growth inhibition of gram-negative bacteria and moulds	Vegetables and fruits, fish, meat, poultry
Ethanol emitters (sachets)	Ethanol/water mixture absorbed onto silicon dioxide powder generating ethanol vapour	Growth inhibition of moulds and yeast	Bakery products (preferably heated before consumption) Dry fish
Antimicrobial preservative releasers (films)	Organic acids, e.g. sorbic acid Silver zeolite Spice and herb extracts Allylisothiocyanate Enzymes, e.g. lyzozyme	Growth inhibition of spoilage and pathogenic bacteria	Meat, poultry, fish, bread, cheese, fruit and vegetables
Sulphur dioxide emitters (sachets)	Sodium metabisulfite incorporated in microporous material	Inhibition of mould growth	Fruits
Antioxidant releasers (films)	BHA BHT Tocopherol Maillard reaction volatiles	Inhibition of oxidation of fat and oil	Dried foodstuffs Fat-containing foodstuffs
Flavouring emitters (films)	Various flavours in polymers	Minimisation of flavour scalping Masking off-odours Improving the flavour of food	Miscellaneous
Pesticide emitters (the outer or inner layer of packaging material)	Imazalil Pyrethrins	Prevention of growth of spoilage bacteria Fungicidal or pest control	Dried, sacked foodstuffs, e.g., flour, rice, grains

Table 2.3 Various other examples of active packaging systems. Adapted from[5]

Packaging type	Examples of working principle/ mechanism/reagents	Purpose	Examples of possible applications
Insulating materials	Special non-woven plastic with many air pore spaces	Temperature control for restricting microbial growth	Various foods to be stored refrigerated
Self-heating aluminium or steel cans and containers	The mixture of lime and water	Cooking or preparing food via built-in heating mechanism	Sake, coffee, tea, ready-to-eat meals
Self-cooling aluminium or steel cans and containers	The mixture of ammonium chloride, ammonium nitrate and water	Cooling of food	Non-gas drinks
Microwave susceptors	Aluminium or stainless steel deposited on substrates such as polyester films or paperboard	Drying, crisping and ultimately browning of microwave food	Popcorn, pizzas, ready-to-eat foods
Modifiers for microwave heating	A series of antenna structures that alter the way microwaves arrive at the food	Even heating, surface browning, crisping and selective heating	As above
Temperature-sensitive films	The gas permeability of the polymer is controlled by filler content, particle size of the filler and degree of stretching of the film	To avoid anaerobic respiration	Vegetables and fruits
UV-irradiated nylon film [24, 25]	The use of excimer laser 193 nm UV irradiation to convert amide groups on the surface of nylon to amines	Growth inhibition of spoilage bacteria	Meat, poultry, fish, bread, cheese, fruit and vegetables
FreshPad[26]	Releasing natural volatile oils, absorbing oxygen and excess juice	Growth inhibition of bacteria Moisture control Self-life improvement	Meat
Surface-treated food packaging materials	Fluorine-based plasmas[27]	Growth inhibition of bacteria	

food, respectively. In this case, the material must be in direct contact with the food (Table 2.3).

More detailed information about active packaging and its application is available in Chapters 3–5 in Part I and the chapters in Parts II and III of this book.

2.3 Intelligent packaging techniques

The definition of intelligent packaging in the Actipak project includes indicators to be used for quality control of packed food (Table 2.4). They can be so-called external indicators, i.e., indicators which are attached outside the package (time-temperature indicators), and so-called internal indicators which are placed inside the package, either to the head-space of the package or attached into the lid (oxygen indicators for indication of oxygen or package leak, carbon dioxide indicators, microbial growth indicators and pathogen indicators).

Time-temperature indicators are discussed in detail in Chapter 6, oxygen and carbon dioxide indicators in Chapter 13 and microbial growth indicators, i.e., freshness indicators and pathogen indicators in Chapter 7. Furthermore, food packaging can be intelligent in ways that give information, e.g., about the origin

Table 2.4 Examples of external and internal indicators and their working principle or reacting compounds to be used in intelligent packaging for quality control of packed food. Adapted from [5,22]

Indicator	Principle/reagents	Gives information about	Application
Time-temperature indicators (external)	Mechanical Chemical Enzymatic	Storage conditions	Foods stored under chilled and frozen conditions
Oxygen indicators (internal)	Redox dyes pH dyes Enzymes	Storage conditions Package leak	Foods stored in packages with reduced oxygen concentration
Carbon dioxide indicator (internal)	Chemical	Storage conditions Package leak	Modified or controlled atmosphere food packaging
Microbial growth indicators (internal/external) i.e. freshness indicators	pH dyes All dyes reacting with certain metabolites (volatiles or non-volatiles)	Microbial quality of food (i.e. spoilage)	Perishable foods such as meat, fish and poultry
Pathogen indicators (internal)	Various chemical and immunochemical methods reacting with toxins	Specific pathogenic bacteria such as *Escherichia coli* O157	Perishable foods such as meat, fish and poultry

of food, authenticity, contents, use, and consumption-date expiration. It can also track a product in the food supply chain, be anti-theft and tamper proof.[3] This book does not cover these technologies, however, some future aspects concerning them are dealt in the Chapter 25.

2.4 Current use of novel packaging techniques

In the USA, Japan and Australia, active and intelligent packaging systems are already being successfully applied to extend shelf-life or to monitor food quality and safety. Despite this, regardless of intensive research and development work on active and intelligent packaging, there are only a few commercially significant systems on the market. Oxygen absorbers added separately as small sachets in the package head-space or attached as labels into the lid probably have the most commercial significance in active food packaging nowadays. Also, ethanol emitters/generators and ethylene absorbers are used, but to a lesser extent than oxygen absorbers. Other commercially significant active techniques include, e.g., absorbers for moisture and off-odour and absorbers/emitters for carbon dioxide. With regard to intelligent packaging, time temperature indicators and oxygen indicators are most used in those countries mentioned above.

In Europe, only a few of these systems have been developed and are being applied. This lag compared to the USA, Japan and Australia is partly due to to the strict European regulations for food-contact materials that cannot keep up entirely with technological innovations and currently prohibit the application of many of these systems. In addition, exiguous knowledge about consumer acceptance, economic aspects and the environmental impact of these novel technologies and, in particular, the exiguous knowledge of hard evidence of their effectiveness and safety demonstrated by independent researchers have inhibited commercial usage.[1] Furthermore, vacuum packaging and protective gas packaging (modified atmosphere packaging) have had an established position in many European countries since 1980. Vacuum packaging, gas packaging and active packaging compete with to each other, at least to some extent. However, all these technologies have their own advantages and disadvantages, and the best package technology should be selected according to individual requirements case by case.[4]

Discussions between VTT Biotechnology and various parties in the food supply chain in Finland and also in other countries have shown that before intelligent and active packaging systems can be launched in greater numbers onto the market, a demonstration of the function and benefits of these systems in the food supply chain is necessary. For this reason, VTT has just started a one-year project 'Demonstration of intelligent packaging as a tool for quality control in the food supply chain' in Finland. The project is financed and supported by Tekes, the National Technology Agency of Finland, packaging companies and franchising groups.

2.5 Current research

Many research institutes in Europe, such as TNO (the Netherlands), Pira International and the Campden and Chorleywood Food Research Association (UK), University of Compostela (Spain), ADRIAC (France), University of Ghent (Belgium), Distam (Italy) Technion-Israel Institute of Technology (Israel), Royal Veterinary and Agricultural University (Denmark), Matforsk and Norconserv (Norway) and VTT Biotechnology (Finland)[1,5] have been working in recent years on shelf-life and quality assurance studies and legislation aspects related to active and intelligent packaging. Outside Europe, probably CSIRO, Australia and the University of Minnesota, Purdue University[6] and Clemson University USA,[7] University of Manitoba, Canada[8] have been the more active.

With regard to the development of new active and intelligent packaging systems, companies in Japan and the USA have innovated and patented most of the active and intelligent systems available.[6,9,10,11] Furthermore, some research institutes are also developing new systems, such as CSIRO in Australia,[12] SIK and Lund University in Sweden,[13] Purdue University and University of California in USA[6,14] and VTT Biotechnology.[15,16] VTT in Finland also started a new five-year project 'Active, communicating package' at the beginning of the year 2002. The aim of this project is to develop an effective logistic system based on wireless communication and active, intelligent and communicating packages for sensitive food.

2.6 The legislative context

At least three types of regulation have an impact on the use of active and intelligent packaging in foods. First, any need for food-contact approval should be established before any form of active and intelligent packaging can be used. Second, environmental regulations of packaging material usage can be expected to increase in the near future. Third, there may be a need for labelling in cases where active or intelligent packaging can give rise to consumer confusion.

Legislative demands regarding food packaging and food contact materials include specific consumer protection and environmental concerns. In various countries, legislation related to food contact materials has been framed. The basic criteria for these regulations differ between countries. Some rules are based on restrictions as to the composition of materials, whereas others regulate mainly migration limits.

In the Actipak project mentioned in Section 2.1, non-European legislation on active and intelligent food packaging concepts were screened. It appeared that there are only a few specific regulations for these innovative concepts. Generally, these new systems should meet the conventional requirements for food contact materials.

2.6.1 USA

In the USA, components directly added/to the food or via packaging are considered as food additives. Consequently, these active substances have to be evaluated as additives according to rigorous toxicological testing prior to use. If a substance is added to the packaging material and has only an indirect effect on foods, then it should be subjected to regulations similar to those for migration of monomers and other polymer components.

2.6.2 Japan

Considering new active and intelligent agents in food packaging materials, it has to be primarily checked whether the agent is on the list of the Ministry of Health and Welfare.[17] New components must be registered as chemicals according to the Guidelines for Screening Toxicity Testing of Chemicals. In addition, for the application of sachets, the packaging must be clearly labelled with the text 'Do not eat contents', including a diagram demonstrating this warning.[18]

2.6.3 Australia

In Australia, standard A12 of the Food Standards Codes up to the amendment of 31 October 1996 sets maximum levels for metals and other contaminants in food, such as monomers and additives from packaging materials. Standard A13 of the Food Standards Codes sets out regulations regarding articles and materials in contact with food. In this standard, the application of sachets containing active agents is explicitly described; the main aspects are listed below.

1 (a) Packages of food may contain sachets of silicon dioxide for the purpose of inhibiting the growth of mould or absorbing moisture.
  (b) Sachets specified in (a) may contain ethanol and flavouring.
  (c) There shall be written on the label on or attached to a sachet to which the clause applies, in standard type, the words 'MOULD INHIBITOR' or 'MOISTURE ABSORBER' or words having the same or a similar intent, immediately followed, in standard type of 3 mm, by the warning 'DO NOT EAT'.
  (d) Sachets to which this clause applies shall have at least one face with a surface area not less than 10 cm^2.

2 (a) Packages of food may contain sachets of reduced iron powder for the purpose of absorbing oxygen.
  (b) Sachets specified in paragraph (a) of this clause shall have at least one face with a surface area of not less than 10 cm^2, and may also contain one or more of the following:

(i) calcium chloride	(v) iron oxide
(ii) calcium hydroxide	(vi) magnesium hydroxide
(iii) carbon, activated	(vii) magnesium stearate
(iv) gypsum	(viii) perlite

(ix) salt
(x) talc
(xi) water
(xii) zeolite
(xiii) magnesium silicate
(xiv) diatomaceous salts

(c) Such sachets should be labelled, in standard type 3 mm, with the words 'OXYGEN ABSORBER', or with words having the same or a similar intent, immediately followed, in standard type of 3 mm, by the warning 'DO NOT EAT'.

In fact, only in Australia is the content of 'active' sachets defined in a specific regulation. In the USA and Japan, active or intelligent components are considered as additives and are subject to FDA approval or must be evaluated according to Japanese law, the Guidelines for Screening Toxicity Testing of Chemicals. Active and intelligent components must be toxicologically safe. Migration behaviour of active and intelligent concepts has not been explicitly described in any of the above-mentioned regulations.

2.6.4 Europe

At the moment, no European regulation currently covers specifically the use of active and intelligent packaging.[1] Furthermore, none of the European countries has a specific regulation concerning active or intelligent packaging. Only France and Spain have an additional list that probably contains components used as active agents in active packaging concepts. In other European countries the use of active agents is restricted to the components on the positive list.

A few years ago, two initiatives were taken to implement active and intelligent packaging within the European regulations. deKruijf and Rijk in Chapter 22 describe the content of these initiatives.

2.7 Consumers and novel packaging

The food industry's main concern about introducing active components to packaging seems to be that consumers will consider the components harmful and will not accept them.[19] Before the food industry can decide on the best available active and/or intelligent packaging technique, studies are needed both in domestic and foreign markets to evaluate consumer attitudes towards these techniques. Even the naming of the 'absorbers' or 'indicators' may not sound familiar to consumers.[10] Dr Liisa Lähteenmäki discusses testing consumer responses to new packaging concepts in Chapter 26. She will also give some examples about the results of current research. Furthermore, Table 2.5 outlines some potential problems and solutions that the food industry should take into account before deciding to use active and/or intelligent packaging techniques.

Table 2.5 Problems and solutions encountered with introducing new products using active and/or intelligent packaging techniques[4]

Problem/fear	Solution
• Consumer attitude	→ Consumer research: education and information.
• Doubts about performance	→ Storage tests before launching. Consumer education and information.
• Increased packaging costs	→ Use in selected, high quality products. Marketing tool for increased quality and quality assurance.
• False sense of security, ignorance of date markings	→ Consumer education and information.
• Mishandling and abuse	→ Active compound incorporated into label or packaging film. Consumer education and information.
• False complaints and return of packs with colour indicators	→ Colour automatically readable at the point of purchase.
• Difficulty of checking every colour indicator at the point of purchase.	→ Barcode labels: intended for quality assurance for retailers only.

2.8 Future trends

2.8.1 Active packaging

Active packaging will probably increase in European countries in the near future due to consumer preferences for minimally processed and naturally preserved foods and the food industry's eagerness to invest in product quality and safety. The future trend in active packaging is to use absorbing or releasing compounds incorporated in the packaging film or in an adhesive label to get rid of separate objects in packaging and thus to avoid consumer resistance towards new packaging techniques. In the near future these invisible active absorbers or emitters will possibly be launched on the market on a larger scale.

The antimicrobial packaging materials are a potential way to decrease the amount of preservatives and focus the function of preservatives more precisely where microbial growth and spoilage mainly occur, on the surface of the food. However, effective materials are still rather rare on the market and need much research and development work. The more significant challenges in developing these materials are, first, to find new physical, chemical and biological methods to add preservatives effectively into packaging materials so that preservatives are still active against the microbes or to treat packaging materials and polymers in such a way that they are converted to antimicrobial. Secondly, to develop antimicrobial materials that are effective against several spoilage and pathogenic microbes. This obviously means that more than one preservative should be incorporated into the same packaging material.

It is self-evident that these novel materials should have proper permeability properties, good appearance, good mechanical properties and they must be reasonable in price, suitable for packaging machines already used in the food industry and suitable for normal sealing procedures. In other words they must have all those properties that traditional packaging materials have.

2.8.2 Intelligent packaging

There are several reasons for the bright future of intelligent packaging.

- The significance of freshness and safety will increase.
- The demands of consumers will increase.
- Globalisation and expansion of the marketing area make logistic chains longer placing more demands on traceability.
- The facilitation of in-house control for industry and retailing in the complete food supply chain. Intelligent packaging can also monitor product quality and trace the critical points in the food supply chain.

Thus, an intelligent product quality control system enables more efficient production, higher product quality and a reduced number of complaints from retailers and consumers. Today the commercially available intelligent concepts are labels reacting with a visible change in response to time and temperature (TTI) or the presence of certain chemical compounds (leak indicators, freshness indicators). In the future it can be expected that the intelligent package can contain more complex invisible messages which can be read at a distance. According to Byrne,[20] this type of electronic labelling could soon be as common as bar coding today. A label could be introduced as a chip but advances in ink technology might enable the use of clever printed circuits as well. The advantages of printed structures include low price and disposability. The security tags, which are already used today, are the first examples of electronic labelling. Pre-programmed miniature radio frequency identity tags have also been used in the identification of containers for military supplies.[20] The tags can have either a built-in battery or can be energised by the external transmitter.

In addition to information on product identification, date of manufacture, price, etc., electronic tags could also function as a time-temperature, leak and/or freshness indicator and as pilferage protection (all these different functions in the same tag).[21] Electronic tags might also be informative labels and give instructions about the use, healthiness, etc., of food. It could be expected that advances in electronics, biotechnology (e.g., biosensors, immunodiagnostics), enzyme technology, analytic methods (e.g., electronic nose), material technology (intelligent materials, modification of polymers), (micro)-electronics (price, printable structures), sensor technology and digital printing would be followed by the emergence of new concepts of intelligent packaging.

With regard to the development of freshness and pathogen indicators, identification and quantification of the most influential volatile and non-volatile metabolic compounds contributing to safety and spoilage with various foods are essential. A lot of information is already available (see Chapter 7), but a lot of new research is still needed, particularly concerning pathogenic microbes.

Last but not least, intelligent packaging systems should be easy in use, low cost, integrated in the packaging and capable of handling tasks. Furthermore, they must correlate well with product quality (not with the environment), be irreversible in colour change, easy to understand, easy to read (i.e. clear and standardised colour changes, particularly if the indicator is to be read by a consumer), and easy to store before use.

2.8.3 Intelligent food supply chain

In order to derive maximum benefit, technologies developed to ensure the safety and quality of food (active and intelligent packages), to track and trace goods through the logistic chain (barcodes, smart tags, RF-ID technologies) and to produce packages just in time on demand and preferably personalised, will be integrated with each other in the future. There are two ways to act, tagging or marking of the goods. Tagging is still quite expensive and different kinds of marking have great potential in consumer goods. The average price of the consumer good is around 1.5 Euro, which means that the price of markings/tags should be low. The tags can be active or passive. The information can be saved in the memory of the tag and also up-dated with the reading device or the tag or marking can act as a link to the information that can be read from the server via network connections. The latter makes it possible to use much cheaper marking methods and also to update the information whenever it is required. The course of action should be selected on the basis of expenses and profits. The technologies to add ambient intelligence in the goods are developing quickly and the IST technologies to create new solutions for the control and communication of the logistic chain are well developed. Mobile communication devices are well accepted in Europe and could be used both for control and consumer purposes.

It is very probable that in the future the management of the food supply chain will be based on wireless communication and active, intelligent and communicating packages. The packages will protect the food without additives, inform about the product quality and history in every stage of the logistic chain, guide the journey of the package, reduce product loss, and will give real-time information to the consumer about the properties/quality/ use of the product.

2.9 Sources of further information and advice

FLOROS, J.D., DOCK, L.L. and HAN, J.H. 1997. 'Active packaging technologies and applications'. *Food, Cosmetics and Drug Packaging* January 1997, pp. 10–17.

ROONEY, M.L. (ed.). 1995. *Active Food Packaging*. Blackie Academic &

Professional, an imprint of Chapman & Hall. Glasgow, UK. 260 pages.

SMOLANDER, M., HURME, E. and AHVENAINEN, R. 1997. 'Leak indicators for modified-atmosphere packages'. *Trends in Food Science & Technology*, 8, No. 4, pp. 101–6.

SMOLANDER M. (2000), 'Principles of smart packaging', *Packaging Technology* March/April 2000, pp. 9–12.

FABECH, B., HELLSTRØM, T., HENRYSDOTTER, G., HJULMAND-LASSEN, M., NILSSON, J., RÜDINGER, L., SIPILÄINEN-MALM, T., SOLLI, E., SVENSSON, K., THORKELSSON, Á.E. and TUOMAALA, V. 2000. 'Active and intelligent food packaging'. *A Nordic report on the legislative aspects*. TemaNord 2000: 584. Copenhagen: Nordic Council of Ministers, 84 pp. ISBN 92-893-0520-7.

2.10 References

1. DE KRUIJF, N., VAN BEEST, M., RIJK, R., SIPILÄINEN-MALM, T., PASEIRO L. and DE MEULENAER, B. 2002. 'Active and intelligent packaging: applications and regulatory aspects'. *Food Additives and Contaminants* Vol. 19, pp. 144–62.
2. RIJK, R., VAN BEEST, M., DE KRUIJF, N., BOUMA, K., MARTIN, C., DE MEULENAER, B. and SIPILÄINEN-MALM, T. 2002. 'Active and intelligent packaging systems and the legislative aspects'. *Food Packaging Bulletin*. Vol. 10, No. 9 & 10, pp. 2–10.
3. DAY, B.P.F. 2000. 'Intelligent packaging for foodstuffs'. *Food, Cosmetics and Drug Packaging* Vol. 23, No. 12, pp. 233–9.
4. HURME, E. and AHVENAINEN, R. 1996. 'Active and smart packaging of ready-made foods'. In: *Minimal Processing and Ready Made Foods*. T. Ohlsson, R. Ahvenainen and T. Mattila-Sandholm (eds). Göteborg, SIK, pp. 169–82.
5. HURME, E., SIPILÄINEN-MALM, T., AHVENAINEN, R. and NIELSEN, T. 2002. 'Active and intelligent packaging'. In: *Minimal Processing Technologies in the Food Industry*. T. Ohlsson and N. Bengtsson (eds), Woodhead Publishing Limited, Cambridge, England, pp. 87–123.
6. FLOROS, J.D., DOCK, L.L. and HAN, J.H. 1997. 'Active packaging technologies and applications'. *Food, Cosmetics and Drug Packaging*, Vol. 20, No. 1, pp. 10–17.
7. COOKSEY, K. 2001. 'Antimicrobial food packaging'. *Food, Cosmetics and Drug* Packaging, 24, No. 7, 133–7.
8. HAN, J.H. 2000. 'Antimicrobial food packaging'. *Food Technology* 54, No. 3, pp. 56–65.
9. ROONEY, M.L. (ed.). 1995. *Active Food Packaging*. Blackie Academic and Professional, an imprint of Chapman & Hall. Glasgow, UK. 260 pages.
10. AHVENAINEN, R. and HURME, E. 1997. 'Active and smart packaging for meeting consumer demands for quality and safety'. *Food Additives and*

Contaminants, **14**, 6–7, pp. 753–63.

11. SMOLANDER, M., HURME, E. and AHVENAINEN, R. 1997. 'Leak indicators for modified-atmosphere packages'. *Trends in Food Science & Technology*, 8, No. 4, pp. 101–6.
12. ROONEY, M.L. 1995. 'Development of active and intelligent packaging systems'. In: *New Shelf-life Technologies and Safety Assessments.* R. Ahvenainen, T. Mattila-Sandholm. and T. Ohlsson, (eds), VTT Symposium 148, Espoo 1995, pp. 75–83.
13. HÖRNSTEN, G. 2001. 'Active packaging: Immobilized enzymes integrated within a laminate that remove oxygen from within the package'. In: *The proceedings of 2nd Nordic Foodpack*, September 5–7, 2001. Norconserv, Stavanger, Norway, 11 pp.
14. ANON. 2002. 'Foodstuff packaging set to use organic sensor to provide ''freshness'' data'. *Food, Cosmetics and Drug Packaging* Vol. 25, No. 6, p. 107.
15. SMOLANDER M. (2000), 'Principles of smart packaging', *Packaging Technology* March/April 2000, pp. 9–12.
16. VARTIAINEN, J., MOTION, R., SKYTTÄ, E., ENQVIST, J., SIPILÄINEN-MALM, T., HURME, E. and AHVENAINEN, R. 2002. 'Antimicrobial Packaging Materials Based on Traditional Food Preservatives'. 13th IAPRI World Conference on Packaging and the 50th Anniversary of the Michigan State University School of Packaging, June 23–28, 2002, East Lansing, Michigan, USA.
17. IGARASHI, S. 1996 TNO Japan Office, personal communications.
18. SMITH, J. P., HOSHINO, J. and ABE, Y. 1995. 'Interactive packaging involving sachet technology'. In: *Active food packaging.* M.L.Rooney (ed.). London, Blackie Academic & Professional, pp. 143–73.
19. MIKKOLA, V., LÄHTEENMÄKI, L., HURME, E., HEINIÖ, R., KÄÄRIÄINEN, T. and AHVENAINEN, R. 1997. 'Consumer attitudes towards oxygen absorbers in food packages'. *VTT Research Notes 1858*, Espoo, 34 pp.
20. BYRME, G. 1997. 'Intelligent packaging'. *Prod. Image Secur.* 1(3), pp. 21–2.
21. ANON. 1994. 'Talking boxes, food temperature sensors and other smart packaging is not far off'. *Quick Frozen Foods International* 3682, 114.
22. FABECH, B., HELLSTRØM, T., HENRYSDOTTER, G., HJULMAND-LASSEN, M., NILSSON, J., RüDINGER, L., SIPILÄINEN-MALM, T., SOLLI, E., SVENSSON, K., THORKELSSON, Á.E. and TUOMAALA, V. 2000. 'Active and intelligent food packaging'. *A Nordic report on the legislative aspects.* TemaNord 2000:584. Copenhagen: Nordic Council of Ministers, 84 pp. ISBN 92-893-0520-7.
23. DAY, B.P.F. 2000. 'Underlying principles of active packaging technology'. *Food, Cosmetics and Drug Packaging* Vol. 23, No. 7, pp. 134–9.
24. OZDEMIR, M. and SADIKOGLU, H. 1998. 'A new emerging technology: laser-induced surface modification of polymers'. *Trends in Food Science & Technology* Vol. 9, pp. 159–67.
25. PAIK, J. S., DHANASEKHARAN, M. and KELLY, M. J. 1998. 'Antimicrobial

activity of UV-irradiated nylon film for packaging applications'. *Packaging Technology and Science* 11, pp. 179–87.

26. ANON. 2001. 'New approaches to packaging are being developed to improve food safety and efficiency'. *Meat International* Vol. 11, No. 8, p. 26.
27. OZDEMIR, M., YURTERI, C. U. and SADIKOGLU, H. 1999. 'Surface treatment of food packaging polymers by plasmas'. *Food Technology* Vol. 53, No. 4, pp. 54–8.

3

Oxygen, ethylene and other scavengers

L. Vermeiren, L. Heirlings, F. Devlieghere and J. Debevere, Ghent University, Belgium

3.1 Introduction

The best known and most widely used active packaging technologies for foods today are those engineered to remove undesirable substances from the headspace of a package through absorption, adsorption or scavenging. To achieve this goal a physical or chemical absorbent or adsorbent is incorporated in the packaging material or added to the package by means of a sachet. In most publications, the term 'absorption' is used loosely to describe any system that removes a substance from the headspace. However, there is a clear difference between absorption and adsorption. Adsorption is a two-dimensional phenomenon while absorption is three-dimensional. According to Mortimer (1993), absorption involves a substance being taken into the bulk of a phase while adsorption involves a substance being taken onto a surface. Both, absorption and adsorption are physical phenomena while scavenging implies a chemical reaction (Brody *et al.*, 2001). This chapter focuses mainly on oxygen and ethylene scavenging and finally also discusses carbon dioxide absorbers and odour removers.

3.2 Oxygen scavenging technology

3.2.1 Introduction

In many cases, food deterioration is caused by the presence of oxygen, as oxygen is responsible for oxidation of food constituents and proliferation of moulds, aerobic bacteria and insects. Modified atmosphere packaging (MAP) and vacuum packaging have been widely adopted to exclude oxygen from the headspace. However, these physical methods of oxygen elimination do not always remove the oxygen completely. Some oxygen (0.1–2%) generally remains in the package and even more when the food is porous. Moreover, the oxygen that permeates through the packaging film during storage cannot be removed by these techniques. In the presence of such amounts of oxygen, many of the oxidation reactions and mould proliferation still proceed. Oxygen scavengers are able to reduce the oxygen concentration to less than 0.01% and can maintain those levels (Rooney, 1995; Hurme and Ahvenainen, 1998; Vermeiren *et al.*, 1999). An oxygen scavenger is a substance that scavenges oxygen chemically or enzymatically and therefore, protects the packaged food completely against deterioration and quality changes due to oxygen.

3.2.2 Role of oxygen scavengers

Preventing oxidation

Oxygen scavengers effectively prevent oxidative damage in a wide range of food constituents such as (i) oils and fats to prevent rancidity, (ii) both plant and muscle pigments and flavours to prevent discolouration (e.g. meat) and loss of taste and (iii) nutritive elements, e.g., vitamins to prevent loss of the nutritional value. Berenzon and Saguy (1998) investigated the effect of oxygen scavengers on the shelf-life extension of crackers packaged in hermetically sealed tin cans which were stored at 15, 25 and 35ºC for up to 52 weeks. Oxygen scavengers reduced the hexanol concentration significantly. Peroxide values were markedly reduced by the presence of oxygen scavengers. In the presence of oxygen scavengers, the lag period before the peroxides started to build up was prolonged to, respectively, 17 and 10 weeks at 25 and 35ºC. Sensory evaluations showed that in the presence of oxygen scavengers and independently of storage temperature, no oxidative rancid odours were observed for up to 44 weeks.

Preventing insect damage

Oxygen scavengers are effective for killing insects and worms or their eggs growing in cereals such as rice, wheat and soybeans. Fumigation treatments using gases such as bromides and methyl disulfide kill insects but their residues can remain in the food. Additionally, insects in the egg or pupal stages can be resistant against fumigation treatments. Oxygen scavengers are very effective against insects because they remove the oxygen the insects need to survive.

Prevention of proliferation of moulds and strictly aerobic bacteria

Oxygen scavenging is effective in preventing growth of moulds and aerobic bacteria. Mould spoilage is an important microbial problem limiting the shelf-life of high and intermediate moisture products. Losses due to mould spoilage are a serious economic concern in the bakery industry. Some moulds, such as *Aspergillus flavus* and *Aspergillus parasiticus*, can also produce highly toxic substances called mycotoxins. In gas packaging aerobic growth can still occur depending on the residual oxygen level in the package headspace. It has been demonstrated that moulds can proliferate in headspaces with oxygen concentrations as low as 1–2% (Smith, 1996). Oxygen levels of 0.1% or lower are required to prevent the growth and mycotoxin production of many moulds (Rooney, 1995). The effects of modified atmosphere packaging involving oxygen scavengers, storage temperature and packaging film barrier characteristics on the growth of and aflatoxin production by *Aspergillus parasiticus* in packaged peanuts was investigated (Ellis *et al.*, 1994). A slight mould growth was visible in air-packaged peanuts using a high gas barrier film (Oxygen Transmission Rate (OTR) of 3–6 cc. m^{-2}. day^{-1} at 23°C and dry conditions) while extensive growth was observed in peanuts packaged under similar air conditions using a low gas barrier film (OTR of 4000 cc m^{-2} day^{-1}). When an oxygen scavenger (Ageless® type S) was incorporated, mould growth was inhibited in peanuts packaged in a high gas barrier film and was reduced when a low barrier film was used. Aflatoxin B_1 production was inhibited in peanuts packaged in a high barrier film with an oxygen scavenger, while a limited amount of aflatoxin less than the regulatory level of 20 ng. g^{-1} was detected in absorbent packaged peanuts using a low barrier film. This study showed that oxygen scavengers are effective for controlling the growth of and aflatoxin production by *Aspergillus parasiticus.* However, the effectiveness of the scavengers will be dependent on the gas barrier properties of the packaging film.

Smith *et al.* (1986) showed that oxygen scavengers are three times more effective than gas packaging for increasing the mould-free shelf-life of crusty rolls. In gas packaged (40% N_2/60% CO_2) crusty rolls with Ageless® the headspace oxygen never increased beyond 0.05% and the product remained mould-free for over 60 days at ambient storage temperature. A similar mould-free shelf-life was obtained in air and N_2 packaged crusty rolls with Ageless®. The mould-free shelf-life of white bread packaged in a polypropylene film could be extended from 4–5 days at room temperature to 45 days by using an Ageless® sachet. Pizza crust, which moulds in 2–3 days at 30°C was mould-free for over 10 days using an appropriate O_2 scavenger (Nakamura and Hoshino, 1983).

It is well known that an oxygen-free atmosphere at a water activity greater than 0.92 can favour the growth of many microbial pathogens including *Clostridium botulinum* (Labuza and Breene, 1989). *Clostridium botulinum* mainly grows under anaerobic conditions but can also have a limited growth under low O_2 conditions. The use of oxygen scavengers could be dangerous if the temperature is not kept close to 0°C. Daifas *et al.* (1999) investigated the growth and toxin production by *Clostridium botulinum* in English-style crumpets, using an Ageless® FX_{200} oxygen scavenger at room temperature. All inoculated crumpets were toxic within 4 to 6 days and were organoleptically acceptable at the time of toxigenesis. Counts of *C. botulinum* increased to approximately 10^5 CFU/g at the time of toxin production. This study confirms that *C. botulinum* could pose a public health hazard in high a_w – high pH crumpets using an oxygen scavenger when stored at non-chilled conditions. Lyver *et al.* (1998) have done challenge studies on raw surimi nuggets, which were inoculated with 10^4 spores/g of *Clostridium botulinum* type E spores. All

Table 3.1 Effects of oxygen scavengers on foods (Abe, 1994; Smith *et al.*, 1990)

Effect	Typical application
Fresh taste and aroma	Various food items, coffee, tea
↔ Mould growth	Bakery products, cheese, processed seafood, pasta
↔ Rancidity	Nuts, fried foods, processed meat, whole milk powder product
↔ Discolouration	Processed meat, green noodle, herbs, tea, dried vegetables
↔ Insect damage	Beans, grain, herbs, spices
Maintaining nutritional value	All kinds of foods

products were packaged in air and air with an Ageless® SS oxygen absorber and stored at 4, 12 and 25ºC. Toxin was not detected in any raw product throughout storage (28 days). The absence of toxigenesis was attributed to the low pH (4.1–4.3) due mainly to the growth of lactic acid bacteria. Whiting and Naftulin (1992) showed that controlling the pH and NaCl concentration of the food product is an important factor in controlling growth of *C. botulinum* under low oxygen concentrations. When oxygen absorbers are used, challenge studies should be done to investigate if *C. botulinum* is able to grow. An overview of the effects of oxygen scavengers and their most important food applications is shown in Table 3.1.

3.3 Selecting the right type of oxygen scavenger

Oxygen scavengers must satisfy several requirements: they must

1. be harmless to the human body. Though the oxygen scavengers themselves are neither food nor food additives, they are placed together with food in a package, and there is therefore the possibility of accidental intake by consumers.
2. absorb oxygen at an appropriate rate. If the reaction is too fast, there will be a loss of oxygen absorption capacity during introduction into the package. If it is too slow, the food will not be adequately protected from oxygen damage.
3. not produce toxic substances or unfavourable gas or odour.
4. be compact in size and are expected to show a constant quality and performance.
5. absorb a large amount of oxygen.
6. be economically priced (Nakamura and Hoshino, 1983; Abe, 1994; Rooney, 1995).

An appropriate oxygen scavenger is chosen depending on the O_2-level in the headspace, how much oxygen is trapped in the food initially and the amount of oxygen that will be transported from the surrounding air into the package during storage. The nature of the food (e.g. size, shape, weight), water activity and desired shelf-life are also important factors influencing the choice of oxygen absorbents. For an oxygen scavenger (sachet) to be effective, some conditions have to be fulfilled (Nakamura and Hoshino, 1983; Abe, 1994; Smith, 1996). First of all, packaging containers or films with a high oxygen barrier must be used, otherwise the scavenger will rapidly become saturated and lose its ability to trap O_2. Films with an oxygen permeability not exceeding 20 ml/m^2.d.atm are recommended for packages in which an oxygen scavenger will be used. Examples of barrier layers used with oxygen scavengers are EVOH (ethylene vinyl alcohol) and PVDC (polyvinylidene chloride) (Nakamura and Hoshino, 1983; Rooney, 1995). If films with high O_2 permeabilities are used (> 100 ml/m^2.d.atm), the O_2 concentration will reach zero within a week but after some days, it will return to ambient air level because the absorbent is saturated. If high-barrier films (e.g. < 10 ml/m^2.d.atm) are used, the headspace O_2 will be reduced to 100 ppm within 1–2 days and remain at this level for the duration of the storage period provided that package integrity is maintained (Rooney, 1995). Secondly, for flexible packaging heat sealing should be complete so that no air invades the package through the sealed part. A rapid, inexpensive and efficient method of monitoring package integrity and ensuring low residual headspace oxygen throughout the storage period is through the incorporation of a redox indicator, e.g. Ageless® Eye®. Ageless® Eye® is a tablet which indicates the presence of oxygen by a colour change. When placed inside the package, the colour changes from blue to pink when the O_2 concentration approaches zero. If the indicator reverts to its blue colour, this is an indication of poor packaging integrity (Smith *et al.*, 1990; Nakamura and Hoshino, 1983; Rooney, 1995). Finally, an oxygen scavenger of the appropriate type and size must be selected. The appropriate size of the scavenger can be calculated using the following formulae (Roussel, 1999; ATCO® technical information, 2002). The volume of oxygen present at the time of packaging (A) can be calculated using the formula:

$$A = (V - P) \times [O_2]/100$$

$V =$ volume of the finished pack determined by submersion in water and expressed in ml;

$P =$ weight of the finished pack in g;

$[O_2] =$ initial O_2 concentration in package ($= 21\%$ if air).

In addition, it is necessary to calculate the volume of oxygen likely to permeate through the packaging during the shelf-life of the product (B). This quantity in ml may be calculated as follows:

$$B = S \times P \times D$$

S = surface area of the pack in m^2;

P = permeability of the packaging in $ml/m^2/24h/atm$;

D = the shelf-life of the product in days.

The volume of oxygen to be absorbed is obtained by adding A and B. Based on these calculations, the size of the scavenger and the number of sachets can be determined.

3.3.1 Oxygen scavenging sachets

In general, O_2 scavenging technologies are based on one of the following concepts: iron powder oxidation, ascorbic acid oxidation, catechol oxidation, photosensitive dye oxidation, enzymatic oxidation (e.g. glucose oxidase and alcohol oxidase), unsaturated fatty acids (e.g. oleic acid or linolenic acid) or immobilised yeast on a solid material (Floros *et al.*, 1997). A summary of the most important trademarks of oxygen scavenger systems and their manufacturers is shown in Table 3.2.

The majority of presently available oxygen scavengers are based on the principle of iron oxidation (Nakamura and Hoshino, 1983; Rooney, 1995; Vermeiren *et al.*, 1999)

$$Fe \rightarrow Fe^{2+} + 2e^-$$

$$\frac{1}{2}O_2 + H_2O + 2e^- \rightarrow 2OH^-$$

$$Fe^{2+} + 2OH^- \rightarrow Fe\,(OH)_2$$

$$Fe\,(OH)_2 + \frac{1}{4}O_2 + \frac{1}{2}H_2O \rightarrow Fe\,(OH)_3$$

The principle behind oxygen absorption is iron rust formation. To prevent the iron powder from imparting colour to the food, the iron is contained in a sachet. The sachet material is highly permeable to oxygen and water vapour. A rule of thumb is that 1 g of iron will react with 300 ml of O_2 (Labuza, 1987; Nielsen, 1997; Vermeiren *et al.*, 1999). The LD_{50} (lethal dose that kills 50% of the population) for iron is 16 g/kg body weight. The largest commercially available sachet contains 7 grams of iron so this would amount to only 0.1 g/kg for a person of 70 kg, or 160 times less than the lethal dose (Labuza and Breene, 1989). Iron-based oxygen scavengers have one disadvantage: they cannot pass the metal detectors usually installed on the packaging line. This problem can be avoided, e.g. by ascorbic acid or enzyme based O_2 scavengers (Hurme and Ahvenainen, 1998).

Some important iron-based O_2 absorbent sachets are Ageless® (Mitsubishi Gas Chemical Co., Japan), ATCO® O_2 scavenger (Standa Industrie, France), Freshilizer® Series (Toppan Printing Co., Japan), Vitalon (Toagosei Chem.

Table 3.2 Some manufacturers and trade names of oxygen scavengers (Ahvenainen and Hurme, 1997; Day, 1998; Vermeiren *et al*., 1999)

Company	Trade name	Type	Principle/Active substances
Mitsubishi Gas Chemical Co., Ltd. (Japan)	Ageless	Sachets and labels	Iron based
Toppan Printing Co., Ltd. (Japan)	Freshilizer	Sachets	Iron based
Toagosei Chem. Ind. Co. (Japan)	Vitalon	Sachets	Iron based
Nippon Soda Co., Ltd. (Japan)	Seaqul	Sachets	Iron based
Finetec Co., Ltd. (Japan)	Sanso-cut	Sachets	Iron based
Toyo Pulp Co. (Japan)	Tamotsu	Sachets	Catechol
Toyo Seikan Kaisha Ltd. (Japan)	Oxyguard	Plastic trays	Iron based
Dessicare Ltd. (US)	O-Buster	Sachets	Iron based
Multisorb technologies Inc. (US)	FreshMax	Labels	Iron based
	FreshPax	Sachets	Iron based
Amoco Chemicals (US)	Amosorb	Plastic film	unknown
Ciba Specialty chemicals (Switzerland)	Shelfplus O_2	Plastic film	Iron based
W.R. Grace and Co. (US)	PureSeal	Bottle crowns	Ascorbate/metallic salts
	Darex	Bottle crowns, bottles	Ascorbate/sulphite
CSIRO/Southcorp Packaging (Australia)	Zero$_2$	Plastic film	Photosensitive dye/ organic compound
Cryovac Sealed Air Co. (US)	OS1000	Plastic film	Light activated scavenger
CMB Technologies (UK)	Oxbar	Plastic bottles	Cobalt catalyst/ nylon polymer
Standa Industrie (France)	ATCO	Sachets	Iron based
	Oxycap	Bottle crowns	Iron based
	ATCO	Labels	Iron based
Bioka Ltd. (Finland)	Bioka	Sachets	Enzyme based

Table 3.3 Types and properties of Ageless oxygen scavenging sachets (Rooney, 1995; Ageless technical information, 2002)

Type	Function	Moisture status	Water activity	Absorption speed[a] (day)
ZP/ZPT	Decreases [O_2]	Self-reacting	< 0.95	1–3
SA	Decreases [O_2]	Self-reacting	0.65–0.95	0.5–1.0
SS	Decreases [O_2]	Self-reacting	0.65–0.95	2–3 (0–4°C) 10 (–25°C)
FX	Decreases [O_2]	Moisture dependent	> 0.85	0.5–1.0
FM	Decreases [O_2]	Moisture dependent also microwaveable products	> 0.80	1.0
E	Decreases [O_2] Decreases [CO_2]	Self-reacting	< 0.3	3–8
G	Decreases [O_2] increases [CO_2]	Self-reacting	0.3–0.5	1–4
GL	Decreases [O2]	Self-reacting	0.3–0.95	2–4

[a] number of days to reduce the oxygen level to less than 0.01% (measured at room temperature)

Industry Co., Japan), Sanso-cut (Finetec Co., Japan), Seaqul (Nippon Soda Co., Japan), FreshPax® (Multisorb technologies Inc., USA) and O-Buster® (Dessicare Ltd., USA). Some of them will be discussed in detail.

Ageless® can reduce the oxygen in an airtight container down to 0.01% (100 ppm) or less to prolong shelf-life of food products. Several types of Ageless® are commercially available and applicable to many types of foods (Labuza and Breene, 1989; Smith *et al.*, 1990; Abe, 1994; Ageless® technical information, 1994; Rooney, 1995; Smith, 1996). The different types and properties of Ageless® oxygen scavenging sachets are shown in Table 3.3.

A self-reacting type contains moisture in the sachet and as soon as the sachet is exposed to air, the reaction starts.. In moisture-dependent types, oxygen scavenging takes place only after moisture has been taken up from the food. These sachets are stable in open air before use because they do not react immediately upon exposure to air therefore they are easy to handle if kept dry.

Toppan Printing Co. developed another type of oxygen scavenging sachet, named Freshilizer®. Two series are commercially available, the F series and C series. Sachets of the F series contain ferrous metal and scavenge oxygen without generating another gas. The C series contain non-ferrous particles and are able to sorb oxygen and generate an equal volume of carbon dioxide to prevent package collapse.

FreshPax™ is a patented oxygen scavenger developed by Multisorb technologies. Four main types of FreshPax are commonly available: type B, D, R and M. Type B is used for moist or semi-moist foods with a water activity above 0.7. Type D is recommended for use with dehydrated and dried foods. To scavenge oxygen at refrigerated or frozen storage temperatures, type R should be used. Type M can be used for moist or semi-moist foods, which are packaged under modified atmospheres containing carbon dioxide.

Another scavenging technology is based on catechol oxidation. As catechol is an organic compound, it passes metal detectors. Tamotsu is the only commercial product in Japan based on this technology (Abe, 1994). Tamotsu type D is used for dry products such as spices, freeze-dried foods, tea. These sachets do not require moisture for their oxygen scavenging reaction.

Another way of controlling the oxygen level in a food package is by using enzyme technology. A combination of two enzymes, glucose oxidase and catalase, has been applied for oxygen removal. In the presence of water, glucose oxidase oxidises glucose, that can be originally present or added to the product, to gluconic acid and hydrogen peroxide (Greenfield and Laurence, 1975; Labuza and Breene, 1989; Nielsen, 1997). The reaction is:

$$2\ \text{glucose} + 2\ O_2 + 2\ H_2O \rightarrow 2\ \text{gluconic acid} + 2\ H_2O_2$$

where glucose is the substrate.

Since H_2O_2 is an objectionable end product, catalase is introduced to break down the peroxide (Rooney, 1995; Vermeiren *et al.*, 1999):

$$2\ H_2O_2 + \text{catalase} \rightarrow 2\ H_2O + O_2 + \text{catalase}$$

Enzymatic systems are usually very sensitive to changes in pH, water activity, temperature and availability of solvents. Most systems require water for their action, and therefore, they cannot be effectively used with low-water content foods (Floros *et al.*, 1997). The enzyme can either be part of the packaging structure or put in an independent sachet. Both polypropylene (PP) and polyethylene (PE) are good substrates for immobilising enzymes (Labuza and Breene, 1989). A commercially available enzyme-based oxygen absorbent sachet is Bioka (Bioka Ltd., Finland). It is claimed that all components of the reactive powder and the generated reaction products are food-grade substances safe for both the user and the environment (Bioka technical information, 1999). The oxygen scavenger eliminates the oxygen in the headspace of a package and in the actual product in 12–48 hours at 20°C and in 24–96 hours at 2–6°C. With certain restrictions, the scavenger can also be used in various frozen products. When introducing the sachet into a package, temperature may not exceed 60°C because of the heat sensitivity of the enzymes (Bioka technical information, 1999). An advantage is that it contains no iron powder, so it presents no problems for microwave applications and for metal detectors in the production line.

Besides glucose oxidase, other enzymes are able to scavenge oxygen. One such enzyme is alcohol oxidase, which oxidises ethanol to acetaldehyde. It could be used for food products in a wide a_w range since it does not require water to operate. If a lot of oxygen has to be absorbed from the package, a great amount of ethanol would be required, which could cause an off-odour in the package. In addition, considerable aldehyde would be produced which could give the food a yoghurt-like odour (Labuza and Breene, 1989).

The Pillsbury Company holds a 1994 patent that utilises ascorbic acid as reducing agent (Graf, 1994). The product, also referred to as Oxysorb, comprises a combination of a reducing agent, ascorbic acid, and a small amount of a transition metal, such as copper. The oxygen removing system may be added in a small oxygen permeable pouch.

The oxidation of polyunsaturated fatty acids (PUFAs) is another technique to scavenge oxygen. It is an excellent oxygen scavenger for dry foods. Most known oxygen scavengers have a serious disadvantage: when water is absent, their oxygen scavenging reaction does not progress. In the presence of an oxygen scavenging system, the quality of the dry food products may decline rapidly because of the migration of water from the oxygen scavenger into the food. Mitsubishi Gas Chemical Co. holds a patent that uses PUFAs as a reactive agent. The PUFAs, preferably oleic, linoleic or linolenic, are contained in carrier oil such as soybean, sesame or cottonseed oil. The oil and/or PUFA are compounded with a transition metal catalyst and a carrier substance (for example calcium carbonate) to solidify the oxygen scavenger composition. In this way the scavenger can be made into a granule or powder and can be packaged in sachets (Floros *et al.*, 1997).

3.3.2 Oxygen scavenging films

It should be noted that the introduction of oxygen scavenger sachets into the food package suffers from the disadvantage of possible accidental ingestion of the contents by the consumer. Another concern is that the sachet could leak out and contaminate the product. When sachets are used, there also needs to be a free flow of air surrounding the sachet in order to scavenge headspace oxygen (Rooney, 1995). To eliminate this problem, oxygen removing agents can be incorporated into the packaging material such as polymer films, labels, crown corks, liners in closures. These oxygen scavenging materials have the additional advantage that they can be used for all products, including liquid products. The oxygen consuming substrate can be either the polymer itself or some easily oxidisable compound dispersed or dissolved in the packaging material (Nielsen, 1997; Hurme and Ahvenainen, 1998).

A problem related to the use of O_2 scavenging films is that the films should not react with atmospheric oxygen prior to use. This problem has been solved by inclusion of an activation system triggering the O_2 consuming capabilities of the film in the packaging system. Activation by illumination or catalysts or reagents, supplied at the time of filling, may be required to start the reaction.

Illumination of a package that contains a photosensitising dye and a singlet oxygen acceptor results in rapid scavenging of oxygen from the headspace. Australian researchers have reported that reaction of iron with ground state O_2 is too slow for shelf-life extension (Hurme and Ahvenainen, 1998). The singlet-excited state of oxygen, which is obtained by dye sensitisation of ground state oxygen using near infra-red, visible or ultraviolet radiation, is highly reactive and so its chemical reaction with scavengers is rapid (Rooney, 1981). The technique involves sealing of a small coil of ethyl cellulose film, containing a dissolved photosensitising dye and a singlet oxygen acceptor, in the headspace of a transparent package. When the film is illuminated with light of the appropriate wavelength, excited dye molecules sensitise oxygen molecules, which have diffused into the polymer, to the singlet state. These singlet oxygen molecules react with acceptor molecules and are thereby consumed. The photochemical reaction can be presented as follows (Rooney, 1981; Vermeiren *et al.*, 1999):

$$\text{photon} + \text{dye} \rightarrow \text{dye}^*$$
$$\text{dye}^* + O_2 \rightarrow \text{dye} + O_2^*$$
$$O_2^* + \text{acceptor} \rightarrow \text{acceptor oxide}$$
$$O_2^* \rightarrow O_2$$

This scavenging technique does not require water as an activator, so it is effective for wet and dry products. Its scavenging action is initiated on the processor's packaging line by an illumination-triggering process. Examples of light-activated oxygen scavenger films are OS1000, developed by Cryovac Sealed air corporation and Zero$_2$™ developed by CSIRO and marketed by Southcorp Packaging (Australia). Cryovac OS1000 is a multi-layer flexible film with a coextruded sealant. The invisible, oxygen scavenging polymer is a component of the sealant. UV lights are used to trigger the scavenging reaction through a patented activation process. These films are activated on the packaging line just before filling and sealing, so light never comes in contact with the food. The active ingredient in the Southcorp technology, named Zero$_2$, is integrated into the polymer backbones of such common packaging materials as PET, polyethylene, polypropylene and EVA. The active ingredient is non-metallic and is activated by UV light once it is incorporated into packaging material (Graff, 1998).

Amoco Chemicals has developed Amosorb® oxygen scavenger, a plastic concentrate that sorbs oxygen in food and beverage packages. Amosorb concentrate is a polymer-based oxygen scavenger that can be incorporated as an inner layer within a multi-layer packaging structure during co-extrusion or lamination. The oxygen scavenger is activated by moisture and can reduce headspace oxygen levels to less than 0.01%. It can be incorporated into several packaging structures such as the sidewall or lid of rigid containers, flexible film and closure liners (Edwards, 1998; Amoco Chemicals bulletin AS-1, 1999a; Amoco Chemicals bulletin AS-3, 1999b). In June 2000, the Amosorb 2000 oxygen absorber technology was acquired by Ciba Specialty Chemicals Corporation. The trade name was changed to Ciba Shelfplus O2-2400 (polyethylene application) and Shelfplus O2-2500 (polypropylene application) (Brody *et al.*, 2001).

Oxyguard, from Toyo Seikan Group, is an oxygen scavenger that uses an iron salt-based additive and is available in the form of a flexible or rigid plastic. It is a multi-layer that consists of an outer layer, a barrier layer, an oxygen scavenging layer and an inner layer. The oxygen sorption is initiated by water (Vermeiren *et al.*, 1999; Oxyguard technical information, 2002).

Oxbar^{TM} is a system developed by Carnaud-Metal Box (UK) and is composed of a PET/MXD6/Co film where the PET serves as the structural material and the active ingredients are MXD6 nylon (polymetaxylylene adipamide or polymetaxylylene diamine-hexanoic acid) and cobalt salt. Cobalt catalyses the oxidation of the nylon polymer (Miltz *et al.*, 1995). It can be used for plastic packaging of beer, wine, sauces and other beverages.

3.3.3 Other scavenging devices

Labels

Other layouts for oxygen scavengers are cards and sheets in or labels on the packaging. In 1991, Multisorb technologies introduced the iron-based oxygen scavenging label FreshMax. FreshMax is designed for adhesion within packages and so the risk for ingestion is minimised. The technology has a printed surface and is acceptable for food contact. It is resistant to fat and moisture and can be used for several food products (Anon., 1991; Rooney, 1995; Silgelac technical information, 1998; Caldic technical information, 2002). Standa Industrie has also developed a self-adhesive oxygen scavenging label, named ATCO® and Ageless® also has a label type and a card type.

Bottle closures

Removal of oxygen from a bottle by a closure requires that a component reacts with gaseous oxygen in the headspace of the bottle. Darex oxygen scavenging technology utilises a material that can be incorporated into barrier packaging such as crowns, cans and a broad variety of plastic and metal closures. The basic reaction of their oxygen scavenging technique is ascorbate oxidising to dehydroascorbic acid and sulphite to sulphate. Darex DarExtend is designed to be incorporated as an integral part of traditional barrier packaging such as the aluminium roll-on closures as well as in plastic closures and crowns. Darex DarEval is an EVOH-based oxygen scavenging barrier resin used as an inner layer in multi-layer PET. DarEval enables multi-layer PET bottles to have a glass-like performance (Rooney, 1995; Vermeiren *et al.*, 1999; Darex Technical Information, 2002). The major use is in crown caps to protect beer from oxidation. Other examples of oxygen scavenging crowns caps are Pure Seal caps (W.R. Grace Co., USA) and Oxycap (Standa Industrie, France).

3.3.4 Economic aspects

Oxygen scavengers have several economic advantages for the food processor (Nakamura and Hoshino, 1983; Rooney, 1995; Smith, 1996):

- increased product shelf-life and distribution radius
- a longer time between deliveries enabling larger deliveries

- increased length of time product can stay in the distribution pipeline
- reduced distribution losses
- reduced evacuation/gas flushing times in gas packaged products thereby increasing product throughput
- reduced costs required for gas flushing equipments.

To save time and labour, oxygen scavenging sachets can be inserted automatically (Ageless® technical information, 1994). Also labels can be applied automatically at conventional line speeds. An iron-based sachet with a capacity of 100 ml O_2 would cost 5.03 euro for 3000 pieces. The same sachet with a capacity of 1000 ml O_2 would cost 20.4 euro for 500 pieces. The price of another iron-based sachet of another trademark, which absorbs 100 ml O_2, is about 40 euro for 300 pieces.

3.4 Ethylene scavenging technology

3.4.1 Introduction

Ethylene acts as a plant hormone that has different physiological effects on fresh fruit and vegetables. It accelerates respiration, leading to maturity and senescence, and also softening and ripening of many kinds of fruit. Furthermore, ethylene accumulation can cause yellowing of green vegetables and may be responsible for a number of specific post-harvest disorders in fresh fruits and vegetables. Although some effects of ethylene are positive, such as degreening of citrus fruit, ethylene is often detrimental to the quality and shelf-life of fruits and vegetables. To prolong shelf-life and maintain an acceptable visual and organoleptical quality, accumulation of ethylene in the packaging of fruits and vegetables should be avoided. A number of ethylene sorbing substances are described. Most of these are supplied as sachets or integrated into films. Many of the claims for ethylene adsorbing or absorbing capacity have been poorly documented so the efficacy of these materials is difficult to substantiate (Vermeiren *et al.*, 1999; Zagory, 1995).

3.4.2 The role of ethylene scavengers

Ethylene is a naturally occurring, chemically simple molecule of the alkene type that regulates numerous aspects of growth, development and senescence of many fruits and vegetables. As it is effective at part-per-million to part-per-billion concentrations and its effects are very dose-dependent, it is considered as a plant hormone (Saltveit, 1999). Environmental ethylene can be produced both biologically and non-biologically. Non-biological sources of ethylene are incomplete combustion of fossil fuels, burning of agricultural wastes and leakage from industrial polyethylene plants (Sawada and Totsuka, 1986). Ethylene is thus a common air pollutant and ambient atmospheric levels are normally in the range of 0,001–0,005 ppm (Abeles *et al.*, 1992). Biological sources of ethylene include higher plant tissues, several species of bacteria and fungi, some algae and mosses (Zagory, 1995).

In higher vascular plants, a relatively simple biosynthetic pathway produces ethylene:

$$\begin{array}{ccccccc} & & & \text{ACC synthase} & & \text{ACC oxidase } (O_2, CO_2) & \\ \text{Methionine} & \rightarrow & \text{S-adenosylmethionine} & \rightarrow & \text{1-amino-cyclopropane-} & \rightarrow & CH_2{=}CH_2 \\ & & \text{(SAM)} & & \text{carboxylic acid (ACC)} & & \\ & & & & \uparrow & & \uparrow \\ & & & & \text{fruit ripening, wounding} & & \text{ripening, ethylene} \end{array}$$

The amino acid methionine (MET) is converted to S-adenosyl methionine (SAM) which is then converted in the next step to 1-amino-cyclopropane carboxylic acid (ACC) by the enzyme ACC synthase. The production of ACC is often the controlling step for ethylene synthesis. A number of intrinsic (e.g. developmental stage) and extrinsic (e.g. wounding) factors influence this pathway. In the final step, ACC is oxidised by the enzyme ACC oxidase to form ethylene. This last step requires the presence of oxygen and low levels of CO_2 to activate ACC oxidase. The ACC oxidase activity can show a dramatic increase in ripening of fruit in response to ethylene exposure (Saltveit, 1999). As in the case of other hormones, ethylene is thought to bind to a receptor, forming an activated complex that in turn triggers a primary reaction. This primary reaction then initiates a chain of reactions leading to a wide variety of physiological responses (Yang, 1985).

Ethylene has since long been recognised as a problem in post-harvest handling of horticultural products (fruits, vegetables and flowers). The diverse physiological effects of ethylene have been extensively reviewed by many authors (Abeles *et al.*, 1992; Hopkins, 1995 and Saltveit, 1999). Some of the effects of ethylene are beneficial and economically useful, such as flowering of pineapples, de-greening of citrus fruits and ripening of tomatoes. However, ethylene is often involved in the decline of the quality and shelf-life of many fruits and vegetables. Only those effects that are deleterious to packaged plant produce will be discussed here.

First of all, ethylene accelerates the respiration of fruits and vegetables. Respiration rate generally is well correlated with perishability of produce. Commodities such as asparagus, broccoli, mushrooms, and raspberries with high respiration rates have short shelf-lives. At the end of growth, climacteric fruit (e.g. banana, avocado) undergoes a large increase in respiration accompanied by marked changes in composition and texture. The ripening of climacteric fruit is associated with a large increase in ethylene production. The increases in respiration and ethylene production can be induced prematurely in climacteric fruit by treating them with a suitable concentration of ethylene. The ripening process is irreversible once endogenous ethylene production increases to a certain level (McGlasson, 1985). As climacteric fruit starts to ripen, this negative feedback inhibition of ethylene on ethylene synthesis changes into a positive feedback promotion in which ethylene stimulates its own synthesis (i.e. autocatalytic ethylene production) and copious amounts of ethylene are produced. Reducing the external concentration of ethylene around bulky ripening climacteric fruit (e.g. apples, bananas, melons, tomatoes) has almost no effect on reducing the internal concentration in these fruit. Internal concentrations of ethylene can exceed 100μ l/l, even when the external concentration is zero. Therefore, reducing the external ethylene concentration generally has no effect on the ripening of fruit that has progressed a few days into its climacteric stage. However, at the initial stages of ripening, when the internal levels are still low, inhibiting the synthesis of ethylene and removal of ethylene can significantly retard ripening (Saltveit, 1999). Non-climacteric fruit show no increase in respiration and ethylene production during ripening. In contrast, an unnaturally climacteric-like respiratory increase can be induced in non-climacteric fruit by treating them with ethylene. Yet this increased respiration is not accompanied by an increase in endogenous ethylene production and is still reversible upon removal of the exogenous ethylene (McGlasson, 1985). In most cases, exposure to a few parts per million of ethylene leads to increased respiration and consequently increased perishability (Zagory, 1995).

Ethylene is often referred to as the ripening hormone because it can accelerate softening and ripening of many kinds of fruit by the direct or indirect stimulation of the synthesis and activity of many enzymes such as pectinases, cellulases, esterases and polygalacturonase. Examples are a reduced firmness of watermelons through ethylene exposure and an increased toughness of asparagus spears after exposure to 100 ppm ethylene for 1 hour, which was associated with increased activity of peroxidases and accelerated lignin biosynthesis. In most cases, for packaged fruits it would be desirable to prevent exposure to ethylene and thereby preventing rapid ripening (Zagory, 1995; Kader, 1985). Ethylene accelerates chlorophyll degradation and induces yellowing of green tissues, thus reducing market quality of leafy green vegetables such as spinach, floral vegetables such as broccoli and immature fruits such as cucumbers (Kader, 1985) and promotes changes that are important to flavour such as starch to sugar conversion, loss of acidity and formation of aroma volatiles in climacteric fruit. Furthermore, ethylene can be responsible for a number of specific post-harvest disorders of fruits and vegetables such as russet spotting of lettuce, sprouting of potatoes and formation of bitter-tasting isocoumarins in carrots (Zagory, 1995; Kader, 1985).

One of the major problems in the post-harvest storage of fruits and vegetables is the proliferation of opportunistic microorganisms that thrive on injured or senescent tissues. By stimulating ripening and senescence, ethylene also enhances the opportunities for pathogenesis. Fruits and vegetables have an epidermal layer that provides a protective barrier against infections but plant pathogenic moulds and bacteria possess mechanisms to penetrate into external tissues (Jacxsens, 2000). The growth of a number of post-harvest pathogens e.g. the development and sporulation of the decay-causing fungi *Penicillium* and *Botrytis cinerea* is directly stimulated by ethylene. In addition, several post-harvest plant pathogens produce ethylene and this ethylene may compromise the natural defences of the plant tissues (Barkai-Golan, 1990; Saltveit, 1999).

Plant organs like stems, roots and leafy parts that are consumed as vegetables are less sensitive to ethylene exposure compared to fruits, but some vegetables such as tomato, cucumber and broccoli are from a morphological point of view 'fruits' and responsive to ethylene (Jacxsens, 2000). These vegetables can benefit from the removal of ethylene as ethylene detrimentally affects their colour by yellowing and texture by promoting unwanted softening in cucumbers and peppers or toughening in asparagus and sweet potatoes (Saltveit, 1999).

Strategies for protecting harvested horticultural products from the detrimental effects of ethylene can be placed into three major categories: avoidance, e.g., through temperature control, removal and inhibition. In the category 'removal', adsorption of ethylene will be discussed below.

3.4.3 Principle of ethylene adsorption

The double bond of ethylene makes it a very reactive compound that can be altered or degraded in many ways. This creates a diversity of opportunities for commercial methodologies for the removal of ethylene. Ethylene can be absorbed or adsorbed by a number of substances, reviewed by Zagory (1995), including activated charcoal, molecular sieves of crystalline aluminosilicates, Kieselguhr, bentonite, Fuller's earth, brick dust, silica gel and aluminium oxide. A number of clay materials such as cristobalite, Oya stone and zeolite have been reported to have ethylene sorbing capacity. Some regenerable sorbents have been shown to have ethylene adsorbing capacity and have the benefit of being reusable after purging. Examples are propylene glycol, hexylene glycol, squalene, phenylmethylsilicone, polyethylene and polystyrene. Some sorbents have been combined with catalysts or chemical agents that modify or destroy the ethylene after adsorption. For example, activated charcoal, used to adsorb ethylene, has been impregnated with bromine or with 15% $KBrO_3$ and 0,5 M H_2SO_4 to eliminate the activity of ethylene. A number of catalytic oxidisers have been combined with adsorbents to remove the adsorbed ethylene such as potassium dichromate, potassium permanganate (KMnO4), iodine pentoxide and silver nitrate, each respectively embedded on silica gel. Electron-deficient dienes or trienes such as benzenes, pyridines, diazines, triazines and tetrazines, having electron-withdrawing substitutes such as fluorinated alkyl groups, sulphones and esters, will react rapidly and irreversibly with ethylene. Such compounds can be embedded in permeable plastic bags or printing inks to remove ethylene from packages of plant produce (Holland, 1992). Metal catalysts immobilised on absorbents such as powdered cupric oxide, will effectively oxidise ethylene, but in many cases the reactions require high temperatures (>180°C). Clearly such systems would be inappropriate for food packaging applications (Zagory, 1995).

Most suppliers offer ethylene adsorbers based on KMnO4. To be effective, KMnO4 must be adsorbed on a suitable inert carrier with a large surface area such as celite, vermiculite, silica gel, alumina pellets, activated carbon, perlite or glass. Typically, such products contain about 4–6% KMnO4. The oxidation of ethylene with potassium permanganate can be thought of as a two-step process. Ethylene (CH_2CH_2) is initially oxidised to acetaldehyde (CH_3CHO), which in turn is oxidised to acetic acid (CH_3COOH). Acetic acid can be further oxidised to carbon dioxide and water:

$$3CH_2CH_2 + 2KMnO_4 + H_2O \rightarrow 2MnO_2 + 3CH_3CHO + 2KOH \quad (1)$$

$$3CH_3CHO + 2KMnO_4 + H_2O \rightarrow 3CH_3COOH + 2MnO_2 + 2KOH \quad (2)$$

$$3CH_3COOH + 8KMnO_4 \rightarrow 6CO_2 + 8MnO_2 + 8KOH + 2H_2O \quad (3)$$

Combining eq. 1–3, we get:

$$3CH_2CH_2 + 12KMnO_4 \rightarrow 12MnO_2 + 12KOH + 6CO_2$$

Potassium permanganate adsorbers change from purple to brown as the MnO_4^- is reduced to MnO_2, indicating the remaining adsorbing capacity. Adsorbent materials containing $KMnO_4$ cannot be integrated into food-contact packaging but are supplied only as sachets because of their toxicity and purple colour (Sherman, 1985; Zagory, 1995). Different studies have shown that these sachets effectively remove ethylene from packages of pears (Scott and Wills, 1974), bananas (Liu, 1970; Jayaraman and Raju, 1992; Chamara *et al.*, 2000), kiwifruit (Ben-arie and Sonego, 1980), diced onions (Howard *et al.*, 1994), apples (Shorter *et al.*, 1992), grapes (Don and Koo, 1996), mango, tomato and other fruits (Jayaraman and Raju, 1992). Examples of suppliers of potassium permanganate based ethylene scavengers are given in Table 3.4. Not only sachets are commercialised but the technique has been transferred to household refrigerators e.g. Mrs. Green's Extra Life cartridges from Dennis Green Ltd. and Fridge Friend box. A special case is the paper Frisspack (Dunapack, Hungary), for manufacture into corrugated fibreboard cases. This paper contains a chemosorbent to bond with ethylene, which is then oxidised by $KMnO_4$ (Brody *et al.*, 2001).

Another type of ethylene scavenger is based on the adsorption of ethylene on activated carbon and subsequent breakdown by a metal catalyst. Use of charcoal with palladium chloride prevented the accumulation of ethylene and was effective in reducing the rate of softening in kiwifruits and bananas and chlorophyll loss in spinach leaves, but not in broccoli (Abe and Watada, 1991). Some Japanese concepts such as Neupalon, Hatofresh System and Sendomate (Table 3.4) are also based on the adsorption of ethylene by activated carbon that is impregnated with different types of substances (palladium catalyst or bromine-type inorganic chemicals) to help the breakdown of ethylene (Zagory, 1995).

Other ethylene absorbing technologies are based on the inclusion of finely dispersed minerals. Typically these minerals are zeolites or local kinds of clays that are embedded in polyethylene (PE) bags that are then used to package fresh produce (Zagory, 1995). The fine pores of these minerals serve to absorb gases such as ethylene. Most of these films are opaque and not

Table 3.4 Some commercialised ethylene scavengers

Company	Trade name	Type	Principle/active substances
Purafil (Georgia, US)	Purafil	Pellets to be used in e.g. sachets	Potassium permanganate-impregnated alumina pellet
DeltaTRAK (US)	Air Repair	Sachets for shipments	Potassium permanganate
Ethylene Control (US)	Fridge Friend	Sachets, box for consumer's refrigerator	Potassium permanganate
International Ripening Company (US)	No specific name	Sachets for shipping boxes	Potassium permanganate
Dennis Green Ltd. (US)	Mrs. Green's Extra Life	Cartridges for consumer's refrigerator	Potassium permanganate
Grofit plastics (Israel)	Biofresh	Zipper bags, bags and films	-
Nippon Container Corporation (Japan)	FAIN	Films for inner surface of cardboard	-
Sekisui Jushi (Japan)	Neupalon	Sachet	Activated carbon
Mitsubishi Chemical Co. (Japan)	Sendomate	Sachet	Activated carbon + Pd-catalyst
Honshu Paper (Japan)	Hatofresh System	Paper bag or corrugated box	Activated carbon + bromine type inorganic chemical
E-I-A Warenhandels GmbH (Austria)	Profresh	Film	Minerals
Evert-fresh Co. (US)	Evert-Fresh Green-Bags	Bags for consumer use	Minerals
Peakfresh products (Australia)	Peakfresh	Film	Minerals
Odja Shoji C. (Japan)	BO film	Film	Crysburite ceramics
Cho Yang Heung San Co. (Korea)	Orega bag	Bags for consumer use	Minerals
OhE Chemicals (Japan)	Crisper SL	Film	-
Marathon products (US)	Ethylene Filter products	Sachet	-
Dessicare (US)	Ethylene EliminatorPak	Sachet	Zeolites
Pacific Agriscience (Singapore)	BI-ON		-

capable of sufficiently absorbing ethylene (Suslow, 1997). Although the incorporated minerals may absorb ethylene, they also alter the permeability of the films, both ethylene and CO_2 diffuse more rapidly and oxygen will enter more rapidly than through pure PE. These changes in permeability can reduce headspace ethylene concentrations and consequently improve shelf-life independently of any ethylene absorption. In fact, any powdered material can be used to reach these effects. However, even if the minerals do absorb ethylene, this capacity is often lost when incorporating these minerals into a polymer matrix (Zagory, 1995). Japanese or Korean companies marketed many of these bags for their internal markets and some of them are now also sold in the USA and Australia but fewer in Europe. The Orega bag based on the US patent of Dr Matsui (Matsui, 1989) contains fine porous material consisting of pumice-tuff, zeolite, active carbon, cristobalite and clinoptilolite mixed with a metal oxide. A similar concept is a sheet, described in the US patent assigned to Nissho and Co. (Japan) (Someyo and Nobuo, 1992), consisting of a synthetic resin film or a fibrous material and containing crushed coral, a stony substance formed from the massed skeletons of marine organisms that has calcium carbonate as the main ingredient. Others are Evert-Fresh Green-Bags, Peakresh™, BO film and Profresh®. An overview of some commercially available ethylene scavengers is given in Table 3.4.

3.4.4 Measuring ethylene sorption

There are many bags and films being sold offering improved post-harvest life of fresh produce due to the sorption of ethylene by minerals finely dispersed into polyethylene bags. The evidence offered in support of this claim is generally based on shelf-life experiments comparing common polyethylene bags with the so-called ethylene sorbing films. Such studies generally show an extension of the shelf-life and/or reduction of the headspace ethylene. However, such data do not support claims of ethylene sorbing capacity as the improved shelf-life and reduced ethylene level could also result from the increased gas permeabilities of these types of films. To evaluate the ethylene sorbing capacity of any ethylene sorbing substance, a direct measurement of ethylene depletion in closed systems containing samples of the bags without any produce is necessary. Furthermore, such studies should be done at low temperature and high relative humidity to mimic the conditions of performance (Zagory, 1995).

A possible method to determine the ethylene sorbing capacity uses closed recipients which include the film in their screw top. These recipients are flushed with ethylene and stored at a low temperature. At regular times the ethylene concentration in the recipients is measured by using a gas chromatograph. As the permeability is influenced by the relative humidity, water is brought in each recipient to reach a high relative humidity.

3.4.5 Economic aspects

It is extremely difficult to assess the economic importance of protecting harvested horticultural products from ethylene. Detrimental effects of ethylene during the normal short-term marketing of fruit and vegetables are not well defined and certainly are secondary to considerations regarding the maintenance of optimum temperature and humidity. However, costs to the individual shippers involved can easily run into tens of thousands of dollars when losses do occur from problems like russet sprouting of lettuce. Losses caused by ethylene are known to occur, but they are usually quantitatively undefined. A conservative estimate for the US would be in the tens of millions of dollars annually (Sherman, 1985). The ethylene sorbing packaging concepts could possibly contribute to an increase in the export of fresh produce.

In the US, ethylene control within packages of fresh and minimally processed fruit and vegetable products remains almost exclusively a reaction of $KMnO_4$ on a porous mineral structure (Brody *et al.*, 2001). The major disadvantage of permanganate scavengers seems to be their expense e.g. 0.33 euro for a 27g sachet, 0.39 euro for a 28g sachet, 0.26 euro for an 8g sachet, 6.25 euro for a Mrs. Green's Extra Life cartridge and 2.2 euro for a Fridge friend box. Prices of mineral-containing bags for consumer use range, e.g., from 1.96 euro to 4.26 euro depending of the size of the sachets. Despite the many Asian claims for the effectiveness of activated carbon and minerals, little proof has been offered to convince packagers to use them. For this reason, there are no strong commercial applications of ethylene-removing films in the US. On the other hand, sachets based on $KMnO_4$ are in widespread commercial use.

3.5 Carbon dioxide and other scavengers

3.5.1 Carbon dioxide scavengers

Role

Carbon dioxide is formed in some foods due to deterioration and respiration reactions. The produced CO_2 has to be removed from the package to avoid food deterioration and/or package destruction. Fresh roasted coffee can release considerable amounts of CO_2 due to the Strecker degradation reaction between sugars and amines (Labuza and Breene, 1989). Unless removed, the generated CO_2 can cause the packaging to burst due to the increasing internal pressure. Another CO_2-producing food product is kimchi, a general term for fermented vegetables such as oriental cabbage, radish, green onion and leaf mustard mixed with salt and spices. Because kimchi cannot be pasteurised for its sensory quality, the fermentation process still continues with the concomitant production of CO_2. The accumulation of CO_2 in the packages causes ballooning or even bursting. Scavengers might therefore be useful.

Principle

The reactant commonly used to scavenge CO_2 is calcium hydroxide, which, at a high enough water activity, reacts with CO_2 to form calcium carbonate:

$$Ca(OH)_2 + CO_2 \rightarrow CaCO_3 + H_2O$$

A disadvantage of this CO_2 scavenging substance is that it scavenges carbon dioxide from the package headspace irreversibly and results in depletion of CO_2, which is not always desired. In the case of packaged kimchi, depletion of CO_2 in the kimchi juices causes loss of the product's characteristic fresh carbonic taste. Therefore reversible absorption or adsorption by physical sorbents such as zeolites and active carbon may be an alternative (Lee *et al.*, 2001).

Commercially available technologies

Carbon dioxide scavengers are often commercialised as a sachet with a dual function, both O_2 and CO_2 scavenging. The O_2 and CO_2 scavenging sachet FreshLock or Ageless® E (Mitsubishi Gas Chemical Company, Japan), containing $Ca(OH)_2$ (Natawa *et al.*, 1982) is used for storing coffee. A similar sachet is Frehilizer type CV of Toppan printing Co. (Japan) (Smith *et al.*, 1995). Multiform Desiccants patented (US 5322701) a CO_2-absorbent sachet including a porous envelope containing CaO and a hydrating agent such as silica gel on which water is adsorbed. The water given off by the supersaturated silica gel combines with the calcium oxide to form calcium hydroxide (Cullen and Vaylen, 1994). Furthermore, a whole range of freshness-retaining mineral-based ethylene scavengers, mentioned in Table 3.4 also claim to scavenge carbon dioxide.

3.5.2 Odour scavengers

Role

As far as food aromas are concerned, plastics are usually considered to have a negative impact on food quality. Flavour scalping, i.e., sorption of food flavours by polymeric packaging materials, may result in loss of flavour and taste intensities and changes in the organoleptic profile of foods. However, flavour sorption could be used in a positive way to selectively absorb unwanted odours or flavours. Odour removers have the potential to scavenge the malodorous constituents of both oxidative and nonoxidative biochemical deterioration. Many foods such as fresh poultry and cereal products develop during product distribution very slight but nevertheless detectable deterioration odours such as sulphurous compounds and amines from protein/amino acid breakdown or aldehydes and ketons from lipid oxidation or anaerobic glycolysis. These odours are trapped within gas-barrier packaging so that, when the package is opened, they are released and detected by consumers. Another reason for incorporating odour removers into packages is to obviate the effect of odours developed in the package materials themselves (Vermeiren *et al.*, 1999; Brody *et al.*, 2001). Although removal of these undesired odours may be attractive from a commercial point of view, care must be taken as in some cases these odours may be a signal indicating that the products are exceeding the microbial or chemical limits.

Principle and commercial applications
An active packaging to reduce bitterness in grapefruit juices has been described. The causes of the bitter taste are glycosidic flavanone naringin and triterpenoid lactone limonin. Naringin is the bitter component found in most fresh citrus fruits and therefore in freshly processed citrus juices. Limonin is formed as a result of heat treatment of the juice during processing and a chemical reaction in the acidic juice medium. To counteract this, an active thin cellulose acetate (CA) layer for application on the inside of the packaging has been developed. This layer contains the fungal-derived enzyme naringinase, consisting of α-rhamnosidase and β-glucosidase, which hydrolyses naringin to naringenin and prunin, both non-bitter compounds. Food-contact approved CA films, which contained immobilised naringinase showed a 60% naringin hydrolysis in grapefruit juice in 15 days at 7ºC and a reduction in the limonin content due to adsorption on the CA film (Soares and Hotchkiss, 1998a,b, Vermeiren *et al.*, 1999).

Malodorous amines, resulting from protein breakdown in fish muscle, include strongly alkaline compounds (Rooney, 1995). A Japanese patent based on the interactions between acidic compounds, e.g., citric acid, incorporated in polymers and the alkaline off-odours, claims amine-removing capabilities. Hence the earliest work involved incorporation of such acids in heat-seal polymers such as polyethylene and extruding them as layers in packaging (Rooney, 1995; Hoshino and Osanai, 1986). Another approach to remove amine odours has been provided by the ANICO Co. (Japan). The ANICO bags made from a film containing ferrous salt and an organic acid such as citric or ascorbic acid are claimed to oxidise the amine or other oxidisable odour-causing compounds as they are absorbed by the polymer (Rooney, 1995).

Aldehydes, formed from the breakdown of peroxides produced during the initial stages of auto-oxidation of fats and oils, can make a wide variety of fat-containing foods, such as potato crisps, biscuits and cereal products, organoleptically unacceptable. Removal of aldehydes such as hexanal and heptanal from package headspaces by means of the layer Bynel IXP101, a HDPE master batch, is claimed by Dupont Polymers (Rooney, 1995). Brody *et al.* (2001) described laboratory tests on peanut butter, coffee and a snack product demonstrating the effectiveness of the aldehyde scavenger. DuPont's materials are compositions of polyalkylene imine (PAI), particularly polyethylene imine and polyolefin polymer. The invention comprises a discontinuous PAI phase and an olefinic polymer continuous phase in a weight ratio of PAI to olefinic polymer of about 0.001 to 30:100 (Brody *et al.*, 2001). DuPont also developed a scavenger for the removal of hydrogen sulphide that could be incorporated into the lid of packaged processed cured poultry (Brody *et al.*, 2001).

Some commercialised odour-absorbing sachets, e.g., MINIPAX® and STRIPPAX® (Multisorb technologies, USA) absorb the odours mercaptanes and H_2S developing in certain packaged foods during distribution (Vermeiren *et al.*,1999).

2-in-1™ from United Desiccants (USA) is a combination of silica gel and activated carbon packaged together for use in controlling moisture, gas and odour within packaged products. Furthermore, a whole range of freshness-retaining mineral-based ethylene scavengers, mentioned in Table 3.4 also claim to absorb ammonia, hydrogen sulphide and other unpleasant odours. Ecofresh and Profresh® (E-I-A Warenhandels GmbH, Vienna) are claimed to be fresh keeping and malodour control master batches (Vermeiren *et al.*, 1999; Brody *et al.*, 2001).

UOP Corporation reported on the odour absorbing properties of a molecular-sieve technology 'Smellrite/Abscents'. This material, a crystalline zeolite, has molecular sized pores that trap odour within its structure (Brody *et al.*, 2001). Vitamin E or alpha-tocopherol has been marketed as a food-grade odour remover in packaging materials. Michigan State University researchers concluded that alpha-tocopherol should be considered for incorporation into package materials for food products such as crackers or potato chips (crisps) in which lipid oxidation is a major concern (Brody *et al.*, 2001).

Flavour incorporation in packaging material might be used to minimise flavour scalping. Flavour release might also provide a means to mask off-odours coming from the food or the packaging. It is of importance that this technology is not misused to mask the development of microbial off-odours thereby concealing the marketing of products that are below standard or even dangerous for the consumer (Nielsen, 1997).

3.6 Future trends

Although more successfully applied in the US, Japan and Australia than in Europe, active packaging is still in its early stages and has a distance to travel before being applied on a large scale. However, the group of the scavengers seems to have the best chance to become popular. The effectiveness of these types of active systems has been studied profoundly and a whole range of scavenging technologies has been patented and/or commercialised. However, consumers are not always very keen on the use of sachets in food packaging. This centres around fear of ingestion of the sachet even though the content is safe. Precautions to minimise the risk have been taken by clearly stating 'Do not eat' on the label and by legislating a minimum size of the sachets. Another concern is that the content of the sachet could leak out and adulterate the product (Smith *et al.*, 1995; Nielsen, 1997; Hurme and Ahvenainen, 1998). To avoid mishandling, abuse and resistance to sachets, scavengers can be incorporated in labels, oxygen scavenging films or crown corks. In Finland a consumer survey conducted in order to determine consumer attitudes towards O_2 scavengers revealed that the new concepts would be accepted if consumers are informed well by using reliable information channels. When the consumers understand the quality improvement and/or the assurance function of the scavengers, they will have more confidence in the safety of the food they buy (Mikkola *et al.*, 1997). In Europe, the introduction of scavenging technologies is limited because of legislative restrictions. Active compounds need to be registered on positive lists and the overall and specific migration limits need to be respected. Moreover, traditional migration testing is not always a realistic simulation of the real use of the scavenging system and could result in a serious overestimation of the migration of the active compound. The solution for this legislative issue is complex and will probably require some more time.

As legislative barriers disappear and more companies become aware of the economic advantages of using absorbent technology, and consumers accept this approach, the technology will be very likely to emerge as an important preservation technology.

3.7 References

ABE K and WATADA A E (1991), 'Ethylene absorbent to maintain quality of lightly processed fruits and vegetables', *Journal of food science*, 56(6), 1589–92.

ABE Y (1994), 'Active packaging with oxygen absorbers', in Ahvenainen R, Nattila-Sandholm T, Ohlsson T, *Minimal processing of foods*, VTT symposium 142, Espoo, 209–33.

ABELES F, MORGAN P and SALTVEIT M (1992), *Ethylene in plant biology*, San Diego, Academic Press Inc.

AGELESS® TECHNICAL INFORMATION (1994), *Ageless® – oxygen absorber preserving product purity, integrity and freshness*', Mitsubishi Gas Chemical Company, inc., Japan, 24p.

AGELESS® TECHNICAL INFORMATION (2002), www.mitsubishiAGELESS.com.

AHVENAINEN R and HURME E (1997), 'Active and smart packaging for meeting consumer demands for quality and safety', *Food additives and contaminants*, 14, 753–63.

AMOCO CHEMICALS BULLETIN AS-1 (1999a), *Amosorb oxygen scavenger concentrates*', Amoco Chemicals, USA.

AMOCO CHEMICALS BULLETIN AS-3 (1999b), *Freshness by Amosorb*, Amoco Chemicals, USA.

ANON. (1991), 'Oxygen absorption reaches new plateau and provides superior product protection', *Food engineering,* 63, 50.

ATCO® TECHNICAL INFORMATION (2002), *ATCO®* oxygen absorbers, Standa industrie, France.

BARKAI-GOLAN R (1990), 'Post harvest diseases suppression by atmospheric modifications' in Calderon M and Barkai-Golan R, *Food preservation by modified atmospheres*, Boca Raton, CRC Press, 237–64.

BEN-ARIE R and SONEGO L (1980), 'Modified-atmosphere storage of kiwifruit with ethylene removal', *Scientia Horticulturae*, 27, 263–73.

BERENZON S and SAGUY I S (1998), 'Oxygen absorbers for extension of crackers shelf-life', *Food science and technology*, 31, 1–5.

BIOKA TECHNICAL INFORMATION (1999), *Bioka oxygen absorber*, Bioka Oy, Finland.

BRODY A L, STRUPINSKY, E R and KLINE L R (2001), *Active packaging for food applications*, Pennsylvania, Technomic publishing company.

CALDIC TECHNICAL INFORMATION (2002), *Freshmax/Freshpax*, Multiform desiccants Inc.

CHAMARA D, ILLEPERUMA K, GALAPPATTY, P T and SARANANDA K H (2000), 'Modified atmosphere packaging of Kolikuttu bananas at low temperature', *Journal of Horticultural Science and Biotechnology*, 75(1), 9296.

CULLEN J S and VAYLEN N E (1994), *Carbon dioxide absorbent packet and process*, United States Patent 5322701.

DAIFAS D P, SMITH J P, BLANCHFIELD B and AUSTIN J W (1999), 'Growth and toxin production by *Clostridium botulinum* in English-style crumpets packaged under modified atmospheres', *Journal of food protection,* 62, 349–55.

DAREX TECHNICAL INFORMATION (2002). *Darex Active packaging technology brochure*.

DAY B P F (1998), 'Active packaging of foods', *CCFRA New technologies bulletin*, 17, 23p.

DON Y S and KOO L S (1996), 'Effect of ethylene removal and sulfur dioxide fumigation on grape quality during MA storage', *Journal of the Korean Society for Horticultural Sciences*, 37 (5), 696–9.

EDWARDS A M (1998), 'Striking back', *Packaging today*, 5, 34–6.

ELLIS W O, SMITH J P, SIMPSON B K, RAMASWAMY H and DOYON G (1994), 'Novel techniques for controlling the growth and aflatoxin production by *Aspergillus parasiticus* in packaged peanuts', *Food microbiology,* 11, 357–68.

FLOROS J D, DOCK L L and HAN J H (1997), 'Active packaging technologies and applications', *Food Cosmetics and Drug Packaging*, 20(1), 10–17.

GRAF E (1994), *Oxygen removal*, US patent 5284871.

GRAFF G (1998), 'O_2 scavengers give "smart" packaging a new lease on shelf-life', *Modern plastics*, 75 (2), 69–72.

GREENFIELD P F and LAURENCE R L (1975), 'Characterization of glucose oxidase and catalase on inorganic supports', *Journal of food science,* 40, 906–10.

HOLLAND R V (1992), ' *Absorbent material and uses thereof*, Australian Patent Application PJ6333.

HOPKINS W G (1995), *Introduction to plant physiology,* New York, John Wiley and Sons.

HOSHINO A and OSANAI T (1986), *Packaging films for deodorization*, Japanese patent 86209612.

HOWARD L R, YOO K S, PIKE L M and MILLER G H (1994), 'Quality changes in diced onions stored in film packages', *Journal of food science*, 59 (1), 110–17.

HURME E and AHVENAINEN R (1998), 'Active and smart packaging of food', *Technical report VTT*, Finland.

JACXSENS L (2000), 'Application of ethylene adsorbers for the storage of fresh fruit and vegetables', *International Conference on active and intelligent packaging*, Campden, CCFRA.

JAYARAMAN K S and RAJU P S (1992), 'Development and evaluation of a permanganate-based ethylene scrubber for extending the shelf-life of fresh fruits and vegetables', *Journal of Food Science and technology India*, 29(2), 77–83.

KADER A A (1985), 'Ethylene-induced senescence and physiological disorders in harvested horticultural crops', *Hortscience*, 20(1), 54–57.

LABUZA T P (1987), 'Oxygen scavenger sachets', *Food research*, 32, 276–7.

LABUZA T P and BREENE W M (1989), 'Applications of 'active packaging' for improvement of shelf-life and nutritional quality of fresh and extended shelf-life foods', *Journal of food processing and preservation*, 13, 1–69.

LEE D S, SHIN D H, LEE D U, KIM, J C and CHEIGH H S (2001), 'The use of physical carbon dioxide absorbents to control pressure build-up and volume expansion, of kimchi packages', *Journal of food engineering*, 48, 183–8.

LIU F W (1970), 'Storage of bananas in polyethylene bags with an ethylene absorbent', *Hortscience*, 5(1), 25–7.

LYVER A, SMITH J P, AUSTIN J and BLANCHFIELD B (1998), 'Competitive inhibition of *Clostridium botulinum* type E by Bacillus species in a value-added seafood product packaged under a modified atmosphere', *Food research international*, 31, 311–19.

MATSUI M (1989), *Film for keeping freshness of vegetables and fruit*, US patent 4847145.

MCGLASSON W B (1985), 'Ethylene and fruit ripening', *Hortscience*, 20(1), 51–4.

MIKKOLA V, LÄHTEENMÄKI L., HURME E, HEINIÖ R, JÄRVI-KÄÄRIÄINEN T and AHVENAINEN R (1997), *Consumer attitudes towards oxygen absorbers in food packages*, VTT research notes 1858, Espoo.

MILTZ J, PASSY N and MANNHEIM C H (1995), 'Trends and applications of active packaging systems', in Ackermann P, Mortimer R G (1993), *Physical chemistry*, California, The Benjamin/Cummings Publishing Co.

NAKAMURA H and HOSHINO J (1983), 'Techniques for the preservation of food by employment of an oxygen absorber', Mitsubishi Gas Chemical Co., Tokyo, Ageless® Division, 1–45.

NATAWA T, KOMATSU T and OHTSUKA M (1982), *Oxygen and carbon dioxide absorbent and process for storing coffee by using the same*', United States Patent 4366179.

NIELSEN T (1997), *Active packaging-a literature review*, SIK-rapport n 631, Sweden.

OXYGUARD TECHNICAL INFORMATION (2002), *Oxyguard technology*, Toyo Seikan, Japan.

ROONEY M L (1981). 'Oxygen scavenging from air in package headspaces by singlet oxygen reactions in polymer media', *Journal of Food Science*, 47,

291–4, 298.

ROONEY M L (1995), 'Active packaging in polymer films', in Rooney M L, *Active food packaging,* London, Blackie Academic and Professional, 74–110.

ROUSSEL (1999), 'Les emballages absorbeurs d'oxygene', in Gontard N, *Les emballages actifs*, Paris, Tec and Doc, 31–7.

SALTVEIT M E (1999), 'Effect of ethylene on quality of fresh fruits and vegetables', *Postharvest biology and technology*, 15, 279–92.

SAWADA S and TOTSUKA T (1986), 'Natural and anthropogenic sources and fate of atmospheric ethylene', *Atmospheric Environment*, 20, 821–31.

SCOTT K J and WILLS R B H (1974), 'Reduction of brown heart in pears by absorption of ethylene from the storage atmosphere', *Australian Journal of Experimental Agriculture and Animal Husbandry*, 14, 266–8.

SHERMAN M (1985), 'Control of ethylene in the post harvest environment', *Hortscience*, 20(1), 57–60.

SHORTER A J, SCOTT, K J, WARD G and BEST D J (1992), 'Effect of ethylene absorption on the storage of Granny Smith apples held in polyethylene bags', *Postharvest Biology and technology*, 1(3), 189–94.

SILGELAC TECHNICAL INFORMATION (1998), *L'emballage actif- une technologie innovante pour vos emballages alimentaires*, Silgelac, France.

SMITH J P (1996), 'Improving shelf life of packaged baked goods by oxygen absorbents', *AIB research department technical bulletin*, 18, 2–7.

SMITH J P, HOSHINO, J and ABE Y (1995), 'Interactive packaging involving sachet technology', in Rooney M L, *Active Food Packaging*, London, Blackie Academic and Professional, 143–73.

SMITH J P, OORAIKUL B, KOERSEN W J, JACKSON E D and LAWRENCE R A (1986), 'Novel approach to oxygen control in modified atmosphere packaging of bakery products', *Food microbiology*, 3, 315–20.

SMITH J P, RAMASWAMY H S and SIMPSON B K (1990), 'Developments in food packaging technology, Part 2: Storage aspects', *Trends in food science and technology*, 11, 111–18.

SOARES N F F and HOTCHKISS J H (1998a), 'Bitterness reduction in grapefruit juice through active packaging', *Packaging technology and science*, 11, 9–18.

SOARES N F F and HOTCHKISS J H (1998b), 'Naringinase immobilisation in packaging films for reducing naringin concentration in grapefruit juice', *Journal of food science*, 63, 61–5.

SOMEYO and NOBUO (1992), *Packaging sheet for perishable goods*, US patent 5084337.

SUSLOW T (1997), 'Performance of zeolite based products in ethylene removal', *Perishables handling quarterly*, 92, 32–3.

VERMEIREN L, DEVLIEGHERE F, VAN BEEST M, DE KRUIJF N and DEBEVERE J (1999), 'Developments in the active packaging of foods', *Trends in Food science and technology,* 10, 77–86.

WHITING R C and NAFTULIN K A (1992), 'Effect of headspace oxygen concentration on growth and toxin production by proteolytic strains of

Clostridium botulinum', *Journal of foods protection*, 55, 23–7.

YANG S F (1985), 'Biosynthesis and action of ethylene', *HortScience*, 20(1), 41–5.

ZAGORY D (1995), 'Ethylene-removing packaging', in Rooney M L, *Active food packaging*, London, Blackie Academic and Professional, 38–54.

4

Antimicrobial food packaging

J. H. Han, The University of Manitoba, Canada

4.1 Introduction

Antimicrobial packaging is one of many applications of active packaging (Floros *et al.*, 1997). Active packaging is the packaging system which possesses attributes beyond basic barrier properties, which are achieved by adding active ingredients in the packaging system and/or using actively functional polymers (Han and Rooney, 2002). Antimicrobial packaging is the packaging system that is able to kill or inhibit spoilage and pathogenic microorganisms that are contaminating foods. The new antimicrobial function can be achieved by adding antimicrobial agents in the packaging system and/or using antimicrobial polymers that satisfy conventional packaging requirements. When the packaging system acquires antimicrobial activity, the packaging system (or material) limits or prevents microbial growth by extending the lag period and reducing the growth rate or decreases live counts of microorganisms (Han, 2000).

Compared to the goals of conventional food packaging such as (i) shelf-life extension, (ii) quality maintenance, and (iii) safety assurance which could be achieved by various methods, antimicrobial packaging is specifically designed to control microorganisms that generally affect the above three goals adversely. Therefore some products, which are not sensitive to microbial spoilage or contamination, may not need the antimicrobial packaging system. However, most foods are perishable and most medical/sanitary devices are susceptible to contamination. Therefore, the primary goals of an antimicrobial packaging system are (i) safety assurance, (ii) quality maintenance, and (iii) shelf-life extension, which is the reversed order of the primary goals of conventional packaging systems. Nowadays food security is a big issue and antimicrobial packaging could play a role in food security assurance.

All antimicrobial agents have different activities which affect microorganisms differently. There is no 'Magic Bullet' antimicrobial agent effectively working against all spoilage and pathogenic microorganisms. This is due to the characteristic antimicrobial mechanisms and due to the various physiologies of the microorganisms. Simple categorisation of microorganisms may be very helpful to select specific antimicrobial agents. Such categories may consist of oxygen requirement (aerobes and anaerobes), cell wall composition (Gram positive and Gram negative), growth-stage (spores and vegetative cells), optimal growth temperature (thermophilic, mesophilic and psychrotropic) and acid/osmosis resistance. Besides the microbial characteristics, the characteristic antimicrobial function of the antimicrobial agent is also important to understand the efficacy as well as the limits of the activity. Some antimicrobial agents inhibit essential metabolic (or reproductive genetic) pathways of microorganisms while some others alter cell membrane/wall structure. For example, lysozyme destroys cell walls without the inhibition of metabolic pathways and results in physical cleavages of cell wall, while lactoferrin and EDTA act as coupling agents of essential cationic ions and charged polymers. Two major functions of microbial inhibition are microbial-cidal and microbial-static effects. In the case of microbial-static effects, the packaging system has to possess the active function of maintaining the concentration above the minimal inhibitory concentration during the entire storage period or shelf-life in order to prevent regrowth of target microorganisms.

Traditional preservation methods sometimes consist of antimicrobial packaging concepts, which include sausage casings of cured/salted/smoked meats, smoked pottery/oak barrels for fermentation, and bran-filled pickle jars. The basic principle of these traditional preservation methods and antimicrobial packaging is a hurdle technology (Fig. 4.1). The extra antimicrobial function of the packaging system is another hurdle to prevent the degradation of total quality of packaged foods while satisfying the conventional functions of moisture and oxygen barriers as well as physical protection. The microbial hurdle may not contribute to the protection function from physical damage. However, it provides tremendous protection against microorganisms, which has never been achieved by conventional moisture and oxygen barrier packaging materials.

Antimicrobial functions which are achieved by adding antimicrobial agents in the packaging system or using antimicrobial polymeric materials show generally three types of mode; (i) release; (ii) absorption; and (iii) immobilisation. Release type allows the migration of antimicrobial agents into foods or headspace inside packages, and inhibits the growth of microorganisms. The antimicrobial agents can be either a solute or a gas. However, solute antimicrobial agents cannot migrate through air gaps or over the space between the package and the food product, while the gaseous antimicrobial agents can penetrate through any space. Absorption mode of antimicrobial system removes essential factors of microbial growth from the food systems and inhibits the growth of microorganisms. For example, the oxygen-absorbing system can prevent the growth of moulds inside packages. Immobilisation system does not

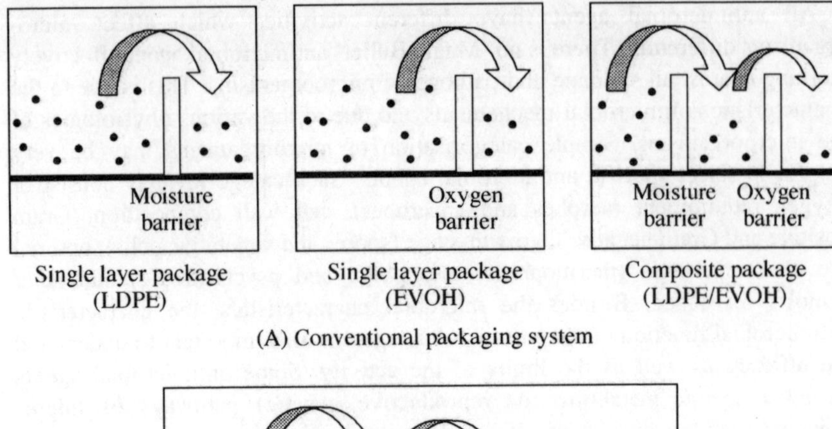

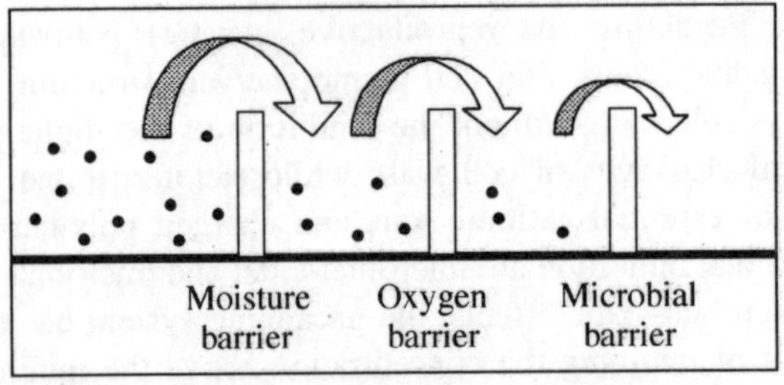

Fig. 4.1 Hurdle technology in antimicrobial packaging system compared to the conventional packaging system.

release antimicrobial agents but suppresses the growth of microorganisms at the contact surface. Immobilisation systems may be less effective in the case of solid foods compared to the liquid foods because there is less possibility for contact between the antimicrobial package and the whole food products.

4.2 Antimicrobial agents

There are many antimicrobial agents that exist and are widely used. To be able to use antimicrobial agents in the foods, pharmaceuticals and cosmetic products, the industry must follow the guidelines and regulations of the country that they are going to use them in, for example, FDA and/or EPA in the United States. This implies that new antimicrobial packaging materials may be developed using only agents which are approved by the authorisation agencies as examples of FDA-approved or notified-to-use within the concentration limits for food safety enhancement or preservation. Various antimicrobial agents may be incorporated in the packaging system, which are chemical antimicrobials, antioxidants, biotechnology products, antimicrobial polymers, natural antimicrobials and gas (Table 4.1).

Chemical antimicrobial agents are the most common substances used in the industry. They include organic acids, fungicides, alcohols and antibiotics.

Table 4.1 Antimicrobial agents and packaging systems

Antimicrobials	Packaging materials	Foods	Microorganisms	References
Organic acids				
Benzoic acids	PE	Tilapia fillets	Total bacteria	Huang *et al.*, 1997
	Ionomer	Culture media	*Pen. spp., Asp. nige*	Weng *et al.*, 1997
Parabens	LDPE	Simulants	Migration test	Dobias *et al.*, 2000
	PE coating	Simulants	Migration test	Chung *et al.*, 2001a
	Styrene-acrylates	Culture media	*S. cerevisiae*	Chung *et al.*, 2001b
Benzoic & sorbic acids	PE-co-met-acrylates	Culture media	*Asp. niger, Pen. spp.*	Weng *et al.*, 1999
Sorbates	LDPE	Culture media	*S. cerevisiae*	Han and Floros, 1997
	PE, BOPP, PET	Water, cheese	Migration test	Han and Flores, 1998a; b
	LDPE	Cheese	Yeast, mould	Devileghere *et al.*, 2000a
	MC/palmitic acid	Water	Migration test	Rico-Pena and Torres, 1991
	MC/HPMC/fatty acid	Water	Migration test	Vojdana and Torres, 1990
	MC/chitosan	Culture media		Chen *et al.*, 1996
	Starch/glycerol	Chicken breast		Baron and Summer, 1993
	WPI	Culture media	*S. cerevisiae,.* *Asp niger, Pen. roqueforti*	Ozdermir, 1999
	CMC/paper	Cheese		Ghosh *et al.*, 1973, 1977
Sorbic anhydride	PE	Culture media	*S. cerevisiae, moulds*	Weng and Chen 1997; Weng and Hotchkiss, 1993
Sorbates & propionates	PE/foil	Apples	Firmness test	Yakovlleva *et al.*, 1999
Acetic, propionic acid	Chitosan	Water	Migration test	Ouattara *et al.* 2000a
Enzymes				
Lysozyme, nisin, EDTA	SPI, zein	Culture media	*E. coli, Lb. plantarum*	Padgett *et al.*, 1998
Lysozyme, nisin, EDTA, propyl paraben	WPI	Culture media	*L. monocytogenes, Sal. typhimurium, E. coli, B. thermosph., S. aureus*	Rodrigues and Han, 2000; Rodrigues *et al.*, 2002

Table 4.1 Continued

Antimicrobials	Packaging materials	Foods	Microorganisms	References
Immobilised lysozyme	PVOH, nylon, cellulose acetate	Culture media	Lysozyme activity test	Appendini and Hotchkiss, 1996; 1997
Glucose oxidase		Fish		Fields *et al.*, 1986
Bacteriocins				
Nisin	PE	Beef	*B. thermosph.*	Siragusa *et al.*, 1999
	HPMC	Culture media	*L. monocytogenes, S. aureus*	Coma *et al.*, 2001
	Corn zein	Shredded cheese	Total aerobes	Cooksey *et al.*, 2000
Nisin, lacticins	Polyamide/LDPE,	Culture media	*M. favus, L. monocytogenes*	An *et al.*, 2000
Nisin, lacticin, salts	Polyamide/LDPE	Culture media	*M. flavus*	Kim *et al.* 2000
Nisin, EDTA	PE, PE-co-PEO	Beef	*B. thermosphacta*	Cutter *et al.*, 2001
Nisin, citrate, EDTA	PVC, nylon, LLDPE	Chicken	*Sal. typhimurium*	Tatrajan and Sheldon 2000
Nisin, organic acids mixture	Acrylics, PVA-co-PE	Water	Migration test	Choi *et al.*, 2001
Nisin, lauric acid	Zein	Simulants	Migration test	Hoffman *et al.*, 2001
Nisin, pediocin	Cellulose casing	Turkey breast, ham, beef	*L. monocytogenes*	Ming *et al.* 1997
Fungicides				
Benomyl	Ionomer	Culture media		Halek and Garg, 1989
Imazalil	LDPE	Bell pepper		Miller *et al.*, 1984
	PE	Cheese	Moulds	Weng and Hotchkiss, 1992

Polymers				
Chitosan	Chitosan/paper	Strawberry	*E. coli*	Yi *et al.*, 1998
Chitosan, herb extract	LDPE	Culture media	*Lb. plantarum, E. coli, S. cerevisiae, Fusarium oxysporum*	Hong *et al.*, 2000
UV/excimer laser irradiated nylon	Nylon	Culture media	*S. aureus, Pseudo. fluorescens, Enterococcus faecalis*	Paik *et al.*, 1998; Paik and Kelly, T995
Natural extract				
Grapefruit seed extract	LDPE, nylon	Ground beef	Aerobes, coli-forms	Ha *et al.*, 2001
	LDPE	Lettuce, soy-sprouts	*E. coli, S. aureus*	Lee *et al.*, 1998
Clove extract	LDPE	Culture media	*L. plantarum, E coli, F. oxysporum, S.cerevisiae*	Hong *et al.*, 2000
Herb extract, Ag-Zirconium	LDPE	Lettuce, cucumber	*E. coli, ,S. aureus, L mesenteroides, S. cerevisiae, Asp. spp, Pen. spp.*	An *et al.*, 1998
	LDPE	Strawberry	*Firmness*	Chung *et al.*, 1998
Eugenol, cinnam aldehyde,	Chitosan	Bologna, ham	Enterobac., lactic acid bacteria,*Lb. sakei Serratia spp.*	Outtara *et al.*, 2000b
Horseradish extract	Paper	Ground beef	*E. coli 0157: H7*	Nadarajah *et al.*, 2002
Allyl isothiocyanate	PE film/pad	Chicken, meats, smoked salmon	*E. coli, S. enteritidis, L. monocytogenes*	Takeuchi and Yuan, 2002

Table 4.1 Continued

Antimicrobials	Packaging materials	Foods	Microorganisms	References
Oxygen absorber				
Ageless	Sachet	Bread	Moulds	Smith *et al.*, 1989
BHT	HDPE	Breakfast cereal		Hoojjatt *et al.*, 1987
Gas				
Ethanol	Silicagel sachet	Culture media		Shapero *et al.*, 1978
	Silicon oxide (Ethicap) sachet	Bakery		Smith *et al.*, 1987
Hinokithiol	Cyclodextrin/plactic (Seiwa) sachet	Bakery		Gontard, 1997
$C10_2$	Plastic films		Migration test	Ozen and Floros, 2001
Others				
Hexamethyleneteh tramine	LDPE	Orange juice	Yeast, lactic acid bacteria	Devlieghere *et al.* 2000b
Silver zeolite, silver nitrate	LDPE	Culture media	*S. cerevisiae, E. coli, S. aureus, Sal. typhimurium, Vibrio parahaemolyticus*	Ishitani, 1995
Antibiotics	PE	Culture media	*E. coli, S. aureus, Sal. typhimurium, Klebsiella neumoniae*	Han and Moon, 2002

MC: methyl cellulose; HPMC: hydroxypropyl methyl cellulose; WPI: whey protein isolate; CMC: carboxyl methyl cellulose; SPI: soy protein isolate

Organic acids such as benzoic acids, parabens, sorbates, sorbic acid, propionic acid, acetic acid, lactic acid, medium-size fatty acids and their mixture possess strong antimicrobial activity and have been used as food preservatives, food contact substances and food contact material sanitisers. Benomyl and imazalil had been incorporated in plastic films and demonstrated antifungal activity. Ethanol has strong antibacterial and antifungal activity, however, it is not sufficient to prevent the growth of yeast. Ethanol may enhance some volatile flavour compounds but also causes a strong undesirable chemical odour in most food products. Some antibiotics can be incorporated into animal feedstuffs for the purpose of disease treatment, disease prevention or growth enhancement as well as human disease curing. The use of antibiotics as package additives is not approved for the purpose of antimicrobial functions and is also controversial due to the development of resistant microorganisms. However, antibiotics may be incorporated for short-term use in medical devices and other non-food products.

Antioxidants are effective antifungal agents due to the restrictive oxygen requirement of moulds. Food grade chemical antioxidants could be incorporated into packaging materials to create an anaerobic atmosphere inside packages, and eventually protect the food against aerobic spoilage (Smith *et al.*, 1990). Since the package did not contain oxygen, the partial pressure difference of oxygen is formed between the outside and inside of packaging materials. Therefore, in order to maintain the low concentration of oxygen inside the package, the packaging system requires high oxygen barrier materials such as EVOH, PVDC or aluminum foil that prevent the permeation of oxygen. Besides the antioxidants, a multi-ingredient oxygen scavenging system, such as commercial oxygen-absorbing sachets, can be used to reduce oxygen concentration inside the package.

Various bacteriocins that are produced by microorganisms also inhibit the growth of spoilage and pathogenic microorganisms. These fermentation products include nisin, lacticins, pediocin, diolococcin, and propionicins (Daeschul, 1989; Han, 2002). These biologically active peptides possess strong antimicrobial properties against various bacteria. Other non-peptide fermentation products such as reuterin also demonstrate antimicrobial activity. Besides the above food grade bacteriocins, other bacteriocins would be utilised for the development of antimicrobial packaging systems.

Some synthetic or natural polymers also possess antimicrobial activity. Ultraviolet or excimer laser irradiation can excite the structure of nylon and create antimicrobial activity. Among natural polymers, chitosan (chitin derivative) exhibits antimicrobial activity. Short or medium size chitosan possesses quite good antimicrobial activity, while long change chitosan is not effective. Chitosan has been approved as a food ingredient from FDA recently; therefore, the use of chitosan for new product development as well as a natural antimicrobial agent would become more popular.

The use of natural plant extracts is desirable for the development of new food products and nutraceuticals, as well as new active packaging systems. Some plant extracts such as grapefruit seed, cinnamon, horseradish and clove have been added to packaging systems to demonstrate effective antimicrobial activity against spoilage and pathogenic bacteria. More use of natural extracts is expected because of the easier regulation process and consumer preference when compared to the chemical antimicrobial agents.

Gaseous antimicrobials have some benefit compared to the solid or solute types of chemical antimicrobial agents. They can be vaporised and penetrated into any air space inside packages that cannot be reached by non-gaseous antimicrobial agents. An ethanol sachet is one example of a gaseous antimicrobial system. Headspace ethanol vapour can inhibit the growth of moulds and bacteria. The use of chlorine dioxide has been permitted with no objection notification from FDA recently and can be incorporated into packaging material. Chlorine dioxide shows effective antimicrobial activity and some bleaching effect. Allyl isothiocyanate, hinokithiol and ozone have been incorporated into packages and demonstrated effective antimicrobial activity. However, the use of these reactive gaseous agents has to be considered after careful studies of their reactivity and permeability through packaging materials.

Since most antimicrobial agents have different antimicrobial mechanisms, the mixture of antimicrobial agents can increase antimicrobial activity through synergic mechanisms when they do not have any interference mechanisms. Therefore, the optimisation study on the combination of various antimicrobials will extend the antimicrobial activity of the mixture and maximise the efficacy and the safety of the antimicrobial packaging system.

4.3 Constructing an antimicrobial packaging system

Antimicrobial agents can be incorporated into a packaging system through simple blending with packaging materials, immobilisation or coating differently depending on the characteristics of packaging system, antimicrobial agent and food. The blended antimicrobial agents can migrate from packaging materials to foods, while the immobilised agent cannot migrate. Fig. 4.2 explains the antimicrobial systems and their releasing profiles. Systems (A) and (B) release antimicrobial agents through diffusion, while systems (C) and (D) release volatile antimicrobial agents by evaporation. Fig. 4.2 presents (A) One-layer system: the antimicrobial agent is incorporated into the packaging material or chemically bound on the packaging material by immobilisation. (B) Two-layer system: the antimicrobial agent (outer layer) is coated on the packaging material (inner layer), or the antimicrobial matrix layer (outer layer) is laminated with the control layer (inner layer) to control the release rate specifically. (C) Headspace system: the volatile antimicrobial agent initially incorporated into the matrix layer releases into the headspace. Headspace antimicrobial agent is partitioned with the food product by equilibrium sorption/isotherm. (D) Headspace system with control layer: the control layer specifically controls the permeation of the volatile antimicrobial agent and maintains specific headspace concentration.

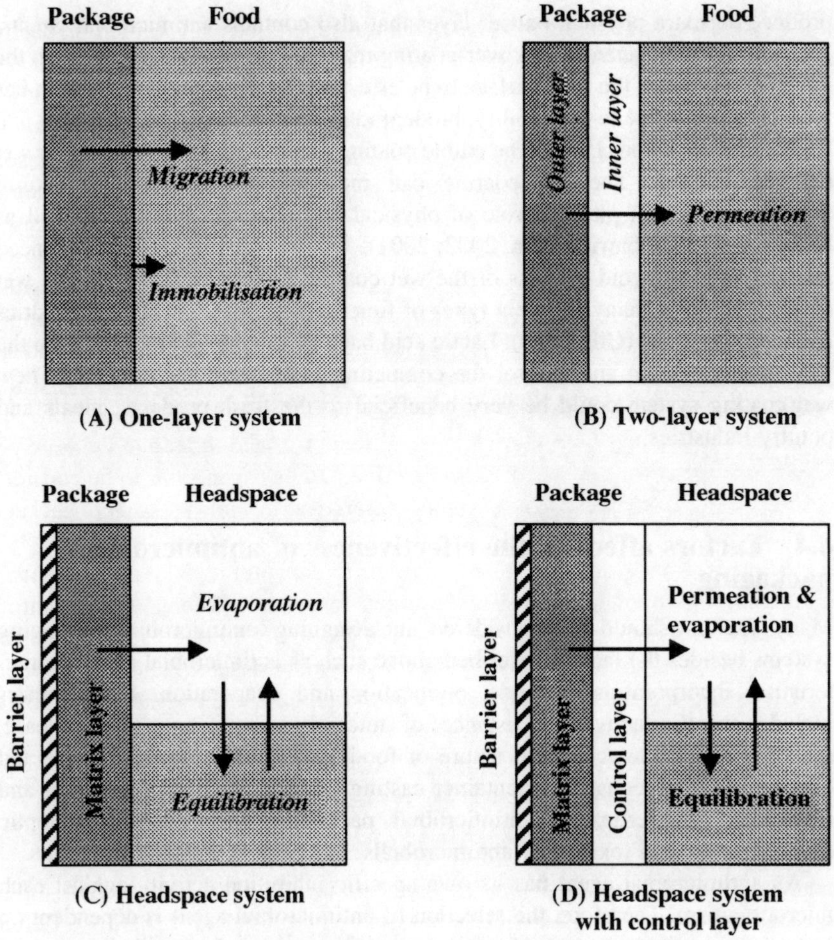

Fig. 4.2 Antimicrobial packaging system

As mentioned above, the gaseous (volatile) agents can evaporate into the headspace of the packaging system and reach the food. As examples of incorporation processes, the antimicrobial agents were impregnated into packaging materials before final extrusion (Han and Floros, 1997; Nam *et al*., 2002), dissolved into coating solvents (Rodrigues and Han, 2000; Rodrigues *et al*., 2002), or mixed into sizing/filling materials of paper and paperboards (Nadarajah *et al*., 2002). Chemical immobilisation utilises the covalent binding of the agents into chemical structures of packaging materials when regulation does not permit migration of the agents into foods (Appendini and Hotchkiss, 1996; 1997; Halek and Garg, 1989; Miller *et al*., 1984). The immobilised antimicrobial agents will inhibit the growth of microorganisms on the contact surface of packaged products.

The coating process can produce an antimicrobial packaging system. Over-coating on the pre-packaged products or edible coating on the food itself can produce an extra physical barrier layer that also contains antimicrobial agents. The antimicrobial agent in the over-coating material has to penetrate through the inner liner to reach the food surface to be effective. An edible coating system has various benefits due to its edibility, biodegradability and simplicity (Krochta and De Mulder-Johnston, 1997). The edible coating may be either dry coating or wet battered coating. The dry coating can incorporate chemical and natural antimicrobials, and play the role of physical and chemical barriers as well as being a microbial barrier (Han, 2002; 2001). The wet coating system may need another wrap to avoid the loss of the wet coating materials. However, the wet system can carry many different types of functional agents as well as probiotics and antimicrobials (Gill, 2000). Lactic acid bacteria can be incorporated into the wet coating system and control the competing undesirable bacteria. This new wet coating system could be very beneficial to the fresh products, meats and poultry industries.

4.4 Factors affecting the effectiveness of antimicrobial packaging

Many factors should be considered in designing antimicrobial packaging systems besides the factors described above such as antimicrobial agent characteristics, incorporation methods, permeation and evaporation. Extra factors include specific activity, resistance of microorganisms, controlled release, release mechanisms, chemical nature of foods and antimicrobials, storage and distribution conditions, film/container casting process conditions, physical and mechanical properties of antimicrobial packaging materials, organoleptic characteristics and toxicity of antimicrobials, and corresponding regulations.

An antimicrobial agent has its own specific inhibition activity against each microorganism. Therefore, the selection of antimicrobial agent is dependent on its activity against a target microorganism. Due to the characteristics of food products such as pH, water activity, compositions and storage temperature, the growth of potential microorganisms that can spoil food products are predictable. The antimicrobial agent has to be selected by the inhibition activity of the agent against the targeted potential microorganisms in the environmental conditions of the packaged foods.

The design of an antimicrobial packaging system requires controlled release technology and microbial growth kinetics. When the migration rate of an antimicrobial agent is faster than the growth rate of the target microorganism, the antimicrobial agent will be depleted before the expected storage period and the packaging system will lose its antimicrobial activity because the packaged food has an almost infinite volume compared to the volume of packaging material and the amount of antimicrobial agent. Consequently the microorganism will start to grow after the depletion of the antimicrobial agent. On the other hand, when the release rate is too slow to control the growth of the microorganism, the microorganism can grow instantly before the antimicrobial

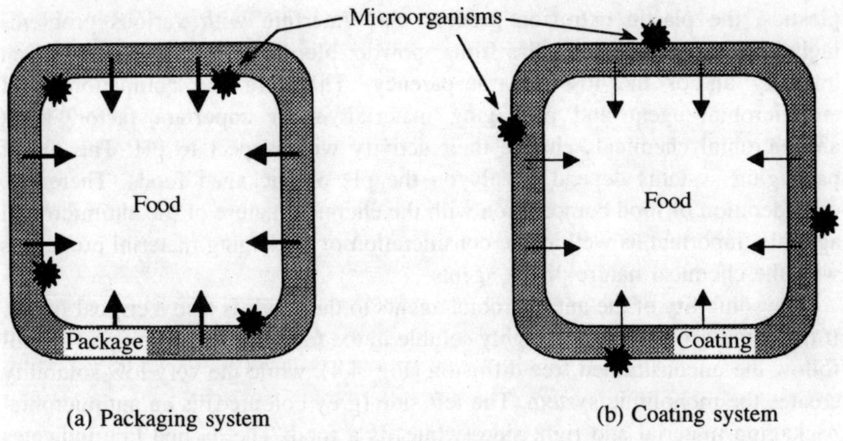

Fig. 4.3 Antimicrobial packaging and edible coating systems

agent is released. Therefore, the release rate of the antimicrobial agent from the packaging material to food is specifically controlled to match the release rate with the growth kinetics of the target microorganism. Figure 4.3 also shows the importance of the mass transfer kinetics and growth kinetics in the cases of a film-packaging system and a coating system. Antimicrobial agents which have been incorporated in the package material (A) or coating material (B) will migrate into the foods during storage and distribution. Packaging system (A) mostly has contaminating microorganisms on the surface of the food product inside the package, while coating system (B), which has been coated by the antimicrobial material, may have the contaminating microorganisms on the surface of coating layer. The migration of antimicrobial agents from the package into the food product is an essential phenomenon to inhibit the growth of microorganisms on the surface of food products. While the concentration of antimicrobial agents is maintained over the m.i.c. (minimal inhibitory concentration) on the food surface, the system actively presents effective antimicrobial activity. However, the migration of the incorporated antimicrobial agents from the coating layer into the food product dilute the concentration in the coating layer. Compared to the volume of the coating layer, the coated food has almost infinite volume. Therefore, the migration will deplete the antimicrobial agent inside the coating layer, reduce the concentration below m.i.c. and eliminate the antimicrobial activity of the coating system. The migration of the incorporated antimicrobial agents contributes the antimicrobial effectiveness in the case of packaging systems. On the other hand no migration is beneficial to the coating system.

The chemical nature of antimicrobial agents is also an important factor. Some agents are soluble in water but some are not. If water-soluble agents are mixed into plastic resins to make antimicrobial films, special consideration of film properties should be involved to obtain high quality films. Due to the hydrophilic nature of the agents compared with the hydrophobic nature of plastics, the plastic extrusion process may interfere with various problems including hole creation in the films, powder-blooming, the loss of physical integrity and/or the loss of transparency. Therefore the compatibility of antimicrobial agent and packaging material is an important factor. Most antimicrobial chemicals change their activity with respect to pH. The pH of packaging systems depend mostly on the pH of packaged foods. Therefore, consideration of food composition with the chemical nature of the antimicrobial agent is important as well as the consideration of packaging material properties with the chemical nature of the agents.

The solubility of the antimicrobial agents to the foods is also a critical factor. If the antimicrobial agent is highly soluble in the food, the migration profile will follow the unconstrained free diffusion (Fig. 4.4), while the very low solubility creates the monolithic system. The left side (grey coloured) is an antimicrobial packaging material and right side (white) is a food. The dashed line indicates m.i.c. of antimicrobial agents. Unconstrained free diffusion model (A) shows the highly soluble antimicrobial agent positioned in the packaging material migrating into the food layer and the concentration of the antimicrobial agent inside the package decreases as migration continues. The concentration of the antimicrobial agent on the surface of the food (C_s) decreases as the concentration inside the package decreases and eventually reduces below the m.i.c. losing the antimicrobial activity. Monolithic system (B) consists of not-very-soluble (or lower affinity) migrants to the food layer. In this system, the concentration of antimicrobial agent on the surface of food (C_s) is much lower than that of soluble migrants. The concentration is highly dependent on the

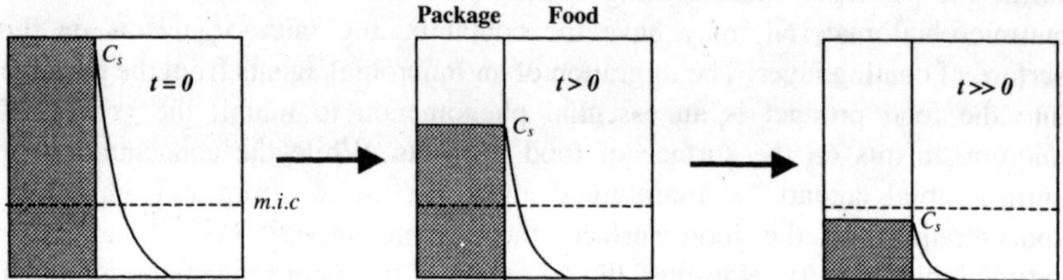

(A) Release of soluble antimicrobial agents through free diffusion

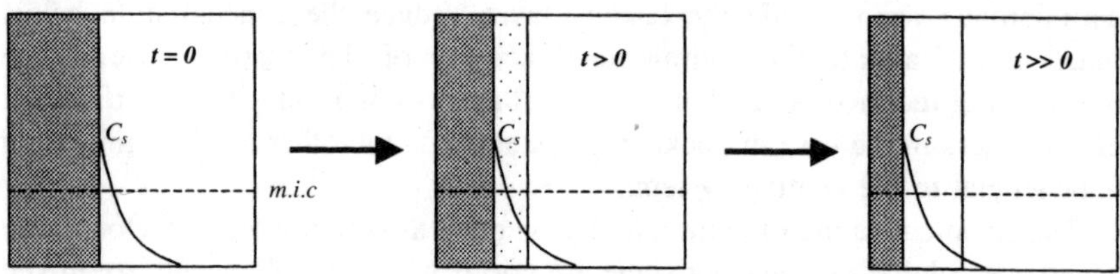

(B) Release of antimicrobial agents from monolithic system

Fig. 4.4 Changes in the concentration of antimictobial agent in two different antimicrobial packaging systems.

solubility of the antimicrobial agent in the food. Until complete depletion of the antimicrobial agent in the package, the surface concentration (C_s) is maintained as a constant concentration (actually maximum solubility) maintaining constant antimicrobial activity, while the total amount of antimicrobial agent inside the package decreases. The migration profile of antimicrobial agents should be fully understood and able to be controlled to achieve predictable antimicrobial effectiveness during entire shelf-life. The importance of controlled release principles is described in Fig. 4.5. The integrated areas ($\Sigma C_s \times \Delta time$) below the concentration lines of both systems have a similar value assuming an equal amount of initial antimicrobial agents. However, system (B) has the longer period of effectiveness in which concentration is above the m.i.c. compared with the period of system (A).

Storage and distribution conditions are important factors. The conditions include storage temperature and time. This time-temperature integration affects the microbial growth profile. To prevent microbial growth, a storage period at the favourable temperature range for microbial growth should be avoided or minimised during the whole period of storage and distribution.

In the case of controlled atmosphere storage or modified atmosphere packaging, active gas permeation through the packaging materials should be controlled to maintain optimum gas composition during the whole period of storage and distribution. When the gas composition is altered through unexpected gas permeation or seal defect, microorganisms that are not considered as target microorganisms may spoil the packaged foods.

Film/container casting methods are important to maintain antimicrobial effectiveness. There are two casting methods; one is extrusion and the other is solvent casting. In the case of extrusion, the critical variables related to residual antimicrobial activity are extrusion temperature and specific mechanical energy input. The extrusion temperature is related to the thermal degradation of the antimicrobial agent, and the specific mechanical energy indicates the severity of the process conditions that also induce the degradation of the agents. In the case of the wet casting method using solvent to cast films and containers such as cellulose films and collagen casing, the solubility and reactivity of the antimicrobial agents and polymers to the solvents are the critical factors. The solubility relates to the homogeneous distribution of the agents in the polymeric materials, and the reactivity connects to the activity loss of the reactive antimicrobial agents.

Physical and mechanical integrity of packaging materials is affected by the incorporated antimicrobial agents. If the antimicrobial agent is compatible with the packaging materials and does not interfere with the polymer-polymer interaction, a fair amount of the antimicrobial agent may be impregnated into the packaging material without any physical and mechanical integrity deterioration (Han, 1996). However, the excess amount of antimicrobial agent that is not capable of being blended with packaging materials will decrease physical strength and mechanical integrity (Cooksey *et al.*, 2000). Polymer morphological studies are very helpful in predicting the physical integrity decrease by

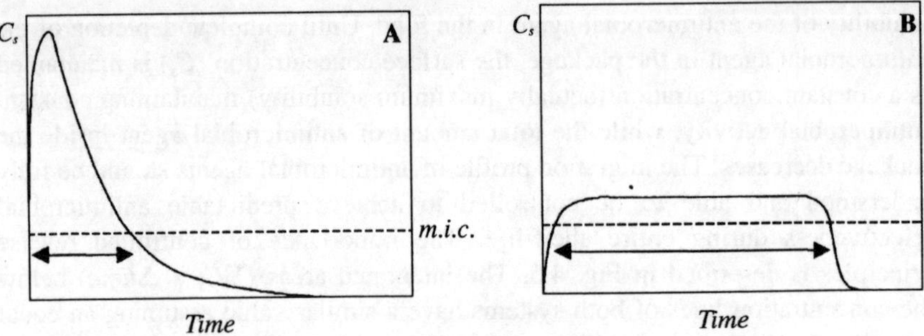

Fig. 4.5 Concentration profile at the surface of foods: (A) unconstrained free diffusion system and (B) monolithic system. Dashed line indicates m.i.c.

adding the antimicrobial agent into the packaging material. Small size antimicrobial agents can be blended with polymeric materials and may be positioned at the amorphous region of the polymeric structure. If the high level of antimicrobial agent is mixed into the packaging materials, the space provided by the amorphous region will be saturated and the mixed agent will start to interfere with the polymer-polymer interactions at the crystalline region. Although there is no physical integrity damage observed after a low level of antimicrobial agent addition, optical properties can be changed by losing transparency or changing colour of the packaging materials (Han and Floros, 1997).

Since the antimicrobial agent is contacting the food or migrating into food, the organoleptic property and toxicity of the antimicrobial agent should be satisfied to avoid quality deterioration and to maintain the safety of the packaged foods. The antimicrobial agents may possess strong taste or flavour, such as a bitter or sour taste as well as an undesirable aroma, that can affect the sensory quality adversely. In the case of antimicrobial edible protein film/coating applications, the allergenicity or chronic disease of the edible protein materials, such as peanut protein, soy protein and wheat gluten, should be considered before use (Han, 2001).

From all the foregoing, the most critical factor that should be considered in designing an antimicrobial packaging system is regulation. The use of an antimicrobial agent is regulated by the various regulatory agencies, for examples, FDA, EPA and USDA in the United States. An antimicrobial agent is an additive of packaging material, not a food ingredient. However, most antimicrobial agents migrate into packaged food therefore the package additive should satisfy all the regulations for food ingredients. The use of an antimicrobial agent should be classified as one of package additive, food contact substance or food ingredient. However, all three categories are applied to a new antimicrobial packaging system in terms of regulatory aspects. The use of natural antimicrobial agents such as plant extracts is a very challenging method because it is simple to deal with the permission process compared with chemical antimicrobial agents.

4.5 Conclusion

Antimicrobial packaging systems can inhibit the growth of spoilage and pathogenic microorganisms, and contribute to the improvement of food safety and the extension of shelf-life of the packaged food. Many factors are involved in designing the antimicrobial packaging system, however, most factors are closely related to the characteristics of antimicrobial agents, packaged foods and target microorganisms.

4.6 References

AN D S, HWANG Y I, CHO S H, and LEE D S (1998), 'Packaging of fresh curled lettuce and cucumber by using low density polyethylene films impregnated with antimicrobial agents', *J Korean Soc Food Sci Nutri*, 27(4), 675–81.

AN D S, KIM Y M, LEE S B, PAIK H D, and LEE D S (2000), 'Antimicrobial low density polyethylene film coated with bacteriocins in binder medium', *Food Sci Biotechnol*, 9(1), 14–20.

APPENDINI P, and HOTCHKISS J H (1996), 'Immobilization of lysozyme on synthetic polymers for the application to food packages', *Book of Abstracts* (1996 IFT Annual Meeting, June 22-26, 1996. New Orleans, LA), Institute of Food Technologists. Chicago, 177.

APPENDINI P, and HOTCHKISS J H (1997), 'Immobilization of lysozyme on food contact polymers as potential antimicrobial films', *Packaging Technol Sci*, 10(5), 271–9.

BARON J K, and SUMNER S S (1993), 'Antimicrobial containing edible films as an inhibitory system to control microbial growth on meat products', *Journal of Food Protection*, 56, 916.

CHEN M-C, YEH H-C, and CHIANG B-H (1996), 'Antimicrobial and physicochemical properties of methylcellulose and chitosan films containing a preservative', *Journal of Food Processing and Preservation*, 20, 379–90.

CHOI J O, PARK J M, PARK H J, and LEE D S (2001), 'Migration of preservation from antimicrobial polymer coating into water', *Food Science and Biotechnology*, 10(3), 327–30.

CHUNG D, PAPADAKIS S E, YAM K L (2001a), 'Release of propyl paraben from a polymer coating into water and food simulating solvents for antimicrobial packaging applications', *J Food Process Preserv*, 25(1), 71–87.

CHUNG D, CHIKINDAS M L, and YAM K L (2001b), 'Inhibition of *Saccharomyces cerevisiae* by slow release of propyl paraben from a polymer coating', *J Food Protect*, 64(9), 1420–4.

CHUNG S K, CHO S H, and LEE D S (1998), 'Modified atmosphere packaging of fresh strawberries by antimicrobial plastic films', *Korean J Food Sci Technol*, 30(5), 1140–5.

COMA V, SEBTI I, PARDON P, DESCHAMPS A, and PICHAVANT F H (2001),

'Antimicrobial edible packaging'based on cellulosic ethers, fatty acids, and nisin incorporation to inhibit *Listeria innocua* and *Staphylococcus aureus*', *J Food Protect*, 64(4), 470–5.

COOKSEY D K, GREMMER A, and GROWER J (2000), 'Characteristics of nisin-containing xorn zein pouches for reduction of microbial growth in refrigerated shredded cheddar cheese', in *Book of Abstracts* (2000 IFT Annual Meeting, Dallas, TX), Institute of Food Technologists, Chicago, 188.

CUTTER C N, WILLETT J L, and SIRAGUSA G R (2001), 'Improved antimicrobial activity of nisin-incorporated polymer films by formulation change and addition of food grade chelator', *Letters in Applied Microbiology*, 33(4), 325–8.

DAESCHUL M A (1989), 'Antimicrobial substances from lactic acid bacteria for use as food preservatives', *Food Technol*, 43(1), 164–7.

DEVLIEGHERE F, VERMEIREN L, BOCKSTAL A, and DEBEVERE J (2000a), 'Study on antimicrobial activity of a food packaging material containing potassium sorbate', *Acta Alimentaria*, 29(2), 137–46.

DEVLIEGHERE F, VERMEIREN L, JACOBS M, and DEBEVERE J (2000b), 'The effectiveness of hexamethylenetetramine-incorporated plastic for the active packaging of foods', *Packaging Technol and Sci*, 13(3), 117–21.

DOBIAS J, CHUDACKOVA K, VOLDRICH M, and MAREK M (2000), 'Properties of polyethylene films with incorporated benzoic anhydride and/or ethyl and propyl esters of 4-hydroxybenzoic acid and their suitability for food packaging', *Food Additive Contaminants*, 17(12), 1047–53.

FIELDS C, PIVARNICK L F, BARNETT S M, and RAND A (1986), 'Utilization of glucose oxidase for extending the shelflife of fish', *Journal of Food Science*, 51(1), 66–70.

FLOROS J D, DOCK L L, and HAN J H (1997), 'Active packaging technologies and applications', *Food Cosmetics and Drug Packaging*, 20(1), 10–17.

GHOSH K G, SRIVATSAVA A N, NIRMALA N, and SHARMA T R (1973), 'Development and application of fungistatic wrappers in food preservation. Part I. Wrappers obtained by impregnation method', *Journal of Food Science and Technology*, 10(4), 105–10.

GHOSH K G, SRIVATSAVA A N, NIRMALA N, and SHARMA T R (1977), 'Development and application of fungistatic wrappers in food preservation. Part II. Wrappers made by coating process', *Journal of Food Science and Technology*, 14(6), 261–4.

GILL A O (2000), Application of Lysozyme and Nisin to Control Bacterial Growth on Cured Meat Products (M.Sc. Dissertation), The University of Manitoba, Winnipeg, MB, Canada.

GONTARD N (1997), 'Active packaging' in Sobral P J do A, and Chuzel G, *Proceedings of Workshop sobre Biopolimeros* (April 22–24, 1997. Universidade de Sao Paulo), Pirassununga, FZEA. Brazil. 23–7.

HA J U, KIM Y M, and LEE D S (2001), 'Multilayered antimicrobial polyethylene films applied to the packaging of ground beef', *Packaging Technology and*

Science, 14(2), 55–62.

HALEK G W, and GARG A (1989), 'Fungal inhibition by a fungicide coupled to an ionomeric film', *Journal of Food Safety*, 9, 215–22.

HAN J H (1996), *Modeling the Inhibition Kinetics and the Mass Transfer of Controlled Releasing Potassium Sorbate to Develop an Antimicrobial Polymer for Food Packaging* (Ph.D. dissertation), Purdue University, West Lafayette, IN.

HAN J H (2000), 'Antimicrobial food packaging', *Food Technol*, 54(3), 56-65.

HAN J H (2001), 'Design edible and biodegradable films/coatings containing active ingredients', in Park H J, Testin R F, Chinnan M S, and Park J W, *Active Biopolymer Films and Coatings for Food and Biotechnological Uses*, Proceedings of Pre-congress Short Course of IUFoST (April 21–22, 2001, Seoul, Korea), 187–98.

HAN J H (2002), 'Protein-based edible films and coatings carrying antimicrobial agents', in Gennadios A, *Protein-based Films and Coatings*, Boca Raton, FL, CRC Press, 485–99.

HAN J H, and FLOROS J D (1997), 'Casting antimicrobial packaging films and measuring their physical properties and antimicrobial activity', *Journal of Plastic Film and Sheeting*, 13, 287–98.

HAN J H, and FLOROS J D (1998a), 'Potassium sorbate diffusivity in American processed and Mozzarella cheeses', *Journal of Food Science*, 63(3), 435–7.

HAN J H, and FLOROS J D (1998b) 'Simulating diffusion model and determining diffusivity of potassium sorbate through plastics to develop antimicrobial packaging film', *Journal of Food Processing and Preservation*, 22(2), 107–22.

HAN J H, and MOON W-S (2002), 'Plastic packaging materials containing chemical antimicrobial agents', in Han J H, *Active Food Packaging*, Winnipeg, Canada, SCI Publication and Communication Services, 11–4.

HAN J H AND ROONEY M L (2002), *personal communications*. Active Food Packaging Workshop, Annual Conference of the Canadian Institute of Food Science and Technology (CIFST), May 26, 2002.

HOFFMAN K L, HAN I Y, and DAWSON P L (2001), 'Antimicrobial effects of corn zein films impregnated with nisin, lauric acid, and EDTA', *J Food Protect*, 64(6), 885–9.

HONG S I, PARK J D, and KIM D M (2000), 'Antimicrobial and physical properties of food packaging films incorporated with some natural compounds', *Food Sci Biotechnol*, 9(1), 38–42.

HOOJJATT P, HONTE B, HERNANDEZ R, GIACIN J, and MILTZ J (1987), 'Mass transfer of BHT from HDPE film and its influence on product stability', *Journal of Packaging Technology*, 1(3), 78.

HUANG L J, HUANG C H, and WENG Y-M (1997), 'Using antimicrobial polyethylene films and minimal microwave heating to control the microbial growth of tilapia fillets during cold storage', *Food Science Taiwan*, 24(2), 263–8.

ISHITANI T (1995), 'Active packaging for food quality preservation in Japan', in

Ackermann P, Jagerstad M, and Ohlsson T, *Food and Food Packaging Materials – Chemical Interactions*, The Royal Society of Chemistry, Cambridge, 177–88.

KIM Y M, LEE N K, PAIK H D, and LEE D S (2000), 'Migration of bacteriocin from bacteriocin-coated film and its antimicrobial activity', *Food Sci Biotechnol*, 9(5), 325–9.

KROCHTA J M, and DE MULDER-JOHNSTON C (1997), 'Edible and biodegradable polymer films: challenges and opportunities', *Food Technology* 51(2), 61–74.

LEE D S, HWANG Y I, and CHO S H (1998), 'Developing antimicrobial packaging film for curled lettuce and soybean sprouts', *Food Sci Biotechnol*, 7(2), 117–21.

MILLER W R, SPALDING D H, RISSE L A, and CHEW V (1984), 'The effects of an imazalil-impregnated film with chlorine and imazalil to control decay of bell peppers', *Proc Fla State Hort Soc*, 97, 108–11.

MING X, WEBER G H, AYRES J W, and SANDINE W E (1997), 'Bacteriocins applied to food packaging materials to inhibit *Listeria monocytogenes* on meats', *J Food Sci*, 62(2), 413–15.

NADARAJAH D, HAN J H, and HOLLEY R A (2002), 'Use of allyl isothiocyanate to reduce Escherichia coli O157:H7 in packaged ground beef patties', in *Book of Abstracts* (2002 IFT Annual Meeting), Chicago, Institute of Food Technologists, 249.

NAM S, HAN J H, SCANLON M G, and IZYDORCZYK M S (2002), 'Use of extruded pea starch containing lysozyme as an antimicrobial and biodegradable packaging', in *Book of Abstracts* (2002 IFT Annual Meeting), Chicago, Institute of Food Technologists, 248.

OUATTARA B, SIMARD R E, PIETTE G, BEGIN A, and HOLLEY R A (2000a), 'Diffusion of acetic and propionic acids from chitosan-based antimicrobial packaging films', *J Food Sci*, 65(5), 768–73.

OUATTARA B, SIMARD R E, PIETTE G, BEGIN A, and HOLLEY R A (2000b), 'Inhibition of surface spoilage bacteria in processes meats by application of antimicrobial films prepared with chitosan', *Int J Food Microbiol*, 62(1/2), 139–48.

OZDERMIR M (1999), *Antimicrobial Releasing Edible Whey Protein Films and Coatings* (Ph.D. dissertation), Purdue University, West Lafayette, IN.

OZEN B F, and FLOROS J D (2001), 'Effects of emerging food processing techniques on the packaging materials', *Trends Food Sci Technol*, 12(2), 60–7, 69.

PADGETT T, HAN I Y, and DAWSON P L (1998), 'Incorporation of food-grade antimicrobial compounds into biodegradable packaging films', *Journal of Food Protection*, 61(10), 1330–5.

PAIK J S, and KELLEY M J (1995), 'Photoprocessing method of imparting antimicrobial activity to packaging film', in *Book of Abstracts* (1995 IFT Annual Meeting, June 3–7, 1996. Anaheim, CA), Institute of Food Technologists, Chicago, IL, 93.

PAIK J S, DHANASSEKHARAN M, and KELLY M J (1998), 'Antimicrobial activity of UV-irradiated nylon film for packaging applications', *Packaging Technol Sci*, 11(4), 179–87.

RICO-PENA D C, and TORRES J A (1991), 'Sorbic acid and potassium sorbate permeability of an edible methylcellulose-palmitic acid film: water activity and pH effects', *Journal of Food Science*, 56, 497–9.

RODRIGUES E T, and HAN J H (2000), 'Antimicrobial whey protein films against spoilage and pathogenic bacteria', in *Book of Abstracts* (2000 IFT Annual Meeting), Chicago, Institute of Food Technologists, 191.

RODRIGUES E T, HAN J H, and HOLLEY R A (2002), 'Optimized antimicrobial edible whey protein films against spoilage and pathogenic bacteria', in *Book of Abstracts* (2002 IFT Annual Meeting), Chicago, Institute of Food Technologists, 252.

SHAPERO M, NELSON D, and LABUZA T P (1978), 'Ethanol inhibition of *Staphylococcus aureus* at limited water activity', *Journal of Food Science*, 43, 1467–69.

SIRAGUSA G R, CUTTER C N, and WILLETT J L (1999), 'Incorporation of bacteriocin in plastic retains activity and inhibits surface growth of bacteria on meat', *Food Microbiol*, 16(3), 229–35.

SMITH J P, OORAIKUL B, KOERSEN W J, VAN DE VOORT F R, JACKSON ED, and LAWRENCE R A (1987), 'Shelf life extension of a bakery product using ethanol vapor', *Food Microbiology*, 4, 329–37.

SMITH J P, VAN DE VOORT F R, and LAMBERT A (1989), 'Food and its relation to interactive packaging', *Canadian Institute of Food Science and Technology Journal*, 22(4), 327–30.

SMITH J P, RAMASWAMY H S, and SIMPSON B K (1990), 'Developments in food packaging technology. Part II: Storage aspects', *Trends in Food Science & Technology*, November, 1990, 111–18.

TAKEUCHI K, and YUAN J (2002), 'Packaging tackles food safety: A look at antimicrobials, Industrial case study', presentation during *IFT Annual Meeting* (June 15–19, 2002, Anaheim, CA).

TATRAJAN N, and SHELDON B W (2000), 'Efficacy of nisin-coated polymer films to inactivate Salmonella typhimurium on fresh broiler skin', *J Food Protect*, 63(9), 1189–96.

VOJDANA F, and TORRES J A (1990), 'Potassium sorbate permeability of methylcellulose and hydroxypropyl methylcellulose coatings: Effect of fatty acid', *Journal of Food Science*, 55(3), 841–46.

WENG Y-M, and CHEN M-J (1997), 'Sorbic anhydride as antimycotic additive in polyethylene food packaging films', *Lebensm Wiss u Technol*, 30, 485–87.

WENG Y-M, and HOTCHKISS J H (1992), 'Inhibition of surface molds on cheese by polyethylene film containing the antimycotic imazalil', *Journal of Food Protection*, 55(5), 367–9.

WENG Y-M, and HOTCHKISS J H (1993), 'Anhydrides as antimycotic agents added to polyethylene films for food packaging', *Packaging Technology and Science*, 6, 123–8.

WENG Y M, CHEN M J, and CHEN W (1997), 'Benzoyl chloride modified ionomer films as antimicrobial food packaging materials', *Int J Food Sci Technol*, 32(3), 229–34.

WENG Y M, CHEN M J, and CHEN W (1999), 'Antimicrobial food packaging materials from poly(ethylene-co-methacrylic acid)', *Lebensl Wiss u Technol*, 32(4), 191–5.

YAKOVLEVA L A, KOLESNIKOV B F, KONDRASHOV G A, and MARKELOV A V (1999), 'New generation of bactericidal polymer packaging material', *Khranenie I Pererabotka Ssel'khozsyr'ya*, 6, 44–5.

YI J H, KIM I H, CHOE C H, SEO Y B, and SONG K B (1998), 'Chitosan-coated packaging papers for storage of agricultural products', *Hanguk Nongwhahak Hoechi*, 41(6), 442–6.

5

Non-migratory bioactive polymers (NMBP) in food packaging

M. D. Steven and J. H. Hotchkiss, Cornell University, USA

5.1 Introduction

Non-migratory bioactive polymers (NMBP) are a class of polymers that possess biological activity without the active components migrating from the polymer to the substrate. This concept has existed for some time (Bachler *et al.*, 1970; Brody and Budny, 1995; Katchalski-Katzir, 1993) and has been applied primarily to immobilised enzyme processing (Katchalski-Katzir, 1993; Mosbach, 1980). It is only now becoming of interest in packaging applications (Appendini and Hotchkiss, 1997; Soares, 1998).

Bioactive materials are based on molecules that elicit a response from living systems. The goal is to use bioactive materials for which the response is desirable from the standpoint of the package or the product, for example inhibition of microbial growth or flavour improvement. Enzymes are classic examples of bioactive substances, as are many peptides, proteins, and other organic compounds. The definition, from the perspective of packaging, is based on function: the way the substance interacts with living systems. Purely physical processes, for example adsorption or diffusion, are excluded from this definition. Bioactive polymers can be formed by attachment of bioactive molecules to synthetic polymers, as in the case of enzyme immobilisation (Appendini and Hotchkiss, 1997; Soares, 1998), or may result from an inherent bioactive effect of the polymer structure, as with chitosan (Collins-Thompson and Cheng-An, 2000; Tanabe *et al.*, 2002). They have potential applications in the packaging of food and other biological materials, in food processing equipment, on biomedical devices (Sodhi *et al.*, 2001; Sun and Sun, 2002) and in textiles (Edwards and Vigo, 2001; Sun and Sun, 2002).

Non-migratory polymers are defined to be those for which the bioactive component does not migrate out of the polymer system into the surrounding

(a)

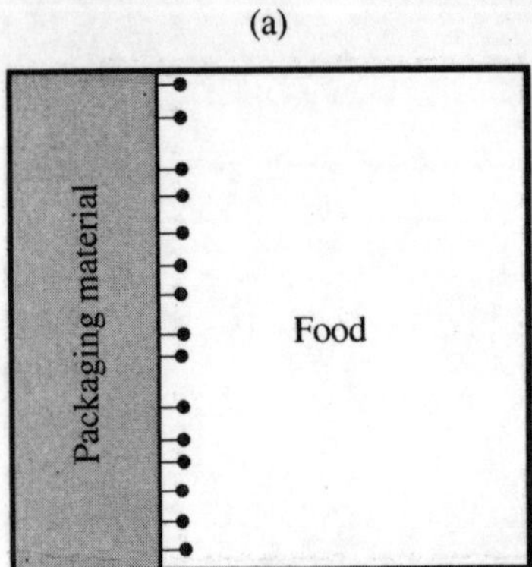

(b)

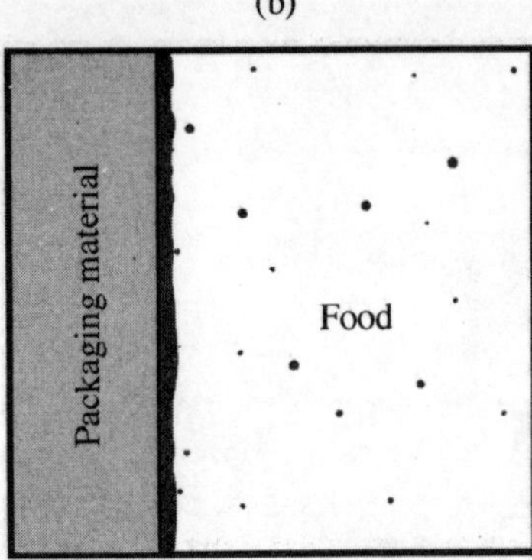

Fig. 5.1 Simplified visual comparison of (a) non-migratory and (b) migratory bioactive packaging. Adapted from Han (2000).

medium (see Fig. 5.1). Typically this is achieved through covalent attachment of the active component to the polymer backbone, inherently bioactive polymer backbones, or entrapment of the active component within the polymer matrix. The first two of these will be discussed in this chapter.

5.2 Advantages of NMBP

In order for any new technology to be considered, it needs to have advantages over existing technologies. Typically, however, these advantages come with certain limitations, in application or utility, and frequently with an increase in cost. Benefits and limitations will apply differently to the different types of NMBP.

The benefits of NP can be divided into four main areas: technical benefits, regulatory advantages, marketing aspects and the food processor's perspective. Note that this list is not exhaustive; particular applications will involve some or all of these plus other considerations specific to that application.

5.2.1 Technical benefits

Technical benefits of NMBP include improved stability of the bioactive substance, and concentration of the bioactive effect at a specific locus. Improved stability is a consideration for covalently immobilised bioactive substances; biological molecules, e.g. enzymes, are typically very sensitive to environmental conditions. They are readily denatured by some solvents, by high, and in some cases low, temperatures; by high pressures, high shear or ionising radiation; by certain levels of pH and in the presence of high concentrations of electrolytes (Richardson and Hyslop, 1985). Conjugation to polymer supports has been shown to enhance dramatically the stability of these molecules. Topchieva and

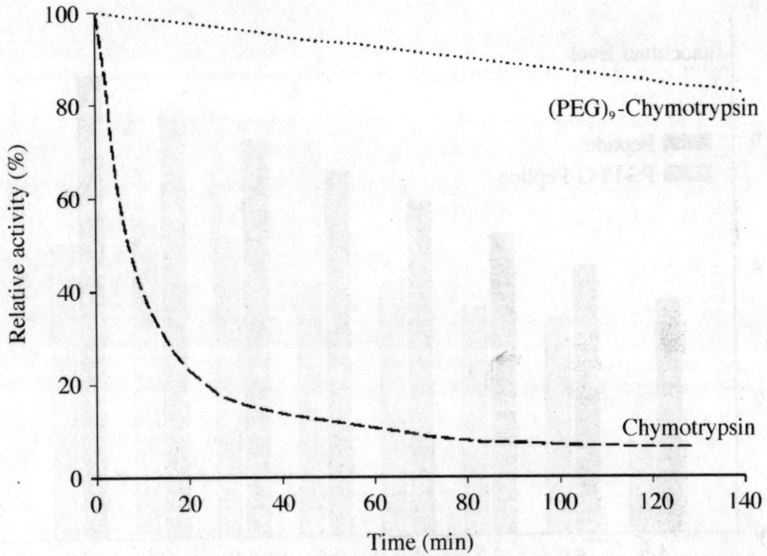

Fig. 5.2 Activity of PEG-conjugated chymotrypsin and native chymotrypsin held at 45°C. Activity is expressed in percent relative to the initial activity of each enzyme preparation. Adapted from Topchieva *et al.* (1995).

colleagues (1995) demonstrated improved thermal stability of chymotrypsin when conjugated to poly(ethylene glycol) (PEG) (see Fig. 5.2). Appendini and Hotchkiss (2001) similarly demonstrated the thermal stability of a small antimicrobial peptide when covalently attached to a PEG-grafted poly(styrene) (PS) support. The immobilised peptide remained active when dry-heated to 200°C for 30 minutes and when autoclaved at 121°C for 15 minutes. Polymers are often processed at temperatures that would denature native proteins; thermally stable protein-polymer conjugates will be resistant to high processing temperatures and suitable for polymer extrusion and other high temperature polymer and food processing applications.

Appendini (1999) also demonstrated the improved activity of the conjugated peptide over a range of pH (see Fig. 5.3). Note that although there is some loss of activity caused by attaching the peptide to the surface (this will be discussed in section 5.3 below), the residual activity is retained over a broader pH range than for the native peptide. Other authors have also reported improved stability of polymer-conjugated enzymes to pH and temperature (Gaertner and Puigserver, 1992; Yang *et al.*, 1996; Yang *et al.*, 1995a; Yang *et al.*, 1995b; Zaks and Klibanov, 1984). The extended range of pH stability will provide activity in a broader range of food products than would be the case for the native compound.

The stability of proteins to inimical media, such as organic solvents, supercritical fluids and gases, is often improved by polymer conjugation and applications have developed to exploit this in non-aqueous enzymology

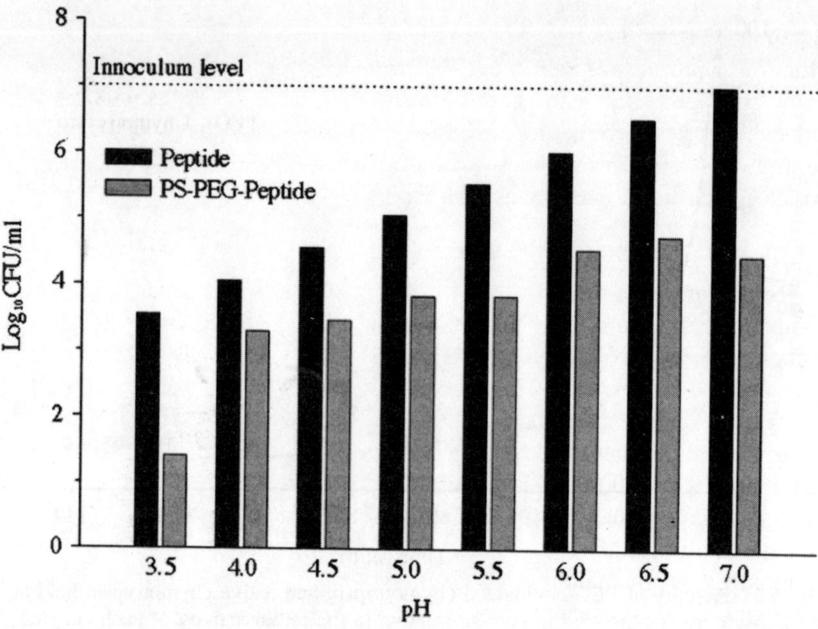

Fig. 5.3 Antimicrobial activity of a small synthetic peptide against *E.coli* 0157:H7 in 0.1M citrate buffer from pH 3.5 to 7.0. Activity is shown for the native peptide (■) and peptide attached to a PS surface through a PEG spacer (■). Equivalent peptide concentrations were used in each determination. Adapted from Appendini (1999).

(Mabrouk, 1997; Panza *et al.*, 1997; Veronese, 2001; Yang *et al.*, 1995a; Yang *et al.*, 1995b; Zaks and Klibanov, 1984). This enhanced stability to organic solvents is useful in allowing a broader range of solvents and chemicals to be used in casting, cleaning/sterilising or treating polymer films prior to package filling without damaging the functional characteristics of immobilised bioactive constituents.

The long-term stability of immobilised peptides and proteins is generally enhanced compared to the native compounds (Katchalski-Katzir, 1993; Panza *et al.*, 1997). The improved stability will help ensure the activity of bioactive packaging is retained for the shelf-life of the packaged food product. Long-term stability is also important in ensuring adequate shelf-life of the NMBP packages before filling; packaging materials are often warehoused for extended periods prior to use; any modifications need to remain active after storage.

The second technical benefit is concentration of the activity at a specific locus within the package and/or the food. This allows the activity to be concentrated where it will be most effective. For many minimally processed food products, such as fresh meat and fresh-cut fruit and vegetables, the majority of contaminating bacteria are located on the surface of the product (Collins-Thompson and Cheng-An, 2000; Hotchkiss, 1995). Concentrating antimicrobials on the surface of the product, as occurs with antimicrobial packaging, allows minimal amounts of the active compounds to be used to maximum effect. Similarly, sampling the headspace of a product for substances indicative of microbial growth using an enzymatic spoilage indicator (de Kruijf *et al.*, 2002), could be accomplished by locating the indicator in the package headspace in a position where it will be most visible to a consumer. This minimises use of expensive materials, e.g. enzymes, and possible undesirable interactions with the food.

5.2.2 Regulatory advantages

Regulations relating to active food packaging are still evolving. As new technologies develop, regulations generally must be modified to encompass them. A detailed discussion of European Union regulations relating to food packaging, with specific discussions of the implications for active and intelligent packaging systems, is presented by de Kruif and Rijk in Chapter 22 of this text (de Kruijf and Rijk, 2003). It is important in interpreting this work from a NMBP standpoint to recall that NMBP do not result in migration of the active components into the food.

As noted by various authors (de Kruijf *et al.*, 2002; de Kruijf and Rijk, 2003; Meroni, 2000; Vermeiren *et al.*, 2002; Vermeiren *et al.*, 1999), there are no specific EU regulations for active or intelligent packaging; rather these packaging systems are subject to the same regulations as traditional packaging. These regulations require that all components used to manufacture food contact materials be on 'positive lists'; active and intelligent agents are not typically included on these lists. Further, the regulations set down migration limits for both overall migration and migration of specific components. For NMBP the migration requirements should not be problematic, although a lack of migration will need to be established as detailed in the appropriate regulations. The compounds used to manufacture NMBP, however, will need to be included on the relevant positive lists. The key Directive (regulation) of concern is 89/109/EEC. De Kruijf and Rijk (2003) indicate that a new Directive, to replace 89/109/EEC, will soon be published and will allow the use of active and intelligent food contact materials. For more information, consult Chapter 22.

In the United States, regulations relating to food contact materials can be found in the Code of Federal Regulations (CFR) Title 21 Parts 170 through 190 (Anon., 2002). The regulations revolve around determining if compounds in packaging materials are food additives. Food additives are defined as substances 'the intended use of which results or may reasonably be expected to result, directly or indirectly, either in their becoming a direct component of food or otherwise affecting the characteristics of food'. Further, 'If there is no migration of a packaging component from the package to the food, it does not become a component of the food and thus is not a food additive' unless it is used 'to give a different flavour, texture of other characteristic in the food', in which case it 'may' be a food additive (21 CFR §170.3 (e) (1)). The regulations also establish guidelines for determining limits below which migration can be considered negligible, negating food additive classification of that substance for that specific application.

These US regulations can be interpreted that any substance for which it can be shown that there is negligible migration into a food product is not classified as a food additive (21 CFR §170.39). This would imply that NMBP needs to meet the regulations required of items for food contact use, but do not need to meet the more stringent food additive regulations, provided lack of migration is proven. However, additive classification may also depend on the intended function of the component. If it was intended that a packaging component be active in the food product, as with an immobilised antimicrobial on a packaging film intended to extend the shelf-life of packaged food, then the component might be classified as a Direct Food Additive and be required to comply with food additive regulations (Brackett, 2002). There is, therefore, some ambiguity as to the status of NMBP materials. If the active components of NMBP are not classified as food additives, then it will be a significant advantage, allowing the use of substances in food packaging that are not currently permitted as food additives, provided, of course, that negligible migration is proven. The process of obtaining food contact approval has been recently reviewed (Heckman and Ziffer, 2001).

The above discussion on US regulations mainly focuses on the issues of attached or entrapped bioactive compounds where the active agent is added to the polymer backbone. For inherently bioactive polymers, which will be discussed in more depth shortly, a different situation may exist. A structural component of the packaging film may not be classified as a direct food additive, for example UV irradiated nylon (Shearer *et al.*, 2000) with antibacterial properties, even if it has a direct effect in the food. This will probably not apply if an edible film is considered, as is often the case in applications involving chitosan (Coma *et al.*, 2002) and other biopolymers. In these cases, food additive regulations will most likely apply.

5.2.3 Marketing aspects

In recent years, consumers have become more aware and concerned about the composition and safety of their food. There have been increasing demands for safe, but minimally processed and preservative-free products (Appendini and Hotchkiss, 2002; Collins-Thompson and Cheng-An, 2000; Vermeiren *et al.*, 1999). This is against a background of recent food-borne microbial disease outbreaks (Appendini and Hotchkiss, 2002; Mead *et al.*, 1999). NMBP may have a key role to play in this area. Incorporating non-migratory antimicrobials in packaging materials may significantly reduce microbial contamination, while providing minimally processed, preservative-free food products. Similarly, immobilised enzyme packaging (Soares, 1998) may provide *in-package* processing opportunities which would not otherwise be possible for 'fresh' products, enhancing the acceptability and shelf-life of minimally processed foods.

5.2.4 The food processor's perspective

From the perspective of the food processor, NMBP would have several advantages. A general benefit would be in achieving a more stable product with a longer shelf-life, but beyond that certain NMBP technologies and applications may offer specific benefits. As an example, consider the production of lactose-free milk. The demand for this product is not high, although there is a place for it in the market. It sells at a high price due to the high cost of production and the low sales volume. Processing requires significant plant down time for cleaning or dedicated production facilities and the high capacity of modern plants means that the minimum production volume may be greater than the demand, leading to product wastage and requiring the use of expensive UHT technology to extend shelf-life. Using lactase-active packaging, however, regular milk could be packed off a normal production run to obtain a lactose-reduced or lactose-free product after a short period of storage. A migratory enzyme, or the direct addition of lactase to milk, cannot be used in this application, due to the strict requirements of the pasteurised milk ordinance (Anon., 1999). Similarly for other products, some of the processing may be accomplished in package, instead of in the processing plant, reducing processing costs and increasing flexibility for the food processor.

5.3 Current limitations

As noted above, any new technology must have benefits over current technologies in order to be successful, but these benefits also typically come with limitations. The most pertinent limitations of NMBP are: a limited locus of activity, specific requirements on the mechanism of activity of the active agent, reduced activity, availability of appropriate technology and an increase in packaging cost.

5.3.1 Limited locus of activity

A significant limitation of NMBP is the need for the reaction constituents to be transported to the package-product interface. This limits the function to areas in intimate contact with the packaging material for solid and viscous liquid foods. For low-viscosity foods this is less of a problem, as agitation during distribution mixes the product and will bring the required constituents in contact with the packaging. With viscous liquids, the high viscosity makes it unlikely that there will be sufficient mixing during distribution to bring all the target constituents into contact with the packaging material. Additionally, the high viscosity will limit diffusive mixing. For solid products, diffusive migration of the target constituents will also be limited and unlikely to be an effective mechanism of ensuring adequate action of the NMBP. Even for applications where the surface of a product is the target, the need for intimate contact with the packaging material may prevent application of the active agent within crevices and folds of the packaged item. This can, however, be alleviated through package design and/or vacuum packaging. Migratory bioactive packaging technologies are often similarly limited in their diffusion and mixing requirements, as the active agent may need to diffuse through the food to achieve the desired effect.

5.3.2 Mechanisms of action

In order for a bioactive agent to be active when covalently anchored to a packaging material, the conformation of the active component in the immobilised state (compared to the free solution form), the location of the covalent link to the polymer, and the mechanism by which the agent interacts with the environment to achieve the desired function must all be considered. If, for example, an antimicrobial ingredient needs to enter the microbial cell to be effective, then it is unlikely to be active in a tethered state, whereas an antimicrobial agent that is active at the microbial surface may maintain activity when tethered. If attachment causes conformational changes in the bioactive compound, or an active site is altered, then activity will be disrupted. Consider also the attachment of an enzyme that requires a co-enzyme for activity. If this coenzyme is not present in the food or otherwise attached along with the primary enzyme, then the primary enzyme will be inactive. Understanding the mechanism of the active agent is a key requirement in creating NMBP.

5.3.3 Reduced activity

One concern in immobilising bioactive compounds is the potential for loss of activity. In many cases, activity is reduced compared to the native compound (Katchalski-Katzir, 1993), and in some cases it is lost completely. With appropriate coupling methodology, however, activity can be retained, albeit normally at a lower level than for the free compound. The activity of a bound bioactive compound can vary drastically compared to the free soluble form. Appendini (1999) compared the activity of a small antimicrobial peptide when immobilised to PEG grafted poly(styrene) (PS) beads and found that it was 200–7000 times less active than the free soluble peptide. It still possessed significant antimicrobial activity, however, and was effective against *E. coli* 0157:H7 at immobilised peptide concentrations of 4μmol/ml in growth media.

In the immobilisation of naringinase, Soares (1998) found that the enzyme retained 23% of its free activity when immobilised. Soares also found that at a pH less than 3.1, the immobilised naringinase possessed higher activity than free naringinase. This often occurs with immobilised enzymes – their increased stability leads to higher activity compared to the free enzyme when conditions depart significantly from the optimum. Mosbach (1980) also suggested that for sequential enzyme pathways, the activity of the immobilised enzymes could be higher than that of the enzymes in free solution if the enzymes were immobilised in close proximity. In other words, although the activity of immobilised enzymes is typically reduced under optimum conditions, they still normally retain sufficient activity to be useful and may show improved activity under extreme conditions.

5.3.4 Technology availability

The commercial availability of technology required to produce NMBP could limit applications. Technologies for functionalising the surface of polymer films are readily available, but newer technologies, which provide controlled surface functionalisation, are still in development; this is especially true with respect to their application in high throughput continuous processes, as required for production of packaging materials. Surface functionalisation is discussed in more detail in section 5.5. Beyond basic surface functionalisation, further modification of film surfaces is not typically practised commercially. Production of most NMBP will require further processing, probably involving wet chemical treatments for immobilisation of active agents; adaptation of existing technologies will be required to implement these treatments. If NMBP becomes widespread, the technology will become more readily available and this will cease to be a limitation.

5.3.5 Cost

The final limitation of NMBP is the likely increase in cost. NMBP will require further steps and additional materials in film manufacturing/converting processes, increasing production costs. Intensive modifications, such as the attachment of proteins, will incur significant cost increases due to the additional processing steps required, the chemicals used in processing, and from the cost of the agent itself. The peptides, proteins and enzymes involved can be quite expensive, although increased demand will undoubtedly result in cost reductions of these components in the long term. Additionally, the need to recover research and development expenses and new equipment requirements will also increase the film cost. Over time, new equipment will become cheaper and more readily available, increased material availability should lead to lower material costs and the overall cost of the films will decrease. This is typical of the cycle involved in introducing new technologies.

5.4 Inherently bioactive synthetic polymers: types and applications

As previously mentioned, there are two main types of NMBP – inherently bioactive polymers and polymers with covalently immobilised bioactive agents. For inherently bioactive polymers, the structural polymer itself is bioactive. For example, polymers containing free amines have been shown to be antimicrobial (Shearer *et al.*, 2000). Included in this definition are structural polymers with modified backbones. These polymers differ from those with immobilised bioactive compounds in that no previously synthesised bioactive compound is attached to the polymer chain. Several materials have been found to have inherent bioactivity (Oh *et al.*, 2001; Ozdemir and Sadikoglu, 1998; Shearer *et al.*, 2000; Vigo, 1999; Vigo and Leonas, 1999) and new ones are currently being developed (Tew *et al.*, 2002). Most examples of inherently bioactive polymers involve antimicrobial activity.

5.4.1 Chitosan

Chitosan is the probably the most studied inherently bioactive NMBP to date (Coma *et al.*, 2002; Oh *et al.*, 2001; Tanabe *et al.*, 2002). It possesses broad spectrum antimicrobial activity in simple media and is available commercially as an antifungal coating for shelf-life extension of fresh fruit (Appendini and Hotchkiss, 2002; Padgett *et al.*, 1998). Chitosan is the deacetylated form of chitin (poly-β-(1→4)-*N*-acetyl-D-glucosamine), a common natural biopolymer extracted from the shells of crustaceans. Production of chitosan from chitin involves demineralisation, deproteinisation, and deacetylation (Oh *et al.*, 2001). The properties of chitosan films, including antimicrobial efficacy, mechanical and barrier properties, are significantly affected by the degree of deacetylation (Oh *et al.*, 2001; Paulk *et al.*, 2002).

Recent research suggests that chitosan disrupts the outer membrane of bacteria (Helander *et al.*, 2001; Tsai and Su, 1999); there were earlier suggestions that the activity was solely due to bacterial adsorption (Appendini and Hotchkiss, 2002), but the weight of evidence now suggests it possesses true antimicrobial activity. Given that chitosan is a large polymeric macromolecule, activity is unlikely to require penetration of the polymer to the intracellular area (Helander *et al.*, 2001). Helander and colleagues (2001) comment that the key feature of the antimicrobial effect of chitosan is probably the positive charge that exists on the amino group at C-2 below pH 6.3. The positive charge on this group creates a polycationic structure, which may interact with the predominantly negatively charged components of the gram-negative outer membrane. They investigated the membrane interactions of chitosan with *E. coli*, *P. aeruginosa* and *S. typhimurium* in microbiological media and determined that the activity was affected by pH, being significant at pH 5.3, but non-existent at pH 7.2; was dependent on the absence of $MgCl_2$ in the media and resulted in increased uptake of a hydrophobic probe (1-*N*-phenylnaphthylamine) from the media, indicating increased membrane permeability. Activity was thought to result from chitosan binding to the outer membrane, and was reduced for mutant *S. typhimurium* strains with cationic outer membranes. Chitosan was found to significantly sensitise the outer membrane to the action of other compounds, for example bile acids and dyes. Tsai and Su (1999) similarly investigated the mechanism of activity of chitosan against *E. coli* and found that higher temperatures and an acidic pH increased chitosan activity, and that divalent cations, such as Mg^{2+}, reduced activity. Chitosan caused leakage of glucose and lactate dehydrogenase from bacterial cells. They also concluded that the activity involves interaction between polycationic chitosan and anions on the bacterial surface, resulting in changes in membrane permeability. A similar mode of action can be assumed against gram-positive bacteria, fungi and yeasts.

The activity of chitosan has been tested against a broad range of microorganisms by researchers in many different fields, including dentistry and pharmaceuticals (Ikinci *et al.*, 2002), textiles (Takai *et al.*, 2002) and food packaging (Oh *et al.*, 2001; Paulk *et al.*, 2002; Tanabe *et al.*, 2002). In microbial growth broths, chitosan has been found effective against gram-positive and gram-negative bacteria, along with some moulds and yeasts (Oh et al., 2001; Tsai and Su, 1999; Tsai *et al.*, 2002). The minimum inhibitory concentration varies with organism and increases as the chitosan degree of deacetylation decreases. Chitosan activity has also been tested in mayonnaise against *Z. bailii* and *L. fructivorans* (Oh *et al.*, 2001). The addition of chitosan to the mayonnaise formulation increased the bacterial inhibition compared to the control mayonnaise, although bacteria numbers also decreased in the control. Higher concentrations of chitosan were required for a significant inhibitory effect in mayonnaise than in growth broth. Tsai and colleagues (2000) investigated the antimicrobial efficacy of chitosan in milk, although strict milk composition regulations and the significant solubility of chitosan in milk mean that this application is unlikely to be commercially viable. At this stage, it is important to note that most research on chitosan activity has been conducted in solution, not with chitosan films, so extrapolation to packaging applications is difficult. Additionally, the high solubility of chitosan makes its use in liquid packaging applications unlikely, as it would dissolve into the food over time, violating the non-migratory principle.

As far as packaging uses go, chitosan activity has been investigated as an edible antimicrobial film for fish fillets (Tsai *et al.*, 2002). The indigenous microflora of fish was inhibited by films formed from 0.5% and 1.0% chitosan solutions. After ten days of storage, mesophilic and psychrotrophic bacteria were reduced compared to control samples. For both types of organism, counts were reduced by approximately 1 log. Additionally, volatile basic nitrogen evolution was decreased and pH increase was suppressed compared to the controls. Coliforms were inhibited throughout a 14-day storage trial, while *Aeromonas* and *Vibrio* species showed negligible initial inhibition but slower growth and decreased numbers during the second half of storage. *Pseudomonas* spp. were initially inhibited on the dipped fillets, but increased after five days to the same levels as the control fillets. Overall, chitosan appeared to be an effective antibacterial coating; it may be suitable for use as an antimicrobial edible film for processed fish products.

In another edible film application, the antimicrobial effect of edible 98% deacetylated chitosan films was investigated against *Listeria* spp. on agar media and cheese. Significant anti-listerial activity was found in an agar plate assay: reductions of 5–8 log cycles were observed. The effect was also significant on the chitosan coated cheese samples. No viable cells were detected three and five days after dipping the cheese samples first in an inoculating solution of 10^4 cells/ml and then in a chitosan film-forming solution. These results are promising for the application of chitosan edible films to help control pathogenic contamination on the surfaces of solid food products. Similarly, the antifungal properties of chitosan may make it suitable for use as an edible film for low moisture applications where mould spoilage is a concern, e.g. bakery applications.

5.4.2 UV irradiated nylon

A recent development is surface modification of polymers leading to antimicrobial activity, for example treatment of nylon with an excimer laser at UV frequencies (193 nm) (Ozdemir and Sadikoglu, 1998; Shearer *et al.*, 2000). This has been described as a physical modification (Appendini and Hotchkiss, 2002), although the actual change which leads to the induced antimicrobial activity is a chemical change: amides on the nylon surface are converted to amines, which remain bound to the polymer chains, as observed with X-ray photoemission spectroscopy (XPS). Antimicrobial nylon-6,6 is prepared by irradiating with an UV excimer laser at 193 nm for a total exposure of 1-3 J/cm^2. This results in conversion of approximately 10% of the surface amides and some etching of the film surface (see Fig. 5.4). The antimicrobial effect is strongly dependent on the wavelength of the laser used, with films treated at 193 nm showing a 5 log reduction in *K. pneumoniae* in one hour, while film treated at 248 nm had no antimicrobial effect (Ozdemir and Sadikoglu, 1998). XPS analysis of the surface of film treated at 193 nm indicated that surface amide groups were converted to amines, while film treated at 248 nm showed no such change. The mechanism of antimicrobial activity is presumably similar to that of chitosan, poly-L-lysine and other cationic polymers, involving interaction with negatively charged microbial membranes leading to membrane disruption and leakage of cellular constituents.

The activity of UV irradiated nylon has been tested against various bacteria (Shearer *et al.*, 2000). In comparisons with untreated nylon, the treated nylon resulted in slight reductions in viable cell numbers for *E. coli* and *S. aureus*. Some bacterial reduction was also observed for the untreated nylon, presumably due to bacterial adsorption. The treated and untreated nylons were not significantly different for reduction of *E. faecalis* and *P. fluorescens*. At least three hours of exposure were required for a significant reduction in cell counts and, for *S.aureus*, the activity of the treated nylon increased with increasing temperature; no effect was observed a 4ºC or 15ºC. Protein (0.1% Bovine Serum Albumin) completely inhibited the antimicrobial activity. Shearer and colleagues (2000) compared the antimicrobial effect of treated film to that of secondary amines (*n*-butyl butyl amine) in solution and found that, to obtain a significant effect, a ten fold higher concentration of soluble secondary amine was required compared to the calculated number of amines formed on the surface of the film. It is not mentioned if the increased surface roughness was factored into this calculation; increased surface roughness results in a significant increase in the absolute surface area of the film, with a consequent increase in the number of active sites.

The results of the antimicrobial assays of UV irradiated nylons are not definitive. The data does not clearly show that the bacteria are killed as opposed to adsorbed on the surface of the film; the increased surface area resulting from

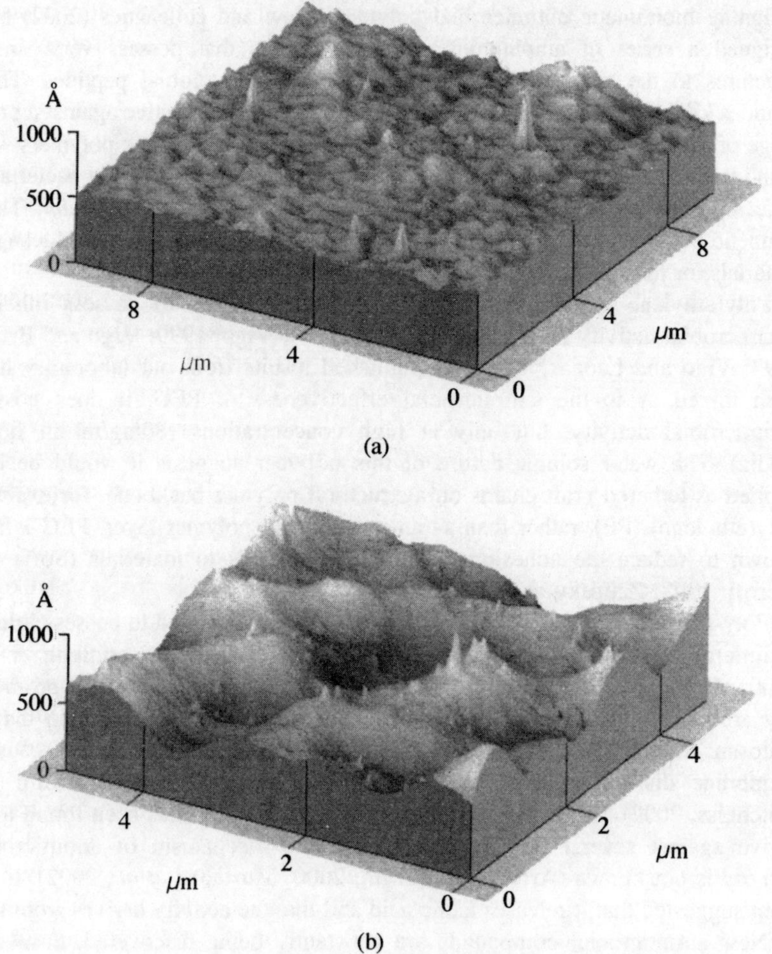

Fig. 5.4 Atomic force microscopy surface profiles of (a) untreated and (b) UV irradiated nylon film (from Shearer *et al.* (2000), reprinted by permission of Wiley-Liss, Inc., a subsidiary of John Wiley and Sons, Inc.)

etching of the film by the laser treatment would result in increased bacterial adsorption compared to the native film. Bacterial adsorption is the most likely cause of the decreased bacterial populations observed for the untreated nylon films, and in most cases the differences observed between the treated and untreated films were small, typically of the order of one log. More investigation is needed to determine whether cells are adsorbed or inactivated.

5.4.3 Others

In addition to the polymers mentioned above, many others have been investigated for potential bioactivity. An interesting recent development is the possibility of designing biomimetic antimicrobial polymers. Tew and colleagues (2002) have designed a series of amphiphilic acrylic polymers that possess very similar structures to the magainin class of amphiphilic antimicrobial peptides. These peptides kill microbes by disrupting cell membranes and are active against a broad range of microorganisms. Similarly, the new synthetic biomimetic polymers were found to disrupt synthetic membranes and to be active against several bacteria: *E. coli*, *K. pneumoniae*, *S. typhimurium*, *P. aeruginosa* and *E. faecium*. These synthetic polymers might be suitable as independent polymer layers in packaging materials, or as a grafted layer on a structural-polymer backbone.

Poly(ethylene glycol) (PEG) has also been reported to possess inherent antimicrobial activity (Jinkins and Leonas, 1994; Vigo, 1999; Vigo and Bruno, 1993; Vigo and Leonas, 1999). Unpublished results from our laboratory have been mixed as to the antimicrobial effectiveness of PEG: it does possess antimicrobial activity, but only at high concentrations (80mg/ml in liquid media). The water soluble nature of this polymer suggests it would be best applied as tethered graft chains on a structural polymer backbone, for example poly(ethylene) (PE), rather than as an independent polymer layer. PEG is also known to reduce the adhesion of proteins and cells to materials (Sofia and Merrill, 1998; Zalipsky and Harris, 1997).

Poly-l-lysine and poly(lactic acid) have both been reported to possess limited antimicrobial activity (Appendini and Hotchkiss, 2002; Ariyapitipun *et al.*, 2000; Mustapha *et al.*, 2002). Poly-l-lysine is a cationic polymer that promotes cell adhesion and is thought to be active by a mechanism similar to that of chitosan; interaction with the negative charges on the cell membrane causing membrane disruption and leakage of cellular constituents (Appendini and Hotchkiss, 2002). Low molecular weight poly(lactic acid) has been found to be active against several organisms, although the mechanism of antimicrobial activity is not known (Ariyapitipun *et al.*, 2000; Mustapha *et al.*, 2002). It has been suggested that it releases lactic acid and that the activity derives from this.

New antimicrobial compounds are constantly being discovered. Some are compounds that are well known, but for which the activity was not recognised, whereas others are novel compounds. Regardless, the possible applications for inherently antimicrobial polymers are significant and research in this area will continue. One important note: many of the abovementioned compounds are water soluble and initial trials are normally conducted on the soluble form of the polymer. Activity of the soluble form does not necessarily carry over to activity of the polymer in a film, either as an independent layer or as a graft layer on another backbone polymer. It is important to test the film activity as well as the solution activity if there is to be any application of these new materials.

5.5 Polymers with immobilised bioactive compounds

The second major type of NMBP is a polymer backbone to which an active agent is covalently attached. The active agent may be a peptide, protein or enzyme; it can be synthesised on the surface, or it can be synthesised or extracted separately and then covalently linked to the polymer. To date there has been more research conducted in this area than in that of inherently bioactive polymers, and a number of examples have been commercialised.

Often, enzymes that are adsorbed on polymers are termed immobilised, however, these compounds are not truly immobilised as they readily migrate out of the polymer in suitable non-reactive solvents (solvents that do not result in significant disruption of the covalent linkages of the system; concentrated acids and bases, for example, will cleave some covalent linkages). The term immobilised, as used here, implies covalent attachment of bioactive molecules to the polymer backbone. In some cases large compounds can also be immobilised by entrapment in the polymer matrix such that they cannot migrate out of the polymer under ambient conditions, but this state is harder to preserve at elevated polymer processing temperatures. This review does not consider entrapment immobilisation.

5.5.1 Developing NMBP by immobilisation

Some key considerations in developing NMBP by immobilising bioactive compounds are the nature of the polymer backbone and whether immobilisation is needed in the bulk polymer or just on the polymer surface. The nature of the polymer backbone is a significant consideration in designing attachment schemes. If the polymer is essentially inert, such as PE, then reactive functional groups need to be created on the polymer backbone to provide sites for attachment. This step is termed functionalisation of the polymer – the development of functional groups on the polymer backbone – and should be tailored to develop the maximum number of target groups for the desired coupling (immobilisation) chemistry while minimising polymer degradation and side reactions. For polymers which already possess suitable functional groups in the polymer backbone, for example poly(acrylic acid), the coupling chemistry needs to be chosen to target the available groups.

Most of the functionalisation and coupling chemistries we will describe are surface centric, but bulk functionalisation can also be achieved with these technologies by treating the polymer as a fine powder and then heat processing, e.g. extruding, moulding or pressing. This will result in the bioactive component being distributed throughout the bulk of the polymer. Polymers can also be bulk modified by dissolution in appropriate solvents, followed by solution modification and then removal of the solvent. However, these solvents may also denature the bioactive compounds of interest, e.g. proteins; proteins can sometimes be protected from denaturation by conjugation with, for example, PEG oligomers prior to dissolution in a solvent. Bulk modification in solution is definitely an option for water-soluble biopolymers such as chitosan, zein and poly(lactic acid).

For bulk covalent immobilisation, the active agent will not be able to migrate from the bulk of the polymer to the surface where it will be active, so the applications of non-migratory bulk immobilisation in food packaging are limited. Covalent immobilisation in the bulk may be of interest for constructing food-processing equipment where surface wear is an issue. Many of the technologies used for surface coupling are also applicable to coupling reagents in solution.

For surface modification of polymers, it is a good idea to have the polymer in as close to the final container form as possible to prevent surface rearrangement burying the active groups in the bulk polymer during heat processing/forming. Surface modification of a polymer film is the simplest situation. For more complex shapes and processes, such as blow moulding, it may be best to perform the covalent attachment with powdered polymer prior to forming the container. Extruding the active polymer as a thin layer in the appropriate location, e.g. an inner layer in a blow-moulding parison, will minimise wastage of the active agent. Similarly, covalently modified polymer powder could be used in co-extrusion of polymer films to provide an inner layer, but given the typically high cost of the active agent, it may be more economical to attach it only to the surface of the container. For both films and powders, the basic processes of surface immobilisation are the same and the active agent will be attached to the surface of either the individual powder grains or the film.

Some organic solvents swell polymers without dissolving them, allowing reagents to penetrate the polymer matrix; polar polymers often swell in water. Polymer swelling allows increased modification of the polymer surface, but may bury some of the active agent in the polymer bulk if the food to be contained does not similarly swell the polymer. Tailoring the solvents used for the reactions can control swelling.

Polymer functionalisation

For inert polymers, such as PE, the polymer backbone requires functionalisation prior to attaching or generating the bioactive agent of interest. The polymer processing literature contains many examples of this. Some of the simplest methods for laboratory use involve wet chemical treatments of the polymers to oxidise the surface, for example concentrated chromic acid in sulphuric acid, potassium permanganate in concentrated sulphuric acid (Eriksson *et al.*, 1984; Larsson *et al.*, 1979) and potassium hypochlorite in concentrated sulphuric acid (Eriksson *et al.*, 1984). Although these methods are relatively simple and do not require overly complex equipment, the hazardous nature of the reagents makes them undesirable in commercial applications. A recent development ameliorates this problem by utilising a microwave catalysed reaction between solid potassium permanganate and powdered polyolefins (Mallakpour *et al.*, 2001a; Mallakpour *et al.*, 2001b). This process reduces some of the problems inherent with wet chemical methods, but still produces waste water containing high concentrations of $KMnO_4$.

Wet chemical oxidations introduce various carbonyl groups, predominantly carboxylic acids, aldehydes and ketones, on polymer surfaces. The reaction can be optimised to produce the maximum concentration of the desired carbonyl function (Eriksson *et al.*, 1984; Holmes-Farley *et al.*, 1985; Rasmussen *et al.*, 1977). Side reactions include incorporation of sulphate groups and surface etching/ablation. Sulphate groups can be removed by nitric acid treatment post-oxidation and surface etching can be controlled by optimising the reaction conditions. For more information, a recent review of polymer surface modification using wet chemical procedures is recommended (Garbassi *et al.*, 1994), as is a second review on the chemical modification of poly(ethylene) surfaces (Bergbreiter, 1994).

Wet chemical modifications have numerous safety and environmental concerns that limit their commercial application. More common in commercial applications are physical surface treatments such as flame treatment and corona discharge. Corona discharge involves applying a high voltage (10–40 kV) at a high frequency (1–4 kHz) between a discharge electrode and an earthed roller carrying the film (Robertson, 1993). This oxidises the surface of the film, introducing a range of oxygen and nitrogen functional groups to the polymer backbone. Careful control is required to prevent excessive etching. Flame treatment also produces an oxidised film surface and introduces a range of oxygen and nitrogen functions, but is more difficult to control than corona treatment. Both treatments require specialised equipment, but this equipment is common in polymer processing and converting operations. The disadvantage of both these methods is that it is very difficult to control the exact chemical nature of the functional groups created on the surface of the film, increasing the difficulty of further coupling. It might be possible to control the chemical groups formed in corona discharge treatment by varying the gas composition of the treatment atmosphere, although the typical installation does not have such capabilities.

Controlling the treatment atmosphere has been a successful strategy to control the chemical groups created in plasma surface treatment of polymers (Groning *et al.*, 2001; Klapperich *et al.*, 2001; Schroder *et al.*, 2001; Terlingen *et al.*, 1995). The disadvantage of classic plasma processing is that it requires a high vacuum to generate a stable plasma. As such, plasma processing is a batch process, which is not suited to high throughput polymer converting operations. A new development in plasma processing is the APNEP system, an atmospheric pressure plasma treatment system developed by EA Technology Ltd in the UK (Shenton and Stevens, 1999). This has been tested with a range of common polymers and various atmospheric compositions (Shenton *et al.*, 2001; Shenton and Stevens, 1999; Shenton and Stevens, 2001; Shenton *et al.*, 2002). Controlled surface functionalisation should be possible similar to that obtainable through vacuum plasmas. One disadvantage of the APNEP system compared to vacuum plasmas is that the plasma is at a very high temperature and care is needed to prevent thermal degradation of polymers during treatment. This is achieved by placing the films in the downstream afterglow region of the plasma rather than in the plasma itself – the distance from the plasma source is an important variable for this system. The APNEP system was also found to cause greater polymer etching than vacuum plasmas.

Plasma treatment technologies, or possibly controlled atmosphere corona discharge treatments, are likely to be the most useful techniques for controlled surface functionalisation of a broad range of polymers for coupling with bioactive compounds. Reviews of physical methods for modification/ functionalisation of polymer surfaces are available (Lane and Hourston, 1993; Ozdemir *et al.*, 1999a; Ozdemir *et al.*, 1999b), although these do not include the novel APNEP technology. One review by Ozdemir and colleagues (1999a) is particularly apt in that it approaches surface functionalisation from the food packaging standpoint, albeit with different intended applications.

Polymeric spacers

One of the difficulties in attaching bioactive agents to polymeric systems is the necessity of maintaining the conformation and structure of the attached compound. The activity of most bioactive compounds is closely related to their structure and is normally lost if this structure is disrupted. The hydrophobic nature of many common polymers will disrupt the structure of a hydrophilic bioactive compound if they are coupled directly. To prevent this, it may be necessary to use a hydrophilic spacer molecule between the bioactive compound and the hydrophobic polymer backbone. A spacer also helps reduce steric hindrances to the activity of the attached compound (Weetall, 1993). Many different oligomers can be used as spacers, although the one most commonly used is PEG. The main considerations in selecting a spacer are that it does not disrupt the structure of the bioactive compound, it is approved for food contact use and suitable chemistry exists for coupling it to both the polymer backbone and the bioactive compound. PEG is safe, well characterised (Zalipsky and Harris, 1997) and approved for food use (Anon., 2002). It has been used extensively for conjugation with peptides and proteins (Zalipsky and Harris, 1997) and a large range of derivatives are available for conjugation with different functional groups (Anon., 2001b). Methods are also well established for grafting it to polymer backbones, typically by attaching preformed PEG chains of defined molecular weight using various coupling chemistries (Bae *et al.*, 1999; Emoto *et al.*, 1998; Kang *et al.*, 2001; Malmsten *et al.*, 1998; Sofia and Merrill, 1998). PEG is water soluble, allowing coupling in aqueous media and increasing the probability that liquid food products will swell the polymer surface, so increasing the interactions between the attached active agent and the target constituents in the food.

Coupling chemistries

There are many coupling chemistries available for covalently linking bioactive compounds to polymers and many different types of linkage can be formed. Amide bonds are formed between an amino group (on either the bioactive agent or the polymer) and a carboxylic acid group. Other common linkages are esters and thioesters, formed by interactions between carboxylic acids and alcohols or thiols, respectively. All these groups are common constituents of peptides, proteins and enzymes. The coupling chemistries which have been explored for polymer conjugation are generally the same as those used for peptide synthesis; texts on peptide synthesis are good sources of information on coupling techniques (Bodanszky, 1993a; Bodanszky, 1993b; Bodanszky and Bodanszky, 1994).

The carbodiimide method is a well-established and relatively simple coupling technique (see Fig. 5.5). It can be used in organic solvents or aqueous systems, depending on the carbodiimide chosen. 1,3-Dicyclohexyl carbodiimide (DCC) is typically the carbodiimide of choice in organic solvents (Bodanszky and Bodanszky, 1994), whereas 1-ethyl-3-(3-dimethylaminopropyl) carbodiimide, often referred to as WSC or EDC, is the most commonly used carbodiimide for aqueous coupling (Bae *et al.*, 1999; Carraway and Koshland, 1972; Hinder *et al.*, 2002; Kang *et al.*, 2001; Nakajima and Ikada, 1995; Plummer and Bohn, 2002; Valuev *et al.*, 1998). The ideal situation is for carboxylic acid groups to be present on the film and amino functions on the bioactive compound. The carboxyl functions are activated by the carbodiimide; if amines are present within the structure of an activated compound, as with a peptide, then the activated carboxyl groups couple with them spontaneously, leading to inter- and intra-molecular cross-linking. The film is immersed in a buffered solution of carbodiimide to activate the carboxyl functions, removed and gently rinsed, then immersed in a buffered solution containing the bioactive compound to be attached. The exact conditions, e.g. time, temperature and pH, require fine tuning for each individual coupling system.

A second common method for coupling bioactive agents to polymers is glutaraldehyde coupling. Glutaraldehyde has long been recognised as an efficient cross-linking agent for use with proteins and other biological molecules (Bigi *et al.*, 2001; Kikuchi *et al.*, 2002; Weissman, 1979) and has been used to immobilise bioactive materials onto polymeric backbones (Appendini and Hotchkiss, 1997; Molday *et al.*, 1975; Soares, 1998). It has also been used clinically for cross-linking biological molecules in the construction of bioprosthetic implants (Bigi *et al.*, 2001). Glutaraldehyde ($OHC-CH_2-CH_2-CH_2-CHO$) is a bifunctional short-chain aldehyde that reacts with amines. It requires amines on both the support and the bioactive molecule, and forms three-dimensional cross-linked aggregates.

A third possibility for coupling reagents are succinimidyl succinate (SS) active esters and their derivatives. These are commercially available (Anon., 2001b) and have been extensively employed for PEG conjugation to peptides and proteins. Succinimide esters react with free amine groups to form stable amide linkages. The most commonly used derivative is the succinimidyl propionic acid ester of PEG (PEG-SPA), which has been used to conjugate PEG with insulin (Caliceti and Veronese, 1999) and human growth hormone antagonist (hGHA) (Olson *et al.*, 1997). The reaction conditions can be modified to suit the system of interest.

A succinimidyl ester, N-hydroxy succinimide (NHS), can also be used in an extension of carbodiimide coupling (Hinder *et al.*, 2002; Plummer and Bohn, 2002). Carbodiimide activated carboxylic acids are relatively unstable and may

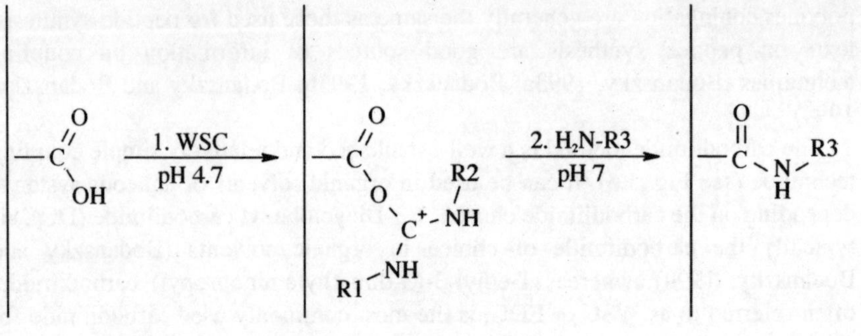

1. WSC pH 4.7 → 2. H_2N-R3 pH 7 →

O, C, OH; O, C, O, R2, C^+-NH, NH, R1; O, C, N, H, R3

Fig. 5.5 Simplified mechanism of attaching peptides to surface carboxyl functions using 1-ethyl-3-(3-dimethylaminopropyl) carbodiimide (WSC). R1: ethyl, R2: 3-dimethylaminopropyl, R3: molecule to be attached. Based on the mechanism proposed by Nakajima and Ikada (1995).

not be suitable for two-step coupling processes, especially if there is a significant delay between the activation and the coupling steps. Instead, the activation can be performed in the presence of NHS, creating a more stable, but still reactive, NHS ester of the carboxylic acid. The NHS-activated polymer film is then immersed in a solution of the amine-containing compound to be coupled and coupling proceeds with the formation of amide linkages. NHS active esters of PEG can also be obtained commercially (Anon., 2001b).

5.6 Applications of polymers with immobilised bioactive compounds

Although there are many possible applications of immobilised bioactives in food packaging, only a few have been explored to date. These can be broken into three main applications: *in-package* processing, antimicrobial packaging/shelf-life extension and intelligent packaging.

5.6.1 In-package processing

In-package processing is the most novel form of NMBP packaging. It involves immobilising an enzyme on the surface of the packaging material so as to perform in the package what would otherwise be a processing step prior to packaging. An example is the hydrolysis of naringin, one of the bitter compounds in citrus juices, with immobilised naringinase (a mixture of α-rhamnosidase and β-glucosidase) (Soares and Hotchkiss, 1998; Soares, 1998). This packaging material was able to reduce significantly the bitter naringin content of grapefruit juice during storage, providing what was perceived as a sweeter product as storage progressed. Another potential application exists for immobilised lactase (β-galactosidase) packaging to produce lactose-reduced or

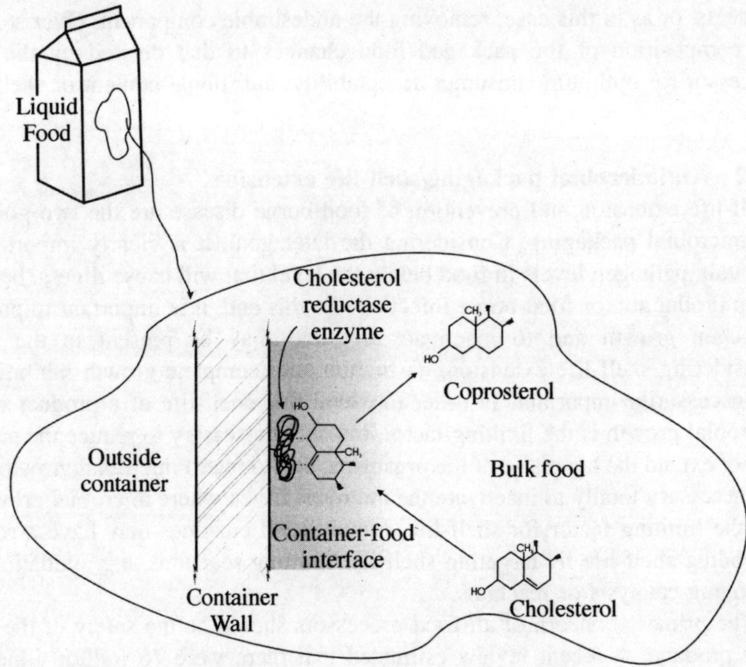

Fig. 5.6 The principle of in-package processing with NMBP, using the example of cholesterol reduction of milk with covalently immobilised cholesterol reductase enzyme. Adapted from Brody and Budny (1995).

lactose-free milk products at a reduced cost compared to current methods. Lactose-free milk products are required for consumers who are lactose-intolerant; these people are unable to consume regular milk products.

Only the number of enzymes with potential processing applications limits the range of potential applications. Other examples include the use of hydrolases to inactivate enzymes which limit shelf-life (many of these are currently inactivated with thermal techniques which also cause loss of product 'freshness'), glucose isomerase to convert glucose to fructose, so increasing the sweetness of products, and cholesterol reductase to reduce the cholesterol content of products. The concepts of lactase and cholesterol reductase immobilised packaging were explored by Pharmacal Biotechnologies in the early 1990s (Brody and Budny, 1995), although no known commercialisation has arisen to date.

The basic principle of in-package processing with immobilised enzymes is shown using the example of cholesterol reductase in Figure 5.6. The interior surface of the packaging material contains immobilised cholesterol reductase in contact with the food. The enzyme substrate, in this case cholesterol, contacts the immobilised enzyme due to natural convective currents or diffusion within the product or mixing caused by product agitation during handling and transportation. The substrate is acted on by the enzyme, producing the desired products, or as in this case, removing the undesirable component. Over storage, the composition of the packaged food changes to that desired by the food processor for optimum consumer acceptability, nutritional content or shelf-life.

5.6.2 Antimicrobial packaging/shelf-life extension

Shelf-life extension and prevention of food-borne disease are the two goals of antimicrobial packaging. Considering the later goal, it is clearly important to maintain pathogen levels in food below the level that will cause illness, be it by toxin production or food-borne infection. To this end, it is important to prevent pathogen growth and to inactivate any that may be present in the food. Considering shelf-life extension, destruction and complete growth inhibition is not necessarily important. In order to extend the shelf-life of a product where microbial growth is the limiting factor, it is only necessary to reduce the growth rate or extend the lag phase of the organisms, i.e. to retard microbial growth. It is not necessary totally to inactivate the microbes. Even where microbial growth is not the limiting factor for shelf-life, immobilised enzymes may have a role in extending shelf-life by targeting shelf-life limiting reactions, e.g. oxidation, by removing catalysts or reactants.

The primary concern of all food processors should be the safety of the food they produce. A recent review estimated that there were 76 million illnesses, 325,000 hospitalisations and 5000 deaths per year due to food-related illness in the United States (Mead *et al.*, 1999). The role of antimicrobial packaging in preventing food-borne illness has two major aspects; sterilisation or sanitation of packaging materials and prevention of pathogen growth in packaged foods. Although less dire than food poisoning, food spoilage due to microbial action is also a problem, resulting in the loss of large quantities of food. Preventing microbial spoilage may allow food, which would otherwise spoil and be wasted, to be transported to places where it is needed. Several reviews of antimicrobial food packaging have recently been published (Appendini and Hotchkiss, 2002; Collins-Thompson and Cheng-An, 2000; Han, 2000; Vermeiren *et al.*, 2002).

Both synthetic and natural compounds have been investigated for attachment to polymers to create non-migratory antimicrobial packaging. Appendini and Hotchkiss (Appendini, 1996; Appendini and Hotchkiss, 1997) investigated the attachment of lysozyme, derived from hen egg white, to poly(vinyl alcohol), nylon and cellulose triacetate (CTA). Although lysozyme was successfully immobilised on all materials, its activity on nylon and poly(vinyl alcohol) was insufficient for commercial use when tested against *M. lysodeikticus*. Greater activity was retained on CTA films, with significant bacterial retardation observed in trypticase soy broth exposed to the modified CTA. There was still, however, a significant reduction in activity compared to free lysozyme. Activity of the lysozyme film also decreased with repeated usage, indicating that the lysozyme was either inactivated over time/use or migrated out of the film.

Another natural antimicrobial system that could be used for antimicrobial packaging is the lactoperoxidase system of milk. The immobilisation of lactase and glucose oxidase enzymes on nylon pellets has been investigated with the goal of producing hydrogen peroxide to activate the lactoperoxidase system naturally present in milk (Garcia-Garibay *et al.*, 1995). The system investigated was designed for use in a bioreactor, rather than in packaging. The advantage of this system is that milk regulations in many countries prevent the addition of preservatives to milk; activating a natural antimicrobial system present in the milk provides antimicrobial effect without contravening regulations. Lactase and glucose oxidase were immobilised onto nylon pellets using glutaraldehyde coupling with a poly(ethyleneimine) spacer. The system resulted in reductions of 0.5–2 log cycles in the natural microflora of raw milk. Milk samples were exposed to the enzymes for only three minutes and microbial counts were taken 24 hours after exposure (storage at 8°C). Modifying this system for use in a package, with prolonged exposure of the milk to the hydroperoxide, would probably result in greater bacterial inhibition, but might also result in oxidation of milk components leading to undesirable sensory characteristics.

A small synthetic antimicrobial peptide has been investigated for non-migratory antimicrobial packaging applications (Appendini, 1999; Appendini and Hotchkiss, 2001; Haynie *et al.*, 1995; Haynie, 1998). The peptide was attached to a PEG grafted PS bead support during peptide synthesis, but this support was not amenable to further processing. A high activity was observed for the immobilised peptide, however, with significant antimicrobial activity against a broad range of microorganisms, gram positive and gram negative bacteria, yeasts and moulds. There is some doubt as to whether the peptide remained bound to the support: some evidence suggested that the PEG spacer was hydrolysing, releasing PEG-peptide conjugates into solution. The conditions that caused the hydrolysis were not investigated. The degree of activity was found to be dependent on the test media; positive activity was found in buffer, trypticase soy broth, apple juice and meat exudate. Further investigation of this system continues, with research focusing on producing a non-migratory antimicrobial film using this peptide.

Other enzymes with indirect antimicrobial activity have also been immobilised onto polymers suitable for food packaging. Glucose oxidase was immobilised in conjunction with catalase for use as an oxygen scavenger (de Kruijf *et al.*, 2002; Labuza and Breene, 1989). The glucose oxidase oxidises glucose to produce glucono-delta-lactone and hydrogen peroxide. Hydrogen peroxide could lead to potential undesirable oxidations of food components, so is degraded to water and oxygen by catalase. The net reduction in oxygen content is half a mole per mole of reactions. A review of the mechanism and kinetics of glucose oxidase can be found in the literature (Labuza and Breene, 1989). Alcohol dehydrogenase has also been immobilised to polymer films and can also be used as an oxygen scavenger (Labuza and Breene, 1989). Reduced oxygen concentration inhibits the growth of aerobic microbes, especially yeasts and moulds, however care must be taken that anaerobic, low acid, high moisture conditions do not result, as these may favour the growth of pathogenic *Clostridium botulinum*.

Immobilising suitable enzymes on packaging films could also control carbon dioxide concentration. Non-enzymatic carbon dioxide emitters and absorbers have been developed and commercialised (Labuza and Breene, 1989), although there has been no commercialisation of enzymatic carbon dioxide-reducing systems to date. Other enzymes may be used to produce ethanol, which has well known antimicrobial activity. Commercial ethanol emitters (Labuza and Breene, 1989) are not typically enzyme based, however, normally using encapsulated slow-release ethanol instead. The ethanol vapours released have been found to inhibit microbial growth.

In addition to their indirect antimicrobial effect, many enzymes that modify the gaseous atmosphere of packaged products can also extend product shelf-life by inhibiting non-microbial degradative mechanisms. The main application of this is in reduced-oxygen packaging to inhibit the oxidation of food components and prevent resulting negative sensory and nutritional effects. Carbon dioxide control can also be important in extending shelf-life, as in the case of packaged coffee. Enzyme systems may be used to prevent taints and off-flavours by metabolising compounds of concern. Care needs to be taken, however, that the taints inhibited are not indicative of microbial spoilage. Off odours and flavours can be key indicators to consumers that food is spoiled and unfit for consumption; removing these indicators may result in consumption of spoiled, possibly pathogenic, foods.

In all cases where indirect enzyme action is used to control microbial growth, and in fact for all immobilised enzyme reactions in food packaging, the by-products of the reaction must be carefully considered. As noted above, glucose oxidase produces hydrogen peroxide, which could cause potentially detrimental oxidations of food components. Other enzymes also produce by-products that may have detrimental effects on the sensory characteristics or shelf-life of the packaged food. Complete understanding of the catalysed reactions is required to ensure undesirable by-products are minimised.

5.6.3 Intelligent packaging

Intelligent packaging, defined as packaging systems that monitor the condition of packaged food and communicate information on food quality during transport and storage (de Kruijf *et al.*, 2002), has recently received a lot of attention and is dealt with in detail in other chapters of this text. The following discussion is a brief overview of potential NMBP applications in intelligent packaging.

Immobilised enzymes and antibodies are common components of intelligent packaging systems, so NMBPs have many potential applications in this area. A range of different indicators involving immobilised bioactive compounds has been developed, including time-temperature integrators (de Kruijf *et al.*, 2002; Labuza and Breene, 1989), spoilage indicators (Anon, 2001a) and indicators of chemical or other contamination (Woodaman, 2001). Time-temperature integrators (TTIs) based on enzyme catalysed reactions are available commercially (de Kruijf *et al.*, 2002; Labuza and Breene, 1989). Although the commercial versions do not include NMBP, this is an area in which NMBP could be effective. For microbial spoilage, enzymatic TTIs indicators may be particularly suited since microbial growth depends on enzyme-catalysed reactions.

For indicators of microbial or toxicant contamination, there are two main methods by which immobilised bioactive compounds can be used: (i) enzyme catalysed reactions requiring microbial metabolites or contaminant chemicals as substrates, or (ii) immobilised antibodies specific to bacterial metabolites and toxins, or contaminating chemicals. Both the enzymes in (i) and the antibodies in (ii) could be used to develop NMBP which indicates contamination.

A final paradigm for the use of immobilised bioactive compounds in intelligent packaging is as detection units on biosensors. The incorporation of biosensor systems in packaging films is an area for future research. Biosensors may allow remote monitoring of package conditions or point-of-sale testing of product condition by interfacing with appropriate electronic devices. These may be useful for both TTIs or for the detection of contaminating bacteria or toxicants. Various systems are in development.

5.7 Future trends

Some clear trends and opportunities exist in the field of non-migratory bioactive polymers for food packaging. The use of immobilised enzymes in food packaging is well established and will become more so, leading to *in-package* enzymatic processing as an extension of the food processing factory. Antimicrobial packaging will become another weapon in the war against foodborne disease and spoilage indicators will help prevent consumption of food that does spoil, leading to fewer food-poisoning incidents. The technology will increasingly involve the combination of biotechnology and materials science leading to a broader range of materials with a bioactive function. Additionally, the understanding of the processes causing food deterioration will increase, as will knowledge of the structure and function of bioactive polymers, enzymes and antibodies, leading to still more opportunities for NMBP in food packaging.

5.8 References

ANON. (1999) *Grade 'A' pasteurized milk ordinance, 1999 revision*, Website: *http://vm.cfsan.fda.gov/~ear/p-nci.html*: US Department of Health and Human Services: Public Health Service and Food and Drug Administration.

ANON. (2001a) Plastic wrap detects spoiled food. *J. Sci. Ind. Res.* **60**, 171.

ANON. (2001b) Shearwater Corporation Catalog 2001: Polyethylene glycol and derivatives for biomedical applications. Huntsville, AL: Shearwater Corporation.

ANON. (2002) Code of Federal Regulations, Title 21, Parts 170-199. *http://www.access.gpo.gov/nara/cfr/waisidx_02/21cfrv3_02.html*. Published: 1 Apr. 2002.

APPENDINI, P. (1996) Immobilization of Lysozyme on synthetic polymers for application to food packages. Masters Thesis. Ithaca, NY: Cornell University.

APPENDINI, P. (1999) Synthetic antimicrobial peptide. Potential application in foods and food packaging materials. PhD Thesis. Ithaca, NY: Cornell University.

APPENDINI, P. and HOTCHKISS, J.H. (1997) Immobilisation of lysozyme on food contact polymers as potential anti-microbial films . *Packag. Technol. Sci.* **10**, 271–9.

APPENDINI, P. and HOTCHKISS, J.H. (2001) Surface modification of poly(styrene) by attachment of an antimicrobial peptide. *J. Appl. Polym. Sci.* **81**, 609–16.

APPENDINI, P. and HOTCHKISS, J.H. (2002) Review of antimicrobial food packaging. *Innovative Food Science and Emerging Technologies* **3**, 113–26.

ARIYAPITIPUN, T., MUSTAPHA, A. and CLARKE, A.D. (2000) Survival of *Listeria monocytogenes* Scott a on vacuum-packaged raw beef treated with polylactic acid, lactic acid, and nisin. *J. Food Prot.* **63**, 131–6.

BACHLER, M.J., STRANDBE, G.W. AND SMILEY, K.L. (1970) Starch conversion by immobilized glucoamylase. *Biotechnol. Bioeng.* **12**, 85.

BAE, J.S., SEO, E.J. and KANG, I.K. (1999) Synthesis and characterization of heparinized polyurethanes using plasma glow discharge. *Biomaterials* **20**, 529–37.

BERGBREITER, D.E. (1994) Polyethylene surface-chemistry. *Prog. Polym. Sci.* **19**, 529–60.

BIGI, A., COJAZZI, G., PANZAVOLTA, S., RUBINI, K. and ROVERI, N. (2001) Mechanical and thermal properties of gelatin films at different degrees of glutaraldehyde crosslinking. *Biomaterials* **22**, 763–68.

BODANSZKY, M. (1993a) *Peptide chemistry: a practical textbook*. Berlin: Springer-Verlag.

BODANSZKY, M. (1993b) *Principles of peptide synthesis*. Berlin: Springer-Verlag.

BODANSZKY, M. and BODANSZKY, A. (1994) *The practice of peptide synthesis*. Berlin: Springer-Verlag.

BRACKETT, R.E. (2002) Regulatory perspective on antimicrobials in packaging . A presentation at IFT 2002 Annual Meeting Session #24: Packaging tackles food safety: a look at antimicrobials, 16 June 2002, Anaheim, CA.

BRODY, A.L. and BUDNY, J.A. (1995) Enzymes as active packaging agents. Rooney, M.L., (Ed.) *Active food packaging* pp.174–92. Glasgow: Blackie Academic & Professional.

CALICETI, P. and VERONESE, F.M. (1999) Improvement of the physicochemical and biopharmaceutical properties of insulin by poly(ethylene glycol) conjugation. *STP Pharma Sci.* **9**, 107–13.

CARRAWAY, K.L. and KOSHLAND, D.E. (1972) Carbodiimide modification of proteins. *Methods Enzymol.* **28**, 616–23.

COLLINS-THOMPSON, D. and CHENG-AN, H. (2000) Chilled storage of foods - packaging with antimicrobial properties. In: Robinson, R.K., Batt, C.A. and Patel, P.D. (Eds.) *Encyclopedia of Food Microbiology*, pp. 416–20. San Diego, CA: Academic Press.

COMA, V., MARTIAL-GROS, A., GARREAU, S., COPINET, A., SALIN, F. and DESCHAMPS, A. (2002) Edible antimicrobial films based on chitosan matrix. *J. Food Sci.* **67**, 1162–9.

DE KRUIJF, N., VAN BEEST, M., RIJK, R., SIPILAINEN-MALM, T., LOSADA, P.P. and DE MEULENAER, B. (2002) Active and intelligent packaging: applications and regulatory aspects. *Food Addit. Contam.* **19**, 144–62.

DE KRUIJF, N. and RIJK, R. (2003) Legislative issues relating to active and intelligent packaging. In: Ahvenainen, R. (Ed.) *Novel Food Packaging Techniques*, Cambridge, UK: Woodhead Publishing Ltd.

EDWARDS, J.V. and VIGO, T.L. (2001) Bioactive fibers and polymers. Washington, DC: American Chemical Society.

EMOTO, K., VAN ALSTINE, J.M. and HARRIS, J.M. (1998) Stability of poly(ethylene glycol) graft coatings. *Langmuir* **14**, 2722–9.

ERIKSSON, J.C., GOLANDER, C.G., BASZKIN, A. and TERMINASSIANSARAGA, L. (1984) Characterization of $KMnO_4/H_2SO_4$ oxidized polyethylene surfaces by means of ESCA and Ca-45(2+) adsorption. *J. Colloid Interface Sci.* **100**, 381–92.

GAERTNER, H.F. and PUIGSERVER, A.J. (1992) Increased activity and stability of poly(ethylene glycol)-modified Trypsin. *Enzyme Microb. Technol.* **14**, 150–5.

GARBASSI, F., MORRA, M. and OCCHIELLO, E. (EDS.) (1994) *Polymer surfaces: from physics to technology*, pp. 243–73. West Sussex, England: John Wiley and Sons.

GARCIA-GARIBAY, M., LUNA-SALAZAR, A. and CASAS, L.T. (1995) Antimicrobial effect of the lactoperoxidase system in milk activated by immobilized enzymes. *Food Biotechnol.* **9**, 157–66.

GRONING, P., COEN, M.C. and SCHLAPBACH, L. (2001) Polymers and cold plasmas. *Chimia* **55**, 171–7.

HAN, J.H. (2000) Antimicrobial food packaging. *Food Technol* **54**, 56–65.

HAYNIE, S.L., CRUM, G.A. and DOELE, B.A. (1995) Antimicrobial activities of amphiphilic peptides covalently bonded to a water-insoluble resin. *Antimicrob. Agents Chemother.* **39**, 301–7.

HAYNIE, S.L. (1998) Antimicrobial composition of a polymer and a peptide forming amphiphilic helices of the magainin-type. USA Patent. 5,847,047.

HECKMAN, J.H. and ZIFFER, D.W. (2001) Fathoming food packaging regulation revisited. *Food Drug Law J.* **56**, 179–95.

HELANDER, I.M., NURMIAHO-LASSILA, E.L., AHVENAINEN, R., RHOADES, J. and ROLLER, S. (2001) Chitosan disrupts the barrier properties of the outer membrane of Gram-negative bacteria. *Int. J. Food Microbio.* **71**, 235–44.

HINDER, S.J., CONNELL, S.D., DAVIES, M.C., ROBERTS, C.J., TENDLER, S.J.B. and WILLIAMS, P.M. (2002) Compositional mapping of self-assembled monolayers derivatized within microfluidic networks. *Langmuir* **18**, 3151–8.

HOLMES-FARLEY, S.R., REAMEY, R.H., MCCARTHY, T.J., DEUTCH, J. and WHITESIDES, G.M. (1985) Acid-base behavior of carboxylic-acid groups covalently attached at the surface of polyethylene - the usefulness of contact-angle in following the ionization of surface functionality. *Langmuir* **1**, 725–40.

HOTCHKISS, J.H. (1995) Safety considerations in active packaging. Rooney, M.L., (Ed.) *Active food packaging* pp. 238–54. Glasgow: Blackie Academic & Professional.

IKINCI, G., SENEL, S., AKINCIBAY, H., KAS, S., ERCIS, S., WILSON, C.G. and HINCAL, A.A. (2002) Effect of chitosan on a periodontal pathogen *Porphyromonas gingivalis*. *Int. J. Pharm.* **235**, 121–7.

JINKINS, R.S. and LEONAS, K.K. (1994) Influence of a polyethylene-glycol treatment on surface, liquid barrier and antibacterial properties. *Text. Chem. Color.* **26**, 25–9.

KANG, I.K., SEO, E.J., HUH, M.W. and KIM, K.H. (2001) Interaction of blood components with heparin-immobilized polyurethanes prepared by plasma glow discharge. *J. Biomat. Sci.-Polym. E.* **12**, 1091–8.

KATCHALSKI-KATZIR, E. (1993) Immobilized enzymes – learning from past successes and failures. *TIBTECH* **11**, 471–8.

KIKUCHI, M., TAGUCHI, T., MATSUMOTO, H.N., TAKAKUDA, K. and TANAKA, J. (2002) Cross-linkage of hydroxyapatite/collagen nano-composite with 3 different reagents. *Key Eng. Mater.* **218**, 449–52.

KLAPPERICH, C., PRUITT, L. and KOMVOPOULOS, K. (2001) Chemical and biological characteristics of low-temperature plasma treated ultra-high molecular weight polyethylene for biomedical applications. *J. Mater. Sci.-Mater. M.* **12**, 549–56.

LABUZA, T.P. and BREENE, W.M. (1989) Applications of active packaging for improvement of shelf-life and nutritional quality of fresh and extended shelf life foods. *J. Food Process. Pres.* **13**, 1–70.

LANE, J.M. and HOURSTON, D.J. (1993) Surface treatments of polyolefins. *Prog. Org. Coat.* **21**, 269–84.

LARSSON, R., ERIKSSON, J.C. and OLSSON, P. (1979) Effects of surface localized sulfate groups on the clotting time in plasma. *Thromb. Res.* **14**, 941–52.

MABROUK, P.A. (1997) The use of poly(ethylene glycol)-enzymes in nonaqueous enzymology. Harris, J.M. and Zalipsky, S., (Eds.) *Poly(Ethylene Glycol): Chemistry and Biological Applications* pp. 118–33. ACS Symposium Series #680. Washington, DC: American Chemical Society.

MALLAKPOUR, S.E., HAJIPOUR, A.R., MAHDAVIAN, A.R., ZADHOUSH, A. and ALI-HOSSEINI, F. (2001a) Microwave assisted oxidation of polyethylene under solid-state conditions with potassium permanganate. *Eur. Polym. J.* **37**, 1199–206.

MALLAKPOUR, S.E., HAJIPOUR, A.R., ZADHOUSH, A. and MAHDAVIAN, A.R. (2001b)

Efficient and novel method for surface oxidation of polypropylene in the solid phase using microwave irradiation. *J. Appl. Polym. Sci.* **79**, 1317–23.

MALMSTEN, M., EMOTO, K. and VAN ALSTINE, J.M. (1998) Effect of chain density on inhibition of protein adsorption by poly(ethylene glycol) based coatings. *J. Colloid Interface Sci.* **202**, 507–17.

MEAD, P.S., SLUTSKER, L., DIETZ, V., MCCAIG, L.F., BRESEE, J.S., SHAPIRO, C., GRIFFIN, P.M. and TAUXE, R.V. (1999) Food-related illness and death in the United States. *Emerg. Infect. Dis.* **5**, 607–25.

MERONI, A. (2000) Active packaging as an opportunity to create package design that reflects the communicational, functional and logistical requirements of food products. *Packag. Technol. Sci.* **13**, 243–8.

MOLDAY, R.S., DREYER, W.J., REMBAUM, A. and YEN, S.P.S. (1975) New immunolatex spheres - visual markers of antigens on lymphocytes for scanning electron microscopy. *J. Cell Biol.* **64**, 75–88.

MOSBACH, K. (1980) Immobilized enzymes. *Trends Biochem. Sci.* **5**, 1–3.

MUSTAPHA, A., ARIYAPITIPUN, T. and CLARKE, A.D. (2002) Survival of *Escherichia coli* O157:H7 on vacuum-packaged raw beef treated with polylactic acid, lactic acid, and nisin. *J. Food Sci.* **67**, 262–7.

NAKAJIMA, N. and IKADA, Y. (1995) Mechanism of amide formation by carbodiimide for bioconjugation in aqueous media. *Bioconjugate Chem.* **6**, 123–30.

OH, H.I., KIM, Y.J., CHANG, E.J. and KIM, J.Y. (2001) Antimicrobial characteristics of chitosans against food spoilage microorganisms in liquid media and mayonnaise. *Biosci., Biotechnol., Biochem.* **65**, 2378–83.

OLSON, K., GEHANT, R., MUKKU, V., O'CONNELL, K., TOMLINSON, B., TOTPAL, K. and WINKLER, M. (1997) Preparation and characterization of poly(ethylene glycol)ylated human growth hormone antagonist. Harris, J.M. and Zalipsky, S., (Eds.) *Poly(ethylene glycol): Chemistry and Biological Applications* pp. 170–81. ACS Symposium Series #680. Washinton, DC: American Chemical Society.

OZDEMIR, M. and SADIKOGLU, H. (1998) A new and emerging technology: laser-induced surface modification of polymers. *Trends Food Sci. Tech.* **9**, 159–67.

OZDEMIR, M., YURTERI, C.U. and SADIKOGLU, H. (1999a) Physical polymer surface modification methods and applications in food packaging polymers. *Crit. Rev. Food Sci.* **39**, 457–77.

OZDEMIR, M., YURTERI, C.U. and SADIKOGLU, H. (1999b) Surface treatment of food packaging polymers by plasma. *Food Technol* **53**, 54–8.

PADGETT, T., HAN, I.Y. and DAWSON, P.L. (1998) Incorporation of food-grade antimicrobial compounds into biodegradable packaging films. *J. Food Prot.* **61**, 1330–5.

PANZA, J.L., LEJEUNE, K.E., VENKATASUBRAMANIAN, S. and RUSSELL, A.J. (1997) Incorporation of poly(ethylene glycol) proteins into polymers. Harris, J.M. and Zalipsky, S., (Eds.) *Poly(Ethylene Glycol): Chemistry and Biological Applications,* pp.134–44. ACS Symposium Series #680. Washington, DC:

American Chemical Society.

PAULK, M., COOKSEY, D., AND WILES, J. (2002) Characterization of Chitosan packaging films with differing levels of % deacetylation. A poster presented at *2002 IFT Annual Meeting* 19 June 2002, Anaheim, CA.

PLUMMER, S.T. and BOHN, P.W. (2002) Spatial dispersion in electrochemically generated surface composition gradients visualized with covalently bound fluorescent nanospheres. *Langmuir* **18**, 4142–9.

RASMUSSEN, J.R., STEDRONSKY, E.R. and WHITESIDES, G.M. (1977) Introduction, modification, and characterization of functional-groups on surface of low-density polyethylene film. *J. Am. Chem. Soc.* **99**, 4736–45.

RICHARDSON, T. and HYSLOP, D.B. (1985) Enzymes. Fennema, O.R., (Ed.) *Food Chemistry* 2nd edn, pp. 371–476. New York: Marcel Dekker.

ROBERTSON, G.L. (1993) *Food Packaging: Principles and Practise*. New York: Marcel Dekker.

SCHRODER, K., MEYER-PLATH, A., KELLER, D., BESCH, W., BABUCKE, G. and OHL, A. (2001) Plasma-induced surface functionalization of polymeric biomaterials in ammonia plasma. *Contrib. Plasm. Phys.* **41**, 562–72.

SHEARER, A.E.H., PAIK, J.S., HOOVER, D.G., HAYNIE, S.L. and KELLEY, M.J. (2000) Potential of an antibacterial ultraviolet-irradiated nylon film. *Biotechnol. Bioeng.* **67**, 141–6.

SHENTON, M.J., LOVELL-HOARE, M.C. and STEVENS, G.C. (2001) Adhesion enhancement of polymer surfaces by atmospheric plasma treatment. *J. Phys. D: Appl. Phys.* **34**, 2754–60.

SHENTON, M.J. and STEVENS, G.C. (1999) Investigating the effect of the thermal component of atmospheric plasmas on commodity polymers. *Thermochim. Acta* **332**, 151–60.

SHENTON, M.J. and STEVENS, G.C. (2001) Surface modification of polymer surfaces: atmospheric plasma versus vacuum plasma treatments. *J. Phys. D: Appl. Phys.* **34**, 2761–8.

SHENTON, M.J., STEVENS, G.C., WRIGHT, N.P. and DUAN, X. (2002) chemical-surface modification of polymers using atmospheric pressure non-equilibrium plasmas and comparisons with vacuum plasmas. *J. Polym. Sci., Part A: Polym. Chem.* **40**, 95–109.

SOARES, N.D.F.F. (1998) Bitterness reduction in citrus juice through naringase immobilized into polymer film. PhD Thesis, Cornell University, Ithaca, NY.

SOARES, N.D.F.F. and HOTCHKISS, J.H. (1998) Naringinase immobilization in packaging films for reducing naringin concentration in grapefruit juice. *J. Food Sci.* **63**, 61–5.

SODHI, R.N.S., SAHI, V.P. and MITTELMAN, M.W. (2001) Application of electron spectroscopy and surface modification techniques in the development of anti-microbial coatings for medical devices. *J. Electron Spectrosc. Relat. Phenom.* **121**, 249–64.

SOFIA, S.J. and MERRILL, E.W. (1998) Grafting of PEO to polymer surfaces using electron beam irradiation. *J. Biomed. Mater. Res.* **40**, 153–63.

SUN, Y. and SUN, G. (2002) Durable and regenerable antimicrobial textile materials prepared by a continuous grafting process. *J. Appl. Polym. Sci.* **84**, 1592–9.

TAKAI, K., OHTSUKA, T., SENDA, Y., NAKAO, M., YAMAMOTO, K., MATSUOKA, J. and HIRAI, Y. (2002) Antibacterial properties of antimicrobial-finished textile products. *Microbiol. Immunol.* **46**, 75–81.

TANABE, T., OKITSU, N., TACHIBANA, A. and YAMAUCHI, K. (2002) Preparation and characterization of keratin-chitosan composite film. *Biomaterials* **23**, 817–25.

TERLINGEN, J.G.A., GERRITSEN, H.F.C., HOFFMAN, A.S. and JEN, F.J. (1995) Introduction of functional-groups on polyethylene surfaces by a carbon dioxide plasma treatment. *J. Appl. Polym. Sci.* **57**, 969–82.

TEW, G.N., LIU, D.H., CHEN, B., DOERKSEN, R.J., KAPLAN, J., CARROLL, P.J., KLEIN, M.L. and DEGRADO, W.F. (2002) De novo design of biomimetic antimicrobial polymers. *Proc. Natl. Acad. Sci. U.S.A.* **99**, 5110–14.

TOPCHIEVA, I.N., EFREMOVA, N.V., KHVOROV, N.V. and MAGRETOVA, N. (1995) Synthesis and physicochemical properties of protein conjugates with water-soluble poly(alkylene oxides). *Bioconjugate Chemistry* **6**, 380–8.

TSAI, G.J. and SU, W.H. (1999) Antibacterial activity of shrimp chitosan against *Escherichia coli*. *J. Food Prot.* **62**, 239–43.

TSAI, G.J., SU, W.H., CHEN, H.C. and PAN, C.L. (2002) Antimicrobial activity of shrimp chitin and chitosan from different treatments and applications of fish preservation. *Fisheries Sci.* **68**, 170–7.

TSAI, G.J., WU, Z.Y. and SU, W.H. (2000) Antibacterial activity of a chitooligosaccharide mixture prepared by cellulase digestion of shrimp chitosan and its application to milk preservation. *J. Food Prot.* **63**, 747–52.

VALUEV, I.L., CHUPOV, V.V. and VALUEV, L.I. (1998) Chemical modification of polymers with physiologically active species using water-soluble carbodiimides. *Biomaterials* **19**, 41–3.

VERMEIREN, L., DEVLIEGHERE, F., VAN BEEST, M., DE KRUIJF, N. and DEBEVERE, J. (1999) Developments in the active packaging of foods. *Trends Food Sci. Tech.* **10**, 77–86.

VERMEIREN, L., DEVLIEGHERE, F. and DEBEVERE, J. (2002) Effectiveness of some recent antimicrobial packaging concepts. *Food Addit. Contam.* **19**, 163–71.

VERONESE, F.M. (2001) Peptide and protein PEGylation: a review of problems and solutions. *Biomaterials* **22**, 405–17.

VIGO, T.L. (1999) Antimicrobial polymers and fibers: retrospective and prospective. *Abstr. Pap. Am. Chem.* **218**, 31–CELL.

VIGO, T.L. and BRUNO, J.S. (1993) Antimicrobial activity of fabrics containing cross-linked polyols. *Abstr. Pap. Am. Chem.* **206**, 50–CELL.

VIGO, T.L. and LEONAS, K.K. (1999) Antimicrobial activity of fabrics containing crosslinked polyethylene glycols. *Text. Chem. Color. Am. D.* **1**, 42–6.

WEETALL, H.H. (1993) Preparations of immobilized proteins covalently coupled through silane agents to inorganic support. *Appl. Biochem. Biotechnol.* **41**, 157–88.

WEISSMAN, R.C. (1979) A microscopical and mechanical study of the gelation of collagen crosslinked with glutaraldehyde. PhD Thesis, Cornell University.

WOODAMAN, J.G., INVENTOR. ASSIGNED TO CALIFORNIA SOUTH PACIFIC INVESTORS, P.C. (2001) Detection of contaminants in food. USA US 6,270,724 B1.

YANG, Z., WILLIAMS, D. and RUSSELL, A.J. (1995a) Synthesis of protein-containing polymers in organic solvents. *Biotechnol. Bioeng.* **45**, 10–17.

YANG, Z., MESIANO, A.J., VENKATASUBRAMANIAN, S., GROSS, S.H., HARRIS, M.J. and RUSSELL, A.J. (1995b) Activity and stability of enzymes incorporated into acrylic polymers. *J. Am. Chem. Soc.* **117**, 4843–50.

YANG, Z., DOMACH, M., AUGER, R., YANG, F.X. and RUSSELL, A.J. (1996) Polyethylene glycol-induced stabilization of subtilisin. *Enzyme Microb. Technol.* **18**, 82–9.

ZAKS, A. and KLIBANOV, A.M. (1984) Enzymatic catalysis in organic media at 100-degrees-C. *Science* **224**, 1249–51.

ZALIPSKY, S. and HARRIS, J.M. (1997) Introduction to chemistry and biological applications of poly(ethylene glycol). Harris, J.M. and Zalipsky, S., (Eds.) *Poly(ethylene glycol): Chemistry and Biological Applications.* ACS Symposium Series #680. Washington, DC: American Chemical Society.

6

Time-temperature indicators (TTIs)

P. S. Taoukis, National Technical University of Athens, Greece and T. P. Labuza, University of Minnesota, USA

6.1 Introduction

The modern food industry is called on to deliver seemingly contradictory market demands. On the one hand consumers want improved safety and sensory quality, together with increased nutritional properties, extended shelf-life and convenience in preparation and use. On the other they want food with a traditional, wholesome image, with less processing and fewer additives.

In achieving safer and better quality food scientists and manufacturers apply intense optimisation and control of all the production and preservation parameters and additionally explore and benefit from innovative techniques to ensure safety and reduce food deterioration. Novel packaging such as active packaging is among such innovative tools. Producers and regulators rely on the development and application of structured quality and safety assurance systems based on prevention through monitoring, recording and controlling of critical parameters through the entire product life cycle. These systems should include the post-processing phase and ideally extend to the consumer's table. The ISO 9001:2000 quality management standard (ISO 9001:2000; ISO 15161: 2001), widely adopted by the food industry, emphasises documented procedures for storage, handling and distribution. The globally recommended Hazard Analysis and Critical Control Point (HACCP) safety assurance system also focuses on this phase (93/43/EEC; Codex, 1997; US Federal Register, 1996). Certain stages of the chill chain are recognised as important critical control points (CCPs) for minimally processed chilled products such as modified-atmosphere packaged and other ready-to-eat chilled products. Monitoring and controlling these CCPs is seen as essential for safety.

Research and industrial studies show that chilled or frozen distribution and handling very often deviate from recommended temperature conditions. Since temperature largely constitutes the determining post-processing parameter for shelf-life under good manufacturing and hygiene practices, monitoring and controlling it is of central importance. The complexity of the problem is highlighted when the variation in temperature exposure of single products within batches or transportation sub-units is considered. Ideally, a cost-effective way to monitor the temperature conditions of food products individually, throughout distribution, is required to indicate their real safety and quality. This requirement could be fulfilled by Time Temperature Integrators or Indicators (TTIs). TTIs can be classified as active packaging. A TTI based system could lead to effective quality control of the chill chain, optimisation of stock rotation and reduction of waste, and provide information on the remaining shelf-life of product units. A prerequisite for the application of TTIs is the systematic study and kinetic modelling of the role of temperature in determining shelf-life. Based on reliable models of food product shelf-life and the kinetics of TTI response, the effect of temperature can be monitored, recorded and translated from production to the consumer's table.

6.2 Defining and classifying TTIs

A time temperature integrator or indicator (TTI) can be defined as a simple, inexpensive device that can show an easily measurable, time-temperature dependent change that reflects the full or partial temperature history of a food product to which it is attached (Taoukis and Labuza, 1989). The principle of TTI operation is a mechanical, chemical, electrochemical, enzymatic or microbiological irreversible change usually expressed as a visible response, in the form of a mechanical deformation, colour development or colour movement. The rate of change is temperature dependent, increasing at higher temperatures. The visible response thus gives a cumulative indication of the storage conditions that the TTI has been exposed to. The extent to which this response corresponds to a real time-temperature history depends on the type of the indicator and the physicochemical principles of its operation. Indicators can thus be classified according to their functionality and the information they convey.

An early classification system introduced by Schoen and Byrne (1972) separated devices into six categories. Byrne (1976) revised this classification, realising that the main functional difference is whether the indicator responds above a preselected temperature, or responds continuously thus giving information on the cumulative time-temperature exposure. He proposed three types:

1. defrost indicators
2. time-temperature integrators
3. time-temperature integrators/indicators.

A similar scheme recognised three categories (Singh and Wells, 1986):

1. abuse indicators
2. partial temperature history indicators
3. full temperature history indicators (an alternative nomenclature for time-temperature integrators).

A three-category classification will be used in this chapter (Taoukis *et al.*, 1991).

6.2.1 Critical temperature indicators (CTI)

CTI show exposure above (or below) a reference temperature. They involve a time element (usually short; a few minutes up to a few hours) but are not intended to show history of exposure above the critical temperature. They merely indicate the fact that the product was exposed to an undesirable temperature for a time sufficient to cause a change critical to the safety or quality of the product. They can serve as appropriate warning in cases where physicochemical or biological reactions show a discontinuous change in rate. Good examples of such cases are the irreversible textural deterioration that happens when phase changes occur (e.g., upon defrosting of frozen products or freezing of fresh or chilled products). Denaturation of an important protein above the critical temperature or growth of a pathogenic microorganism are other important cases where a CTI would be useful. The 'critical temperature' term is preferred rather than the used alternative 'defrost' that is too limiting. The term 'abuse' might be misleading as undesirable changes can happen at temperatures which are not as extreme or abusive as the term implies and which are within the acceptable range of normal storage for the product in question.

6.2.2 Critical temperature/time integrators (CTTI)

CTTI show a response that reflects the cumulative time-temperature exposure above a reference critical temperature. Their response can be translated into an equivalent exposure time at the critical temperature. They are useful in indicating breakdowns in the distribution chain and for products in which reactions, important to quality or safety, are initiated or occur at measurable rates above a critical temperature. Examples of such reactions are microbial growth or enzymatic activity that are inhibited below the critical temperature.

6.2.3 Time temperature integrators or indicators (TTI)

TTI give a continuous, temperature dependent response throughout the product's history. They integrate, in a single measurement, the full time-temperature history and can be used to indicate an 'average' temperature during distribution and possibly be correlated to continuous, temperature dependent quality loss reactions in foods. In the remainder of this chapter, the term TTI will refer to this type of indicator, unless otherwise noted. A different method of classification sometimes used is based on the principle of the indicators' operation. Thus, they can be categorised as mechanical, chemical, enzymatic, microbiological, polymer, electrochemical, diffusion based, etc.

6.3 Requirements for TTIs

The requirements for an effective TTI are that it shows a continuous change, the rate of which increases with temperature and which does not reverse when temperature is lowered. There are a number of other desirable attributes for a successful indicator. An ideal TTI would have all the following properties:

- It exhibits a continuous time-temperature dependent change.
- The change causes a response that is easily measurable and irreversible.
- The change mimics or can be correlated to the food's extent of quality deterioration and residual shelf-life.
- It is reliable, giving consistent responses when exposed to the same temperature conditions.
- It has low cost.
- It is flexible, so that different configurations can be adopted for various temperature ranges (e.g., frozen, refrigerated, room temperature) with useful response periods of a few days as well as up to more than a year.
- It is small, easily integrated as part of the food package and compatible with a high-speed packaging process.
- It has a long shelf-life before activation and can be easily activated.
- It is unaffected by ambient conditions other than temperature, such as light, humidity and air pollutants.
- It is resistant to normal mechanical abuse and its response cannot be altered.
- It is non-toxic, posing no safety threat in the unlikely situation of product contact.
- It is able to convey in a simple and clear way the intended message to its target, be that distribution handlers or inspectors, retail store personnel or consumers.
- Its response is both visually understandable and adaptable to measurement by electronic equipment for easier and faster information, storage and subsequent use.

6.4 The development of TTIs

The drive for development of an effective and inexpensive indicator dates from the time when the importance of temperature variations to final food quality during distribution became apparent. Initially, the interest was focused on frozen foods. The first application of a 'device' to indicate handling abuse dates from World War II when the US Army Quartermaster Corps used an ice cube placed inside each case of frozen food. Disappearance of the cube indicated mishandling (Schoen and Byrne, 1972). The first patented indicator goes back to 1933 (Midgley, 1933). Over a hundred US and international patents relevant to time-temperature indicators have been issued since. During the last 30 years numerous TTI systems have been proposed of which only few reached the prototype and even fewer the market stage. Patents dating up to 1990 are tabulated in the literature (Byrne, 1976; Taoukis, Fu and Labuza, 1991). Byrne (1976) gives an overview of the early indicators and Taoukis (1989) presents a detailed history of TTI. Table 6.1 lists significant recent TTI patents classified according to type and principle of operation.

The first commercially available TTI was developed by Honeywell Corp (Minneapolis, MN) (Renier and Morin, 1962). The device never found commercial application, possibly because it was costly and relatively bulky. In the early 1970s, the US government considered mandating the use of indicators on certain products (OTA, 1979). This generated a flurry of research and development. Researchers at the US Army Natick Laboratories developed a TTI that was based on the colour change of an oxidisable chemical system controlled by the temperature dependent permeation of oxygen through a film (Hu, 1972). Field testing over a two-year period with the TTI attached to rations showed their potential for use (Killoran, 1976). The system was contracted to Artech Corp (Falls Church, VA) for commercial development. By 1976 six companies were making temperature

Table 6.1 List of recent TTI patents and classification according to type and mode of response.

Date	*Inventor*	*Principle of operation*	*Patent No*
1991	Jalinski, T.J.	Chemical (TTI)	US5,182,212
1991	Jalinski, T.J.	Chemical (TTI)	US5,085,802
1991	Thierry, A.	Chemical (CTI)	US5,085,801
1991	Swartzel, K.R.	Physicochemical (TTI)	US5,159,564
1992	Jalinski, T.	Chemical (CTI)	EP497459A1
1993	Veitch, R.J.	Physicochemical (CTI)	EP563769A1
1993	Loustaunau, A.	Physical (CTI)	EP615614A1
1994	Loustaunau, A.	Physical (CTI)	US5,460,117
1994	Veitch, R.J.	Physicochemical (CTI)	US5,490,476
1995	Prusik, T.	Physicochemical (TTI)	US5,709,472
1996	Cannelongo, J.F.	Physical (CTI)	US5,779,364
1996	Veitch, R.J.	Physical (CTI)	EP835429A1
1997	Arens R. *et al.*	Physicochemical (TTI)	US5,667,303
1997	Schneider, N.	Physical (CTI)	US6,030,118
1999	Simons, M.J.	Physicochemical (CTI)	EP930488A2
2000	Schaten, B.B.	Physical (CTI)	EP1053726A2
2000	Prusik, T.	Physical (CTTI)	US6,042,264
2000	Ram, A.T.	Chemical (TTI)	US6,103,351
2000	Bray, A.V.	Physical (TTI)	US6,158,381
2001	Simons, M.J.	Physicochemical (TTI)	US6,214,623
2001	Qiu, J.	Physicochemical (TTI)	US6,244,208
2002	Qiu, J.	Physicochemical (TTI)	US6,435,128

indicators at least at the prototype stage (Kramer and Farquhar, 1976). The Artech, the Check Spot Co (Vancouver, WA) (US patent 2,971,852) and the Tempil (S Plainfield, NJ) indicators could be classified as CTI. The I-Point (Malmö, Sweden), the Bio-Medical Sciences (Fairfield, NJ) (US patents 3,946,611 and 4,042,336) and the 3M Co (St Paul, MN) indicators were TTI. The Tempil indicator could function as a CTTI. It involved a change to a red colour and subsequent movement when exposed above the critical temperature. The I-Point was an enzymatic TTI, and the 3M, a diffusion based TTI.

By the end of the 1970s, however, very little commercial application of the TTI had been achieved. Research and development activity subsided temporarily, noted by a decrease in the relevant publications and in the new TTI models introduced. However, the better systems remained available and development continued, aiming at fine tuning and making performance more consistent. In the early 1980s, there were four systems commercially available including the I-Point and the 3M TTI. Andover Labs (Weymouth, MA) marketed the Ambitemp and Tempchron devices up to 1985. Both were for use in frozen food distribution and could be classified as CTTI. Their operation was based on the displacement of a fluid along a capillary.

6.5 Current TTI systems

In the last fifteen years three types of TTI have been the focus of both scientific and industrial trials. They claim to satisfy the requirements of a successful TTI and have evolved as the major commercial types on the market. They are described in detail in the following sections.

6.5.1 Diffusion-based TTIs

The 3M Monitor Mark® (3M Co., St Paul, Minnesota) (US Patent, 3,954,011, 1976) is a diffusion based indicator. One of the first significant applications of TTI was the use of this indicator by the World Health Organization (WHO) to monitor refrigerated vaccine shipments. The response of the indicator is the advance of a blue dyed ester diffusing along a wick. The useful range of temperatures and the response life of the TTI are determined by the type of ester and the concentration at the origin. Thus the indicators can be used either as CTTI with the critical temperature equal to the melting temperature of the ester or as TTI if the melting temperature is lower than the range of temperatures the food is stored at, e.g., below 0ºC for chilled storage. The same company has marketed the successor to this TTI: the Monitor Mark Temperature Monitor (Fig. 6.1) and Freshness Check, based on diffusion of proprietary polymer materials (US patent 5,667,303).

A viscoelastic material migrates into a diffusely light-reflective porous matrix at a temperature dependent rate. This causes a progressive change of the light transmissivity of the porous matrix and provides a visual response. The response rate and temperature dependence is controlled by the tag configuration,

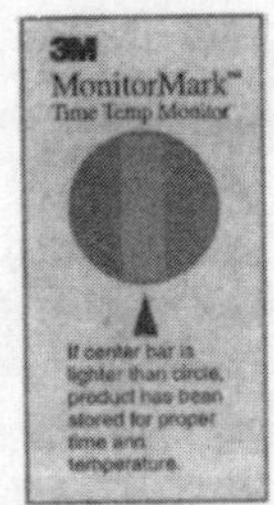

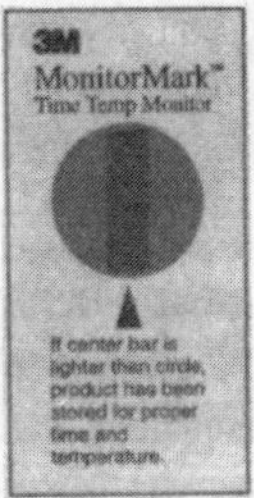

Fig. 6.1 Diffusion based TTI.

the diffusing polymer's concentration and its glass transition temperature and can be set at the desirable range (Shimoni, Anderson and Labuza, 2001). The TTI is activated by adhesion of the two materials. Before use these materials can be stored separately for a long period at ambient temperature.

6.5.2 Enzymatic TTIs

The VITSAB Time Temperature Indicator is an enzymatic indicator. It is the successor of the I-Point Time Temperature Monitor (VITSAB A.B., Malmö, Sweden). The indicator is based on a colour change caused by a pH decrease which is the result of a controlled enzymatic hydrolysis of a lipid substrate (US Patents 4,043,871 and 4,284,719). Before activation the indicator consists of two separate compartments, in the form of plastic mini-pouches. One compartment contains an aqueous solution of a lipolytic enzyme, such as pancreatic lipase. The other contains the lipid substrate absorbed in a pulverised PVC carrier and suspended in an aqueous phase and a pH indicator mix. Glycerine tricapronate (tricaproin), tripelargonin, tributyrin and mixed esters of polyvalent alcohols and organic acids are included in substrates.

Different combinations of enzyme-substrate types and concentrations can be used to give a variety of response lives and temperature dependencies. At activation, enzyme and substrate are mixed by mechanically breaking the barrier that separates the two compartments. Hydrolysis of the substrate (e.g., tricaproin) causes acid release (e.g., caproic acid) and the pH drop is translated in a colour change of the pH indicator from deep green to bright yellow. Reference starting and end point colours are printed around the reaction window to allow easier visual recognition and evaluation of the colour change (Fig. 6.2). The continuous colour change can also be measured instrumentally (Taoukis and Labuza, 1989). The TTI is claimed to have a long shelf-life if kept chilled before activation.

6.5.3 Polymer-based TTIs

The Lifelines Freshness Monitor® and Fresh-Check® indicators (Lifelines Inc, Morris Plains, NJ) are based on a solid state polymerisation reaction (US Patent, 3,999,946 and 4,228,126) (Fields and Prusik, 1983). The TTI function

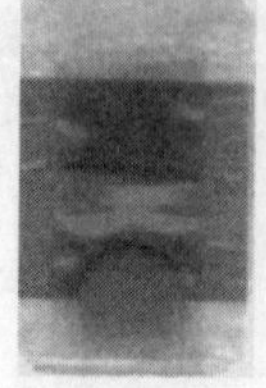
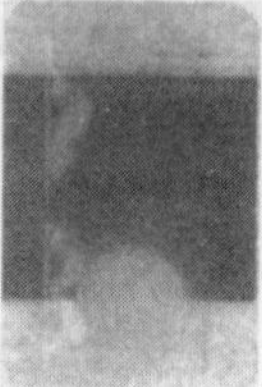

Fig. 6.2 Enzymatic TTI.

is based on the property of disubstituted diacetylene crystals (R–C = C–C = C–R) to polymerise through a lattice-controlled solid-state reaction proceeding via 1,4-addition polymerisation and resulting in a highly coloured polymer. During polymerisation, the crystal structure of the monomer is retained and the polymer crystals remain chain aligned and are effectively one dimensional in their optical properties (Patel and Yang, 1983). The response of the TTI is the colour change measured as a decrease in reflectance.

The Freshness Monitor consists of an orthogonal piece of laminated paper the front face of which includes a strip with a thin coat of the colourless diacetylenic monomer and two barcodes, one about the product and the other identifying the model of the indicator. The Fresh-Check® version, for consumers, is round, and the colour of the 'active' centre of the TTI is compared to the reference colour of a surrounding ring (Fig. 6.3). The laminate has a red or yellow colour so that the change is perceived as a change from transparent to black. The reflectance of the Freshness Monitor can be measured by scanning with a laser optic wand and stored in a hand-held device supplied by the TTI producer. The response of Fresh Scan can be visually evaluated in comparison to the reference ring or continuously measured by a portable colorimeter or an optical densitometer. Before use, the indicators, active from the time of production, have to be stored deep frozen where change is very slow.

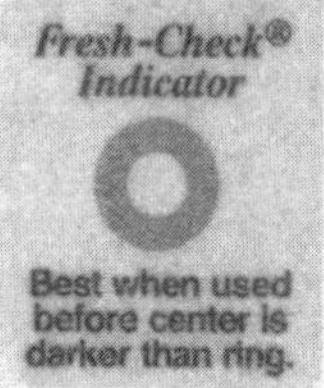

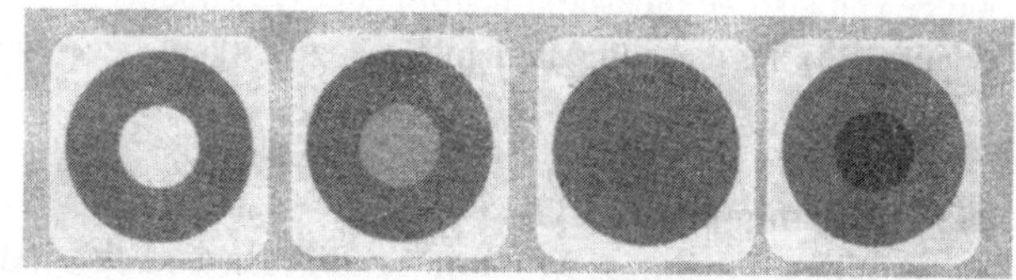

Fig. 6.3 Polymer based TTI.

6.6 Maximising the effectiveness of TTIs

Despite the potential of TTIs to contribute substantially to improved food distribution, reduce food waste and benefit the consumer with more meaningful shelf-life labelling, their application up to now has not lived up to the initial expectations. The main reasons for the reluctance of food producers to adopt the TTI have been:

- cost
- reliability
- applicability.

Cost is volume dependent, ranging from 2 to 20 US cents per unit. If other questions are resolved, cost-benefit analysis should favour use of TTIs. The reliability question has its roots in the history of indicators, due partly to lack of sufficient data, both from studies and from the suppliers. Initial attempts at using TTI as quality monitors were not well designed and hence unsuccessful. Re-emerging discussions by regulatory agencies to make TTI use mandatory, before the underlying concepts were understood and their reliability demonstrated, resulted in resistance by the industry and may have hurt TTI application up to the present time. Current TTI systems have achieved high standards of production quality assurance and provide reliable and reproducible responses according to the specifications stated. Testing standards have been issued by the BSI and can be used by TTI manufacturers as well as TTI users (BS 7908:1999).

The question of applicability, however, has been the most substantial hurdle to TTI use. Earlier studies have been ineffective in establishing a clear methodology on how the TTI response can be used as a measure of food quality. The initial approach was to assume an overall temperature dependence curve (or zone) for the shelf-life of a general class of foods, e.g., frozen foods, and aim for an indicator that has a similar temperature dependence curve for the time to reach a specific point on its scale. Such a generalisation has proved insufficient, as even foods of the same type differ significantly in the temperature dependence of the deterioration in their quality. What is needed is a thorough knowledge of the shelf-life loss behaviour of the food system through accurate kinetic models.

It has been widely assumed that the behaviour of a TTI should strictly match that of the particular food to be monitored at all temperatures. This approach, even if feasible, is impractical, and requires an unlimited number of TTI models. Instead of a TTI exactly mimicking quality deterioration behaviour of the food product, a meaningful, general scheme of translating TTI response to food status is needed. This should be based on systematic modelling of both the TTI and the food. Advances in modelling are now making this possible. Current developments in this area are reviewed by Taoukis (2001). The following sections discuss how modelling contributes to the practical use of TTIs.

6.7 Using TTIs to monitor shelf-life during distribution

TTIs can be used to monitor the temperature exposure of food products during distribution, from production up to the time they are displayed at the supermarket. Attached to individual cases or pallets they give a measure of the preceding temperature conditions at selected control points. Information from TTIs can be used for continuous, overall monitoring of the distribution system, leading to recognition and correction of weak links in the chain. Furthermore, it serves as a proof of compliance with contractual requirements by the producer and distributor. It can guarantee that a properly handled product was delivered to the retailer, thus eliminating the possibility of unsubstantiated rejection claims by the latter. The presence of the TTI itself would probably improve handling, serving as an incentive and reminder to distribution employees throughout the distribution chain of the importance of proper temperature storage.

The same TTIs can be used as shelf-life end point indicators readable by the consumer and attached to individual products. Tests using continuous instrumental readings to define the end point under constant and variable temperatures showed that such end points could be reliably and accurately recognised by panellists (Sherlock *et al.*, 1991). However, for a successful application of this kind there is a much stricter requirement that the TTI response matches the behaviour of the food. To achieve this the TTI end point should coincide with the end of shelf-life at one reference temperature and the activation energy should differ by less than 10kJ/mol from that of the food. In this way the TTI attached to individually packaged products can serve as active shelf-life labelling instead of, or in conjunction with, open date labelling. The TTI assures the consumer that the product was properly handled and indicates the remaining shelf-life. Consumer surveys have shown that consumers can be very receptive to the idea of using these TTI on dairy products along with the date code (Sherlock and Labuza, 1992). Use of TTI can thus also be an effective marketing tool. Diffusion-based TTIs have been used in this way by the Cub Foods Supermarket chain in the USA and polymer-based TTIs by the Monoprix chain in France and the Continent stores in Spain.

A number of experimental studies have sought to establish correlations between the response of specific TTIs and quality characteristics of specific products. Foods have been tested at different temperatures, plotting the response of the TTI v. time and the values of selected quality parameters of the foods before testing the statistical significance of the TTI response correlation to the quality parameters. Foods correlated to TTI include:

- pasteurised whole milk (Mistry and Kosikowski, 1983; Grisius *et al.*, 1987; Chen and Zall, 1987)
- ice cream (Dolan *et al.*, 1985)
- frozen hamburger (Singh and Wells, 1985)
- chilled cod fillets (Tinker *et al.*, 1985)
- refrigerated ready to eat salads (Cambell, 1986)

- frozen bologna (Singh and Wells, 1986)
- UHT milk (Zall *et al.*, 1986)
- refrigerated orange juice (Chen and Zall, 1987b)
- pasteurised cream (Chen and Zall, 1987a)
- cottage cheese (Chen and Zall, 1987; Shellhammer and Singh, 1991)
- frozen strawberries (Singh and Wells, 1987)
- chilled lettuce and tomatoes (Wells and Singh, 1988)
- chilled fresh salmon (Ronnow *et al.*, 1999).

Such studies offer useful information but do not involve any modelling of the TTI response as a function of time and temperature. They are thus applicable only for the specific foods and the conditions that were used. Extrapolation to other similar foods or quality loss reactions, or even use of the correlation equations for the same foods at other temperatures or for fluctuating conditions is not accurate.

A kinetic modelling approach allows the potential user to develop an application scheme specific to a product and to select the most appropriate TTI without the need for extensive testing of the product and the indicator. This approach emphasises the importance of reliable shelf-life modelling of the food to be monitored. Shelf-life models must be obtained with an appropriate selection and measurement of effective quality indices and be based on efficient experimental design at isothermal conditions covering the range of interest. The applicability of these models should be further validated at fluctuating, non-isothermal conditions representative of the real conditions in the distribution chain. Similar kinetic models must be developed and validated for the response of the suitable TTI. Such a TTI should have a response rate with a temperature dependence, i.e., activation energy E_{AI}, in the range of the E_A of the quality deterioration rate of the food. The total response time of the TTI should be at least as long as the shelf-life of the food at a chosen reference temperature. TTI response kinetics should be provided and guaranteed by the TTI manufacturer as specifications of each TTI model they supply.

The above concepts have been applied in studying the suitability of TTIs in monitoring the seafood chill chain within the FAIR-CT96-1090 research project funded by the European Commission entitled 'Development, Modelling and Application of Time-Temperature Integrators to monitor Chilled Fish Quality'. The fish chill chain is noted for substantial losses by spoilage. As part of this programme the shelf-life of different fresh and minimally processed fish products was systematically studied and modelled. Shelf-life analysis requires establishing a time correlation between measured chemical/biochemical changes, microbiological activity and sensory quality. Although each type of fish product, depending on the particular intrinsic and extrinsic factors, has its own specific spoilage pattern, investigation of the influence of each of these factors provides the fundamentals for understanding the spoilage phenomenon and for reliable shelf-life predictive modelling (Dalgaard, 1995; Dalgaard and Huss, 1995). Models of sensory quality and growth of spoilage microflora were developed and validated in dynamic temperature conditions for a variety of different fish. In this context the natural microflora of different Mediterranean fish of commercial interest such as boque, seabass, seabream and red mullet was studied and growth of the specific spoilage bacteria *Pseudomonas* spp. and *Shewanella putrefaciens* was modelled and correlated to organoleptic shelf-life (Taoukis *et al.*, 1999; Koutsoumanis and Nychas, 2000; Koutsoumanis *et al.*, 2000). Arrhenius and square root functions were used to model temperature dependence of maximum growth rates. For example experimental data for growth of the different measured constituents of the boque natural microflora showed that, at all temperatures, growth of Pseudomonads and *Shewanella putrefaciens* followed closely the decrease of average sensory score of the cooked fish. End of shelf-life coincided with an average level of 10^7 for these two bacteria from 0° to 15°C. At 0°C it was determined at 174 hr. The Arrhenius temperature dependence of the rate of sensory degradation and Pseudomonads and *Shewanella putrefaciens* exponential growth rate was determined in terms of activation energy (E_A) as 86.6, 81.6 and 82.7kJ/mol respectively.

Based on this kinetic data the effect of the difference in the activation energies of TTI response and spoilage rate of the monitored fish on the shelf-life predictive ability of TTI can be assessed. The actual effective temperature (based on growth kinetics) of variable temperature profiles is compared to the one calculated from the response of the TTI. The total shelf-life at 0°C is 174 hr based on the Pseudomonads growth with $N_0 = 1000$ and $N_{max} = 10^7$. This coincides with the sensory shelf-life. Setting these limits allows the estimation of remaining shelf-life at 0°C after the 'abusive' storage conditions of the first 24 hr. Table 6.2 shows T_{eff} for the fish, after exposure for 24 hours at the variable temperature profiles (shown in Fig. 6.4), is given.

It can be seen that, for the first temperature profile TD_1, the enzymatic TTI Model C has an activation energy more than 40kJ/mol different from the fish spoilage and gives a T_{eff} error of more than 1°C. This results in a prediction of remaining shelf-life of 74 hrs compared to the 90 hrs of actual remaining shelf-life. T_{eff} and prediction of remaining shelf-life from the enzymatic TTI Model M and the polymer-based TTI Model A6 (92 and 91 hr respectively) are very close to the actual. It should be noted, however, that even the erroneous estimations

Table 6.2 Effective temperature and remaining shelf-life (t_r) of boque at chilled conditions of 0°C, for variable storage temperatures (TD1, TD2) during the initial 24 hr estimated by different TTI

	TD1		TD2	
	T_{eff} (°C)	t_r(hr)	T_{eff} (°C)	t_r(hr)
'ACTUAL'	8.93	90	9.8	81
Enzymatic TTI – Model C	10.50	74	10.0	79
Enzymatic TTI – Model M	8.74	92	9.8	82
Polymer-based TTI – Model A6	8.95	91	9.8	91

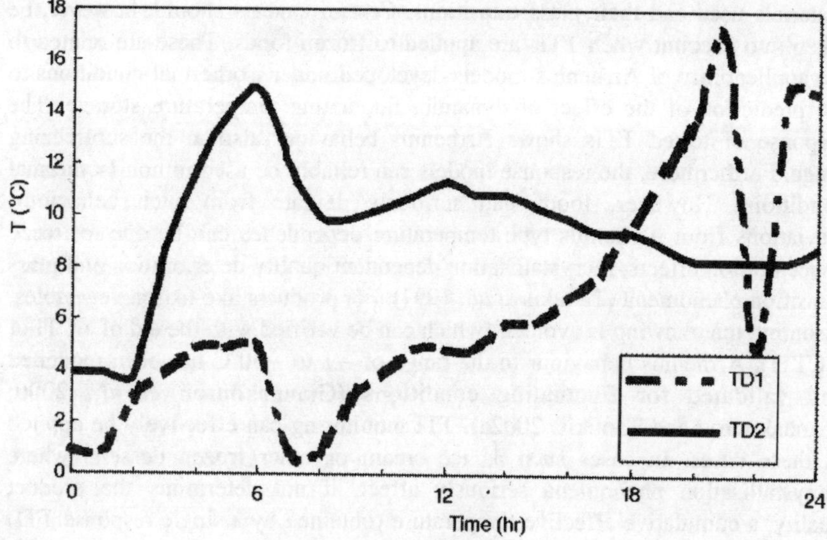

Fig. 6.4 Variable temperature scenarios of storage of boque during 24 hours.

from the TTIs with different activation energies are in practice much better than the 150 hr that would be presumed for shelf-life if no indication of improper storage was available. For the second temperature profile, TD_2, predictions from all TTIs are good. This illustrates the fact that the error depends on the actual temperature distribution. TD_1 and TD_2 are qualitatively different in that the first represents a profile with more abrupt changes than the second. The problem is that the temperature profile in a real situation is unknown. It is therefore advisable to select the TTI that, in addition to other requirements, has an activation energy close to the one of the quality loss rate of the food. Alternatively, response of two or three TTIs (i.e. a multiple TTI) with different E_{AI}s could provide a corrected estimate of T_{eff} giving a reliable estimate of the food quality even when these E_{AI}s differ substantially from the E_A of the food (Stoforos and Taoukis, 1998).

The example shows the potential and demonstrates the methodology of monitoring the chill chain based on a continuous scale TTI response, translatable to an effective temperature history. This methodology can be applied to chilled products other than fish if appropriate quality loss models are available. Long shelf-life chilled foods can benefit from the ability to monitor their temperature history by the introduction of a TTI based distribution control and stock rotation system. Such a system will be described and evaluated in the next section.

Frozen foods can also be monitored based on the same approach. Diffusion-based and enzymatic TTIs have been tested and modelled at temperatures in the range of −1 to −30°C (Giannakourou *et al.*, 2000; Giannakourou and Taoukis, 2002). Diffusion-based TTIs can respond above a temperature at which diffusion commences. This temperature can also be set, based on the type of polymer materials used and their glass transitions. Certain caveats should, however, be taken into account when TTIs are applied to frozen foods. These are related to the applicability of Arrhenius models developed under isothermal conditions to the prediction of the effect of dynamic, fluctuating temperature storage. The response of tested TTIs shows Arrhenius behaviour also at the subfreezing range. Furthermore, the response models can reliably be used in non-isothermal conditions. However, foods can seriously deviate from such behaviour. Deviations from Arrhenius type temperature dependence can be due to freeze concentration effects, recrystallisation dependent quality deterioration and glass transition phenomena (Taoukis *et al.*, 1997). For products like frozen vegetables, assuming that thawing is avoided (which can be verified with the aid of a CTI or a CTTI), Arrhenius behaviour in the range of −1 to −30ºC has been modelled and validated for fluctuating conditions (Giannakourou *et al.*, 2000; Giannakourou and Taoukis, 2002a). TTI monitoring can effectively be applied in these cases. In cases such as ice cream or other frozen desserts where recrystallisation phenomena seriously affect, if not determine, the product quality, a cumulative effective temperature (obtained by a single response TTI) might not be sufficient to predict accurately the quality loss.

6.8 Using TTIs to optimise distribution and stock rotation

The information provided by a TTI, translated to remaining shelf-life at any point of the chill chain, can be used to optimise distribution control and apply a stock rotation system. Such an inventory management and stock rotation tool at the retail level was initially proposed by Labuza and Taoukis (1990). The approach currently used is the First In First Out (FIFO) system according to which, products received first and/or with the closest expiration date on the label are shipped, displayed and sold first. This approach aims in establishing a 'steady state' with all products being sold at the same quality level. The assumption is that all products have gone through uniform handling. Quality is basically seen as a function of time. The use of the indicators can help establish a system that does not depend on this unrealistic assumption. This approach is called LSFO (Least Shelf-life First Out). The LSFO system would reduce the number of rejected products and largely eliminate consumer dissatisfaction since the fraction of product with unacceptable quality at consumption time can be minimised.

The development of LSFO system is based on validated shelf-life modelling of the controlled food product, specification of the initial value of the quality index, A_0, and the value A_s at the limit of acceptability (end of shelf-life), and temperature monitoring in the chill chain with TTI. The above elements for the core of integrated software that allows the calculation of the actual remaining shelf-life of individual product units (e.g. small pallets, 5–10 kg boxes or even single product units) at strategic control points of the chill chain. Based on the distribution of the remaining shelf-life, decisions can be made for optimal

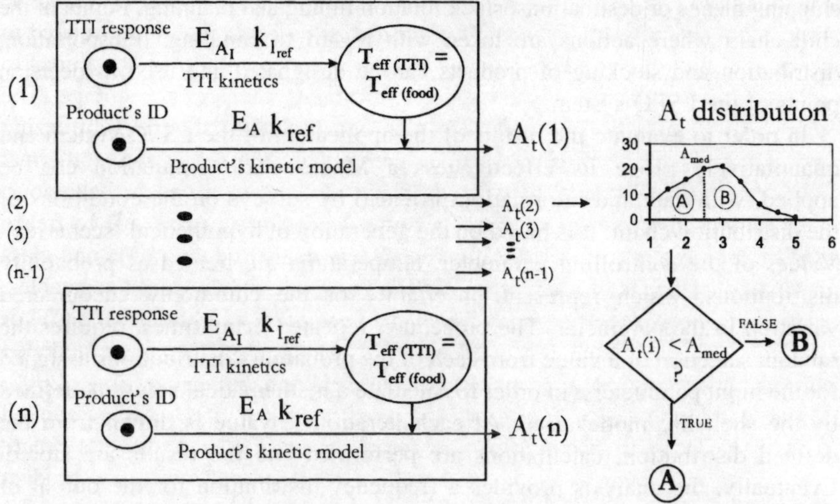

Fig. 6.5 Logical diagram of the decision-making routine of LSFO system at important control points of the distribution chain. Quality at time t (A_t) is computed for all n product units. The computation is based on the response of the TTI, translated to the effective storage temperature (T_{eff}) of the product. The distribution function of quality is constructed and decision for the further handling of each unit is taken based on its value within this function.

handling, shipping destination and stock rotation, aimed at obtaining a narrow distribution of quality at the point of consumption. The diagram of the decision-making process is illustrated in Fig. 6.5. For example, at a certain point, e.g. at the supermarket, one half of a shipment could be forwarded to retail display immediately, the other half the next day. The split could be random according to conventional FIFO practice or it can be based on the actual individual product quality and LSFO. For all units the response of the TTI, cumulatively expressing the temperature exposure of the product, is put in either electronically as a signal to a suitable optical reader or keyed in manually based on visual readings. This information directly fed into a portable unit with the LSFO programme, is translated to quality status, A_t, based on the kinetics of the used TTI, which integrates the time-temperature history of each product into an effective temperature value, T_{eff}, and the shelf-life model of the product. Having estimated A_t for all the n product units, the actual quality distribution for the products at the decision point is constructed. Based on the quality of each product unit relative to this distribution, decisions about its further handling are made.

For the scenario illustrated in Fig. 6.5, products B with less remaining shelf-life, i.e., higher A_t, will be displayed first in the retail display cabinets of the supermarket and will therefore be consumed sooner whereas products with longer remaining shelf-life (lower A_t) will be displayed later. The decision process can involve more options with regard to, e.g., handling methods, shipping means or destinations, stock rotation timing and planning. Points of the chill chain where actions are taken with regard to handling, transportation, distribution and stocking of products can be designated and used as decision points of the LSFO system.

In order to evaluate the results of the application of the LSFO system and quantitatively prove its effectiveness a Monte Carlo simulation can be applied, with data and information provided by surveys on the conditions of the distribution chain. It is based on the generation of hypothetical 'scenarios'. Values of the controlling parameter, temperature, are treated as probability distributions, which represent uncertainty or the commonly encountered variation in the parameter. The procedure, repeated many times, requires the random selection of a value from each of the probability distributions assigned for the input parameters, in order to calculate a mathematical solution, defined by the shelf-life model used. At each iteration, a value is drawn from the defined distribution, calculations are performed and the results are stored. Eventually, the analysis provides a frequency distribution for the output of interest (quality status and remaining shelf-life), that has taken into account the probability distribution of temperature conditions, instead of using a single-point estimate.

The results for the simulated application of LSFO in the cases of two long shelf-life chilled products are shown in Fig. 6.6. These products have both a shelf-life of three months at 4°C and their quality loss rate shows a low and high temperature dependence (low and high E_A respectively). Russian salad, a chilled product widely consumed in Greece with a shelf-life of three months, was used as a case study (Taoukis *et al.*, 1998). For this microbiologically stable, complex food, modelling of shelf-life was based on overall organoleptic deterioration and development of rancidity. Use of Weibull Hazard Analysis facilitated shelf-life determination and modelling of sensory evaluation data. The activation energy of shelf-life loss was estimated at 31.5 kJ/mol. For a realistic estimate of the storage temperature conditions at the different stages, data of chilled product temperatures previously collected at the commercial level and from a survey of home refrigerators was used. The temperature condition distributions are illustrated in Fig. 6.7.

To demonstrate the effectiveness of the LSFO approach as compared to FIFO, a 60-day cycle, from production to consumption, was used. This consisted of three stages:

- Stage 1: 30 days at local distribution centres
- Stage 2: 15 days at the supermarket storage
- Stage 3: 15 days at the domestic refrigerator

Based on a Monte Carlo simulation approach, 2000 temperature scenarios were run, using a program code written in FORTRAN 77. The temperatures used were obtained at random from the distributions of Fig. 6.7 (distribution 10a for stages 1 and 2 and distribution 10b for stage 3). The results of this simulation are illustrated in Fig. 6.6 which shows the probability for the product to be

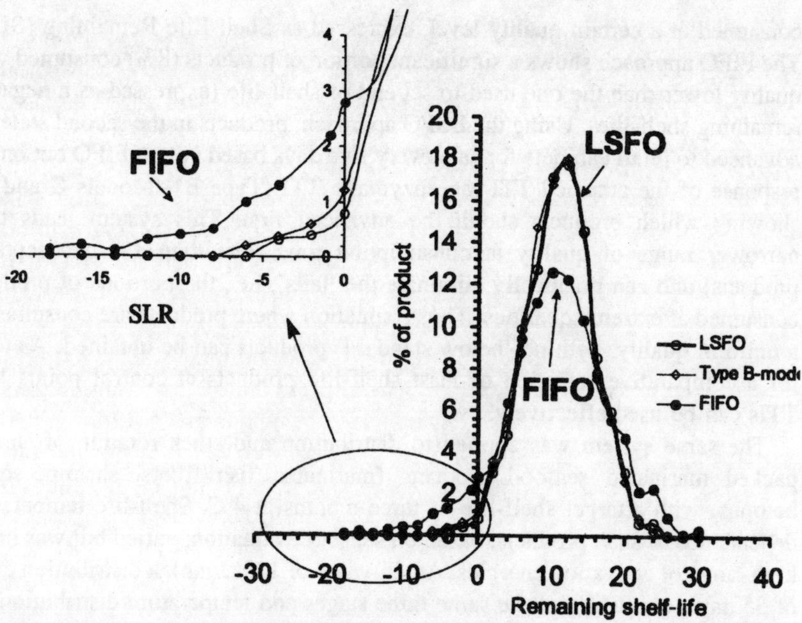

Fig. 6.6 Distribution of quality of Russian salad products after 60 days distribution, retail and domestic storage. For each point the percentage of the products that have a remaining shelf-life in the range of ±2.5 days of the abscissa value can be read on the vertical axis. The line with solid circles corresponds to the FIFO and with open circles to the LSFO system based on actual temperature monitoring or a TTI with $E_{A_I} = E_A$. Open diamonds line is the LSFO line based on the TTI, Type B (practically coincides with the actual LSFO line).

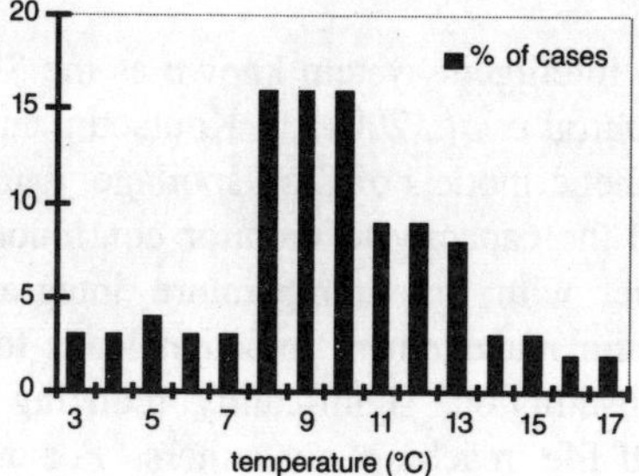

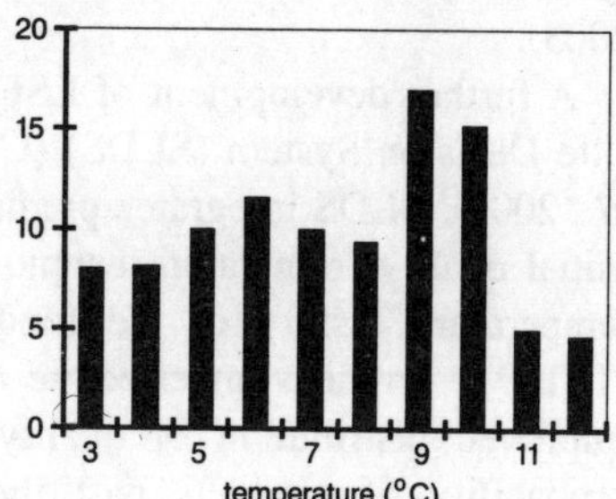

Fig. 6.7 Left: Temperature distribution in commercial chilled storage. (Measurements in 150 supermarkets in the metropolitan area of Athens). Right: Temperature distribution in domestic refrigerators. (Based on measurements in 40 households). (Adopted from Taoukis *et al.*, 1998.)

consumed at a certain quality level, expressed as Shelf-Life Remaining (SLR). The FIFO approach shows a significant portion of products (8%) consumed with quality lower than the one used to set end of shelf-life (expressed as a negative remaining shelf-life). Using the LSFO approach, products in the second state are advanced to retail cabinets for sale every five days based not on FIFO but on the response of the attached TTI (an enzymatic TTI (Type B)–Models C and M) showing which products should be advanced first. This system leads to a narrower range of quality at consumption time (less than 1% unacceptable products) and can practically eliminate the 'tails', i.e., the portions of products consumed at extreme qualities. Thus a situation where products are consumed at a uniform quality, with no 'below standard' products can be obtained. As tools for a comparative selection of least shelf-life products at control points both TTIs can be used effectively.

The same system was applied to distribution and stock rotation of shrink-packed marinated seafood products (marinated fish fillets, shrimp, squid, octopus) with a target shelf-life of three months at 4°C. Shelf-life temperature dependence of such products, based on sensory evaluation, varied but was in the high range of activation energies. An E_A value of 110kJ/mol, a distribution cycle of 35 days, consisting of the same three stages and temperature distributions as above (10 days at stage 1, 15 days at stage 2 and 10 days at stage 3) and enzymatic TTIs (Models C, S and L with activation energies 48.3, 102 and 160 kJ/mol respectively) were used in the Monte Carlo simulation to assess application of LSFO. Results are shown in Fig. 6.8. It can be seen that in products with high activation energies the distribution of quality at consumption time is much wider as temperature variation affects more intensely the rates of quality loss. Application of the LSFO system reduces the percentage of unacceptable products to less than 5% compared to 22% with the FIFO approach. It can also be seen that even TTIs that differ from the food in terms of E_A approximately 50 kJ/mol can serve as tools for the relative comparison of the shelf-life of the products at the control points of the LSFO system. The LSFO system can also be applicable for frozen foods (Giannakourou and Taoukis, 2003).

A further development of LSFO is an intelligent system known as the Shelf Life Decision System (SLDS) (Giannakourou *et al.*, 2001a,b; Koutsoumanis *et al.*, 2002). SLDS integrates predictive kinetic models of food spoilage, data on initial quality from rapid techniques and the capacity to monitor continuously temperature history of the food product with time temperature integrators (TTIs). It provides an effective chill chain management tool that leads to an improved distribution of quality at consumption, significantly reducing the probability of products past their shelf-life reaching consumers. For most processed food products, 'zero time' post-processing parameters, including a target range of initial microbial load, can be fixed and achieved by proper design and control of the processing conditions. This is the working assumption of LSFO. However, initial microflora in fresh foods such as fish or meat can fluctuate significantly, depending on a number of extrinsic factors at slaughter or

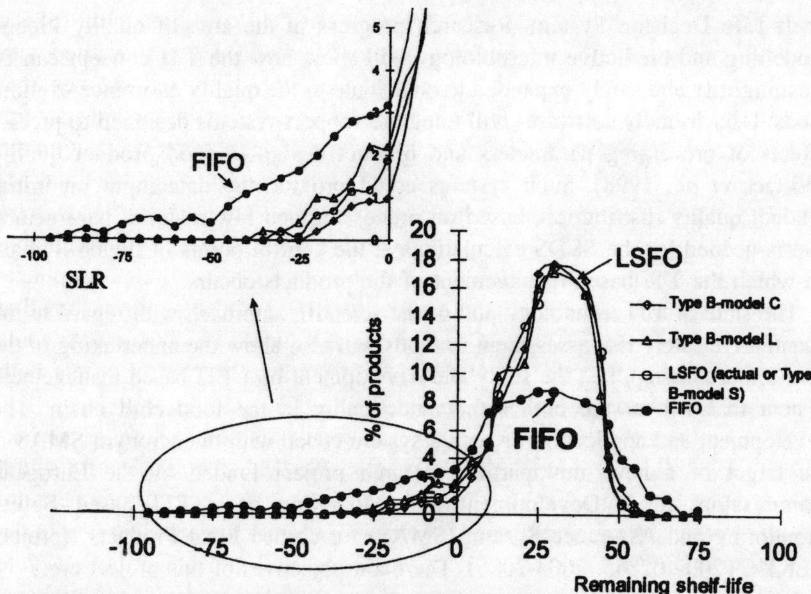

Fig. 6.8 Distribution of quality of shrink-packed products after 30 days distribution, retail and domestic storage. For each point the percentage of the products that have a remaining shelf-life in the range of ±5 days of the abscissa value can be read on the vertical axis. The line with solid circles corresponds to the FIFO and with open circles to the LSFO system based on actual temperature monitoring or the TTI B- Model S with $E_{A_I} \approx E_A$. Open diamonds and open triangles lines are the LSFO lines based on the TTI, Type B–Models L and C.

catch, and subsequent handling and processing (Eisel *et al.*, 1997; Huss, 1995). SLDS takes not only the history of the product in the distribution chain into account but also this variability of initial contamination. Rapid methods of microbial enumeration can be employed to provide such information as input. The Shelf Life Decision System can incorporate other parameters in the calculation of the quality distribution at each control point. Such parameters can include variation of initial pH, water activity, packaging gases composition, provided the shelf-life predictive models can account for the effect of these parameters on the microbiological and chemical reactions responsible for the loss of quality.

6.9 Future trends

TTIs will inevitably find wider application as tools to monitor and control distribution as their potential is thoroughly understood by the food industry. Progress on both the variety, reliability and flexibility of TTIs, and on better quantitative shelf-life characterisation of food products, will allow successful application of chill chain optimisation tools such as the LSFO and the intelligent

Shelf Life Decision System. Research progress in the area of quality kinetic modelling and predictive microbiology will show how the TTI concept can be meaningfully and safely expanded to contribute to the quality assurance of more foods. User friendly softwares will integrate support systems designed to predict effects of processing parameters and product design to food product quality (Wijtzes *et al.*, 1998). Such systems could provide the data input on initial product quality distribution, based on processing and raw material parameters, that is needed for the SLDS calculations at the control points of the chill chain on which the TTI based management of the products occurs.

The state of TTI technology and of the scientific approach with regard to the quantitative safety risk assessment in foods will also allow the undertaking of the next important step, i.e., the study and development of a TTI based management system that will assure both safety and quality in the food chill chain. The development and application of such a system coded with the acronym SMAS is the target of a new, multipartner research project funded by the European Commission titled 'Development and Modelling of a TTI based Safety Monitoring and Assurance System (SMAS) for chilled Meat Products' (project QLK1-CT2002-02545, 2003-2005). The main objectives of this project are:

- modelling the effect of food structure, microbial interactions and dynamic storage conditions on meat pathogens and spoilage bacteria.
- combination of validated pathogen growth models with data on prevalence/ concentration, dose response and chill chain conditions for risk assessment with and without SMAS application.
- development, modelling and optimisation of TTI with accuracy to monitor microbiological safety of meat products.
- development of SMAS into a user-friendly computer software.
- evaluation of the applicability and effectiveness of SMAS in real conditions of meat distribution.
- assessment of the industry acceptance of the TTI and the concept of chill chain management and
- evaluation of the consumer attitude on use of TTI and correlation to quality.

Information on the outputs of this project will be available on the Web (www.cordis.lu/life/src/pub_qol.htm).

6.9 References

93/43/EEC, 1993, 'Hygiene of foodstuffs', *General Directive,* 1993.

BS 7908: 1999, 'Packaging–temperature and time–temperature indictors – performance specification and reference testing', British Standard, 1999.

BYRNE C H, 1976, 'Temperature indicators – the state of the art', *Food Technology,* **30** (6) 66–8.

CAMBELL L A, 1986, *Use of a time–temperature indicator in monitoring quality of refrigerated salads,* M. S. Thesis, Michigan State University.

CHEN J H and ZALL R R, 1987a, 'Packaged milk, cream and cottage cheese can be monitored for freshness using polymer indicator labels', *Dairy Food Sanitation,* **7** (8) 402–5.

CHEN J H and ZALL R R, 1987b, 'Refrigerated Orange Juice can be monitored for freshness using a polymer indicator label', *Dairy Food Sanitation,* **7** 280–3.

CODEX COMMITTEE ON FOOD HYGIENE, 1997, 'Revised International Code of Practice – General Principles of Food Hygiene', *WHO/FNU/97,* 1997.

DALGAARD P and HUSS H H, 1995, 'Mathematical modelling used for evaluation and prediction of microbial fish spoilage', in: *Seafood safety, processing and biotechnology,* Kramer, D, Shahidi, F and Jones, Y (eds) Technomic Pub Co Inc, Lancaster, Canada, 73–89.

DALGAARD P, 1995, 'Modelling of microbial activity and prediction of shelf–life for packed fresh fish', *Int J of Food Microbiol,* **26** 305–17.

DOLAN K D, SINGH R P and WELLS J H, 1985, 'Evaluation of time–temperature related quality changes in ice cream during storage', *J Food Proc Preserv,* **9** 253–71.

EISEL W G, LINTON R H and MURIANA P M, 1997, 'A survey of microbial levels for incoming raw beef and ground beef in a red meat processing plant', *Food Micro,* **14** 273–82.

FIELDS S C and PRUSIK T, 1983, 'Time–Temperature monitoring using solid–state chemical indicators', *Intl Inst Refrig Commission C2 Preprints,* 16th Intl Cong Refrig, 636–40.

GIANNAKOUROU M C, TAOUKIS P S, 2002. Systematic application of Time Temperature Integrators as tools for control of frozen vegetable quality. *J.Food Sci*. **67**(6): 2221–8.

GIANNAKOUROU M C, TAOUKIS P S. 2003. 'Application of a TTI–based distribution management system for quality optimisation of frozen vegetables at the consumer end'. *J.Food Sci*. **68**(1): 201–9.

GIANNAKOUROU M C, SKIADOPOULOS A, POLYDERA A and TAOUKIS P S, 2000, 'Shelf–life modelling of frozen vegetables for quality optimisation with Time–Temperature indicators', *Eighth International Congress on Engineering and Food,* Puebla, Mexico.

GIANNAKOUROU M C, KOUTSOUMANIS K, DERMESONLOUOGLOU E, and TAOUKIS P S, 2001a, 'Applicability of the intelligent Shelf Life Decision system for control of nutritional quality of frozen vegetables'. *Acta Horticulturae,* **566**: 275–80.

GIANNAKOUROU M, KOUTSOUMANIS K, NYCHAS G–J E and TAOUKIS P S, 2001b, 'Development and assessment of an intelligent Shelf Life Decision System for quality optimisation of the food chill chain', *J Food Protection,* **64**(7): 1051–7.

GRISIUS R, WELLS J H, BARRET E L and SINGH R P, 1987, 'Correlation of time–temperature indicator response with microbial growth in pasteurised milk', *J Food Proc Preserv,* **11** 309–24.

HU K H, 1972, 'Time–Temperature indicating system "writes" status of product shelf–life', *Food Technology,* **26** (8) 56–62.

HUSS H H, 1995, 'Quality and quality changes in fresh fish', FAO Fisheries Technical Paper, No. 348, FAO, Rome.

ISO 9001:2000, 'Quality management systems – Requirements', *International Standard*, 2000.

ISO 15161:2001, 'Guidelines on the application of ISO 9001: 2000 for the food and drink industry', *International Standard,* 2001.

KILLORAN J, 1976, 'A time–temperature indicating system for foods stored in the nonfrozen state', *Activities Report, US Army Natick Labs,* 137–42.

KOUTSOUMANIS K and NYCHAS G-J E, 2000, 'Application of a systematic experimental procedure to develop a microbial model for rapid fish shelf–life procedure', *Int J Food Microbiol,* **60** 171–84.

KOUTSOUMANIS K, GIANNAKOUROU M C, TAOUKIS P S and NYCHAS G-J E, 2002. 'Application of Shelf Life Decision System (SLDS) to marine cultures fish quality', *Int J Food Micro.* **73**: 375–82.

KOUTSOUMANIS K, TAOUKIS P S, DROSINOS E and NYCHAS G-J E, 2000, 'Applicability of an Arrhenius model for the combined effect of temperature and CO_2 packaging on the spoilage microflora of fish', *Appl Env Microbiol,* **66** 3528–34.

KRAMER A and FARQUHAR J W, 1976, 'Testing of time–temperature indicating and defrost devices', *Food Technology,* **32** (2) 50–6.

LABUZA T P and TAOUKIS P S, 1990, 'The relationship between processing and shelf–life', in: *Foods for the 90's,* Birch, G G, Campbell–Platt, G and Lindley, M G (eds), Developments Series, Elsevier Applied Science Publishers, London and New York, Ch. 6, 73–105.

MIDGLEY T, 1933, 'Telltale means', *US Patent 1917048.*

MISTRY V V and KOSIKOWSKI F V, 1983, 'Use of time–temperature indicators as quality control devices for market milk', *J Food Protection,* **46** (1) 52–7.

OTA, 1979, *Open Shelf Life Dating of Food,* Office of Technology Assessment, Congress of the US, Library of Congress Cat. No. 79–600128.

PATEL G N and YANG N, 1983, 'Polydiacetylenes: An ideal system for teaching polymer science', *J Chem Ed,* **60** (3) 181–5.

RENIER J J and MORIN W T, 1962, 'Time–temperature indicators', *Intl Inst Refrig Bull Annex,* **1** 425–35.

RONNOW P, SIMPSON R and OTWELL S, 1999, 'The use of an enzymatic TTI to monitor time–temperature exposure in distribution of chilled seafood', in: *Predictive Microbiology Applied to Chilled Food Preservation.* C M Bourgeois, and T A Roberts (eds) Refrigeration Science and Technology Proceedings Series, International Institute of Refrigeration (IIR), Paris, France, 308–15.

SCHOEN H M and BYRNE C H, 1972, 'Defrost indicators: many designs have been patented yet there is no ideal indicator', *Food Technol,* **26** (10) 46–50.

SHELLHAMMER T H and SINGH R P, 1991, 'Monitoring chemical and microbial changes of cottage cheese using a full history time–temperature indicator', *J Food Sci,* **56** 402–5.

SHERLOCK M and LABUZA T P, 1992, 'Consumer perceptions of consumer time–temperature indicators for use on refrigerated dairy foods', *J Dairy Science*, **75**, 3167–76.

SHERLOCK M, FU B, TAOUKIS P S and LABUZA T P, 1991, 'Systematic evaluation of time temperature indicators for use as consumer tags', *J Food Protection*, **54** (11) 885–9.

SHIMONI E, ANDERSON E M and LABUZA T P, 2001, 'Reliability of Time Temperature Indicators Under Temperature Abuse' *J. Food Sci.* **66**(9): 1337–1340.

SINGH R P and WELLS J H, 1985, 'Use of time–temperature indicators to monitor quality of frozen hamburger', *Food Technology,* **39** (12) 42–5.

SINGH R P and WELLS J H, 1986, 'Keeping track of time and temperature: new improved indicators may become industry standards', *Meat Processing,* **25** (5) 41–2, 46–7.

SINGH R P and WELLS J H, 1987, 'Monitoring quality changes in stored frozen strawberries with time–temperature indicators', *Int J Refrig,* **10** 296–9.

STOFOROS N G and TAOUKIS P S, 1998, 'A theoretical procedure for using multiple response time–temperature integrators for the design and evaluation of thermal processes', *Food Control,* **9** (5) 279–87.

TAOUKIS P S and LABUZA T P, 1989, 'Applicability of Time Temperature Indicators as shelf–life monitors of food products', *J Food Sci,* **54** 783–8.

TAOUKIS P S, 1989, *Time–temperature indicators as shelf life monitors of food products,* PhD Thesis, University of Minnesota, St Paul.

TAOUKIS P S, 2001, 'Modelling the use of time temperature indicators in distribution and stock rotation', in Tijskens L, Hertog M and Nicolai B (eds), *Food process modelling,* Woodhead Publishing Ltd, Cambridge.

TAOUKIS P S, BILI M and GIANNAKOUROU M, 1998, 'Application of shelf–life modelling of chilled salad products to a TTI based distribution and stock rotation system', in: *Proceedings of the International Symposium on Application of Modelling as an Innovative Technology in the Agri–Food–Chain,* Tijiskens, L M M and Hertog, M L A T M (eds), *Acta Horticulturae,* **476** 131–40.

TAOUKIS P S, LABUZA T P and SAGUY I, 1997, 'Kinetics of food deterioration and shelf-life prediction', in: *Handbook of Food Engineering Practice*, Valentas K J, Rotstein E and Singh R P (eds), CRC Press, Boca Raton, Ch. 10, 361–403.

TAOUKIS P S, FU B and LABUZA T B, 1991, 'Time–Temperature Indicators', *Food Technology,* **45** (10) 70–82.

TAOUKIS P S, KOUTSOUMANIS K and NYCHAS G–J E, 1999, 'Use of Time Temperature Integrators and Predictive Modelling for Shelf Life Control of Chilled Fish under Dynamic Storage Conditions', *Int J Food Microbiol,* **53** 21–31.

TINKER J H, SALVIN J W, LEARSON R J and EMPOLA V G, 1985, 'Evaluation of automated time–temperature monitoring system in measuring the freshness of chilled fish', *IIF-IIR Commissions C2, D3 4,* 286–90.

US FEDERAL REGISTER, 1996, 'Pathogen Reduction; Hazard analysis and Critical Control Point (HACCP) Systems; Final Rule', Vol. **61**, No. 144, July 25, 1996.

WELLS J H and SINGH R P, 1988, 'Application of time-temperature indicators in monitoring changes in quality attributes of perishable and semi-perishable foods', *J Food Sci,* **53** 148–52.

WIJTZES T, VAN'T RIET K, HUIS IN'T VELDAND J H J and ZWIETERING M H, 1998, 'A decision support system for the prediction of microbial food safety and food quality', *Int J Food Microbiol,* **42** 79–90.

ZALL R R, CHEN J and FIELDS S C, 1986, 'Evaluation of automated time–temperature monitoring systems in measuring freshness of UHT milk', *Dairy and Food Sanitation,* **6** (7) 285–90

7

The use of freshness indicators in packaging

M. Smolander, VTT Biotechnology, Finland

7.1 Introduction

As we know from the definition, intelligent or smart packaging monitors and gives information about the quality of the packed food. According to Huis in't Veld (1996) the changes taking place in the fresh food product can be categorised as (i) microbiological growth and metabolism resulting in pH-changes, formation of toxic compounds, off-odours, gas and slime formation, (ii) oxidation of lipids and pigments resulting in undesirable flavours, formation of compounds with adverse biological reactions or discoloration. The focus of this chapter is on intelligent concepts indicating the changes mainly belonging to the first category.

The intelligence of a package can be based on the package's ability to give information about the requirements of the product quality like package integrity (leak indicators) and time-temperature history of the product (time-temperature indicators). Intelligent packaging can also give information on product quality directly (see Fig. 7.1). A freshness indicator indicates directly the quality of the product. The indication of microbiological quality is, for example, based on a reaction between the indicator and the metabolites produced during growth of microorganisms in the product. Of the indicators mentioned, time-temperature indicators and leak indicators are already commercially available and their use is increasing constantly. An indicator that would show specifically the spoilage or the lack of freshness of the product, in addition to temperature abuse or package leaks, would be ideal for the quality control of packed products. The number of concepts of package indicators for contamination or freshness detection of food is still very low, however new concepts of freshness indicators are patented and new commercially available products are likely to become available in the near future.

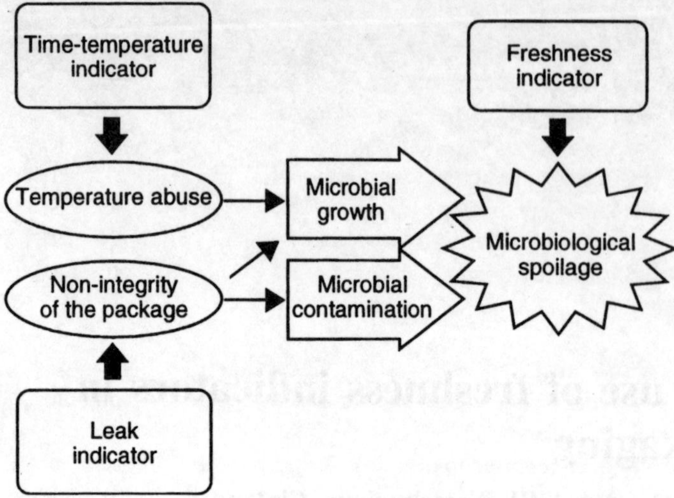

Fig. 7.1 Quality indicators for packaged food products can be either on direct or indirect freshness evaluation.

In this chapter the potential microbial metabolites and other compounds indicating the quality of packaged food are presented. Subsequently, freshness indicator concepts, which are commercially available or have been described in literature are reviewed. Finally, the possibilities for the future are discussed.

7.2 Compounds indicating the quality of packaged food products

An essential prerequisite in the development of freshness indicators is knowledge about the quality indicating metabolites. These metabolites have also been studied because they offer a possibility to replace time-consuming sensory and microbiological analyses traditionally used in the quality evaluation of food products (Dainty, 1996). The formation of the different metabolites depends on the nature of the packaged food product, spoilage flora and the type of packaging. The chemical detection of spoilage has been extensively reviewed by Dainty (1996). Chemical changes in stored meat have been discussed by Nychas *et al.* (1998). They propose that some chemical compounds do indeed indicate the microbiological quality of food products but also that more information is needed about the correlation between the sensory quality and the concentration of the metabolites. In this chapter some of the quality-indicating metabolites and other compounds representing potential target molecules for the quality-indicating freshness indicators are discussed in detail.

7.2.1 Glucose

Glucose is an initial substrate for many spoilage bacteria in air, vacuum packages and modified atmosphere packages (Dainty, 1996). As bacterial growth takes place glucose is depleted from meat surface and it has been proposed by Kress-Rogers (1993) that the measurement of the glucose gradient could be utilised to predict the remaining shelf-life. However glucose is not among the most promising quality-indicating compounds since the concentration decreases during storage and it would be more beneficial to have a quality-indicating compound with non-existent or low intitial concentration.

7.2.2 Organic acids

Organic acids like lactic acid and acetic acid are the major compounds having a role in glucose fermentation by lactic acid bacteria. The amount of L-lactic acid has generally been reported to decrease during storage of fish and meat (Kakouri *et al.*, 1997; Drosinos and Nychas, 1997; Nychas *et al.*, 1998). On the contrary, the concentration of D-lactate has been reported to increase during storage of meat and therefore D-lactate seems to be a more promising freshness indicator (Shu *et al.*, 1993).

Acetate concentrations have been reported to increase during storage of fresh fish (Kakouri *et al.*, 1997). In our studies with modified atmosphere packaged poultry meat (Smolander *et al.* in preparation) we have also found that the concentration of acetic acid in the tissue fluid and homogenised meat increased as a function of storage time and temperature. We also studied the formation of formic acid. At the beginning of storage some decrease in the concentration in meat was seen but later the concentration increased, however the increase was not as clearly dependent on storage temperature as in the case of acetic acid.

7.2.3 Ethanol

In addition to lactic and acetic acid, ethanol is another major end product of fermentative metabolism of lactic acid bacteria. It has been postulated that an increase in ethanol concentration in meat and fish indicates an increase of total viable count of the product. For instance, Rehbein (1993) studied the formation of ethanol in iced fish. He found that on an average the concentration of ethanol was increased as a function of storage time. At sensory rejection point an average of 1.77 mg ethanol was found in 100 g fish. Rehbein (1993) also studied the formation of ethanol in smoked, vacuum-packed salmon. The concentrations in stored fish samples were considerably higher than in fresh, iced fish. At the end of shelf-life concentrations as high as 10 mg/100 g fish were observed. Randell *et al.* (1995) studied the effect of storage time and package integrity on the formation of ethanol in modified atmosphere packaged, marinated rainbow trout slices. They found that the amount of ethanol in package headspace was increased together with storage time and size of the package leakage. In our unpublished studies analogous to Randell *et al.* (1995) we found also that storage temperature had a remarkable effect on the ethanol concentration. The higher the storage temperature was, the higher was the concentration of ethanol.

Randell *et al.* (1995) also studied the volatile compounds in packages containing marinated chicken pieces in modified atmosphere (40%CO_2 + 60%N_2). They found that ethanol concentration in the package headspace increased as a function of storage time. In our own unpublished studies we also found some correspondence between sensory quality of modified atmosphere (40%CO_2 + 60%N_2) packaged, unmarinated poultry meat and ethanol concentration in the package headspace. However, in unmarinated poultry meat packaged in MAP with higher (80%) CO_2 concentration we did not observe a clear trend in the formation of ethanol as a function of storage time and temperature.

7.2.4 Volatile nitrogen compounds

It is well known that high levels of basic volatile nitrogen compounds like ammonia, dimethylamine and trimethylamine give an indication about microbiological spoilage of fish (Ohlenschläger, 1997). The European Commission has even fixed TVB-N (total volatile basic nitrogen contributed by ammonia, dimethylamine and trimethylamine) limits for some fish species (95/149/EEC). Trimethylamine, formed by microbial actions in fish muscle is generally considered as a major metabolite responsible for the spoilage odours of seafood. A drawback of using trimethylamine as a quality indicator for seafood is the variation in the concentration of its precursor trimethylamine N-oxide according to the species and season (Dainty, 1996; Rodríguez *et al.*, 1999).

7.2.5 Biogenic amines

Biogenic amines (e.g. tyramine, cadaverine, putrescine, histamine) are especially widely considered as indicators of hygienic quality of meat products. In addition to this indicative nature, they can have pharmacological, physiological and toxic effects. Due to the health risks, a tolerance level of 100 mg/kg of fish has been established for histamine by FDA (Kaniou, *et al.* 2001). Even if biogenic amines indicate the quality of food products they do not themself contribute to the sensory quality of the product.

Putrescine and cadaverine are formed from ornithine and lysine, respectively in enzymatic decarboxylation. It has been widely suggested that these diamines are indicators of the initial stage of decomposition of meat products. Okuma *et al.* (2000) describe an increase in the diamine concentration together with the increase of the total viable counts in aerobically stored chicken. Kaniou *et al.* (2001) reported the formation of putrescine and cadaverine in unpacked beef. In addition to putrescine and cadaverine, histamine was also produced during the storage of vacuum-packed beef. Also in our studies we have found a clear correspondence between the microbiological quality of modified atmosphere packaged poultry and the total amount of biogenic amines. Tyramine formation took place evenly during the storage period, the formation being however dependent on temperature. Storage temperature had a more striking effect on the formation of putrescine and cadaverine which were accumulated especially at the end of storage period (Rokka *et al.*, submitted).

Ruiz-Capillas and Moral (2002) studied the effect of controlled and modified atmosphere on the production of biogenic amines during storage of hake. They found that controlled and modified atmosphere generally restricted the formation of biogenic amines and that cadaverine was a major biogenic amine formed in most of the studied atmospheres. Rodríguez *et al.* (1999) proposed that biogenic amines cadaverine and putrescine could indicate the freshness of freshwater rainbow trout either stored in air or in a vacuum package.

7.2.6 Carbon dioxide

Carbon dioxide (CO_2) is generally known to be produced during microbial growth. CO_2 is also typically added as a protecting gas to modified atmosphere packages, together with inert nitrogen, since it has bacteriostatic effects. A modified atmosphere package for non-respiring food typically has a CO_2 concentration as high as 20–80 % and, e.g., Fu *et al.* (1992) reported a further increase of CO_2 during storage in modified atmosphere packages containing beef. Even if the indication of microbial growth by CO_2 may be difficult in these modified atmosphere packages already containing a high concentration of CO_2, it is possible to use the increase in CO_2 concentration as a means of determining microbial contamination in other types of product. For instance Mattila *et al.* (1990) found a correlation between CO_2 concentration and the growth of microbes in pea and tomato soup which were packaged aseptically either in air or in a mixture of O_2 (5%) and nitrogen.

7.2.7 ATP degradation products

K-value is defined as the ratio of the sum of hypoxanthine and inosine and the total concentration of ATP-related compounds (Henehan *et al.*, 1997). This value, indicating the extent of ATP-degradation, correlates with the sensory quality of fish and also other types of meat and has been used as a freshness indicating parameter (Watanabe *et al.*, 1989; Yano *et al.*, 1995a). For fresh meat the value is low since the concentration of ATP-degradation products is low as compared to the concentration of all ATP-related compounds. The correlation between ATP-degradation products and fish quality has been extensively studied, e.g., by Hattula (1997).

7.2.8 Sulphuric compounds

Some sulphuric compounds have a remarkable effect on the sensory quality of meat products due to their typical odour and low odour threshold. Hydrogen sulphide (H_2S) is produced from cysteine and triggered by glucose limitation (Borch *et al.*, 1996). H_2S forms a green pigment, sulphmyoglobin, when it is bound to myoglobin (Paine and Paine, 1992, Egan *et al.*, 1989). However, sulphmyoglobin is not formed in anaerobic conditions.

H_2S and other sulphuric compounds have been found to be produced during the spoilage of poultry by pseudomonas, *Alteromonas* sp. and psycrotrophic anaerobic clostridia (Freeman *et al.*, 1976; Lea *et al.*, 1969; Russell *et al.*, 1997, Vieshweg *et al.*, 1989; Arnaut-Rollier *et al.*, 1999; Kalinowski and Tompkin, 1999). According to Dainty (1996), production of H_2S can be used as an indication of Enterobacteriacae and hence also of hygienic problems in aerobically stored meat. H_2S production by *Alteromonas putrefaciens*, *Enterobacter liquefaciens* and pseudomonas was discovered in high ultimate pH beef from stressed animals (Gill and Newton, 1979; Nicol *et al.*, 1970). It has also been found that in vacuum packed meat H_2S indicates the growth of particular strains of lactic acid bacteria (Egan *et al.*, 1989). Also in fish the volatile sulphuric compounds have been suggested as the main cause of putrid spoilage aromas (Olafsdottir and Fleurence, 1997).

7.3 Freshness indicators

A variety of different concepts for freshness indicators have been presented in the scientific literature. Most of these concepts are based on a colour change of the indicator tag due to the presence of microbial metabolites produced during spoilage (e.g. Smolander *et al.*, 2002; Wallach and Novikov, 1998; Kahn, 1996; Namiki, 1996), but also concepts for indicators relying on more advanced technology have been presented. For instance, a miniaturised gas detector based on conducting polymers has been patented by Aromascan, a manufacturer of electronic nose equipment (Payne and Persaud, 1995). Fibre optics can also be used to construct indicators for the volatile compounds produced in microbial spoilage (Honeybourne, 1993; Wolfbeis and List, 1995). Freshness indicator or detector concepts have been proposed, for example, for CO_2, diacetyl, amines, ammonia and hydrogen sulphide (see Table 7.1). These concepts are discussed in detail in following chapters.

7.3.1 Indicators sensitive to pH change

The majority of concepts described in the literature are based on the use of pH-dyes, which change colour in the presence of volatile compounds produced during spoilage. As early as the 1940s Clark (1949) filed a patent application describing 'an indicator which exhibits an irreversible change in visual appearance upon an appreciable multiplication of bacteria in the indicator'. The idea was that if microbiological growth inducing a pH change has been possible in the indicator, the conditions might have been such that the food product itself may also have been subject to deterioration. A direct

Table 7.1 Examples of freshness and contamination indicator systems for food packages

Author/ Patent applicant or holder/Trade name	Metabolite detected	Principle of the indicator
Freshness indicators		
Holte (1993)	CO_2	Colour change of e.g. bromothymol blue
Visual Spoilage Indicator Company (Eaton et al. 1977)	CO_2	Colour change
Mattila et al. (1990)	CO_2	Colour change of bromothymol blue
Horan (1998, 2000)	E.g. CO_2, SO_2, NH_4	Colour change of the indicator (e.g. xylenol blue, bromocresol purple, bromocresol green, cresol red, phenolphalein, bromothymol blue, neutral red) incorporated in the packaging material
Neary (1981)	CO_2, H_2, NH_4	Colour change of liquid crystal/liquid crystal + indicator
AVL Medical Instruments (Wolfbeis and List, 1995)	CO_2, NH_4, amines, H_2S	Colour change of CO_2–, NH_4–, and amine-sensitive dyes, formation of colour of heavy-metal sulfides (H_2S)
Biodetect Corporation (Wallach and Novikov, 1998)	E.g. acetic acid, lactic acid, acetaldehyde, ammonia, amines	Visually detectable colour change of a pH-dye (e.g. phenol red, cresol red, *m*-cresol purple)
Mattila and Auvinen (1990a, b)	Not specified	Colour change of methylene blue or 2,6-dichlorophenol indophenol
Miller *et al.* (1999)	Volatile amines	Colour change of a food dye
Loughran and Diamond (2000)	Volatile amines	Colour change of calix[4]arene-based dye
VTT Biotechnology (Ahvenainen *et al.*, 1997)	H_2S	Colour change of myoglobin
Cameron and Talasila (1995)	Ethanol	Alcohol oxidase-peroxidase-chromogenic substrate system
Honeybourne (1993)	Diacetyl	Detection of optical changes in aromatic orthodiamine
DeCicco and Keeven (1995)	Microbial enzymes	Colour change of chromogenic substrates of the microbial enzymes

Table 7.1 *(continued)*

Author/ Patent applicant or holder/Trade name	Metabolite detected	Principle of the indicator
Namiki (1996)	Micro-organisms	Degradation of lipid membrane by micro-organisms and subsequent diffusion of coloured compound
Aromascan (Payne and Persaud, 1995)	Not specified	Miniaturised electronic component with electrical properties affected by volatile compound associated with spoilage
Pathogen indicators		
Food Sentinel System (Sira Technologies); Goldsmith (1994)	Various microbial toxins	Bar code detector comprising a toxin printed onto a substrate and indicator colour irreversibly bound to the toxin, in the presence of toxin the bar code is illegible
Toxin Guard (Toxin Alert); Bodenhamer (2000)	Various pathogens	Formation of a coloured pattern when a target analyte is first bound to labelled antibody and subsequently to capture antibody
Lawrence Berkeley National Laboratory (Quan and Stevens, 1998)	*E. coli* 0157 enterotoxin	Colour change of polydiacetylene-based polymer

determination of CO_2 from the microbiologically spoiling product itself with a pH-dye (e.g. fuchsine acid) based indicator was proposed by Lawdermilt (1962).

The use of the pH-dye bromothymol blue as an indicator for the formation of CO_2 by microbial growth has been suggested in many studies (Holte, 1993; Mattila *et al.*, 1990). As mentioned before, an increase in CO_2 can be used to determine microbial contamination in certain types of products. On the other hand, Balderson and Whitwood (1994) proposed a CO_2-sensitive packaging indicator for correct filling and subsequent opening of the package. In addition to the detection of spoilage and package filling, pH-dyes reacting to the presence of CO_2 have also been used to construct intelligent packaging concepts indicating the ripeness of traditional fermented vegetable foods in Korea (*kimchi*) (Hong and Park, 2000).

In addition to the most frequently used pH-dye bromothymol blue, many other reagents e.g. xylenol blue, bromocresol purple, bromocresol green, cresol red, phenol red, methyl red and alizarin, among others, have been proposed for the same purpose (Horan 2000). Besides CO_2, other metabolites like SO_2, NH_4, volatile amines and organic acids have been proposed to be suitable target molecules for pH-sensitive indicators (Mattila and Auvinen, 1990a, b; Horan, 2000).

7.3.2 Indicators sensitive to volatile nitrogen compounds

A concept reacting to volatile amines with a colour change, hence indicating freshness of seafood, has been proposed by Miller *et al.* (1999). This concept was marketed by COX Recorders (USA) with the trade name FreshTag®. The concept consisted of a plastic chip incorporating a reagent-containing wick. As the label was attached to the package the sharp barb on the back of the label penetrated the packaging film thus enabling contact between the package headspace gases and the reagent. As volatile amines passed through the wick a bright pink colour was developed along the wick.

Loughran and Diamond (2000) propose a simple method for the determination of volatile nitrogen compounds (NH_3, DMA, TMA) with the aid of chromogenic dye calix[4]arene which has been impregnated on a paper disc. They show that the reflectance spectrum of the indicator disc is changed due to the volatile compounds emitted from cod stored on ice. They also propose that the system could form a base for an intelligent packaging concept.

7.3.3 Indicators sensitive to hydrogen sulphide

In our own studies we have utilised a reaction between hydrogen sulphide and myoglobin in a freshness indicator for the quality control of modified-atmosphere-packed poultry meat (Ahvenainen *et al.*, 1997; Smolander *et al.*, 2002). Freshness indication is based on the colour change of myoglobin by hydrogen sulphide (H_2S), which is produced in considerable amounts during the ageing of packaged poultry during storage. The indicators were prepared by applying commercial myoglobin dissolved in a sodium phosphate buffer on small squares of agarose. These indicators were tested in the quality control of MA-packaged fresh, unmarinated broiler cuts. It was found that the colour change of the myoglobin-based indicators corresponded with the deterioration of the product quality, hence it could be concluded that the myoglobin-based indicators seem to be promising for the quality control of packaged poultry products.

7.3.4 Indicators sensitive to miscellaneous microbial metabolites

Cameron and Talasila (1995) have explored the potential of detecting the unacceptability of packaged, respiring products by measuring ethanol in the package headspace with the aid of alcohol oxidase, peroxidase and a chromogenic substrate. Honeybourne and co-workers (Honeybourne, 1993; Shiers and Honeybourne, 1993) have developed a diamine dye-based sensor system responding to the presence of diacetyl vapour. Diacetyl is a volatile compound evolving from meat as spoilage takes place and Honeybourne and co-workers have proposed that the dye could be printed onto the surface of a gas permeable meat package. Diacetyl migrating through the packaging material would react with the dye and induce a colour change.

7.3.5 Other principles for freshness indicating systems

In addition to indicators based on reactions caused by microbial metabolites, other concepts for contamination indicators have been proposed. DeCicco and Keeven (1995) describe an indicator based on a colour change of chromogenic substrates of enzymes produced by contaminating microbes. The indicator can be applied for contamination detection in liquid health care products.

Besides the products of microbial metabolism, the consumption of certain nutrients could also be used as a measure of freshness. Kress-Rogers (1993) has developed a knife-type freshness probe for meat. Freshness detection is based on the formation of a glucose gradient as glucose from the surface of meat is consumed during microbial growth and the bulk concentration of glucose remains stable. The detection of microorganisms as such was proposed by Namiki (1996). The principle of the indicator is the degradation of lipid membrane by microorganisms and subsequent diffusion of coloured compound.

7.4 Pathogen indicators

In addition to the above-mentioned indicators reacting to the normal spoilage of food products, systems for the detection of a certain contaminant in the food product have also been presented. Commercially available Toxin Guard™ by Toxin Alert Inc. (Ontario, Canada) is a system to build polyethylene-based packaging material, which is able to detect the presence of pathogenic bacteria *(Salmonella, Campylobacter, Escherichia coli* O157 and *Listeria)* with the aid of immobilised antibodies. As the analyte (toxin, microorganism) is in contact with the material it will be bound first to a specific, labelled antibody and then to a capturing antibody printed as a certain pattern (Bodenhamer, 2000). The method could also be applied for the detection of pesticide residues or proteins resulting from genetic modifications. Another commercial system for the detection of specific microorganisms like *Salmonella sp., Listeria sp.* and *E. coli* is Food Sentinel System™. This system is also based on immunochemical reaction, the reaction taking place in a bar code (Goldsmith, 1994). If the particular microorganism is present the bar code is converted unreadable.

Specific indicator material for the detection of *Escherichia coli* O157 enterotoxin has been developed at Lawrence Berkeley National Laboratory (Kahn, 1996; Quan and Stevens, 1998). This sensor material, which can be incorporated in the packaging material, is composed of cross-polymerised polydiacetylene molecules and has a deep blue colour. The molecules specifically binding the toxin are trapped in this polydiacetylene matrix and as the toxin is bound to the film, the colour of the film changes from blue to red.

7.5 Other methods for spoilage detection

In addition to the indicators undergoing a visual colour change it can be expected that in the future intelligent packaging concepts will resemble more and more small analytical tools that are incorporated in the package. Some existing technologies like biosensors and electronic noses are likely to serve as a technological basis for the development of new miniaturised concepts. These technologies and their application in the determination of food quality and freshness are described in the following chapters.

7.5.1 Biosensors

A biosensor is an analytical device consisting of a biological component specific to the analyte and a physical component, which is able to transduce the biological signal to a physical one (Turner *et al.*, 1987). For instance enzymes, antibodies and cells can be used as the biological component of biosensors. The signal can also be detected in many ways e.g. with amperometric, potentiometric, optic and calorimetric methods.

Several types of enzymatic biosensors have been developed for the detection of biogenic amines. The increase in the amount of diamines (putrescine, cadaverine and spermidine) in poultry meat was detected with a putrescine oxidase reactor combined with amperometric hydrogen peroxide electrode by Okuma *et al.* (2000). Niculescu *et al.* (2000) constructed an amperometric bienzyme (grass pea amine oxidase, horseradish peroxidase) electrode and used it for the determination of biogenic amines from fish muscle. Frébort *et al.* (2000) presented a flow system based on spectrophotometric detection of hydrogen peroxide generated in the oxidation of biogenic amines by amine oxidase. They applied their system for the determination of histamine in rainbow trout meat. Yano *et al.* (1995b) developed a tyramine oxidase based sensor for the quality control of beef. ATP degradation products have also been measured with biosensors. For instance, Yano *et al.* (1995a) and Mulchandani *et al.* (1990) developed an enzyme based electrochemical sensor for ATP degradation products and applied it for the quality control of beef.

7.5.2 Electronic nose

Electronic nose is an analytical tool composed of an array of sensors (e.g. metal oxide, polymer) responding to volatile compounds with changes in their electrical properties. The combined pattern of these changes from all the sensors forms a fingerprint for the sample. Individual compounds are not separated/identified with electronic nose, but the samples can be classified to acceptable or unacceptable according to the fingerprint of the volatile compounds and the results obtained with a reference method (e.g. sensory evaluation or microbiological analysis). On the basis of existing data on the correlation between sensor responses and the quality of the sample the system can be adjusted to classify samples of unknown quality.

We have obtained very promising data on the determination of the quality of broiler chicken cuts with electronic nose. The response of electronic nose was found to be consistent with sensory and microbiological quality of the products as well as with the concentration of volatile compounds in the package headspace (Rajamäki *et al.,* submitted). Also Boothe and Arnold (2002) evaluated electronic nose as a promising tool for the quality evaluation of poultry meat. Electronic noses have also been successfully used for the quality evaluation of other products like fresh yellowfin tuna (Du *et al.*, 2001) and vacuum packaged beef (Blixt and Borch, 1999).

An interesting system working analogously with electronic nose has been developed by Suslick and co-workers (Suslick and Rakow, 2001; Rakow and Suslick, 2000). The system is a simple optical chemical sensing method (colorimetric nose) that utilises the colour change taking place in an array of metalloporphyrin dyes upon ligand binding. The liganding vapours (alcohols, amines, ethers, phospines, phosphites, thioethers, thiols, arenes, halocarbons and ketones) can be visually identified. Food quality control has been identified as one of the potential applications of the system.

7.6 Future trends

Today the commercially available intelligent concepts are labels responding with a visible change to time and temperature (TTI) or the presence of certain chemical compounds (leak indicators, freshness indicators). It will however be very likely that the visible labels will be replaced with more developed types of indicator. The replacement of traditional barcodes by electronic tags in a supply chain is becoming more and more common. Security tags, already in use today, are the first examples of electronic labelling. However, it can be expected that in future the tags will be used not only as information carriers but also as miniaturised analytical tools. These quality indicating electronic labels could be introduced, e.g., as chips. The advances in ink technology might enable the use of intelligent printed circuits as well. The advantages of printed structures include a low price and disposability. It is very likely that the advances in electronics and biotechnology (e.g., biosensors, immunodiagnostics) would be followed by the emergence of new concepts of intelligent packaging. However, one has to bear in mind that the basic principles underlying all the indicators are common and knowledge about food quality and safety is required in the development work of all types of freshness indicators. Profound understanding is needed about the formation of quality indicating compounds (both volatile and non-volatile) and also about their correlation with product quality. On the basis of this information new intelligent concepts, e.g., reacting to more than one quality indicating compound could be developed.

7.7 References

AHVENAINEN R, PULLINEN T, HURME E, SMOLANDER M. and SIIKA-AHO M. (VTT BIOTECHNOLOGY AND FOOD RESEARCH, ESPOO, FINLAND) (1997) 'Package for decayable foodstuffs' PCT International patent application WO 9821120.

ARNAUT-ROLLIER I., DE ZUTTER L and VAN HOOF L (1999) 'Identities of *Pseudomonas* spp. in flora from chilled chicken', *International Journal of Food Microbiology* 48, 87–96.

BALDERSON S N and WHITWOOD R J (TRIGON INDUSTRIES LIMITED, AUCKLAND, NEW ZEALAND) (1994) 'Gas Indicator for a Package' *EP* 0627363 A1.

BLIXT Y and BORCH E (1999) 'Using an electronic nose for determining the spoilage of vacuum-packaged beef', *International Journal of Food Microbiology* 46, 123–34.

BODENHAMER W T (TOXIN ALERT, INC., CANADA) (2000) 'Method and apparatus for selective biological material detection', *US Patent* 6051388.

BOOTHE D D H and ARNOLD J W (2002) 'Electronic nose analysis of volatile compounds from poultry meat samples, fresh and after refrigerated storage', *Journal of the Science of Food and Agriculture* 82, 315–22.

BORCH E, KANT-MUERMANS M-L and BLIXT Y (1996) 'Bacterial spoilage of meat and cured meat products' *International Journal of Food Microbiology* 33, 103–20.

CAMERON, A C and TALASILA P C (1995) 'Modified-atmosphere packaging of fresh fruits and vegetables', *IFT Annual Meeting/ Book of Abstracts,* p. 254.

CLARK J D'A (1949) 'Method and device for indicating spoilage' *US Patent* 2485566.

DAINTY H (1996) 'Chemical/ biochemical detection of spoilage', *International Journal of Food Microbiology* 33, 19–33.

DECICCO B T and KEEVEN J K (1995) 'Detection system for microbial contamination in health-care products', *US Patent* 5443987.

DROSINOS E H and NYCHAS G-J E (1997) 'Production of acetate and lactate in relation to glucose content during modified atmosphere storage of gilt-head seabream (*Sparus aurata*) at 0±1°C', *Food Research International*, 30(9), 711–17.

DU W-X, KIM J, CORNELL J A, HUANG T-S, MARSHALL M and WEI C-I (2001) 'Microbiological, sensory and electronic nose evaluation of yellowfin tuna under various storage conditions' *Journal of Food Protection,* 64(12), 2027–36.

EATON J, KILGORE M B and LIVINGSTON R B (1977) 'Visual spoilage indicator for food containers', *US Patent* 4003709.

EGAN A F, SHAY B J and ROGERS P J (1989) 'Factors affecting the production of hydrogen sulphide by *Lactobacillus sake* L13 growing on vacuum-packaged beef', *Journal of Applied Bacteriology* 67, 255–62.

FRÉBORT I, SKOUPA L and PEC P (2000) 'Amine oxidase-based flow biosensor for the assessment of fish freshness', *Food Control* 11, 13–18.

FREEMAN L R, SILVERMAN G J, ANGELINI P, MERRITT C JR. and ESSELEN W B (1976) 'Volatiles produced by microorganisms isolated from refrigerated chicken at spoilage' *Applied Environmental Microbiology* 32, 222–31.

FU A-H MOLINS R A and SEBRANEK J G (1992) 'Storage quality characteristics of beef rib eye steaks packaged in modified atmosphere', *Journal of Food Science* 57(2), 283–301.

GILL C O and NEWTON K G (1979) 'Spoilage of vacuum-packaged dark, firm, dry meat at chill temperatures', *Applied Environmental Microbiology* 37, 362–64.

GOLDSMITH R M (1994) 'Detection of contaminants in food', *US patent* 5 306 466.

HATTULA, T. (1997) 'Adenosine triphosphate breakdown products as a freshness indicator of some fish species and fish products', *VTT Publications* 297, VTT (Technical Research Centre of Finland), Espoo, 48 p. + app. 31 p.

HENEHAN G, PROCTOR M, ABU-GHANNAN N, WILLS C and MCLOUGHLIN J V (1997) 'Adenine nucleotides and their metabolites as determinants of fish freshness' in Ólafsdottir G, Luten J, Dalgaard P, Careche M, Verrez-Bagnis V, Martinsdóttir E and Heia K *Methods to determine the freshness of fish in research and industry EU-FAIR project AIR3CT94 2283,* Paris, International Institute of Refrigeration, 266–72.

HOLTE B (1993) 'An Apparatus for Indicating the Presence of CO_2 and a Method of Measuring and Indicating Bacterial Activity within a Container or Bag', *PCT International Patent Application* WO 93/15402.

HONEYBOURNE, C.L. (BRITISH TECHNOLOGY GROUP LTD., LONDON, UK) (1993) 'Food Spoilage Detection Method', *PCT International Patent Application* WO 93/15403.

HONG S-I and PARK W-S (2000) 'Use of color indicators as an active packaging system for evaluating kimchi fermentation', *Journal of Food Engineering*, 46, 67–72.

HORAN T J (1998) 'Method for determining bacterial contamination in food package', *US Patent* 5753285.

HORAN T J (2000) 'Method for determining deleterious bacterial growth in packaged food utilizing hydrophilic polymers', *US Patent* 6149952.

HUIS IN'T VELD J H J (1996) 'Microbial and biochemical spoilage of foods: an overview' *International Journal of Food Microbiology* 33, 1–18.

KAHN J (1996) New sensor provides first instant test for toxic *E.coli* organism, Berkeley Lab- Research News December 10, 1996 (http://www.lbl.gov/Science-articles/archive/E-coli-sensor.html)

KAKOURI A, DROSINOS E H, NYCHAS G-J E (1997) 'Storage of mediterranean fresh fish (*Boops boops*, and *Sparus aurata*) under modified atmospheres or vacuum at 3 and 10ºC' in *Seafood from producer to Consumer, Integrated Approach to quality*, Elsevier Science, 171–8.

KALINOWSKI R M AND TOMPKIN R B (1999) 'Psychrotrophic clostridia causing spoilage in cooked meat and poultry products', *Journal of Food Protection* 62, 766–772.

KANIOU I, SAMOURIS G, MOURATIDOU T, ELEFTHERIADOU A and ZANTOPOULOS N (2001) 'Determination of biogenic amines in fresh unpacked and vacuum-packed beef during storage at 4ºC', *Food Chemistry* 74, 515–19.

KRESS-ROGERS, E. (1993) 'The marker concept: Frying oil monitor and meat freshness sensor', in Kress-Rogers E, *Instrumentation and Sensors for the food industry*, Stoneham, Mass., Butterworth-Heinemann, 523.

LAWDERMILT R F (1962) 'Spoilage indicator for food containers' *US Patent* 3067015.

LEA C H, PARR L J and JACKSON H F (1969) 'Chemical and organoleptic changes in poultry meat resulting from the growth of psychrophylic spoilage bacteria at 1ºC. 3. Glutamine, glutathione, tyrosine, ammonia, lactic acid, creatine, carbohydrate, haem pigment and hydrogen sulphide', *British Poultry Science* 10(3), 229–38.

LOUGHRAN M and DIAMOND D (2000) 'Monitoring of volatile bases in fish sample headspace using an acidochromic dye', *Food Chemistry*, 69, 97–103.

MATTILA T and AUVINEN M (1990a) 'Headspace Indicators Monitoring the Growth of *B. cereus* and *Cl. perfringens* in Aseptically Packed Meat Soup (Part I)' *Lebensmittel-Wissenschaft und – Technologie* 23, 7–13.

MATTILA T and AUVINEN M (1990b) 'Indication of the Growth of *Cl. perfringens* in Aseptically Packed Sausage and Meat Ball Gravy by Headspace Indicators (Part II)' *Lebensmittel-Wissenschaft und -Technologie* 23, 14–19.

MATTILA T, TAWAST J and AHVENAINEN R (1990) 'New Possibilities for Quality Control of Aseptic Packages: Microbiological Spoilage and Seal Defect Detection using Head-Space Indicators' *Lebensmittel-Wissenschaft und -Technologie.* 23, 246–51.

MILLER D W, WILKES J G and CONTE E D (1999) 'Food quality indicator device', *PCT International patent application WO* 99/04256.

MULCHANDANI A, MALE K B and LUONG J H T (1990) 'Development of a biosensor for assaying postmortem nucleotide degradation in fish tissues' *Biotechnology & Bioengineering*, 35, 739–45.

NAMIKI H (1996) 'Food freshness indicators containing filter paper, lipid membranes and pigments', *JP* 08015251.

NEARY M P (1981) 'Food spoilage indicator', *US Patent* 4 285 697.

NICOL D, SHAW M and LEDWARD D (1970) 'Hydrogen sulfide production by bacteria and sulfmyoglobin formation in prepacked chilled beef', *Applied Microbiology* 19, 937–39.

NICULESCU M, NISTOR C, FRÉBORT I, PEC P, MATTIASSON B and CSÖREGI E (2000) 'Redox hydogel-based amperometric bienzyme electrode for fish freshness monitoring', *Analytical Chemistry* 72, 1591–7.

NYCHAS G-J E, DROSINOS E H and BOARD R G (1998) 'Chemical changes in stored meat' in Davies A and Board R *The Microbiology of Meat and Poultry, London,* Blackie Academic & Professional, 288–326.

OHLENSCHLÄGER J (1997) 'Suitability of ammonia-n, dimethylamine-n, trimethylamine-n, trimethylamine oxide-n and total volatile basic nitrogen as freshness indicators in seafoods' in Ólafsdottir G, Luten J, Dalgaard P, Careche M, Verrez-Bagnis V, Martinsdóttir E and Heia K *Methods to determine the freshness of fish in research and industry EU-FAIR project AIR3CT94 2283,* Paris, International Institute of Refrigeration, 92–99.

OKUMA H, OKAZAKI W, USAMI R and HORIKOSHI K (2000) 'Development of the enzyme reactor system with an amperimetric detection and application to estimation of the incipient stage of spoilage of chicken', *Analytica Chimica Acta* 411, 37–43.

OLAFSDÓTTIR G and FLEURENCE J (1997) 'Evaluation of fish freshness using volatile compounds – Classification of volatile compounds in fish' in Olafsdóttir G, Luten J, Dalgaard P, Careche M, Verrez-Bagnis V, Martinsdóttir E and Heia K *Methods to determine the freshness of fish in research and industry EU-FAIR project AIR3CT94 2283,* Paris, International Institute of Refrigeration, 55–69.

PAINE F A and PAINE H Y (1992) *Handbook of food packaging*, 2nd edition, Blackie Academic & Professional , University Press, Cambridge, UK. pp. 212–14.

PAYNE P A and PERSAUD K C (AROMASCAN PLC) (1995) 'Condition indicator', *PCT International patent application WO* 95/33991

QUAN C and STEVENS R (1998) 'Protein coupled colorimetric analyte detectors', *PCT International patent application WO* 98/36263.

RAJAMÄKI T, ALAKOMI H-L, RITVANEN T, SKYTTÄ E, SMOLANDER M and AHVENAINEN R 'Application of electronic nose for quality assessment of modified atmosphere packaged poultry meat', submitted to *International Journal of Food Microbiology.*

RANDELL K, AHVENAINEN R, LATVA-KALA K, HURME E, MATTILA-SANDHOLM T and HYVÖNEN L (1995) 'Modified atmosphere-packed marinated chicken breast and rainbow trout quality as affected by package leakage', *Journal of Food Science* 60, 667–84.

RAKOW N A and SUSLICK K S (2000) 'A colorimetric sensor array for odour visualization', *Nature* 406, 710–13.

REHBEIN H (1993) 'Assesment of fish spoilage by enzymatic determination of ethanol', *Archiv für Lebensmittelhygiene* 44, 1–24.

RODRÍGUEZ C J, BESTEIRO I AND PASCUAL C (1999) 'Biochemical changes in freshwater rainbow trout (Oncorhynchus mykiss) during chilled storage', *Journal of the Science of Food and Agriculture*, 79, 1473–80.

ROKKA M, EEROLA S, SMOLANDER M and AHVENAINEN R, 'The formation of biogenic amines in modified atmosphere packaged broiler chicken cuts during storage', submitted to *Food Control.*

RUIZ-CAPILLAS C and MORAL A (2002) 'Effect of controlled and modified atmospheres on the production of biogenic amines and free amino acids during storage of hake' *European Food Research and Technology*, 214, 476–81.

RUSSELL S M, FLETCHER D L and COX NA (1997) 'Spoilage bacteria of fresh chicken carcasses', *Poultry Science* 74, 2041–7.

SHIERS V P and HONEYBOURNE C L (1993) 'An optical fibre based vapour sensor for spoilage in meat' in Gratta K T V and Augousti A T, *Sensors VI*, Bristol, Inst. Phys. 45–50.

SHU H-C, HÅKANSON H AND MATTIASSON B (1993) 'D-lactic acid in pork as a freshness indicator monitored by immobilized D-lactate dehydrogenase using sequential injection analysis', *Analytica Chimica Acta* 283, 727–37.

SMOLANDER M, HURME E, LATVA-KALA K, LUOMA T, ALAKOMI H-L and AHVENAINEN R (2002) 'Myoglobin-based indicators for the evaluation of freshness of unmarinated broiler cuts', *Innovative Food Science and Emerging Technologies* 3(3), 277–85.

SMOLANDER M, ALAKOMI H-L, RAJAMÄKI T, RITVANEN T, MOKKILA M, VAINIONPÄÄ J and AHVENAINEN R, 'Assesment and comparison of different quality evaluation methods for modified atmosphere packaged broiler chicken cuts' in preparation.

SUSLICK K S and RAKOW N A (2001) 'A colorimetric nose: "Smell-seeing" ', in Stetter J R and Pensrose W R, *Artificial Chemical Sensing, Olfaction and Electronic nose,* Pennington, Electrochemical Society, 8–14.

TURNER A P F, KARUBE I and WILSON G S (1987) Biosensors, Fundamentals and Applications, Oxford, Oxford University Press.

VIESHWEG S H, SCHMITT R E and SCHMIDT-LORENZ W (1989) 'Microbial spoilage of refrigerated fresh broilers. VII: Production of off odours from poultry skin by bacterial isolates', *Lebensmittel-Wissenschaft und-Technologie* 22, 356–66.

WALLACH D F H and NOVIKOV A (BIODETECT CORPORATION) (1998) 'Methods and devices for detecting spoilage in food products', *PCT International patent application WO* 98/20337.

WATANABE A, TSUNEISHI A and TAKIMOTO Y (1989) 'Analysis of ATP and its breakdown products in beef by reversed-phase HPLC', *Journal of Food Science*, 54(5), 1169–72.

WOLFBEIS O S and LIST H (AVL MEDICAL INSTRUMENTS, AG, SCHAFFHAUSEN, SWITZERLAND) (1995) 'Method for Quality Control of Packaged Organic Substances and Packaging Material for use with this Method' *US Patent* 540782.

YANO Y, KATAHO N, WATANABE M, NAKAMURA T and ASANO Y (1995a) 'Evaluation of beef ageing by determination of hypoxanthine and xanthine contents: application of a xanthine sensor', *Food Chemistry* 52, 439–45.

YANO Y, KATAHO N, WATANABE M, NAKAMURA T and ASANO Y (1995b) 'Changes in the concentration of biogenic amines and application of tyramine sensor during storage of beef', *Food Chemistry* 54, 155–9.

8

Packaging-flavour interactions

J. P. H. Linssen, R. W. G. van Willige and M. Dekker, Wageningen University, The Netherlands

8.1 Introduction

Interactions within a package system refer to the exchange of mass and energy between the packaged food, the packaging material and the external environment. Food-packaging interactions can be defined as an interplay between food, packaging, and the environment, which produces an effect on the food, and/or package (Hotchkiss, 1997).

Mass transfer processes in packaging systems are normally referred to as permeation, migration and absorption (Fig. 8.1). Permeation is the process resulting from two basic mechanisms: diffusion of molecules across the package wall, and absorption/desorption from/into the internal/external atmospheres. Migration is the release of compounds from the plastic packaging material into the product (Hernandez and Gavara, 1999). The migration of compounds from polymer packaging materials to foods was the first type of interaction to be investigated due to the concern that human health might be endangered by the leaching of residues from the polymerisation (e.g., monomers, oligomers, solvents), additives (e.g., plasticisers, colourants, UV-stabilisers, antioxidants) and printing inks. Later, absorption or scalping of components originally contained in the product by the packaging material attracted attention. Product components may penetrate the structure of the packaging material, causing loss of aroma, or changing barrier and/or mechanical properties, resulting in a reduced perception of quality (Johansson, 1993).

The fundamental driving force in the transfer of components through a package system is the tendency to equilibrate the chemical potential (Hernandez and Gavara, 1999). Mass transport through polymeric materials can be described as a multistep process. First, molecules collide with the polymer surface. Then they

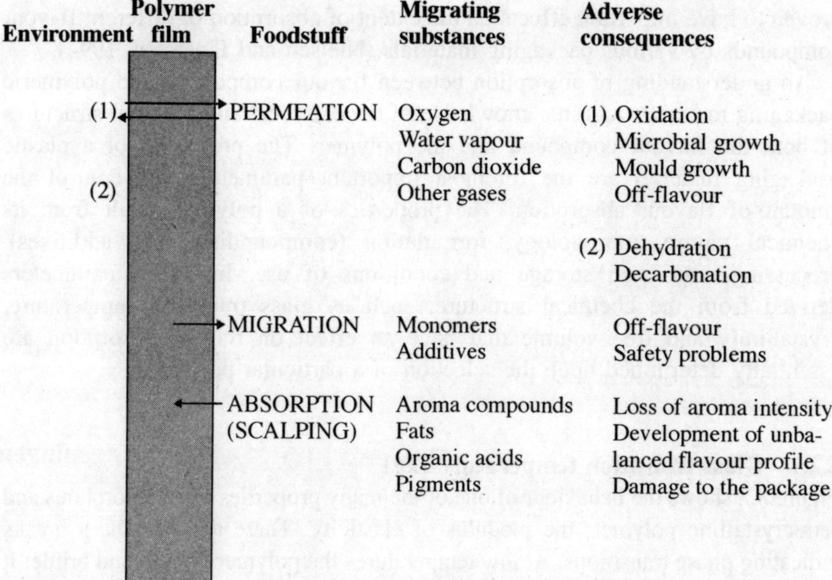

Fig. 8.1 Possible interactions between foodstuff, polymer film and the environment, together with the adverse consequences (Nielsen and Jägerstad, 1994)

adsorb and dissolve into the polymer mass. In the polymer film, the molecules 'hop' or diffuse randomly as their own kinetic energy keeps them moving from vacancy to vacancy as the polymer chains move. The movement of the molecules depends on the availability of vacancies or 'holes' in the polymer film. These 'holes' are formed as large chain segments of the polymer slide over each other due to thermal agitation. The random diffusion yields a net movement from the side of the polymer film that is in contact with a high concentration or partial pressure of permeant to the side that is in contact with a low concentration of permeant. The last step involves desorption and evaporation of the molecules from the surface of the film on the downstream side (Singh and Heldman, 1993). Absorption involves the first two steps of this process, i.e. adsorption and diffusion, whereas permeation involves all three steps (Delassus, 1997).

8.2 Factors affecting flavour absorption

As polymer packaging is more and more widely used for direct contact with foods, product compatibility with the packaging material must be considered. Flavour scalping, or the absorption of flavour compounds, is one of the most important compatibility problems. The problem of aroma absorption by plastic packages has been recognised for many years (Johansson, 1993). Several research groups throughout the world investigated flavour absorption phenomena extensively. It is a complex field, and several factors have been proven to have important effects on the extent of absorption of different flavour compounds by various packaging materials (Nielsen and Jägerstad, 1994).

An understanding of absorption between flavour compounds and polymeric packaging materials requires knowledge of the chemical and physical structures of both the flavour compound and the polymer. The properties of a plastic packaging material are the foremost important parameters that control the amount of flavour absorption. The properties of a polymer result from its chemical nature, morphology, formulation (compounding with additives), processing, and even storage and conditions of use. Important parameters derived from the chemical structure, such as glass transition temperature, crystallinity and free volume that have an effect on flavour absorption are essentially determined upon the selection of a particular polymer.

8.2.1 Glass transition temperature (*Tg*)

Figure 8.2 shows the behaviour of one of the many properties of an amorphous and semicrystalline polymer: the modulus of elasticity. There are two sharp breaks indicating phase transitions. At low temperatures the polymer is rigid and brittle: it forms a 'glass'. At the glass transition temperature *Tg* the modulus of elasticity drops dramatically. Many of the properties of the polymer change a little at this temperature. Above *Tg* the polymer becomes soft and elastic; it forms a '*rubber*'. At high temperatures, the polymer may melt, to form a viscous liquid (Wesselingh and Krishna, 2000). The polymers that we know as glassy polymers, such as the polyesters polyethylene terephthalate (PET), polycarbonate (PC) and polyethylene nafthalate (PEN), have a *Tg* above ambient temperature. At room temperature, glassy polymers will have very stiff chains and very low diffusion coefficients for flavour molecules at low concentrations. Rubbery polymers, such as the polyolefins polyethylene (PE) and polypropylene (PP), have a *Tg* below ambient temperature. Rubbery polymers have high diffusion coefficients for flavour compounds and steady-state permeation is established quickly in such structures (Giacin and Hernandez, 1997). Stiff-chained polymers that have a high glass transition temperature generally have low permeability, unless they also have a high free volume (Miller and Krochta, 1997).

8.2.2 Free volume

The free volume of a polymer is the molecular 'void' volume that is trapped in the solid state. The permeating molecule finds an easy path in these voids. Generally, a polymer with poor symmetry in the structure, or bulky side chains, will have a high free volume and a high permeability (Salame, 1989).

8.2.3 Crystallinity

The importance of crystallinity to absorption has been recognised for many years. All polymers are at least partly amorphous; in the amorphous regions the

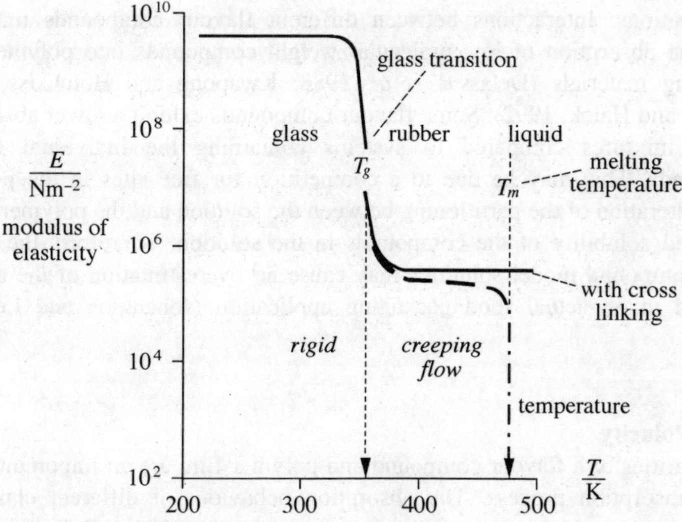

Fig. 8.2 Modulus of elasticity against temperature, showing the glass transition and melting temperatures (Wesselingh and Krishna, 2000).

polymer chains show little ordering. However, polymers often contain substantial 'crystalline' parts, where the polymer chains are more or less aligned. The crystalline areas are typically a tenth denser than the amorphous parts; for many permeants they are practically impermeable. So, diffusion occurs mainly in the amorphous regions in a polymer, where small vibrational movements occur along the polymer chains. These micro Brownian motions can result in 'hole' formation as parts of the polymer chains move away from each other. It is through such 'holes' that permeant molecules can diffuse through a polymer (Johansson, 1993; Wesselingh and Krishna, 2000). Therefore, the higher degree of crystallinity in a polymer, the lower the absorption.

8.2.4 Concentration and mixtures of flavour compounds

There are relatively few reports relating flavour absorption to the relative concentrations of the sorbants in a liquid or vapour. Mohney *et al.* (1988) reported that low sorbant concentrations will affect the polymer only to a very limited extent and the amount of absorbed compounds will be directly proportional to the concentration of the sorbants. At higher concentrations, however, the absorption of compounds into a polymer material may alter the polymer matrix by swelling (Charara *et al,* 1992; Sadler and Braddock, 1990). Consequently, to avoid overestimation of the amounts of absorbed compounds or swelling of the polymer, it is advisable to use a mixture of compounds in the concentration range that can be expected to be found in a food application (Johansson and Leufvén, 1997). However, to generate reliable and reproducable analytical data, experimental procedures are usually carried out with enhanced concentrations. Interactions between different flavour compounds may also affect the absorption of low molecular weight compounds into polymer food packaging materials (Delassus *et al,* 1988; Kwapong and Hotchkiss, 1987; Letinski and Halek, 1992). Some flavour compounds exhibit a lower absorption rate in mixtures compared to systems containing the individual flavour compounds. This may be due to a competition for free sites in the polymer and/or alteration of the partitioning between the solution and the polymer due to an altered solubility of the compounds in the solution. Therefore, the use of single compound model solutions may cause an overestimation of the amount absorbed in an actual food packaging application (Johansson and Leufvén, 1997).

8.2.5 Polarity

The polarities of a flavour compound and polymer film are an important factor in the absorption process. The absorption behaviour of different classes of flavour compounds depends to a great extent on their polarity. Different plastic materials have different polarities; hence their affinities toward flavour compounds may differ from each other (Gremli, 1996). Flavour compounds are absorbed more easily in a polymeric film if their polarities are similar (Quezada Gallo *et al.,* 1999). Polyolefins are highly lipophilic and may be inconvenient for packaging products with non-polar substances such as fats, oils, aromas etc., since they can be absorbed and retained by the package (Hernandez-Muñoz *et al.,* 2001). The polyesters, however, are more polar than the polyolefins and will therefore show less affinity for non-polar substances.

8.2.6 Molecular size and structure

The size of the penetrant molecule is another factor. Smaller molecules are absorbed more rapidly and in higher quantities than larger molecules. Very large molecules plasticise the polymer, causing increased absorption into the newly available absorption sites (Landois-Garza and Hotchkiss, 1987). Generally, the absorption of a series of compounds with the same functional group increases with an increasing number of carbon atoms in the molecular chain, up to a certain limit. Shimoda *et al.* (1987) reported that absorption of aldehydes, alcohols and methyl esters increased with increasing molecular weight up to about ten carbon atoms. For even larger molecules the effect of molecular size overcomes the effect of the increased solubility of the compounds in the polymer, and the solubility coefficient decreased. Linssen *et al.* (1991a) reported that compounds with eight or more carbon atoms were absorbed from yoghurt drinks by HDPE, while shorter molecules remained in the product. They also observed that highly branched molecules were absorbed to a greater extent than linear molecules.

8.2.7 Temperature

Temperature is probably the most important environmental variable affecting transport processes. The permeability of gases and liquids in polymers increases with increasing temperature according to the Arrhenius relationship. Possible reasons for increased flavour absorption at higher temperatures are (Gremli, 1996):

- increased mobility of the flavour molecules
- change in polymer configuration, such as swelling or decrease of crystallinity
- change in the volatile solubility in the aqueous phase.

8.2.8 Relative humidity

For some polymers, exposure to moisture has a strong influence on their barrier properties. The presence of water vapour often accelerates the diffusion of gases and vapours in polymers with an affinity for water. The water diffuses into the film and acts like a plasticiser. Generally, the plasticising effect of water on a hydrophilic film, such as ethylene-vinyl alcohol (EVOH) and most polyamides, would increase the permeability by increasing the diffusivity because of the higher mobility acquired by the polymer network (Johansson, 1993). Absorbed water does not affect the permeabilities of polyolefins and a few polymers, such as PET and amorphous nylon, show a slight decrease in oxygen permeability with increasing humidity. Since humidity is inescapable in many packaging situations, this effect cannot be overlooked. The humidity in the environment is often above 50%RH, and the humidity inside a food package can be nearly 100%RH (Delassus, 1997).

8.3 The role of the food matrix

The quality and the shelf-life of the packaged food depend strongly on physical and chemical properties of the polymeric film and the interactions between food components and package during storage. Several investigations have shown that considerable amounts of aroma compounds can be absorbed by plastic packaging materials, which can cause loss of aroma intensity or an unbalanced flavour profile (Hotchkiss, 1997; Arora *et al.*, 1991; Lebossé *et al.*, 1997; Linssen *et al.*, 1991a; Nielsen *et al.*, 1992; Paik, 1992).

The composition of a food matrix is of great importance (besides other factors, see Fig. 8.3) in determining the amount of flavour absorption by plastic packaging materials. There is only limited information available in literature about the influence of the food matrix on flavour absorption by polymers. Linssen *et al.* (1991b) and Yamada *et al.* (1992) showed that the presence of juice pulp in orange juice decreased absorption of volatile compounds into polymeric packaging materials. They suggested that pulp particles hold flavour compounds (e.g., limonene) in equilibrium with the

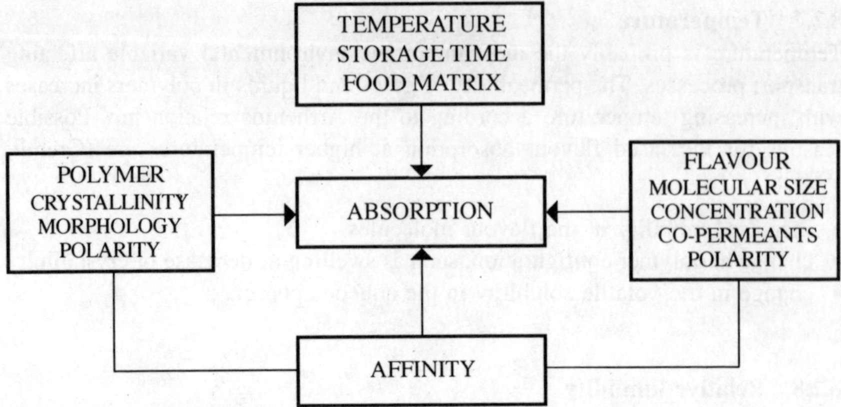

Fig. 8.3 Factors influencing flavour absorption by plastic polymers (Van Willige, 2002c).

watery phase, which could be responsible for the decrease of absorption of these compounds by the plastics.

Fukamachi *et al.* (1996) studied the absorption behaviour of flavour compounds from an ethanolic solution as a model of alcoholic beverages. The absorption of a mixture of homologous volatile compounds (esters, aldehydes and alcohols with carbon chain length 4-12) into LDPE film first increased with a maximal absorption at 5–10% (v/v) aqueous ethanol and then decreased remarkably with increasing ethanol concentration. EVOH film showed similar absorption behaviour, with maximal absorption at 10–20% (v/v) aqueous ethanol. Nielsen *et al.* (1992) investigated the effects of olive oil on flavour absorption into LDPE. Olive oil and, thereby, the flavours dissolved in the oil, were absorbed in large amounts by the plastic. The partition coefficients for alcohols and short-chained esters in an oil/polymer system were higher than in a water/polymer system, while the partition coefficients for aldehydes and long-chained esters were lower in an oil/polymer system than in a water/polymer system. Not only the type of plastic used is of importance for the uptake of aroma compounds, but also possible interactions between flavour and food components. Flavour components may be dissolved, adsorbed, bound, entrapped, encapsulated or retarded in their diffusion through the matrix by food components. The relative importance of each of these mechanisms varies with the properties of the flavour chemical (functional groups, molccular size, shape, volatility, etc.) and the physical and chemical properties of the components in the food (Kinsella, 1989; Le Thanh *et al.* 1992).

Knowledge of the binding behaviour of flavour components to non-volatile food components and their partitioning between different phases (food component/water and water/polymer) is of great importance in estimating the rate and amount of absorption by polymers. Because many food products are emulsions of fat and water, such as milk and milk products, the fat content is an important variable in the food matrix. Fat/oil content is often reduced in order to decrease caloric intake to make food healthier. Removal or reduction of lipids can lead to an imbalanced flavour, often with a much higher intensity than the original full fat food (Widder and Fischer, 1996; Ingham *et al.,* 1996).

De Roos (1997) reported that in products containing aqueous and lipid phases, a flavour compound is distributed over three phases: fat (or oil), water, and air. Flavour release from the oil/fat phase of a food proceeded at a lower rate than from the aqueous phase. This was attributed, first to the higher resistance to mass transfer in fat and oil than in water and, second to the fact that in oil/water emulsions flavour compounds had initially to be released from the fat into the aqueous phase before they could be released from the aqueous phase to the headspace. Kinsella (1989) reported that several mechanisms might be involved in the interaction of flavour compounds with food components. In lipid systems, solubilisation and rates of partitioning control the rates of release. Polysaccharides can interact with flavour compounds mostly by non-specific adsorption and formation of inclusion compounds. In protein systems, adsorption, specific binding, entrapment, encapsulation and covalent binding may account for the retention of flavours.

Oil and fatty acids can also be absorbed by polymers (Arora and Halek, 1994; Riquet *et al.,* 1998) resulting in increased oxygen permeability (Johansson and Leufvén, 1994) and delamination of laminated packaging material (Olaffson and Hildingson, 1995; Olaffson *et al.,* 1995) However, the availability of data about the influence of oil on the absorption of flavour compounds by plastic packaging materials is limited. Nielsen *et al.* (1992) found that some apple aroma compounds added to and stored in pure olive oil were lost to a greater extent to LDPE than from an aqueous solution, probably due to differences in polarity of the aromas, polymer and solutions. Thus, oil/fat has a major influence on flavour compounds (perception, intensity, volatility, etc.) and on the properties of packaging material.

Van Willige *et al.* (2000a, b) did a more detailed study on the influence of the composition of the matrix on food products. These authors used a model system, consisting of limonene, decanal, linalol and ethyl 2-methylbutyrate to study flavour scalping in LLDPE from different models representing differences in food matrices. The proteins, β-lactoglobuline (β-lg) and caseine were able to suppress absorption of decanal and limonene, because β-lg interacted irreversibly with decanal and caseine was capable of binding limonene and decanal by hydrophobic and covalent interactions. Dufour and Haertlé (1990) and Charles *et al.* (1996) reported that β-lg does not bind terpenes as limonene and linalol. The behaviour of ethyl 2-methylbutyrate could not be fully explained and needs further investigation.

The presence of carbohydrates also affected the absorption of flavour compounds by LLDPE. Absorption rates of limonene and to a lesser extent of decanal were decreased in the presence of pectine and carboxymethylcellulose. Increasing viscosity slowed down diffusion of flavour compounds from the matrix to LLDPE. Roberts *et al.* (1996) also reported that thickened solutions of similar viscosity did not show the same flavour release. Their results showed an influence of both viscosity and binding interactions with the thickener on the release of flavour. Binding interactions with carbohydrate-based thickeners are often due to adsorption, entrapment in microregions, complexation, encapsulation and hydrogen bonding between appropriate functional groups (Kinsella, 1989; Damodaran, 1996). The presence of disaccharides, lactose and saccharose was able to bind water and cause a salting out effect of the lesser polar flavour compounds, linalol and ethyl 2-methylbutyrate, resulting in an increased absorption in the polymer. Also Godshall (1997) reported that disaccharides can lower the amount of bulk water due to hydration, which increases the effective concentration of flavour compounds and therefore can enhance their absorption into polymers.

The main effect of the influence of the food matrix on flavour scalping, however, is the presence of oil or fat. Even a small amount of oil (50g/l) had a major effect on the amount of flavour absorption. Absorption of limonene and decanal is reduced to approximately 5%. A quantity of oil as low as 2 g/l results in a decrease of about 50% of absorption, meaning that the presence of oil very strongly influences the level of absorption of flavour compounds in polymeric packaging material (Fig. 8.4).

The composition of the food matrix plays a major role in the absorption of flavour compounds by LLDPE. Several studies have already revealed that flavour compounds interact with oil, carbohydrates and proteins, but the

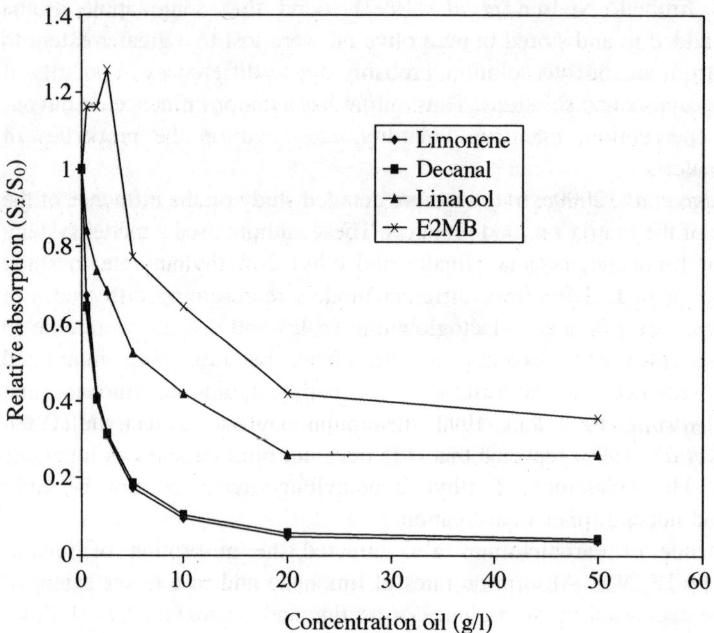

Fig. 8.4 Influence of oil on the relative absorption of limonene, decanal, linalool and ethyl-2-methylbutyrate (E2MB) by LLDPE after one day of exposure at 4° (Van Willige *et al.*, 2000a)

influence on flavour absorption by plastic packaging materials in different food matrices has been unclear for a long time. Van Willige *et al.* (2000a,b) showed that food components can affect the quantity of absorbed flavour compounds by LLDPE in the following order: oil or fat >> polysaccharides and proteins > disaccharides. Because of the lipophilic character of many flavour compounds, food products with a high oil or fat content will lose less flavour by absorption into LLDPE packaging than food products containing no or a small quantity of oil.

8.4 The role of differing packaging materials

An important requirement in selecting food-packaging systems is the barrier properties of the packaging material. Barrier properties include permeability of gases (such as O_2, CO_2, N_2 and ethylene), water vapour, aroma compounds and light. These are vital factors for maintaining the quality of foods. A good barrier to moisture and oxygen keeps a product crisp and fresh, and reduces oxidation of food constituents. Plastics are widely used for food packaging due to their flexibility, variability in size and shape, thermal stability, and barrier properties. PE and PP have been used for many years because of their good heat sealability, low costs and low water vapour permeability. However, poor gas permeability makes laminating of PE with aluminium foil and paper necessary. During the last decades, PET and, to a lesser extent, PC have found increased use for food packaging. PET has good mechanical properties, excellent transparency and relatively low permeability to gases. PC is tough, stiff, hard and transparent, but has poor gas permeability properties and is still quite expensive.

Unlike glass, plastics are not inert allowing mass transport of compounds such as water, gases, flavours, monomers and fatty acids between a food product, package and the environment due to permeation, migration and absorption. The quality and shelf-life of plastic-packaged food depend strongly on physical and chemical properties of the polymeric film and the interactions between food components and package during storage. Several investigations showed that considerable amounts of aroma compounds can be absorbed by plastic packaging materials, resulting in loss of aroma intensity or an unbalanced flavour profile (Van Willige *et al.*, 2000a,b; Arora *et al.*, 1991; Lebossé *et al.*, 1997; Linssen *et al.*, 1991b; Nielsen *et al.*, 1992; Paik, 1992) Absorption may also indirectly affect the food quality by causing delamination of multilayer packages (Olafsson and Hildingsson, 1995; Olafsson *et al.*, 1995) or by altering the barrier and mechanical properties of plastic packaging materials (Tawfik *et al.*, 1998). Oxygen permeability through the packaging is an important factor for the shelf-life of many packed foods. Little information is available in literature about the influence of absorbed compounds on the oxygen permeability of packaging materials. Hirose *et al.* (1988) reported that the oxygen permeability of LDPE and two types of ionomer increased due to the presence of absorbed d-limonene. Johansson and Leufvén (1994) studied the effect of rapeseed oil on the oxygen barrier properties of different polymer packaging materials. They found that amorphous PET remained an excellent oxygen barrier even after storage in rapeseed oil for 40 days. The polyolefins (PP and high-density PE) showed an increased oxygen transmission rate (OTR) after being in contact with rapeseed oil for 40 days. This was attributed to swelling of the polymer matrix. However, the increase in OTR was not proportional to the amount of absorbed oil.

Sadler and Braddock (1990) showed that the oxygen permeability of LDPE was proportional to the mass of absorbed limonene. In another paper, they concluded that oxygen permeability of LDPE and the diffusion coefficients of citrus flavour volatiles in LDPE were related to the solubility of these compounds in the polymer (Sadler and Braddock, 1991). The increased oxygen permeability of LDPE could only be explained by absorption. Attachment of volatile molecules at the polymer surface (adsorption) might hinder oxygen permeation, which would lower the oxygen permeation, or leave it unchanged. An increased oxygen permeability of LDPE indicated that absorption of volatiles must be responsible for structural changes in the polymer. Flavour absorption can have a major influence on the oxygen permeability of plastic packaging materials, and consequently on the shelf-life of a food product, making it necessary to investigate this important aspect more thoroughly.

Van Willige *et al.*, (2002b) investigated the influence of oxygen permeation on absorption of several flavour compounds (limonene, decanal, hexyl acetate and 2-nonanone) into LDPE, PP, PC and PET packaging materials. They measured the oxygen permeability of the exposed polymer specimens with a set-up based on the isostatic continuous flow technique. In the isostatic method the pressure differential across the test film remains constant during the total permeation process. Whereas the high-pressure side (oxygen chamber) remains constant at a certain value, the low-pressure side (nitrogen chamber) is maintained by sweeping the permeated molecules with a continuous flow of carrier gas (Hernandez and Gavara, 1999).

Figure 8.5 gives a good picture of the influence of the total amount of flavour absorption on the oxygen permeability of all four investigated polymers. PP and LDPE showed an increase in oxygen permeability after absorption of flavour compounds. This increase in oxygen permeability indicated that molecular changes occurred in the polymer network. Several researchers reported that swelling of a polymer by a permeant (i.e. plasticising) greatly increased the diffusivity. During the absorption process molecules are absorbed in the free volume ('holes') which is always present in the amorphous regions. Diffusion and a slow relaxation of the polymer, reducing the intercatenary forces and even promoting polymer swelling control the rate of absorption. This further enhances the rate of diffusion, which further influences the relaxation. As a result, the permeation of one component affects the permeation of another component, i.e. the plasticising effect within the polymer matrix becomes apparent (Halek, 1988; Hernandez-Muñoz *et al.,* 1999).

Absorbed water has a similar effect on the permeability of some hydrophilic polymers, such as ethylene vinyl alcohol (EVOH) and most polyamides. Water

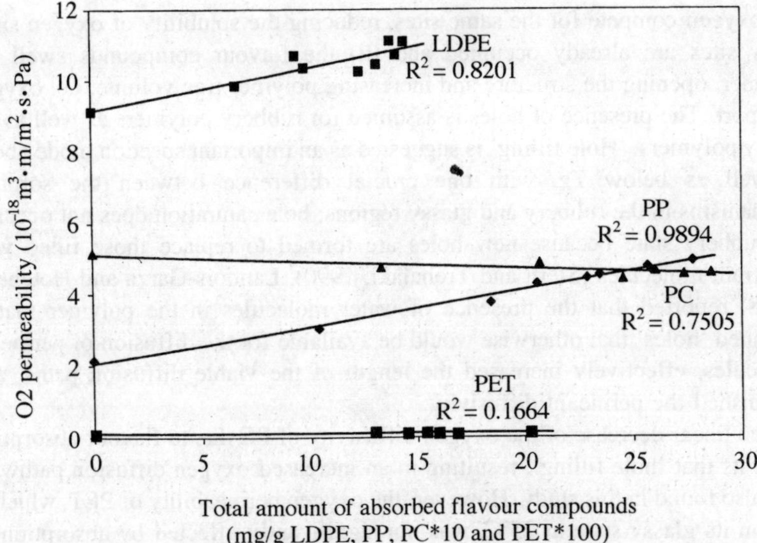

Fig. 8.5 Influence of total flavour absorption on oxygen permeability of PP, LDPE, PC and PET at 25ºC (Van Willige *et al.*, 2002b)

molecules absorbed at high relative humidities are believed to combine with hydroxyl groups in the polymer matrix and weaken the existing hydrogen bounds between polymer molecules. As a result, the interchain distances increase and thus free volume, facilitating the diffusion of oxygen and perhaps other gases. The presence of water in the hydrophilic polymer matrix not only influences how a permeant is sorbed and diffused, it also leads to depression of the glass transition temperature (*Tg*) of the polymer due to the plasticising effect of water. When the *Tg* drops below storage temperature, a substantial increase in oxygen permeability is expected (Zhang *et al.*, 1999; Delassus *et al.*, 1988).

Krizan *et al.* (1990) reported that free volume in a polymer is the dominant factor in determining the permeation properties. A plot of the log of the oxygen permeability coefficients versus the reciprocal of the specific free volume showed a good linear relationship. Also Sadler and Braddock (1990) reported that the oxygen permeability was proportional to the mass of absorbed limonene. The specific molecular composition of a flavour compound seems to play a more important role than the mass of absorbed flavour compounds. Each individual absorbed flavour compound caused swelling of PP; i.e. increased the specific free volume. Rubbery polymers (LDPE and PP) have very short relaxation times and respond very rapidly to stresses that tend to change their physical conditions. Glassy polymers (PC and PET) have very long relaxation times. Penetrant (molecules) can therefore potentially be present in 'holes' or irregular cavities with very different intrinsic diffusional mobilities (Stern and Trohalaki, 1990).

Hernandez-Muñoz *et al.* (1999) reported that there are two possible effects of absorbed flavour compounds on oxygen mass transport: (i) flavour compounds and oxygen compete for the same sites, reducing the solubility of oxygen since many sites are already occupied and (ii) the flavour compounds swell the polymer, opening the structure and increasing polymer free volume, i.e. oxygen transport. The presence of holes is assumed for rubbery polymers as well as for glassy polymers. 'Hole filling' is suggested as an important sorption mode above as well as below *Tg*, with one crucial difference between the sorption mechanisms in the rubbery and glassy regions; hole saturation does not occur in the rubbery state because new holes are formed to replace those filled with penetrant molecules (Stern and Trohalaki, 1990). Landois-Garza and Hotchkiss (1988) reported that the presence of water molecules in the polymer matrix occupied 'holes' that otherwise would be available for the diffusion of permeant molecules, effectively increased the length of the viable diffusion paths, and diminished the permeant diffusivity.

The linear decrease of the oxygen diffusivity of PC due to flavour absorption suggests that 'hole filling', resulting in an increased oxygen diffusion pathway, was also found in this study. However, the oxygen permeability of PET, which is also in its glassy state at 25°C, was not significantly affected by absorption of flavour compounds. Because of the low oxygen permeability of PET, which was close to the detection limit of the oxygen analyser, a significant effect of flavour absorption on oxygen permeability cannot be ruled out. A more sensitive oxygen analyser or a smaller permeation cell should be used in order to investigate the influence of absorption of flavour absorption on the oxygen permeability of PET. Van Willige *et al.* (2002b) concluded that flavour absorption increased oxygen permeability of PP and LDPE by 130% and 21% after 8 hours of exposure to various flavour compounds.

Because of the higher oxygen permeability a reduction in the shelf-life of oxygen sensitive products, which are packed in LDPE or PP and contain the tested flavour compounds (such as orange juice and apple juice) can be expected. Furthermore, flavour absorption has probably a positive effect on the shelf-life of oxygen-sensitive products packed in PC, because of the reduction in oxygen permeability of 11% after 21 days of exposure to various flavour compounds. Oxygen permeability of PET was not influenced by the presence of flavour compounds, meaning that PET remained a good oxygen barrier. One should realise that the concentrations of flavour compounds in real food products are usually substantially lower, with the exception of limonene, than the concentrations used in this study. Therefore, the observed effects may be less or even not significant in foods and beverages.

8.5 Flavour modification and sensory quality

One of the main aspects of flavour scalping is how this phenomenon is able to affect the quality of foods. In this field a lot of research has been carried out on fruit juices. During the last decades the quality of juices, aseptically packed in laminated cartons, has been investigated extensively. Loss of organoleptic characteristics during storage has been commonly observed (Marshall *et al.*, 1985, Moshonas and Shaw 1989a, b). LDPE laminated carton packs, such as Tetra Brik® and Combibloc®, are commonly used for packaging aseptically filled juices. LDPE is able to absorb considerable amounts of flavour compounds (Arora *et al.*, 1991, Nielsen *et al.*, 1992, Van Willige *et al.*, 2002a). Therefore, food industries often correct this absorptive effect by adding excess flavour compounds to the food for keeping taste and flavour acceptable for consumers until the end of the product's shelf-life (Lebossé *et al.*, 1997).

Although instrumental analysis indicates that considerable amounts of flavour compounds are absorbed into polymeric packaging and that the loss of flavours may be high enough to affect the sensory quality of a packaged food, only few authors have conducted sensory tests to go along with the analytical results (Dürr *et al.*, 1981, Kwapong and Hotchkiss, 1987, Mannheim *et al.*, 1987, Moshonas and Shaw, 1989b, Sharma *et al.*, 1990). Dürr *et al.* (1981) reported that absorption of d-limonene up to 40% did not affect the sensory quality of orange juice during three months storage at 20ºC. d-Limonene was suggested scarcely to contribute to the flavour of orange juice. Moreover, they even considered limonene absorption as an advantage, since limonene is known as a precursor to such off-flavour compounds as α-terpeniol. They also reported that the storage temperature was the main quality parameter for the shelf-life of orange juice.

Kwapong and Hotchkiss (1987) found that assessors were able to detect significant differences in odour profile due to absorption of citrus essential oils from aqueous model solutions by LDPE strips. Moshonas and Shaw (1989a) reported significant reduced flavour scores using a sensory panel in a commercial aseptically packaged orange juice stored for six weeks at 21ºC and 26ºC. The detected flavour changes were caused by the combined loss of limonene due to absorption together with the increase of potential off-flavour compounds. Mannheim *et al.*, (1987) found that the product shelf-life of orange and grapefruit juices was significantly shorter in LDPE laminated cartons than in glass jars. A loss of ascorbic acid and an increase in brown colour was observed and a 40% decrease of limonene was found; other volatiles were not assayed. After ten weeks of storage at 25ºC they revealed a difference in taste. Sharma *et al.* (1990) reported that PE and PP contact did not cause perceptible changes in sensory quality of fruit squash (orange and lemon) and beverages (mango, orange and blue grapes). Pieper *et al.*, (1992) stored orange juice in glass bottles and in LDPE laminated cardboard packages at 4ºC for 24 weeks. Absorption of d-limonene up to 50% and small amounts of aldehydes and alcohols by the packaging materials did not affect the sensory quality of orange juice significantly. A reason could be the low storage temperature. Sadler *et al.*, 1995 reported that no evidence was found that flavour absorption directly altered sensory characteristics of orange juice through general or selective absorption of volatile compounds by LDPE, PET and EVOH after three weeks of storage at 4.5ºC. Marin *et al.* (1992) exposed orange juice to LDPE and an ionomer (i.e. Surlyn). The polymers absorbed more than 70% of the limonene content in 24 hours at 25°C. However, results from gaschromatography-olfactometry (GCO) analysis indicated that limonene possessed only trace odour activity. Furthermore, the polymers did not alter the odour-active components present in orange juice substantially.

Van Willige *et al.* (2003) investigated three types of packaging materials (LDPE, PET and PC) on flavour scalping in a model system (consisting of the compounds octanal, decanal, ethylbutyrate and 2-nonanone) and a reconstituted orange juice. Sensory evaluation was carried out by a sensory panel of 27 assessors. From the model system valencene was almost completely absorbed by LDPE, followed to a lesser extent by decanal, hexylacetate, octanal and nonanone. Much less flavour compounds were absorbed by PC and PET. In contrast to LDPE valencene was absorbed to the lowest extent and decanal to the highest. From the orange juice limonene was readily absorbed by LDPE, while myrcene, valencene, pinene and decanal were absorbed in smaller quantities. Only three flavour compounds were absorbed from orange juice by PC and PET in very small amounts: limonene, myrcene and decanal. Although instrumental analysis showed a substantial decrease of the flavour content between control and polymer treated sample, meaning that the polymers absorbed a substantial amount of the flavour compounds, the sensory panel was not able to detect any significant differences caused by flavour scalping. Van Willige (2002b) stated that flavour scalping is not the main reason for a possible change of flavour perception during storage of food products into polymer packaging materials. It is more likely that other mechanisms play a more important role, such as chemical degradation, resulting in a development of off-flavour compounds.

Sizer *et al.* (1988) stated that storage temperature remains the single most important factor in delaying flavour loss and achieving satisfactory shelf-life and quality. From all the published data, Gremli (1996) stated that there is ample evidence that flavour compounds migrate from beverages and foods into plastic packaging materials. However, investigations about the relevance of the loss of flavour compounds for the sensory quality of a product are insufficient and sometimes contradictory because flavour alteration depends on many parameters, such as storage temperature and type of packaging material. Therefore, investigations regarding the effect of flavour absorption on sensory quality of a product should be carried out at ambient temperature (i.e. usual storage conditions of aseptic packs), because the rate and amount of flavour absorption by packaging materials increases with increasing temperature (Van Willige *et al.*, 2002a). Moreover, it is important that the polymer treated and untreated (=control) samples are comparable. That means similarly packed and using a sensory procedure that evaluates complete packaging systems with similar properties, e.g. oxygen permeability (i.e. glass-glass, and not glass-laminated carton).

8.6 Case study: packaging and lipid oxidation

Lipid oxidation is an important chemical process in food products containing (poly)unsaturated fats. It is a process that leads eventually to the formation of volatile off-flavours by further reaction of the formed hydroperoxides in secondary lipid oxidation reactions (Nawar, 1996). Dekker *et al.* (2002) studied the primary oxidation process in a defined model system for a packed food using sunflower oil. The package consisted of a PE/EVOH/PE tub, with a top-film of either LLDPE or PE/EVOH/PE. At regular time intervals measurements were made of: oxygen concentrations in headspace and oil (sensor technology), peroxide value of the oil and the volume of the headspace. Storage of the packages was in the dark at three different temperatures (25, 37 and 50ºC). The headspace was either air or nitrogen, and for some experiments oxygen scavengers (Ageless) were used glued to the top-film. The degree of lipid oxidation was studied as a function of temperature, top-film material, initial headspace composition and the presence or absence of an oxygen scavenger. In Fig. 8.6 the effect of temperature is clearly visible, at 50ºC the peroxide values obtained are about tenfold the values at 37ºC. The effect of initial headspace composition and scavenger is as expected, although the effects of these packaging concepts is limited under the present conditions, especially due to the low barrier properties of the used LLDPE film. When the high barrier EVOH film is used, oxidation stops after all oxygen from the package headspace is consumed. For the description of product quality the level of primary oxidation is only an indicator value. The real problem is the subsequent secondary oxidation process that leads to the formation of off-flavours. In section 8.7.2. the process of lipid oxidation in different packaging concepts is translated into a mathematical model.

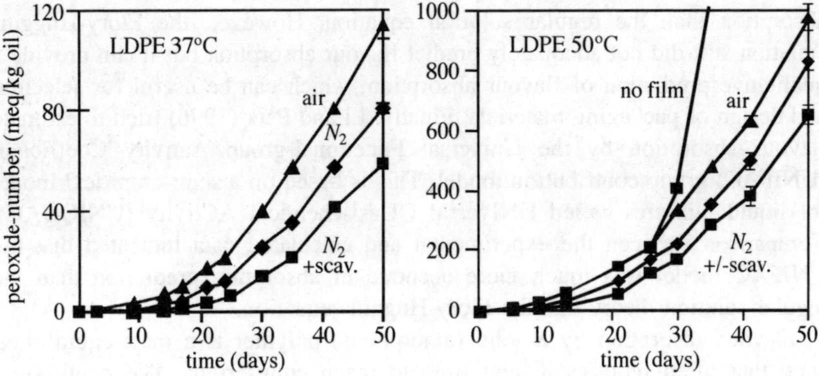

Fig. 8.6 Peroxide value for different temperatures and packaging concepts for LLDPE top-film.

8.7 Modelling flavour absorption

8.7.1 Modelling of flavour scalping

Knowledge of solubility and binding behaviour of flavour compounds to non-volatile food components and their partitioning behaviour between different phases (component/water, component/oil or component/oil/water on one side and water/polymer, oil/polymer or water/oil/polymer on the other side) is of great importance in estimating the rate and amount of absorption from real food products by polymers. Enormous amounts of different flavour compounds are used in foods. It is impossible to study them all. A determination of the relationship between flavour compounds and polymeric packaging materials for predicting flavour absorption would save research time for the packaging industry. Prediction of flavour absorption in relation to the packed food and the packaging material would be a valuable tool in product development. It can help the food industry in choosing packaging material or in determining product formulation.

In the literature little information is available on the prediction of flavour absorption. Attempts were made by using several theories. Tigani and Paik (1993) used the dielectric constants of polymers and flavour compounds to predict flavour absorption. They concluded that the dielectric constant might not encompass all of the factors for an accurate prediction of absorption by polymers. Paik and Tigani (1993) examined the application of Hildebrand's regular solution theory for predicting the equilibrium absorption of flavour compounds by polymer packaging materials. However, they found a poor correlation between the regular solution theory and flavour absorption values, indicating that the entropy contribution could not be assumed negligible.

Paik and Writer (1995) applied the Flory-Huggins equation for prediction of flavour absorption. The Flory-Huggins theory is based on the entropy contribution due to molecular size and shape differences of molecules. They showed that the Flory-Huggins equation gave much better estimations of flavour absorption than the regular solution equation. However, the Flory-Huggins equation still did not adequately predict flavour absorption but it can provide a qualitative prediction of flavour absorption, which can be useful for selection and design of packaging materials. Finally, Li and Paik (1996) tried to estimate flavour absorption by the Universal Functional-group Activity Coefficient (UNIFAC) group contribution model. This is based on a semi-empirical model for liquid mixtures called UNIversal QUasi-chemical ACtivity (UNIQUAC). Comparison between the experimental and calculated data indicated that the UNIFAC model was much more accurate in absorption prediction than the regular solution theory and the Flory-Huggins equation.

Flavour absorption by a solid (amorphous) polymer is a meta-equilibrium state that often requires a long time to reach equilibrium. The equilibrium distribution of flavour compounds will depend on their partitioning behaviour of compounds between different phases in the system: polymeric packaging and food matrix. The properties of the package, such as polarity and crystallinity, as well as the composition of the food matrix (presence of oil, proteins, carbohydrates) are extremely important factors. Modelling of flavour absorption could be based upon a set of equations describing these equilibrium distributions together with mass balances of the flavour compounds.

The final goal is to make predictions of flavour absorption for other food matrices and other compounds based upon their characteristics, such as polarity and molecular weight. In the future, a fitting model could be extended with the dynamics of the absorption phenomena (including mass transfer effects as a consequence of product texture, viscosity, etc.) and also for different packaging materials. In a food-packaging material system flavour molecules will strive for a thermodynamic equilibrium situation in which their chemical activities in all phases of the system will be equal. The time it will take to reach this equilibrium will depend on the composition of the food matrix. It can be assumed that in liquid foods this equilibrium will be reached well before consumption of the product. Experimental data of flavour absorption confirm this (van Willige *et al.*, 2000a, b). In solid or highly viscous food products this equilibrium might take longer, in that case only the outer part of the food will be affected by the flavour absorption effect. Dekker *et al.* (2003) proposed a formula based on equilibrium in a food-package system. The equilibrium concentration in the packaging material can be calculated from equation 8.1:

$$C_{P,\infty} = \frac{M_{FP} \cdot C_{FP,t=0}}{M_O \cdot \frac{K_{O/A}}{K_{P/A}} + \frac{M_A}{K_{P/A}} + M_P} \tag{8.1}$$

in which C is concentration (mg/g), M is mass (g) and K is the partitioning coefficient (–). The indices F, P, O and A refer to food, polymer, oily and aqueous phase, respectively.

To make predictions of the extent of flavour absorption, information is required about the value of the partition coefficients for the flavours of interest. The partitioning will depend largely on the nature of the flavour, especially on its polarity. Experimental determination of all partition coefficients is a very laborious task, therefore models describing the relation of the partition coefficients with known quantitative information on the nature of the flavour molecules are valuable. Dekker *et al.* (2003) made an attempt to do this based upon the log P value of the flavours, which is a good measure of their polarity. These values are reported for many molecules and can also be calculated from their chemical structure. Figure 8.7 shows the relationship between log P values and partitioning coefficients of the flavour compounds limonene, decanal and linalool.

Figure 8.7 also shows that a linear relationship between the log P values and the partition coefficients is obtained (R^2 is 0.95 and 0.99 for $K_{O/A}$ and $K_{P/A}$ respectively). In equations 8.2 and 8.3 the relations are given:

$$\text{Log}(K_{O/A}) = -0.85 + 0.88 \cdot \log P \tag{8.2}$$

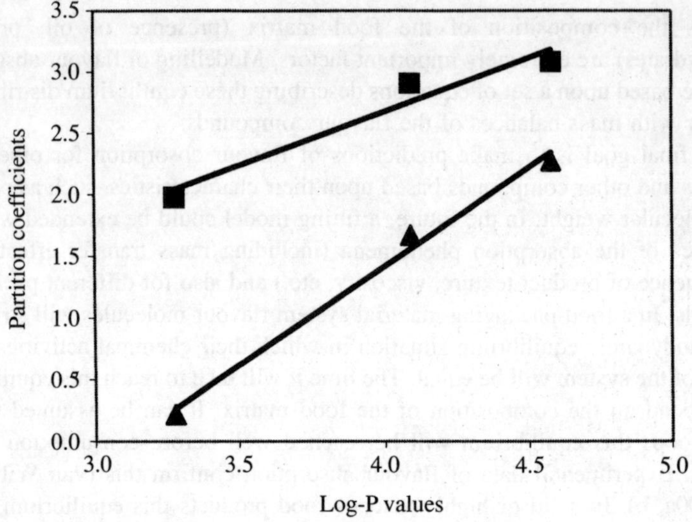

Fig. 8.7 Relationship between log-*P* values of the flavour compounds and their partition coefficients $K_{O/A}$ (■) *and* $K_{P/A}$ (▲) *at 4ºC (Van Willige, 2002d).*

$$\text{Log}(K_{P/A}) = -5.00 + 1.60 \cdot \log P \tag{8.3}$$

With equations 8.2 and 8.3 a prediction can be made of the partition coefficient of other flavour compounds, which have log *P* values in the range of the studied compounds (log *P* 3 to 5). In this way the amount of flavour absorption is predicted and this can be used for selection of packaging concepts for giving indications for adjustment of the formulation of the product accordingly. The modelling of the absorption of flavour molecules into LLDPE based on the partitioning behaviour between the different phases in the systems enables the prediction of this phenomenon based on the polarity of the flavour compounds involved. This can limit the amount of work that would be required for experimental determination of the amount of absorption in product development. Future research could focus on the extension of this modelling approach for other polymer packaging materials and other conditions.

8.7.2 Modelling of packaging and lipid oxidation

To enable predictions of the degree of lipid oxidation in a given situation mathematical models can be useful. Dekker *et al.*, (2002) developed a model for primary oxidation that included the effect of packaging material and geometry, temperature by using data on diffusion and reaction rates for sunflower oil. To model the oxidation reaction rates in the packed product one has to consider the barrier properties of the packaging material used. This was described by equation 8.4:

$$V_h \frac{dO_{2,headspace}}{dt}\bigg|_{permeation} = P \cdot A_{film} \cdot \Delta p_{O_2} \tag{8.4}$$

In which V_h is the headspace volume, P the permeation coefficient of the packaging material, A the surface area, and Δp_{O2} the oxygen pressure difference between air and the headspace. It was assumed that at the interface between the product and the headspace an equilibrium exists between the oxygen concentrations in both phases, as described by equation 8.5:

$$O_{2,headspace} = O_{2,oil}\big|_{x=O} \cdot H \tag{8.5}$$

In which H is the partitioning coefficient for oxygen.

In the product one has to calculate the rates of the primary lipid oxidation reactions. The formation of radicals was assumed to be mainly due to an oxygen induced initiation reaction. This led to the following equations describing the reaction rates in the product phase (8.6–8.9):

$$\frac{dO_2}{dt}\bigg|_{reaction} = k_1 \cdot [RH] \cdot [O_2] + k_2 \cdot [R\cdot] \cdot [O_2] \tag{8.6}$$

$$\frac{d[R\cdot]}{dt} = k_1 \cdot [RH] \cdot [O_2] - k_2 \cdot [R\cdot] \cdot [O_2] + k_3 \cdot [ROO\cdot] \cdot [RH] \tag{8.7}$$

$$\frac{d[ROO\cdot]}{dt} = k_2 \cdot [R\cdot] \cdot [O_2] - k_3 \cdot [RH] \cdot [ROOH] \tag{8.8}$$

$$\frac{d[ROOH]}{dt} = k_3 \cdot [ROO\cdot] \cdot [RH] \tag{8.9}$$

Diffusion will take place because of the concentration difference of oxygen and all lipid components involved in the reactions. The diffusion process was described by the second law of Fick:

$$\frac{\partial C_i}{\partial t} = D_i \cdot \frac{\partial^2 C_i}{\partial x^2} \tag{8.10}$$

In which D_i is the diffusion coefficient of component i, and x is the space-coordinate. All reaction, diffusion and permeation rates will depend on temperature. To describe this the Arrhenius equation was used:

$$k_n = k_0 \cdot e^{\frac{-E_a}{R \cdot T}} \tag{8.11}$$

In which k_O is the pre-exponential factor, E_a is the activation energy, R the gas constant, and T the absolute temperature. With similar experiments to the one presented in section 8.6 using additional temperatures, the parameters used in equations 8.4–8.11 were estimated. An example of the description of the measured values of peroxide concentration, oxygen concentration and headspace volume is shown in Fig. 8.8. The parameters obtained enable prediction of the lipid oxidation for different packaging geometry, temperature, and initial headspace composition. It should be realised that primary lipid oxidation is not

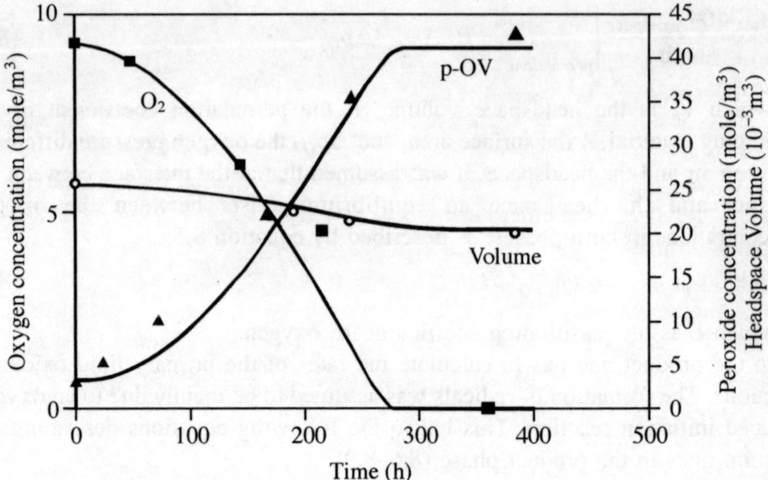

Fig. 8.8 Experimental data and model description of the lipid oxidation of sunflower oil with EVOH top-film.

the final quality indicator for edible oils. For the sensory quality of oils, the further (secondary) reactions of the fatty acid hydroperoxides results in the formation of volatile compounds that are responsible for the perceived off-flavour (rancidity) of the product by consumers (see section 8.6.).

The predictive modelling approach enables an efficient way of assessing the performance of different packaging concepts for oxidation sensitive products. Before conducting actual shelf-life experiments it is possible to get good estimates of what will happen with the product as a response to a different packaging condition (inclusion of a scavenger, modified atmosphere, or various material properties). Further research has to focus on the modelling of secondary oxidation in relation to sensory quality and the modelling of more complex food products like emulsions or structured foods.

8.8 Packaging–flavour interactions and active packaging

How important are packaging–flavour interactions? It depends on the extent they can affect the quality of the packed food. The food matrix is one of the main aspects which determine how important packaging–flavour interactions are. Food products containing fat or oil are able to keep the flavour compounds in the food itself and the loss caused by flavour scalping will be diminished. Some proteins are able to bind some flavour compounds that are no longer available for absorption into a plastic polymer. Aqueous food products have less ability to bind flavour compounds in the food matrix and therefore these foods are more susceptible to losing flavour compounds in the packaging polymer, which can result in quality defects. Many flavour compounds have a lipophilic character and therefore a good affinity to apolar polymers such as PE and PP. It has been shown that the highest amounts of absorbed flavour compounds are found in these types of polymers.

Polyesters, such as PET, PC and PEN, have a more polar character and therefore they show less affinity to the common flavour compounds. This means that polyesters absorb fewer flavour compounds and these polymers are therefore better packaging materials in the context of loss of flavour compounds due to flavour scalping. On the other hand, generally, there is less evidence that flavour scalping influences the taste and odour of a product. Although flavour compounds can be absorbed in substantial amounts, sensory defects are rarely found. Another factor is the way the polymer properties are affected. There is evidence that oxygen permeation can be enhanced due to absorption of flavour compounds. This means that as a secondary aspect, food quality can be affected due to oxidative chemical reactions, e.g., lipid oxidation can influence the quality of the product.

A second parameter of importance is the mechanical properties of a polymer. Some rare research work could be found dealing with the way flavour absorption affects the mechanical strength of a polymer (Tawfik *et al.*, 1998). On the other hand packaging and flavour interactions may act in a possitive way. Attempts to use food packaging and flavour interactions in a positive sense is also mentioned in literature.

We are now entering the field of active packaging applications. Polymer packaging material can be used to remove selectively undesirable compounds, causing off flavours, from the packed food. In certain orange varieties a bitter compound, limonin, is developed during the extraction and pasteurisation process of the juice. Inclusion of an absorbent might remove such a compound selectively (Chandler *et al.*, 1968; Chandler and Johnson, 1979). Also an active packaging concept has been described for reducing bitterness of grapefruit juices, caused by the presence of naringin. A thin cellulose acetate layer is applied to the inside of the packaging as an absorbent. Such a layer contains the enzyme naringinase, which hydrolyses naringin to non-bitter compounds (Soares and Hotchkiss, 1998a,b).

Other types of compounds responsible for off flavours are amines and aldehydes. Such compounds could also be removed by applying active packaging. Amines are formed from protein breakdown in fish muscle and include strongly alkaline compounds (Rooney, 1995). Vermeiren *et al.* (1999) reported that a Japanese patent claimed the removal of amines from food by interaction between acids incorporated in the polymer and the off-flavour compounds. ANICO Company Ltd (Japan) introduced another approach to remove amine odours. Bags made from a polymer containing ferrous salt and an organic acid claimed to oxidise the odours as they are absorbed by the polymer (Rooney, 1995). Aldehydes are formed in the lipid autoxidation reaction and they can reduce the quality of food products considerably. Dupont polymers claimed the selective removal of aldehydes from packaging headspaces by using a layer of Bynel IXP 101, which is a HDPE masterbatch (Rooney, 1995).

In conclusion, although it is very clear that packaging and flavour interactions exist, this phenomenon does not influence the food quality to the extent that it causes insuperable problems in practical situations. Moreover, in some cases packaging and flavour interactions can help to maintain the desired quality of food products.

8.9 References

ARORA A P and HALEK G W, 1994, Structure and cohesive energy density of fats and their sorption by polymer films. *J Food Sci,* **59**, 1325–27.

ARORA D K, HANSEN A P and ARMAGOST M S, 1991, Sorption of flavor compounds by low density polyethylene film. *J Food Sci*, **56**, 1421–23.

CHANDLER B V and JOHNSON R L, 1979, New sorbent gel forms of cellulose esters for debittering citrus juices. *J Sci Food Agric,* **30**, 825–32.

CHANDLER B V, KEFFORD J F and ZIEMELIS G, 1968, Removal of limonin from bitter orange juice. *J Sci Food Agric,* **19**, 83–86.

CHARARA Z N, WILLIAMS J W, SCHMIDT R H and MARSHALL M R, 1992, Orange flavor absorption into various polymeric packaging materials, *J Food Sci,* **57**, 963–66, 972.

CHARLES M, BERNAL B and GUICHARD E, 1996, Interactions of β-lactoglobulin with flavour compounds, in *Flavour Science*, ed. by Taylor A J and Mottram D S. The Royal Society of Chemistry, Cambridge, pp. 433–36.

DAMODARAN S, 1996, Amino acids, peptides, and proteins, in *Food Chemistry*, ed. by Fennema O R. Marcel Dekker, Inc, New York, pp. 321–429.

DEKKER M, KRAMER, M, VAN BEEST, M. and LUNING P 2002, Modelling oxidative quality changes in several packaging concepts, in *Proceedings Worldpak 2002*, Michigan, USA, CRC Press, Boca Raton, Volume 1, 297–303.

DEKKER M, VAN WILLIGE R W G, LINSSEN J P H and VORAGEN A G J 2003, Modelling the effect of oil/fat content in food systems on flavour absorption by LLDPE, *Food Add Contam*, 20, 180–85.

DELASSUS P T, 1997, Barrier polymers, in *The Wiley Encyclopedia of packaging technology*, ed. by Brody A L and Marsh K S. John Wiley & Sons, Inc., New York, pp. 71–77.

DELASSUS P T, TOU J C, BABINEE M A, RULF D C, KARP B K and HOWELL B A, 1988, Transport of apple aromas in polymer films, in *Food and Packaging Interactions*, ed. by Hotchkiss J H. ACS Symposium Series 365, American Chemical Society, Washington, DC, pp. 11–27.

DE ROOS K B, 1997, How lipids influence food flavor. *Food Technol,* **51**, 60–62.

DUFOUR E and HAERTLÉ T, 1990, Binding affinities of β-ionone and related flavor compounds to β-lactoglobulin: Effects of chemical modifications, *J Agric Food Chem*, **38**, 1691–95.

DÜRR P, SCHOBINGER U and WALDVOGEL R, 1981, Aroma quality of orange juice after filling and storage in soft packages and glass bottles. *Alimenta*, **20**, 91–3.

FUKAMACHI M, MATSUI T, HWANG Y H, SHIMODA M and OSAJIMA Y, 1996, Sorption behavior of flavor compounds into packaging films from ethanol solution. *J Agric Food Chem*, **44**, 2810–13.

GIACIN J R and HERNANDEZ R J, 1997, Permeability of aromas and solvents in polymeric packaging materials, in *The Wiley Encyclopedia of packaging technology*, ed. by Brody A L and Marsh K S. John Wiley & Sons, Inc., New York, pp. 724–33.

GODSHALL M A, 1997, How carbohydrates influence food flavor. *Food Technol,* **51**, 63–67.

GREMLI H, 1996, Flavor changes in plastic containers: a literature review. *Perfumer & Flavorist,* **21**, 1–8.

HALEK W H, 1988, Relationship between polymer structure and performance in food packaging applications, in *Food and Packaging Interactions*, ed. by Hotchkiss J H. ACS Symposium Series 365, American Chemical Society, Washington, DC, pp. 195–202 .

HERNANDEZ R J and GAVARA R, 1999, *Plastics packaging – methods for studying mass transfer interactions.* Pira International, Leatherhead, UK, pp. 53.

HERNANDEZ-MUÑOZ P, CATALÁ R and GAVARA R, 1999, Effects of sorbed oil on food aroma loss through packaging materials. *J Agric Food Chem,* **47,** 4370–74.

HERNANDEZ-MUÑOZ P, CATALÁ R and GAVARA R, 2001, Food aroma partition between packaging materials and fatty food simulants. *Food Addit Contam,* **18**, 673–82.

HIROSE K, HARTE B R, GIACIN J R, MILTZ J and STINE C, 1988, Sorption of d-limonene by sealant films and effect on mechanical properties, in *Food and Packaging Interactions*, ed. by Hotchkiss J H. ACS Symposium Series 365, American Chemical Society, Washington, DC, pp. 28–41.

HOTCHKISS J H, 1997, Food-packaging interactions influencing quality and safety, *Food Addit Contam,* **14**, 601–7.

INGHAM K E, TAYLOR A J, CHEVANCE F F V and FARMER L J, 1996, Effect of fat content on volatile release from foods, in *Flavour Science, Recent Developments*, ed. by Taylor A J and Mottram D S. The Royal Society of Chemistry, Cambridge, pp 386–91.

JOHANSSON F, 1993, *Polymer packages for food – materials, concepts and interactions, a literature review*. SIK – The Swedish Institute for Food and Biotechnology, Göteborg, pp 118.

JOHANSSON F and LEUFVÉN A, 1994, Influence of sorbed vegetable oil and relative humidity on the oxygen transmission rate through various polymer packaging films. *Packag Technol Sci,* **7,** 275–81.

JOHANSSON F and LEUFVÉN A, 1997, Concentration and interactive effects on the sorption of aroma liquids and vapors into polypropylene. *J Food Sci,* **62**, 355–58.

KINSELLA J E, 1989, Flavor perception and binding to food components, in *Flavour chemistry of lipid foods*, ed. by Min D B and Smouse T H. American Oil Chemists' Society, Champaign, pp. 376–403.

KRIZAN T D, COBURN J C and BLATZ P S, 1990, Structure of amorphous polyamides: effect on oxygen permeation properties, in *Barrier polymers and structures*, ed. by Koros W J. ACS Symposium Series 423, American Chemical Society, Washington, DC, pp 111–25.

KWAPONG O Y and HOTCHKISS J H, 1987, Comparative sorption of aroma compounds by polyethylene and ionomer food-contact plastics. *J Food Sci*, **52**, 761--63, 785.

LANDOIS-GARZA J and HOTCHKISS J H, 1987, Aroma sorption. *Food Eng*, **4**, 39, 42.

LANDOIS-GARZA J and HOTCHKISS J H, 1988, Permeation of high-barrier films by ethyl esters, in *Food and Packaging Interactions*, ed. by Hotchkiss J H. ACS Symposium Series 365, American Chemical Society, Washington, DC, pp 42–58.

LE THANH M, THIBEAUDEAU P, THIBAUT M A and VOILLEY A, 1992, Interaction between volatile and non-volatile compounds in the presence of water. *Food Chem*, **43**, 129–35.

LEBOSSE R, DUCRUET V and FEIGENBAUM A, 1997, Interactions between reactive aroma compounds from model citrus juice with polypropylene packaging film. *J Agric Food Chem*, **45**, 2836–42.

LETINSKI J and HALEK G W, 1992, Interactions of citrus flavor compounds with polypropylene films of varying crystallinities. *J Food Sci,* **57,** 481–4.

LI S and PAIK J S, 1996, Flavor sorption estimation by UNIFAC group contribution model. *Transactions of the ASAE*, **39**, 1013–17.

LINSSEN J P H, VERHEUL A, ROOZEN J P and POSTHUMUS M A, 1991a, Absorption of flavour compounds by packaging material: Drink yoghurts in Polyethylene bottles. *Int Dairy Journal* **1**, 33–40.

LINSSEN J P H, REITSMA J C E and ROOZEN J P 1991b, Influence of pulp particles on limonene absorption into plastic packaging material, in *Food packaging: a burden or an achievement?* Vol. 2. Symposium Rheims France 5–7 June, pp. 24.18–.20.

MANNHEIM C H, MILTZ J and LETZTER A, 1987, Interaction between polyethylene laminated cartons and aseptically packed citrus juices. *J Food Sci*, **52**, 737–40.

MARIN A B, ACREE T E, HOTCHKISS J H and NAGY S, 1992, Gas chromatography-olfactometry of orange juice to assess the effects of plastic polymers on aroma character. *J Agric Food Chem*, **40**, 650—54.

MARSHALL M R, ADAMS J P and WILLIAMS J W, 1985, Flavor absorption by aseptic packaging materials. Aseptipak 85: *Proceedings of the 3rd international conference and exhibition on aseptic packaging* (Princeton, NJ: Schotland Business Research, Inc.), 299–312.

MILLER K S and KROCHTA J M, 1997, Oxygen and aroma barrier properties of edible films: A review. *Trends in Food Sci & Technol*, **8**, 228–37.

MOHNEY S M, HERNANDEZ R J, GIACIN J R, HARTE B R and MILTZ J, 1988, Permeability and solubility of d-limonene vapor in cereal package liners. *J Food Sci*, **53**, 253–57.

MOSHONAS, M. G. and SHAW, P. E., 1989a, Changes in composition of volatile components in aseptically packaged orange juice during storage. *J Agric Food Chem*, **37**, 157–61.

MOSHONAS, M G and SHAW P E, 1989b, Flavor evaluation and volatile flavor constituents of stored aseptically packaged orange juice. *J Food Sci*, **54**, 82–5.

NAWAR W W, 1996, Lipids, in: *Food Chemistry*, ed. O R Fennema. M. Dekker, New York.

NIELSEN T J and JÄGERSTAD I M, 1994, Flavour scalping by food packaging. *Trends in Food Sci & Technol*, **5**, 353–6.

NIELSEN T J, JÄGERSTAD I M and ÖSTE R E, 1992, Study of factors affecting the absorption of aroma compounds into low-density polyethylene. *J Sci Food Agric*, **60**, 377–81.

OLAFSSON G and HILDINGSSON I, 1995, Sorption of fatty acids into ldpe and its effect on adhesion with aluminium foil in laminated packaging material. *J Agric Food Chem*, **43**, 306–12.

OLAFSSON G, HILDINGSSON I and BERGENSTAHL B, 1995, Transport of oleic and acetic acids from emulsions into low density polyethylene; Effects on adhesion with aluminum foil in laminated packaging. *J Food Sci*, **60**, 420–5.

PAIK J S, 1992, Comparison of sorption in orange flavor components by packaging films using the headspace technique. *J Agric Food Chem*, **40**, 1822–5.

PAIK J S and TIGANI M A, 1993, Application of regular solution theory in predicting equilibrium sorption of flavor compounds by packaging polymers. *J Agric Food Chem*, **41**, 806–8.

PAIK J S and WRITER M S, 1995, Prediction of flavor sorption using the Flory-Huggins Equation. *J Agric Food Chem*, **43**, 175–8.

PIEPER G, BORGUDD L, ACKERMANN P and FELLERS P, 1992, Absorption of aroma volatiles of orange juice into laminated carton packages did not affect sensory quality. *J Food Sci*, **57**, 1408–11.

QUEZADA GALLO J A, DEBEAUFORT F and VOILLEY A, 1999, Interactions between aroma and edible films. 1. Permeability of methylcellulose and low-density polyethylene films to methyl ketones. *J Agric Food Chem*, **47**, 108–13.

RIQUET A M, WOLFF N, LAOUBI S, VERGNAUD J M and FEIGENBAUM A, 1998, Food and packaging interactions: determination of the kinetic parameters of olive oil diffusion in polypropylene using concentration profiles. *Food Addit Contam*, **15**, 690–700.

ROBERTS D D, ELMORE J S, LANGLEY K R and BAKKER J, 1996, Effects of sucrose, guar gum, and carboxymethylcellulose on the release of volatile flavor compounds under dynamic conditions. *J Agric Food Chem*, **44**, 1321–26.

ROONEY M L, 1995, Active packaging in polymer films. in *Active Food Packaging*, ed. by Rooney M L pp. 74–110. London, Blackie Academic and Professional.

SADLER G D and BRADDOCK R J, 1990, Oxygen permeability of low density polyethylene as a function of limonene absorption: An approach to modeling flavor scalping. *J Food Sci*, **55**, 587–8.

SADLER G D and BRADDOCK R J, 1991, Absorption of citrus flavor volatiles by low density polyethylene. *J Food Sci*, **56**, 35–8.

SADLER G, PARISH M, DAVIS J, and VAN CLIEF D, 1995, *Flavor-Package Interaction. Fruit Flavors*, ed. by R. L. Rouseff and M. M. Leahy (Washington DC: American Chemical Society), pp. 202–10.

SALAME M, 1989, The use of barrier polymers in food and beverage packaging, in *Plastic film technology, volume one – high barrier plastic films for packaging*, ed. by Finlayson K M. Technomic Publishing Company, Inc., Lancaster, Pennsylvania, U.S.A., pp 132–45.

SHARMA G K, MADHURA C V, and ARYA S S, 1990, Interaction of plastic films with foods. I. Effect of polypropylene and polyethylene films on fruit squash quality. *J Food Sci Technol*, **27**, 127–32.

SHIMODA M, MATSUI T and OSAJIMA Y, 1987, Effects of the number of carbon atoms of flavor compounds on diffusion, permeation and sorption with polyethylene films. *J Jpn Soc Nutr Food Sci*, **34**, 535–9.

SINGH R P and HELDMAN D R, 1993, *Introduction to food engineering*. Academic Press, Inc., New York, pp 499.

SIZER C E, WAUGH P L, EDSTAM S, and ACKERMANN P, 1988, Maintaining flavor and nutrient quality of aseptic orange juice. *Food Technol*, June, 152–9.

SOARES N N F and HOTCHKISS J H, 1998a, Bitterness reduction in grape fruit juices through active packaging. *Pack. Technol. Sci,* **11**, 9–18.

SOARES N N F and HOTCHISS J H, 1998b, Naringnase immobilisation in packaging films for reducing naringin concentration in grape fruit juice. *J Food Sci,* **63**, 61–5.

STERN S A and TROHALAKI S, 1990, Fundamentals of gas diffusion in rubbery and glassy polymers, in *Barrier polymers and structures*, ed. by Koros W J. ACS Symposium Series 423, American Chemical Society, Washington, DC, pp. 22–59.

TAWFIK M S, DEVLIEGHERE F and HUYGHEBAERT A, 1998, Influence of D-limonene absorption on the physical properties of refillable PET. *Food Chem*, **61**, 157–62.

TIGANI M A and PAIK J S, 1993, Use of dielectric constants in the prediction of flavour scalping. *Packag Technol Sci*, **6**, 203–9.

VAN WILLIGE R W G, LINSSEN J P H and VORAGEN A G J, 2000a, Influence of food matrix on absorption of flavour compounds by linear low-density polyethylene: proteins and carbohydrates. *J Sci Food Agric*, **80**, 1779–89.

VAN WILLIGE R W G, LINSSEN J P H and VORAGEN A G J, 2000b, Influence of food matrix on absorption of flavour compounds by linear low-density polyethylene: oil and real food products. *J Sci Food Agric*, **80**, 1790–7.

VAN WILLIGE R W G, SCHOOLMEESTER D N, VAN OOIJ A N, LINSSEN J P H and VORAGEN A G J, 2002a, Influence of storage time and temperature on absorption of flavour compounds from solutions by plastic packaging materials. *J Food Sci*, **67**, 2023–31.

VAN WILLIGE, R W G, LINSSEN J P H, MEINDERS M B J, VAN DER STEGE H J and VORAGEN A G J, 2002b, Influence of flavour absorption on oxygen permeation through LDPE, PP, PC and PET plastics packaging materials. *Food Addit Contam*, **19**, 303–13.

VAN WILLIGE R W G, 2002c, Effects of flavour absorption of foods and their packaging materials, PhD thesis, Wageningen University, The Netherlands.

VAN WILLIGE R W G, LINSSEN J P H, LEGGER-HUYSMAN A and VORAGEN A G J, 2003, Influence of flavour absorption by food packaging materials (low-density polyethylene, polycarbonate and polyethylene terephthalate) on taste perception of a model solution and orange juice, *Food Add Contam*, **20**, 84–91.

VERMEIREN L, DEVLIEGHERE F, VAN BEEST M, DE KRUIJF N and DEBEVERE J, 1999, Developments in the active packaging of foods. *Trends in Food Sci & Technol.*, **10**, 77–86.

WESSELINGH J A and KRISHNA R, 2000, *Mass transfer in multicomponent mixtures*. Delft. University Press, Delft, NL, pp. 329.

WIDDER S and FISCHER N, 1996, Measurement of the influence of food ingredients on flavour release by headspace gas chromatography-olfactometry, in *Flavour Science*, ed. Taylor A J and Mottram D S. The Royal Society of Chemistry, Cambridge, pp 405–12.

YAMADA K, MITA K, YOSHIDA K and ISHITANI T, 1992, A study of the absorption of fruit juice volatiles by the sealant layer in flexible packaging containers (The effect of package on quality of fruit juice, part IV). *Packag Technol Sci*, **5**, 41–7.

ZHANG Z, BRITT I J and TUNG M A, 1999, Water absorption in EVOH films and its influence on glass transition temperature. *J Polymer Sci: Part B: Polymer Physics*, **37**, 691–9.

9

Moisture regulation

T. H. Powers and W. J. Calvo, Multisorb Technologies, USA

9.1 Introduction

Drying is probably the oldest form of preservation. Wrapping things that have been dried to protect them from moisture may well have been the earliest form of packaging. Even today a lot of technological development resources are expended to find new ways to package things to keep them dry. Some of the oldest materials used to control moisture are still used today: clay, salt, minerals, and plant extracts that have a greater affinity for water than the material being protected. Clay has been used for centuries; moist clay to keep things moist and dried clay to keep things dry. Likewise the importance of salt is legendary, whether added to foodstuffs and plant materials to bind moisture or used in the dry form to adsorb moisture.

Economic losses due to moisture – not to mention the threat to life – in some areas of the world (due in part to spoilage of foodstuffs) attest to the importance of keeping things dry. It has been estimated that up to 25% of the world's food supply is lost each year due to spoilage mostly from failure of packaging, ravages of moisture and lack of refrigeration. One of the earliest sorbents, still widely used today, is clay. It is inexpensive, widely available and requires a minimum of processing. Silica gel is the most popular sorbent due to its availability and purity as well as its whiteness, which connotes purity. Other silicates are likewise widely used in the form of natural zeolites and the synthesised forms called molecular sieves. These are used for their selectivity and their ability to keep things very dry. Many other minerals and salts are also described below.

9.2 Silica gel

9.2.1 Origins

The origins of silica gel lie on every beach and river bottom in the world. Sand is the raw material. Sand is relatively pure crystalline silicon dioxide. In order to manufacture silica gel, sand is first put into solution with a strong alkali. Then after filtration, precipitation, neutralisation, repeated rinsings and drying, amorphous silica is obtained. This is silica gel, needing only to be milled and classified to make it ready for use.

9.2.2 Composition

Silica gel manufactured in this way is completely amorphous, lacking any detectable fraction of crystalline silica, which is of concern as an irritant. There is still some residual salt, typically about 0.5% and mostly Na_2SO_4. The pH is near neutrality. The usual specification is pH 4–8. There is little if any titratable acidity. Evaluation may routinely be accomplished by preparing a 10% slurry of silica gel in distilled water, extracting for two hours, and measuring the supernatant for conductivity, pH, and titratable acidity.

9.2.3 Purity and compliance: EU, FCC, USP

Silica gel is permitted for use as a desiccant with foods and pharmaceuticals under EU regulations. The US Food Chemicals Codex contains a monograph specifying silica gel for food use and the US Pharmacopoeia describes silica gel for pharmaceutical use.

9.2.4 Adsorption profile

Silica gel adsorption, as with any sorbent, is proportional to the equilibrium relative humidity (ERH) and the temperature of its environment. In order to view adsorption characteristics it is customary to plot an adsorption isotherm at 25°C as in Fig. 9.1. As may be seen, the adsorptive capacity of silica gel is only 3–4% at an ERH of 10% rising to a capacity of over 30% at an ERH of 90%.

The rate at which silica gel approaches its capacity at differing ERH is illustrated in Fig. 9.2. Though capacity varies greatly, the rate at which silica gel approaches its capacity does not.

9.2.5 Regeneration

Silica gel may be regenerated and used indefinitely. With repeated adsorption and regeneration, some particle attrition occurs which eventually diminishes its usefulness. Complete regeneration is possible to < 2% moisture at 150°C for three hours. Moreover, 75–80% of capacity may be regained at 115–120°C for six hours. Microwave regeneration at low power (<400W) is also possible.

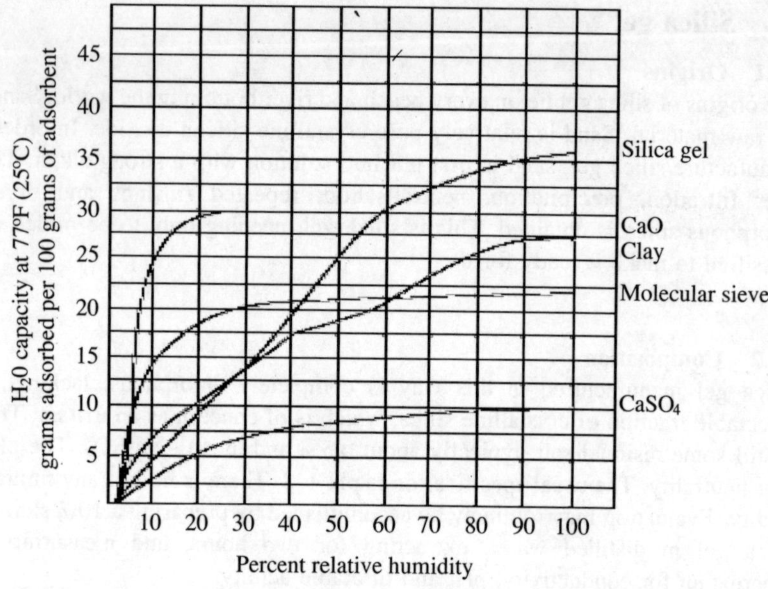

Fig. 9.1 Adsorption isotherms @ 25ºC.

Regeneration of packaged silica gel may be limited by the temperature tolerance of the package material itself.

9.2.6 Packaging and applications

Silica gel, as with other desiccants, is often packed in pouches or sachets. Materials range from adhesive coated papers and paper based laminates to non-wovens (coated and uncoated) to permeable or microperforated films. Semi-rigid capsules of many constructions are available with varying degrees of porosity and permeability. Occasionally, silica gel is filled into injection-moulded packages or incorporated directly into resins for moulding or extrusion. The uses of silica gel are too numerous to list. They include foodstuffs, pharmaceuticals, medical and diagnostic devices, textiles, leather goods, sealed electronics and many more.

9.3 Clay

9.3.1 Nomenclature and sources

Nearly all sources and types of clay, when fully dried, have some adsorptive properties. Clays used commercially fall into the category of Bentonite. The most frequently used is Montmorillonite. Clays are principally composed of metal silicates with some sulfates and phosphates present.

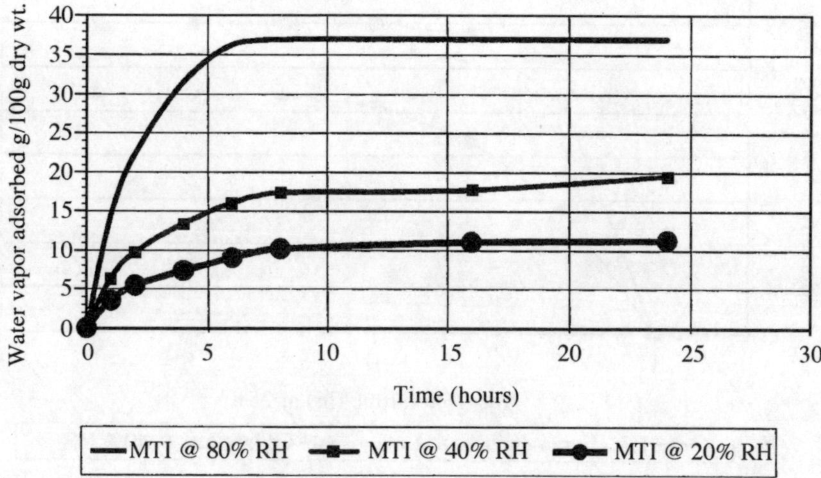

Fig. 9.2 Adsorption rate and capacity at 20, 40, and 80% RH.

9.3.2 Capacity and conditions of use

Mined clay is activated for use through careful drying. Adsorption capacities are in the range of 25–30% of dry weight at normal room temperature and below. Above 35°C clay will begin to desorb moisture. As a result the utility of clay is greatest under temperate conditions.

9.3.3 Adsorption/desorption

Adsorptive capacity varies with composition and the source. Fig. 9.3 illustrates the adsorptive characteristics of clay from a particular mine located in Oklahoma known as Oklahoma 1. Adsorption of moisture by clay is relatively rapid even at low relative humidity. As can be seen in Fig. 9.3 clay will adsorb its full complement of moisture in a matter of hours.

9.3.4 Packaging and applications

Clay is the desiccant of choice for many industrial and particularly bulk applications. Packaging may be in adhesive coated, reinforced kraft paper, non-wovens, or perforated film laminates. Larger sized packages of 1kg or more may be in fabric bags, filled, then sewn closed. As noted above, the adsorption of moisture is more rapid than necessary for some applications. For operational expediency and to permit more open or working time, a package material with somewhat restricted permeability may be chosen.

In order to account for the differences between one source of desiccant clay and another and indeed from one desiccant to another, the US military established a standard unit of adsorption. In November 1963, the Department of Defence

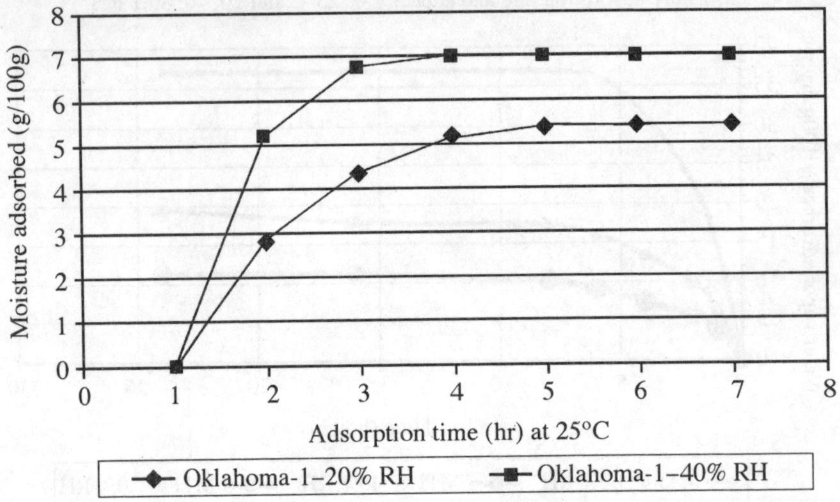

Fig. 9.3 Adsorption vs. time.

released MIL-D-3464C, covering the use of bagged desiccants for packaging and static dehumidification. Three years later, MIL-D-3464D served to update the original specification, creating a uniform standard of comparison in a wide variety of areas: adsorption capacity and rate, dusting characteristics of the package, strength and corrosiveness of the package and particle size of the desiccant.

In 1973, the DOD followed with specifications for cleaning, drying, preserving, and packaging of items, equipment and materials for protection against corrosion, mechanical and physical damage and other forms of deterioration. MIL-D-3464 and MIL-P-116 have long been the only objective source for packaging engineers. The strength of these specifications lies in their determination of a uniform unit of drying capacity, enabling one to compare desiccant effectiveness on a common scale.

9.4 Molecular sieve

9.4.1 Composition and purity

The composition of molecular sieves are sodium-, potassium-, calcium-, and magnesium-, aluminum silicates. These form orderly macrostructures with rigid pores of a consistent size.

9.4.2 Common types and nomenclature

Molecular sieves are usually designated by their pore size expressed in Angstrom units (10^{-10}m). The most frequently encountered is type 4A with a nominal pore size of 4 Angstroms. Other common grades are 3A and 5A. A speciality grade is type 13X with a nominal pore size of ~10 Angstroms. Figures 9.4 and 9.5 illustrate adsorption of types 4A and 13X at selected relative humidity conditions.

9.4.3 Selective and preferential adsorption

Retention and selectivity among various adsorbed compounds is proportional to their polarity and effective molecular size. This is the rationale for the name 'molecular sieve'. The ability of a particular type of molecular sieve selectively to adsorb a particular molecular species depends on what other compounds are present, e.g., more polar will displace less polar compounds in the same size range.

9.4.4 Moisture adsorption

Thus in a moist environment, water will displace nearly every other compound. The adsorption capacity of a molecular sieve in a moist environment will therefore be mostly taken up by water. As a result, desiccant applications most often utilise a type 4A molecular sieve since its pore size is suitable for water and its polarity favours water. These properties combine to give type 4A a significant affinity for moisture even at low ERH. As can be seen in Figs 9.4 and 9.5, a type 4A molecular sieve can adsorb 10% of its weight in moisture at less than 10% relative humidity. This permits the moisture of a closed environment to be maintained at less than a fraction of a percent with a suitable quantity of molecular sieve.

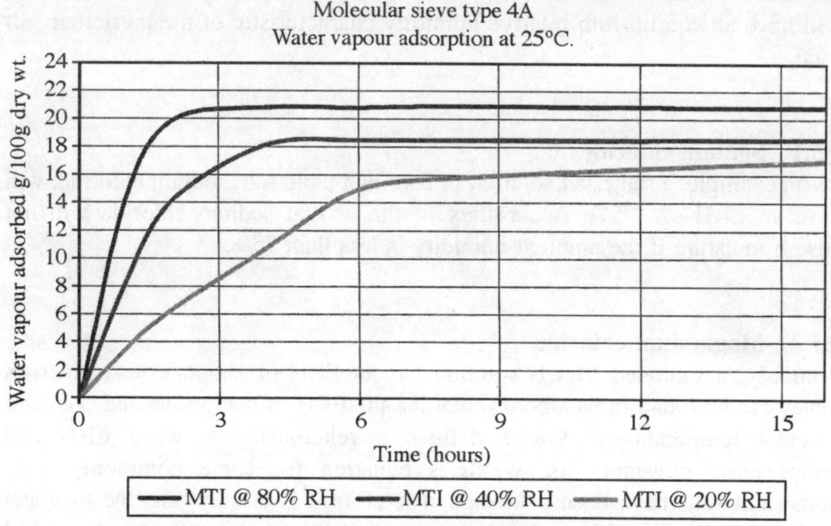

Fig. 9.4 Adsorption isotherms of common types of molecular sieve.

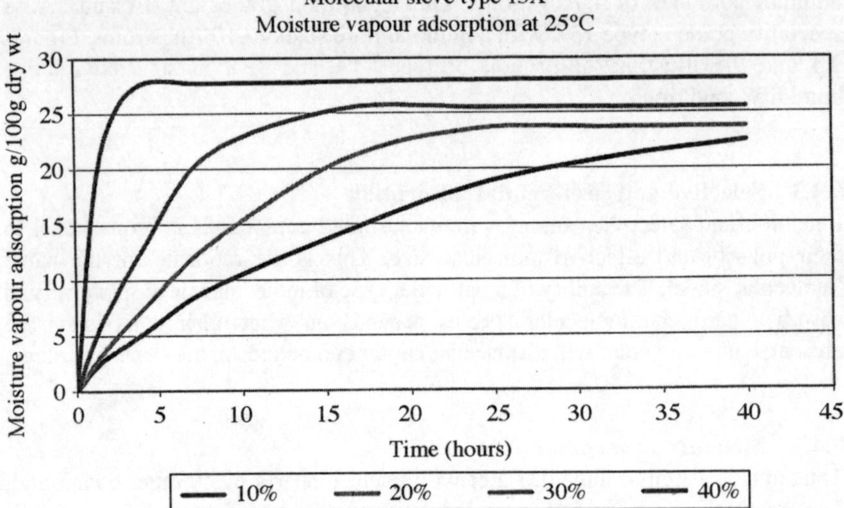

Fig. 9.5 Adsorption isotherms of common types of molecular sieve.

9.5 Humectant salts

A common and economical method of controlling humidity in moist environments is the use of humectant salts. Such salts will adsorb moisture until they go completely into solution. As this occurs, what will be seen is a mixture of solid salt and a salt solution. This solution will be saturated and will have an equilibrium relative humidity characteristic of the particular salt used.

9.5.1 Sodium chloride

As an example, a saturated solution of common table salt, sodium chloride, will have an ERH of ~75%. A corollary of this is that sodium chloride will not adsorb moisture if the ambient humidity is less than 75%.

9.5.2 Magnesium chloride

Similarly, a saturated $MgCl_2$ solution has an ERH of about 35%. Numerous tables can be found in handbooks, that list the ERH of many salts and common ambient temperatures. Wexler defined a relationship between ERH and temperature; constants for Wexler's equation for some commonly used humectant salts have been published. Therefore it is usually possible to select a salt that will control humidity in a range suitable for the system or packaged product being protected.

9.5.3 Calcium sulfate

Certain salts, sulfate salts in particular, will take on water of hydration in fixed mole proportions. Some salts such as calcium sulfate have multiple hydration states. Anhydrous calcium sulfate can take on ½ mole of water which becomes the commonly known 'Plaster of Paris'. Likewise it can take on two moles of water; this is known as gypsum. Many salts then can take on water of hydration. Their utility as moisture sorbents depends on the kinetics of adsorption under the conditions of use. Some experimentation may be required to make the right choice.

9.6 Irreversible adsorption

Closely related to water of hydration is the addition of water to alkali metal and transition metal oxides. Here water reacts with the oxide to form a separate compound. The reaction may be reversed but only with an input of energy sufficient to decompose the compound. Since decomposition occurs at a temperature well above any possible conditions of use, adsorption is considered irreversible.

9.6.1 Calcium oxide

The most frequently encountered example of this is calcium oxide. It is the product of high temperature decomposition of limestone, $CaCO_3$. CaO is principally used as agricultural lime and is known as quicklime. Calcium oxide reacts with water as follows:

$$CaO + H_2O \rightarrow Ca(OH)_2$$

The resulting product, calcium hydroxide, is likewise used for agricultural purposes and is known as slaked lime. To reverse this reaction requires raising the temperature to nearly 600ºC. Its use as a desiccant depends on the same reaction. The great amount of energy required to reverse the reaction makes this irreversible in a practical sense. As may be inferred, the product is quite alkaline. Although it is technically a corrosive product and irritating to the skin, it is not particularly hazardous in use due to its very low solubility (about 0.06g/100cc of water).

Calcium oxide is used as a desiccant principally where extremely low residual moisture is required. In a closed system it is possible to reduce moisture to a few parts per million with a suitable amount of calcium oxide. And it is a fairly efficient desiccant as well, capable of absorbing about 28% of its dry weight as water. In addition it is specific for water. Calcium oxide is used for very demanding packaging applications but, more particularly, for sealed electronic devices that are expected to last for years.

It must be kept in mind that calcium hydroxide is also a good carbon dioxide absorber. Therefore if calcium oxide is used as a desiccant in a carbon dioxide environment it will absorb carbon dioxide as well, releasing water in the process.

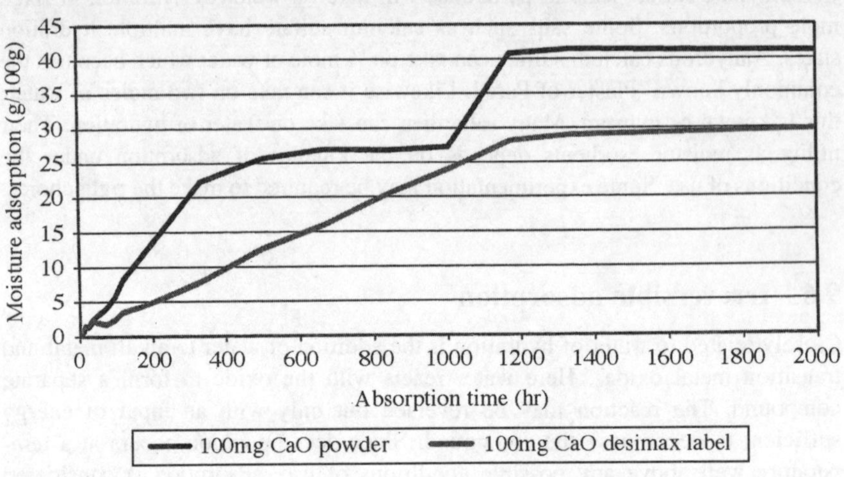

Fig. 9.6 Adsorption profile of calcium oxide.

In most applications this is not an issue since the package or device is typically sealed in a normal atmosphere, containing no more than 0.03% carbon dioxide. Nevertheless, in Fig. 9.6, the calcium oxide powder line can be seen to rise to about 27% moisture absorbed, then display an additional weight gain to over 40%. This second step corresponds to carbon dioxide absorption. Note also that CaO absorbs its full complement of moisture even in a 10% RH environment.

The companion curve represents calcium oxide in a self-adhesive label structure that is sealed but which has a microperforated face that allows diffusion of moisture through to the calcium oxide. The restriction of the microperforated material slows diffusion resulting in a constant slope, rather than the more typical asymptotic absorption curve characteristic of a desiccant in an open environment.

9.6.2 Magnesium oxide and barium oxide

Similar behaviour may be seen with magnesium oxide and barium oxide as well. Both are somewhat slower to react but barium oxide has the capability of reducing moisture in a closed system to less than 1 ppm. Its use is limited to these very special circumstances.

9.7 Planning a moisture defence

Moisture trapped within a food product package or leaking into it during storage and shipping can cause many harmful effects. If not removed, this moisture will be adsorbed by the product or condensate will form, causing growth of mould, mildew and fungus. For example, if a solid is very water soluble (such as a sugar coating), dissolution into the adsorbed layer can trigger irreversible water uptake and subsequent deliquescence, given the appropriate conditions. Selection of the proper desiccant can be inexpensive insurance for protecting packaged food products, thus resulting in improved quality.

9.7.1 Sources of moisture in packaging

Those involved in food packaging applications face a confusing array of variables when selecting moisture adsorbents (desiccants), as moisture regulation is a multi-faceted challenge. More specifically, sources of water permeation into a closed package or container can be attributed to moisture from (i) the product itself, (ii) any material (such as felt, foam, paper, etc.) used to support or retain the product, and (iii) permeation through the protective barrier of the package. With the goal of selecting the appropriate desiccant, moisture contributed by the product environment (ambient moisture) and the package (bound moisture) must be considered independently.

Ambient moisture

Temperature and relative humidity are two of the most influential environmental factors affecting product integrity and must be controlled to match the conditions of optimum product preservation and performance. Before selecting the correct desiccant, it is imperative to know the conditions surrounding the shipment and storage of the product. Furthermore, at the time of packaging, it must be noted that the product is sealed in the conditions of the packaging room.

The moisture content of the air can be defined by its relative humidity, equal to the ratio (expressed as a percentage) of the partial pressure of water vapour present in the air to the saturated vapour pressure. The most useful combined measure of temperature and relative humidity is the dewpoint, that is, the temperature at which the actual vapour pressure equals the saturated vapour pressure. As the temperature drops, the saturation water vapour pressure decreases. Any additional drop in temperature will give rise to condensation, as the amount of water in the air has then exceeded the saturation point. Condensation provides the most dramatic visual observation of the effects of moisture damage. An effective desiccant will adsorb water vapour from the air in a package, lowering the relative humidity to the point where condensation will no longer occur or the threshold relative humidity is never exceeded under the conditions to which the package will be exposed.

As a general rule of thumb, designing the package and the desiccant to maintain an internal relative humidity of 10–12% at normal room temperature conditions (70ºF) will provide adequate protection. It is strongly suggested that the desiccant supplier be contacted to discuss the elements of the package and the level of protection required.

Bound moisture

The moisture content of the packaged product can be defined as the ratio of water present in the product to its dry weight. Under equilibrium conditions (no exchange of product moisture with the environment), the vapour pressure generated from the moisture content is termed 'water activity' and denoted A_w. More specifically, water activity can be defined by the equation

$$A_w = P/P_{sat}$$

where P denotes the partial pressure of water vapour at the product's surface and P_{sat} denotes the saturation pressure at the product temperature. Equilibrium relative humidity (ERH) is the water activity value expressed as a percentage.

The advantage of this definition of A_w is that it defines moisture, which can be 'actively' exchanged between the product and its environment. Water activity can provide better information than a product's total moisture content when considering the stability of foods. It can be directly compared with the relative humidity of the ambient air to prevent hygroscopic powders (such as powdered sugars or salt) from caking, for example. In a closed package, the desiccant will work to adsorb moisture from all sources. Some plastics (such as nylon), foams, paper, wood, felt, cotton and polyester can all contain moisture. Wood, cotton and paper can hold 14% or more, and certain foams up to 10%. The moisture contained in these materials can be released into the air as the desiccant dries the air space around it or as the temperature increases. The amount of water adsorbed by the desiccant depends upon (i) how strongly the moisture is bound by that source (chemisorption or physical adsorption); (ii) the type of desiccant and quantity used; (iii) how much water has already been adsorbed by the desiccant, and (iv) temperature.

The main purpose of moisture regulation in foods is to lower water activity, thereby reducing the growth of microorganisms such as moulds and yeast on foods with high water activity (such as meals ready-to-eat, or MREs). Moreover, any change in the temperature of a hygroscopic food product will trigger the product to exchange moisture with the air or gas that surrounds it. Moisture will be exchanged as such until the partial water vapour pressures at the product surface and in the air or gas are equal.

Moisture ingress: theoretical development

Desiccants, which address moisture concerns, are usually chosen by directly running tests under the intended application. However, product stability testing can become time consuming and expensive. Consequently, it can be beneficial to know how much moisture theoretically permeates into the package, either prior to or in conjunction with product stability tests.

The moisture vapour transmission rate or MVTR is defined as the amount of water vapour passing through one square metre of test material under set conditions (temperature and relative humidity) in 24 hours. It is inversely proportional to the material thickness. If the MVTR (typically in units of grams of water per square metre per hour), the total surface area of the barrier (package surface area) and the length of time in storage are all known, the moisture ingress can be determined. This value represents the amount removed by the desiccant to protect against possible condensation within the package or damage due to prolonged exposure in a high humidity environment.

The permeability coefficient is a material constant which specifies the volume of gas which will pass through a test material of known surface area and thickness in a fixed time, with a given partial pressure difference. The coefficient varies with temperature. Permeability coefficients for packaging materials can usually be obtained from the material supplier or, more appropriately, permeability data on the actual package may be available. Published material permeability values can be found not only for the packaging material, but also for desiccant packaging (a sachet, for example) as well.

Permeability coefficients can be converted into permeability rates defined by the origin of the driving force involved (whether it be water concentration, partial pressure or mole fraction). Water vapour partial pressure data are acquired through thermodynamic steam tables. These rates may have to be adjusted to reflect actual shelf-life conditions. Moisture ingress (mg water/day, for example) across a package wall for given relative humidity conditions inside the package is determined by the use of a general material balance for water, namely

$$\text{Flow of } H_2O = A_{pkg} P_{pkg} (C_{out} - C_{in})$$

where A_{pkg} = surface area of package
P_{pkg} = permeability rate of package material
C_{out} = concentration of water
C_{in} = concentration of water in headspace.

The permeability rate of the package material needs to be adjusted both for the thickness of the material and the partial pressure differential. It is important to note that the reaction rates are assumed to be relatively rapid compared to the rate of diffusion, so that the process is diffusion controlled. The amount of water permeating into a package during shelf-life depends largely on the type of packaging (e.g., flexible, rigid) and the packaging materials that are used. For example, various flexible packaging materials exhibit different MVTRs. Water ingress into flexible packaging can only be roughly estimated from such data. This is because the contribution of moisture ingress from the seal is not included in material permeability data. Desiccants, which address moisture concerns, are usually chosen by directly running tests under the intended application. However, more rigorous testing can become time-consuming and expensive.

9.7.2 Desiccant selection process

In order to maximise success in the selection of a desiccant, the properties of both the commercially available desiccant and the package material(s) must be considered. The conditions involving optimum product preservation and performance are analysed experimentally and/or by seeking published literature values.

In a porous desiccant, water is removed from the headspace either by multilayer adsorption, where thin layers of water molecules are attracted to the surface of the desiccant or by capillary condensation, where the smaller pores become filled with water. In multilayer adsorption, the surface area is high due to the extensive porosity and significant amounts of water can be attracted and adsorbed. In contrast, however, capillary condensation occurs because the saturation water vapour pressure in a small pore is reduced by the effect of surface tension. Capillary condensation also may occur in any pores within the packaged product. The harmful results of moisture condensation may occur in some products at humidity levels below those that would be predicted from examination of the bulk package headspace. Moreover, the container in which the product will be packaged, shipped and stored is vital in determining how much of a particular desiccant is needed and in what packaging form. Before the adsorbent selection process itself, the size of the container based on the flexibility of the package wall structure must be determined.

The three most important parameters to consider in the desiccant selection process are relative humidity, adsorbent capacity and adsorbent rate.

1 Relative humidity

Knowledge of the maximum and minimum humidity levels that should be maintained can further enhance the correct choice of desiccant. Product stability tests by the food manufacturer usually establish the maximum moisture level that should not be exceeded during shelf-life. In addition, the maximum relative humidity in the packaging allowed by the product manufacturer during shelf-life is an important consideration. This is because dividing the total amount of moisture to be adsorbed by the adsorption capacity of pure desiccant (at the maximum allowable relative humidity) yields the minimum amount of pure desiccant required.

The effect of residual moisture, which is usually specified by the desiccant supplier, also needs to be taken into account. There is usually a minimum moisture level to be maintained during shelf-life in order to avoid problems arising as a result of moisture loss in the product.

2 Adsorption capacity

Adsorption capacity is calculated by a mass balance on the various sources of water involved. The total moisture (W_{tot}) is calculated by taking the sum of headspace air humidity (W_{hs}), available residual moisture in the packaged product (W_{rm}) and the amount of water ingress into the package during shelf-life (W_{in}).

$$W_{tot} = W_{hs} + W_{rm} + W_{in}$$

W_{hs} is the product of headspace volume and absolute air humidity in the environment during the packaging operation. The contribution of W_{hs} to W_{tot} is usually negligible, because headspace is minimized during packaging design and when packaging is done under controlled low-humidity conditions. W_{rm} is the amount of water that can be desorbed from the packaged product by the desiccant during shelf-life. This amount depends on several factors including moisture desorption thermodynamics and kinetics (originating from the packaged product) as well as the overall moisture content of the product after manufacturing.

3 Adsorption rate
The third parameter to note when considering desiccants is adsorption rate. For example, silica gel reduces the relative humidity in a closed container from 20% to virtually 0% in about one hour. Under actual manufacturing conditions, the product being packaged contains residual moisture. This moisture has to be adsorbed by the desiccant, thereby increasing the reaction time. Note also that as residual moisture increases, adsorption capacity decreases. In most cases, the desorption kinetics for water from the product are slower than the adsorption kinetics of the desiccant. Consequently, the desorption reaction is the rate-limiting step for overall dehumidification. This phenomenon eliminates the difference between the desiccant reaction rates and leads to more prolonged reaction times.

9.8 Future trends

Moisture management is taking on a broader meaning as the whole field of active packaging expands. It is no longer adequate simply to keep things 'dry'. They must be dry enough to attain stability but not so dry as to affect the structure on a macro or on a molecular level. Therefore moisture control becomes water activity control. There is a growing recognition that in some sealed systems moisture is not the only volatile substance which can degrade the active components of the system. Volatile organics, particularly tri-halo methanes as well as volatile acids, aldehydes and alcohols may be present as well. Desiccants then may need to be blended with other sorbents or impregnated with compounds that can selectively bind molecules other than water.

Finally, greater importance is being placed on the selectivity of sorbents. A wider selection of molecular sieves is becoming available. New synthetic techniques are being developed; some mimicking biosynthetic mechanisms are being studied.

Developments are moving in the direction of more specialised sorbents and multifunctional sorbents. This trend is sure to continue and will most likely accelerate.

Part II

Developments in modified atmosphere packaging (MAP)

10

Novel MAP applications for fresh-prepared produce

B.P.F. Day, Food Science Australia

10.1 Introduction

During recent years there has been an explosive growth in the market for fresh prepared fruit and vegetable (i.e. produce) products. The main driving force for this market growth is the increasing consumer demand for fresh, healthy, convenient and additive-free prepared product items. However, fresh prepared produce items are highly perishable and prone to the major spoilage mechanisms of enzymic discoloration, moisture loss and microbial growth. Good manufacturing and handling practices along with the appropriate use of modified atmosphere packaging (MAP) are relatively effective at inhibiting these spoilage mechanisms, thereby extending shelf-life. Shelf-life extension also results in the commercial benefits of less wastage in manufacturing and retail display, long distribution channels, improved product image and the ability to sell convenient, added-value, fresh prepared produce items to the consumer with reasonable remaining chilled storage life.

The application of novel high oxygen (O_2) MAP is a new approach for the retailing of fresh prepared produce items and is capable of overcoming the many inherent shortcomings of current industry-standard air packaging or low O_2 MAP. The results from an extensive European Commission and industry funded project have shown that high O_2 MAP is particularly effective at inhibiting enzymic discolorations, preventing anaerobic fermentation reactions and moisture losses, and inhibiting aerobic and anaerobic microbial growth. Independent research undertaken in the Netherlands, Belgium, Australia, USA and Spain has also shown many interesting and mainly beneficial effects of high O_2 MAP and references to this research are listed. This chapter highlights how extended shelf-life can be achieved by using high O_2 MAP. Practical guidance on issues such as safety, optimal high O_2 mixtures, produce volume/gas volume ratios, packaging materials and chilled storage temperatures will be outlined so as to facilitate the commercial exploitation of this new technology. Brief reference in this chapter has been made with respect to novel argon (Ar) and nitrous oxide (N_2O) MAP, but in light of the variable results obtained for these novel MAP treatments, the majority of the text concentrates on the applications of novel high O_2 MAP.

Unlike other chilled perishable foods that are modified atmosphere (MA) packed, fresh produce continues to respire after harvesting, and any subsequent packaging must take into account this respiratory activity. The depletion of O_2 and enrichment of carbon dioxide (CO_2) are natural consequences of the progress of respiration when fresh produce is stored in hermetically sealed packs. Such modification of the atmosphere results in a respiratory rate decrease with a consequent extension of shelf-life (Kader *et al.*, 1989). MAs can passively evolve within hermetically air-sealed packs as a consequence of produce respiration. If a produce item's respiratory characteristics are properly matched to film permeability values, then a beneficial equilibrium MA (EMA) can be passively established. However, in the MAP of fresh produce, there is a limited ability to regulate passively established MAs within hermetically air-sealed packs. There are many circumstances when it is desirable to rapidly establish the atmosphere within produce packs. By replacing the pack atmosphere with a desired mixture of O_2, CO_2 and nitrogen (N_2), a beneficial EMA may be established more rapidly than a passively generated EMA. For example, flushing packs with N_2 or a mixture of 5–10% O_2, 5–10% CO_2 and 80–90% N_2 is commercial practice for inhibiting undesirable browning and pinking on prepared leafy green salad vegetables (Day, 1998).

The key to successful retail MAP of fresh prepared produce is currently to use packaging film of correct permeability so as to establish optimal EMAs of typically 3–10% O_2 and 3–10% CO_2. The EMAs attained are influenced by produce respiration rate (which itself is affected by temperature, produce type, variety, size, maturity and severity of preparation); packaging film permeability; pack volume, surface area and fill weight; and degree of illumination. Consequently, establishment of an optimum EMA for individual produce items is very complex. Furthermore, in many commercial situations, produce is sealed in packaging film of insufficient permeability (Betts, 1996) resulting in development of undesirable anaerobic conditions (e.g. <2% O_2 and >20% CO_2). Recently developed, microperforated films, which have very high gas transmission rates, are now commercially used for maintaining aerobic EMAs (e.g. 5–15% O_2 and 5–15% CO_2) for highly respiring prepared produce items such as broccoli and cauliflower florets, baton carrots, beansprouts, mushrooms and spinach. However, microperforated films are relatively expensive, permit moisture and odour losses, and may allow for the ingress of microorganisms into sealed packs during wet handling situations (Day, 1998).

10.2 Novel MAP gases

10.2.1 High O_2 MAP

Information gathered by the author during 1993–1994 revealed that a few prepared produce companies had been experimenting with high O_2 (e.g. 70–100%) MAP and had achieved some surprisingly beneficial results. High O_2 MAP of prepared produce was not exploited commercially during that period, probably because of the inconsistent results obtained, a lack of understanding of the basic biological mechanisms involved and concerns about possible safety implications. Intrigued by the concept of high O_2 MAP, the Campden and Chorleywood Food Research Association (CCFRA) carried out limited experimental trials on prepared iceberg lettuce and tropical fruits, in early 1995. The results of these trials confirmed that high O_2 MAP could overcome the many disadvantages of low O_2 MAP. High O_2 MAP was found to be particularly effective at inhibiting enzymic discolorations, preventing anaerobic fermentation reactions and inhibiting microbial growth. In addition, the high O_2 MAP of prepared produce items within inexpensive hermetically sealed plastic films was found to be very effective at preventing undesirable moisture and odour losses and ingress of microorganisms during wet handling situations (Day, 1998).

The experimental finding that high O_2 MAP is capable of inhibiting aerobic and anaerobic microbial growth can be explained by the growth profiles of aerobes and anaerobes (Fig. 10.1). It is hypothesised that active oxygen radical species damage vital cellular macromolecules and thereby inhibit microbial growth when oxidative stresses overwhelm cellular protection systems (Gonzalez Roncero and Day, 1998; Amanatidou, 2001). Also intuitively, high O_2 MAP inhibits undesirable anaerobic fermentation reactions (Day, 1998).

Polyphenol oxidase (PPO) is the enzyme primarily responsible for initiating discoloration on the cut surfaces of prepared produce. PPO catalyses the oxidation of natural phenolic substances to colourless quinones which subsequently polymerise to coloured melanin-type compounds (McEvily *et al.*, 1992). It is hypothesised that high O_2 (and/or high Ar) levels may cause substrate inhibition of PPO or alternatively, high levels of colourless quinones subsequently formed (Fig. 10.2) may cause feedback product inhibition of PPO.

10.2.2 Argon and nitrous oxide MAP

Argon (Ar) and nitrous oxide (N_2O) are classified as miscellaneous additives and are permitted gases for food use in the European Union (EU). Air Liquide S.A. (Paris, France) has stimulated recent commercial interest in the potential MAP applications of using Ar and, to a lesser extent, N_2O. Air Liquide's broad range of patents claim that in comparison with N_2O, Ar can more effectively inhibit enzymic activities, microbial growth and degradative chemical reactions in selected perishable foods (Brody and Thaler, 1996; Spencer, 1999). More specifically, an Air Liquide patent for fresh produce applications claims that Ar

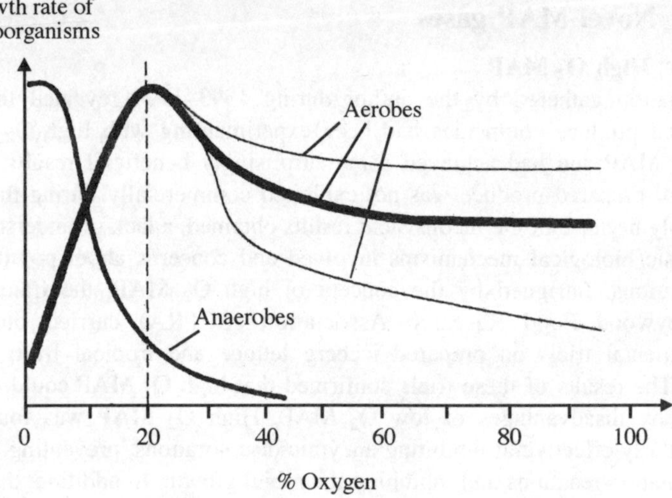

Fig. 10.1 Hypothesised inhibition of microbial growth by high O_2 MAP.

and N_2O are capable of extending shelf-life by inhibiting fungal growth, reducing ethylene emissions and slowing down sensory quality deterioration (Fath and Soudain, 1992). Of particular relevance is the claim that Ar can reduce the respiration rates of fresh produce and hence have a direct effect on extension of shelf-life (Spencer, 1999).

Although Ar is chemically inert, Air Liquide's research has indicated that it may have biochemical effects, probably due to its similar atomic size to molecular O_2 and its higher solubility in water and density compared with N_2 and O_2. Hence, Ar is probably more effective at displacing O_2 from cellular sites and enzymic O_2 receptors with the consequence that oxidative deterioration reactions are likely to be inhibited. In addition, Ar and N_2O are thought to sensitise microorganisms to antimicrobial agents. This possible sensitisation is not well understood but may involve alteration of the membrane fluidity of microbial cell walls with a subsequent influence on cell function and performance (Thom and Marquis, 1984). Clearly, more independent research is needed to better understand the potential beneficial effects of Ar and N_2O (Day, 1998).

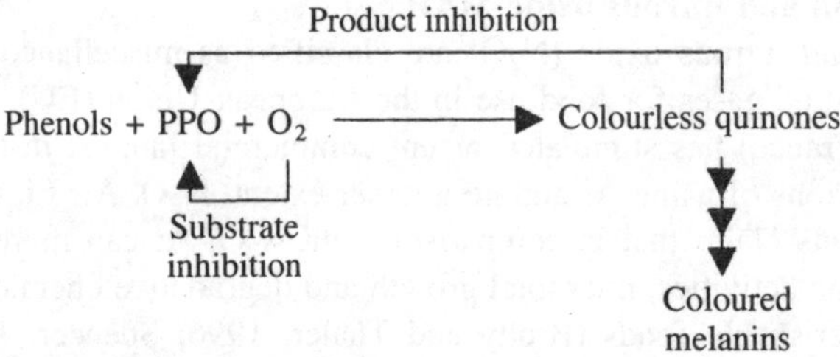

Fig. 10.2 Hypothesised inhibition of enzymic discoloration by high O_2 MAP.

10.3 Testing novel MAP applications

Two industrially funded research Clubs were set up at CCFRA to investigate in detail the interesting effects of novel MAP on fresh prepared produce. A High O_2 MAP Club ran from April, 1995 to September, 1997 and as a follow-up, a Novel Gases MAP Club ran from January, 1998 to December, 1999. These Clubs were supported by a total of nine prepared produce suppliers, five gas companies, four packaging film suppliers, three retailers, two suppliers of non-sulphite dips, two manufacturers of MAP machinery and two gas instrument companies.

In addition, further investigations were carried out during a three-year EU FAIR funded project, which started in September 1996. The overall objective of this project was to develop safe commercial applications of novel MAP for extending the quality shelf-life of a wide range of fresh prepared produce items. Other aims included investigations of the effects of novel MAP on non-sulphite dipped prepared produce, labile nutritional components, and microbial and biochemical spoilage mechanisms. The major focus of this research was on high O_2 MAP, followed by Ar MAP, and to a minor extent, N_2O MAP.

In summary, the following major results and achievements were made during the course of CCFRA's Club and EU-funded novel MAP research:

- High O_2 compatible MAP machines were used safely and successfully during the course of the project's experimental trial work. A non-confidential guidelines document on the safe use of high O_2 MAP was published (BCGA, 1998).
- Enzymic discolorations of prepared non-sulphite dipped potatoes and apples were generally more effectively inhibited by anaerobic (<2% O_2) MAP combinations of N_2, Ar and CO_2, compared with high O_2 MAP. However, high O_2 MAP was found to have certain odour and textural benefits for prepared potatoes and apples. Also, high O_2 MA packed non-sulphite dipped prepared potatoes and bananas were found to have longer achievable shelf-lives, in comparison with equivalent low O_2 (8%) MA packed control samples.
- For most prepared produce items, under defined storage and packaging conditions, high O_2 MAP was found to have beneficial effects on sensory quality in comparison with industry-standard air packing and low O_2 MAP. High O_2 MAP was found to be effective for extending the achievable shelf-lives of prepared iceberg lettuce, sliced mushrooms, broccoli florets, Cos lettuce, baby-leaf spinach, raddichio lettuce, lollo rossa lettuce, flat-leaf parsley, cubed swede, coriander, raspberries, strawberries, grapes and oranges (Tables 10.1 and 10.2).
- Ar-containing and N_2O-containing MAP treatments were found to have negligible, variable or only minor beneficial effects on the sensory quality of several prepared produce items, in comparison with equivalent N_2-containing MAP treatments.
- High O_2 MAs were found to inhibit the growth of several generic groups of bacteria, yeasts and moulds, as well as a range of specific food pathogenic

Table 10.1 Overall achievable shelf-life obtained from fresh prepared iceberg lettuce trial

MAP treatments	Storage days at 8ºC to drop to quality grade C			Shelf-life limiting quality attribute(s)	Overall achievable shelf-life
	Appearance	Odour	Texture		
5% O_2/95% N_2	4	7	4	Appearance/texture	4 days
5% O_2/10% CO_2/85% N_2	7	7	8	Appearance/odour	7 days
80% O_2/ 20% N_2	11	11	11	Appearance/odour/ texture	11 days

and spoilage microorganisms, namely *Aeromonas hydrophila*, *Salmonella enteritidis*, *Pseudomonas putida*, *Rhizopus stolonifer*, *Botrytis cinerea*, *Penicillium roqueforti*, *Penicillium digitatum* and *Aspergillus niger* (e.g. Figs 10.3 and 10.4). High O_2 MAs alone were not found to inhibit or stimulate the growth of *Pseudomonas fragi*, *Bacillus cereus*, *Lactobacillus sake*, *Yersinia enterocolitica* and *Listeria monocytogenes*, but the addition of 10–30% CO_2 inhibited the growth of all these bacteria. Ar-containing and N_2O-containing MAs were found to have negligible antimicrobial effects on

Table 10.2 Overall achievable shelf-life obtained from several fresh prepared produce trials

Prepared produce items	Overall achievable shelf-life (days) at 8ºC	
	Industry standard air/low O_2 MAP	High O_2 MAP
Iceberg lettuce	2–4	4–11
Dipped sliced bananas	2	4
Broccoli florets	2	9
Cos lettuce	3	7
Strawberries	1–2	4
Baby leaf spinach	7	9
Lolla Rossa lettuce	4	7
Radicchio lettuce	3	4
Flat leaf parsley	4	9
Coriander	4	7
Cubed swede	3	10
Raspberries	5–7	9
Little Gem lettuce	4–8	6–8
Dipped potatoes	2–3	3–6
Baton carrots	3–4	4
Sliced mushrooms	2	6

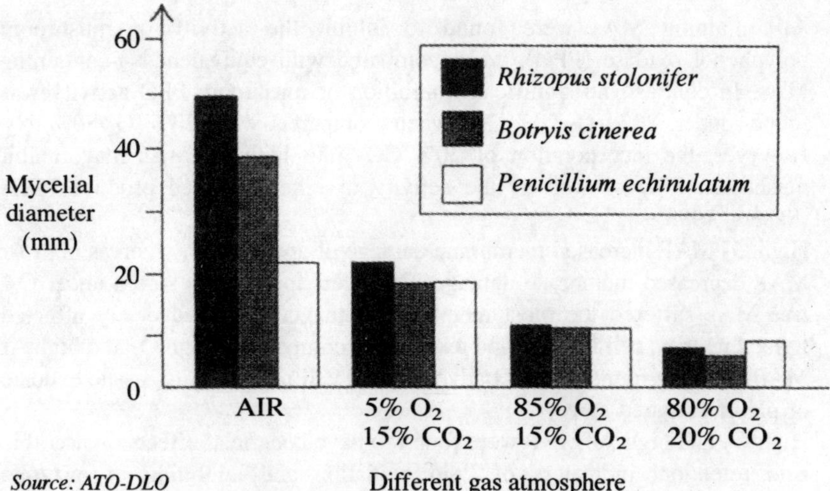

Fig. 10.3 Inhibition of fungal growth by different MAs.

a range of microorganisms, when compared with equivalent N_2-containing MAs.

- Respiration rates of selected prepared produce items were not found to be significantly affected by high O_2 or high Ar MAs, but were substantially reduced by the addition of 10% CO_2.
- High O_2 and high Ar MAP did not prevent the enzymic browning of non-sulphite dipped apple slices, but no further browning took place after pack opening.

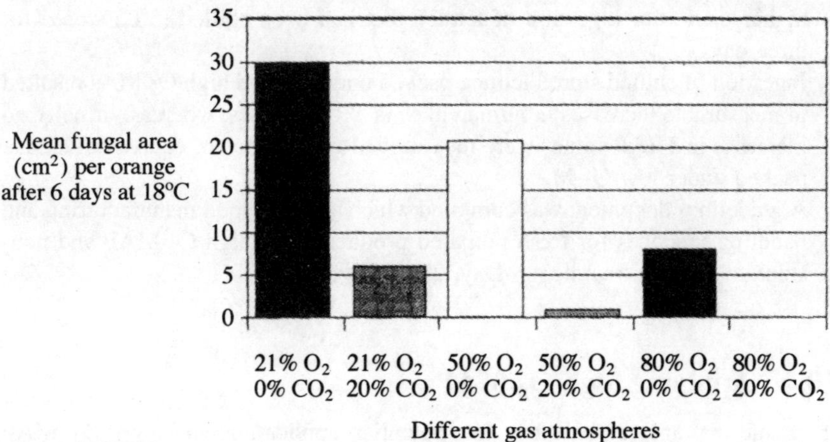

Fig. 10.4 Inhibition of fungal growth on *Penicillium digitatum* infected oranges under different MAs.

- Ar-containing MAs were found to inhibit the activity of mushroom polyphenol oxidase (PPO), when compared with equivalent N_2-containing MAs. In contrast, no significant inhibition of mushroom PPO activity was found under 80% O_2/20% N_2 when compared with 20% O_2/80% N_2. However, the incorporation of 20% CO_2 into high O_2 MAs may inhibit mushroom PPO as well as the activity of other prepared produce PPOs (Sapers, 1993).
- High O_2 MAP increased membrane damage of apple slices, whereas high Ar MAP decreased membrane damage. However, apple slices stored under O_2-free MAs suffered the most membrane damage, which adversely affected tissue integrity, cell leakage and texture. By comparison, high O_2 and high Ar MAP were not found to affect adversely the cell permeability, tissue exudate or pH of prepared carrots.
- High O_2 and high Ar MAP were found to have beneficial effects on ascorbic acid retention, indicators of lipid oxidation and inhibition of enzymic browning on prepared lettuce.
- High O_2 MAs increased the peroxidase activity of *Botrytis cinerea*, but the addition of 10% CO_2 substantially reduced this activity.
- In comparison with air packing and low O_2 MAP, high O_2 MAP was not found to decrease preferentially single antioxidant (ascorbic acid, β-carotene and lutein) levels in prepared lettuce but did induce the loss of certain phenolic compounds, even though desirable total antioxidant capacity (TRAP) values after chilled storage were increased.
- Extracts from high O_2 MA packed prepared lettuce and onions did not have any cytotoxic effects on human colon cells.
- Ingestion of fresh lettuce resulted in an increase in human plasma TRAP values through the absorption of phenolic compounds and single antioxidant molecules. This increase in human plasma TRAP values was significantly higher than after ingestion of lettuce that had been chilled (5°C) stored for three days.
- Ingestion of chilled stored lettuce packed under air and high O_2 MAs resulted in measurable increases in human plasma TRAP values, whereas virtually no increases in TRAP values were measured after ingestion of equivalent lettuce packed under low O_2 MAs.
- A guidelines document was compiled which outlines good manufacturing and handling practices for fresh prepared produce using high O_2 MAP and non-sulphite dipping treatments (Day, 2001a).

10.4 Applying high O_2 MAP

It should be appreciated that the potential applications of high O_2 MAP technology are a recent innovation and new knowledge will evolve in the future. Hence, the following guidance provided only reflects the current status of available knowledge and experience of high O_2 MAP for fresh prepared produce. Potential applications of high O_2 MAP to chilled combination food items (e.g. chilled ready meals, pizzas, kebabs, etc.) have been the subject of recent research (Day, 2001b), but are outside the scope of this chapter.

10.4.1 Safety

A specific guideline document on *The safe application of oxygen enriched atmospheres when packaging food* has been published and is publicly available (BCGA, 1998). This document contains clear and concise advice and recommendations on how to control safely the hazards of utilising O_2-rich gas mixtures for the MAP of food.

Food companies and related industries (e.g. gas companies and MAP machinery manufacturers) are strongly encouraged to purchase this safety guidelines document and to follow closely the advice and recommendations given before undertaking any pre-commercial trials using high O_2 MAP. Further advice and help on the safety aspects of high O_2 MAP can be sought from qualified gas safety engineers and the BCGA.

10.4.2 Optimal gas levels

Based on CCFRA's practical experimental trials, the recommended optimal headspace gas levels immediately after fresh prepared produce package sealing are:

80-95% O_2/5–20% N_2

After package sealing, headspace O_2 levels will decline whereas CO_2 levels will increase during chilled storage due to the intrinsic respiratory nature of fresh prepared produce. As previously explained, the levels of O_2 and CO_2 established within hermetically sealed packs of produce during chilled storage are influenced by numerous variables, i.e. the intrinsic produce respiration rate (which itself is affected by temperature; atmospheric composition; produce type, variety, cultivar and maturity; and severity of preparation); packaging film permeability; pack volume, surface area and fill weight; produce volume/gas volume ratio and degree of illumination (Kader *et al.*, 1989; Day, 1994; O'Beirne, 1999).

To maximise the benefits of high O_2 MAP, it is desirable to maintain headspace levels of O_2 > 40% and CO_2 in the range of 10–25% during the chilled shelf-life of the product. This can be achieved by lowering the temperature of storage, by selecting produce having a lower intrinsic respiration rate, by minimising cut surface tissue damage, by reducing the produce volume/gas volume ratio by either decreasing the pack fill weight or increasing the pack headspace volume, by using a packaging film which can maintain high levels of O_2 whilst selectively allowing excess CO_2 to escape, or by incorporating an innovative active packaging sachet that can adsorb excess CO_2 and emit an equal volume of O_2 (McGrath, 2000).

Also, in order to maintain levels of $O_2 > 40\%$ and CO_2 in the range 10–25% during the chilled shelf-life of the product, it is desirable to introduce the highest level of O_2 (balance, N_2) possible just prior to fresh prepared produce package sealing. Generally, it is not necessary to introduce any CO_2 in the initial gas mixture since levels of CO_2 will build up rapidly within sealed packages during chilled storage. However, for fresh prepared produce items that have low intrinsic respiration rates packaged in a format with a low produce volume/gas volume ratio, are stored at low chilled temperatures, or have an O_2 emitter/CO_2 adsorber sachet incorporated into the sealed package, then the incorporation of 5–10% CO_2 into the initial gas mixture may be desirable. Based on the results of controlled atmosphere storage experiments, the most effective high O_2 gas mixtures were found to be 80–85% O_2/15–20% CO_2. This had the most noticeable sensory quality and antimicrobial benefits on a range of fresh prepared produce items (Day, 2001a).

The type of MAP machinery used will greatly influence the maximum achievable O_2 level that can be introduced just prior to fresh prepared produce package sealing. Most light prepared salad items are commercially MA packed on vertical form-fill-seal (VFFS) and horizontal form-fill-seal (HFFS) machines (Hartley, 2000). These machines use a gas flushing or air dilution technique to introduce gas in MA pillow-packs just prior to sealing. Since these machines do not use an evacuation step, then c.80% O_2 would be the highest practical level that could be achieved within sealed fresh prepared produce packs by initially flushing with 100% O_2. Higher levels of in-pack O_2 could be achieved by substantially increasing the flow rate of O_2 through the gas flushing lance of these machines, but this is not recommended for economic and safety reasons (BCGA, 1998).

In contrast to VFFS and HFFS machines, thermoform-fill-seal (TFFS), preformed tray and lidding film (PTLF), vacuum chamber (VC) and snorkel type (ST) machines use a compensated vacuum technique to evacuate air and then introduce gas into tray and lidding film and/or flexible MA packs (BCGA,1998). Since these machines use an evacuation step prior to gas (i.e. 100% O_2) introduction, much higher levels of headspace O_2 (85–95%) can be achieved within such sealed fresh prepared produce packs. Also, all compensated vacuum machines (except VC machines) are intrinsically safer for high O_2 MAP applications, compared with gas flushing VFFS and HFFS machines, since O_2 is introduced directly into the MA packs after air evacuation and prior to sealing, and consequently O_2 levels in the air surrounding these machines are not enriched (BCGA, 1998).

10.4.3 Produce volume/gas volume ratio

In order to maintain headspace O_2 levels > 40% and CO_2 levels in the range 10–25% during the chilled shelf-life of the product, it is desirable to minimise the produce volume/gas volume ratio of fresh prepared produce MA packs. This can be achieved by either decreasing the pack fill weight or increasing the pack headspace volume. Decreasing the pack fill weight of fresh prepared produce will have the effect of reducing the overall respiratory load or activity within MA packs and hence the rate of O_2 depletion will be reduced. Increasing the pack headspace volume will have the effect of increasing the reservoir of O_2 for respiratory purposes and hence the rate of O_2 depletion will also be reduced. Consequently, low produce volume/gas volume ratios are conducive to maintaining headspace O_2 levels > 40% and CO_2 levels in the range 10–25%.

The important influence of the produce volume/gas volume ratio, in addition to the intrinsic produce respiration rate and packaging film permeability, is well illustrated by the results from CCFRA's bulk iceberg lettuce trial (Day, 2001a). Depletion of O_2 and elevation of CO_2 levels within the high O_2 MA bulk packs of this trial were very rapid because these packs contained 2kg of fresh prepared iceberg lettuce as opposed to only 200g for retail MA packs. Consequently, the produce volume/gas volume ratio and overall respiratory load were much higher in these MA bulk packs compared with MA retail packs. Also, the iceberg lettuce used for this bulk pack trial was shredded (10mm cut) and hence had a much higher intrinsic respiration rate compared with retail salad cut (40–70 mm) iceberg lettuce. In addition, the thicker (60μm compared with 30μm for retail) and less permeable bulk OPP/LDPE bags exacerbated the depletion of O_2 and elevation of CO_2. Hence, it was not surprising that the achievable shelf-life at 8°C for high O_2 MA bulk packed fresh shredded iceberg lettuce was found to be only two days, even though the shelf-life of equivalent low O_2 MA bulk packed iceberg lettuce was even shorter (Day, 2001a).

It should be appreciated that there are practical and commercial limits to the reduction of produce volume/gas volume ratios for fresh prepared produce MA packs. Obviously, retail consumers will not readily accept MA packs of fresh prepared produce that appear to be underfilled with too much headspace gas. Therefore it is recommended that potential users of high O_2 MAP technology should carry out pre-commercial trials with fresh prepared produce packs having different but practical produce volume/gas volume ratios.

10.4.4 Packaging materials

Based on the results of CCFRA's practical experimental trials, the recommended packaging material for high O_2 MA retail packs of fresh prepared produce is 30μm orientated polypropylene (OPP) with anti-mist coating. It should be noted that initial experimental trials carried out at CCFRA on high O_2 MAP of fresh prepared produce used an O_2 barrier film, i.e. 30μm polyvinylidene chloride (PVDC) coated OPP with anti-mist coating, because it was considered at the time to be important to maintain the highest levels of O_2 within high O_2 MA packs. However, extensive experimental trials on high O_2 MAP of fresh prepared iceberg lettuce using 30μm PVDC coated OPP film clearly demonstrated that excess and potentially damaging levels of CO_2 (30–40%) could be generated within such O_2 barrier film packs, particularly at higher chilled storage temperatures (i.e. 6–8°C). Consequently, 30μm OPP film was used for subsequent high O_2 MAP experimental trials, instead of 30μm PVDC coated OPP film, and for the majority of fresh prepared produce items, was found to have sufficient O_2 barrier properties to maintain high in-pack O_2 levels (>40%) and be sufficiently permeable to ensure that in-pack CO_2 levels did not rise above 25%, after 7–10 days storage at 5–8°C (Day, 2001a).

It should be appreciated that other packaging materials, apart from 30μm OPP, may be suitable for high O_2 MAP of fresh prepared produce (Air Products, 1995; Day and Wiktorowicz, 1999). For example, laminations or extrusions of OPP with low density polyethylene (LDPE), ethylene-vinyl acetate (EVA) or polyvinyl chloride (PVC) or other medium to very high O_2 permeability films, may be more suitable for high O_2 MAP of fresh prepared produce items that have a higher respiration rate than iceberg lettuce. Also, the produce volume/gas volume ratio of different retail MA pack formats (e.g. pillow packs or tray and lidding film systems), the intrinsic fresh prepared produce respiration rate and chilled temperature of storage will influence the selection of the most suitable packaging film for high O_2 MAP applications (Day, 2001a).

It is recommended that potential users of high O_2 MAP for fresh prepared produce should initially carry out pre-commercial shelf-life trials using 30μm OPP with anti-mist coating as the packaging film for flexible pillow packs or as a tray lidding film. Regular gas analyses of the in-pack atmospheres during chilled storage will reveal whether the packaging film is not permeable enough (resulting in build-up of excess levels of CO_2 to >25%) or too permeable (resulting in depletion of O_2 to < 40% and slow build-up of CO_2 to <10%). If the in-pack O_2 levels fall < 40% and CO_2 levels lie outside the range 10–25% by the end of the chilled shelf-life of the product, then adjustments to the produce volume/gas volume ratio, chilled temperature of storage, pack format and/or permeability of the package film will need to be made and further shelf-life trials carried out.

It should also be noted that O_2 barrier films could be used for high O_2 (or low O_2) MAP of fresh prepared produce items if an O_2 emitter/CO_2 adsorber sachet is incorporated into sealed packages. Appropriate transparent O_2 barrier films (with anti-mist coatings) include PVDC coated OPP, and coextrusions or laminations containing ethylene-vinyl acetate (EVOH), polyester (PET), polyamide (nylon) and/or PVDC (Air Products, 1995; Day and Wiktorowicz, 1999).

Whatever packaging material is used for high O_2 MAP applications, all of them must comply with statutory legal requirements. In the UK, these requirements include the Materials and Articles in Contact with Food Regulations 1987, Plastic Materials and Articles in Contact with Food Regulations 1998, Producer Responsibility Obligations (Packaging Waste) Regulations 1997 and Packaging (Essential Requirements) Regulations 1998.

All packaging materials should be purchased to an agreed specification that includes details of technical properties and performance. Quality assurance on all incoming packaging materials should be subject to an agreement between the packaging supplier and user. Each delivery or batch should be given a reference code to identify it in storage and use, and the documentation should allow any batch of packaged product to be correlated with deliveries of respective packaging materials. All packaging materials should be stored off the floor in separate and dry areas of the factory and should be inspected at regular intervals to ensure that they remain in acceptable condition. Authorised procedures and documentation should be established and followed for the issue of packaging materials from store (Day, 1992). Further advice on the technical requirements, properties, performance and handling of packaging materials should be sought from reliable suppliers.

10.4.5 Temperature control

The importance of proper temperature control to retard quality deterioration and assure the microbial safety of fresh prepared produce cannot be overemphasised. For high O_2 MA packed fresh prepared produce, it is recommended that the temperature be maintained below 8°C, and ideally in the range 0–3°C, throughout the entire chill chain.

The important influences of storage temperature and packaging film permeability on the quality of high O_2 MA packed fresh prepared produce can be illustrated by the results from CCFRA's fresh prepared iceberg lettuce trials (Day, 2001a). The results from these trials clearly demonstrated that temperature and packaging film permeability are critical factors in determining the development of O_2 and CO_2 levels within high O_2 MA packs, during chilled storage. Higher temperatures of storage correlate to high respiratory rates and hence greater depletion of O_2 and elevation of CO_2 within sealed high O_2 MA barrier (i.e. 30μm PVCD coated OPP) pillow packs of fresh prepared iceberg lettuce. The most beneficial sensory effects of high O_2 MAP were obtained when the temperature of storage was 3–5°C and the O_2 levels dropped from 70% to 55% and the CO_2 levels reached only 15% after ten days' storage. In contrast, largely negative sensory effects were obtained when an elevated chill temperature of storage regime (8°C) was employed. Under this elevated chilled temperature of storage regime, O_2 levels dropped from 80% to 35–40% whereas CO_2 levels reached 35–40% after ten days' storage. These high levels of generated CO_2 within the high O_2 MA barrier pillow packs of fresh prepared iceberg lettuce were responsible for the undesirable 'CO_2 damage' discoloration observed. Later high O_2 MAP experimental trials used more permeable OPP film whereby high O_2 (> 40%) levels were generally maintained and CO_2 levels did not rise above 25% after 7–10 days' storage at 5°C and 8°C. Under these high O_2 MAP conditions, beneficial sensory effects were observed for the majority of the fresh prepared produce items studied, in comparison with industry standard air and/or low O_2 MAP (Day, 2001a).

10.4.6 Fresh prepared produce applications

High O_2 MAP has been found to have beneficial effects on the sensory quality of the vast majority of fresh prepared produce items studied. Under defined storage and packaging conditions and in comparison with industry-standard air packing and/or low O_2 MAP, high O_2 MAP was found to be effective for extending the achievable shelf-lives of retail packs of fresh prepared iceberg lettuce, sliced mushrooms, potatoes, sliced bananas, little gem lettuce, cos lettuce, baby-leaf spinach, raddichio lettuce, lollo rosso lettuce, flat-leaf parsley, cubed swede, coriander, raspberries and strawberries. In addition, the results from trials carried out prior to September 1997, showed beneficial sensory effects of high O_2 MAP for fresh prepared tomato slices, baton carrots, pineapple cubes, broccoli florets, honeydew melon cubes, sliced mixed peppers and sliced leeks. Also, high O_2 controlled atmospheres were found to extend the shelf-life of table grapes and oranges (Day, 2001a).

It should be noted that in comparison with industry-standard air and/or low O_2 MAP, high O_2 MAP was found not to have beneficial effects on the sensory quality of retail packs of fresh prepared apple slices, curly parsley, red oak leaf lettuce and Galia melon cubes, and bulk packs of shredded iceberg lettuce. However, it is probable that beneficial effects of high O_2 MAP on the above fresh prepared produce items would have been achieved if the chilled storage temperature, high O_2 gas level, packaging film permeability, produce volume/gas volume ratio and/or preparation procedures had been optimised adequately. Consequently, it is recommended that potential users of high O_2 MAP for specific fresh prepared produce items or combinations, carry out pre-commercial optimisation trials by utilising the advice given previously.

10.5 Future trends

High O_2 MAP has become a fertile area of research during the last three years. Partly as a result of the interest stimulated by CCFRA's Club and EU funded novel MAP research, several research studies and reviews have recently appeared in the scientific literature (e.g. Amanatidou, 2001; Gözükara, 2000; Kader and Ben-Yehoshua, 2001; Perez and Sanz, 2001; Wszelaki and Mitcham, 2001; and Jacxsens *et al.*, 2002). These studies have shown some interesting and mainly beneficial effects of high O_2 MAP and pointed in the direction of future research needs. Novel MAP (particularly, high O_2) has the potential to maintain the quality and assure the microbial safety of fresh prepared produce. The commercial implementation and success of this new technology may encourage greater consumption of conveniently packed fresh prepared produce and help towards improving the health and well-being of consumers. The publication of practical guidance on high O_2 MAP and non-sulphite dipping has already facilitated commercial exploitation of this new technology (Day, 2001a).

Arun Foods Limited (Littlehampton, West Sussex, UK) has produced a wide range of salads and stir-frys for the commercial retail market using high O_2 MAP technology (Day, 2002). These high O_2 MA packed products have been presented in a tray and lidding film format and were assigned a chilled shelf-life of 7–8 days in comparison with only 3–4 days in control air packs (Dr Steve

Yeo, Arun Foods Limited, personal communication, June 2002). A soft fruit supplier in Belgium is also using high O_2 MAP for extending the chilled shelf-life of its product range (Dr Frank Devlieghere, Universiteit Gent, Belgium, personal communication, June, 2002). In addition, the author is aware of several other companies who are actively trialling high O_2 MAP for fresh prepared produce and chilled ready meal applications.

With specific regard to the high O_2 MAP of fresh prepared produce, the following future research directions are suggested:

- Further investigate the potential applications of an innovative dual-action O_2 emitter/CO_2 scavenger active packaging sachet that has been developed by Standa Industrie (Caen, France) and marketed by EMCO Packaging Systems (Worth, Kent, UK). Initial trials carried out by CCFRA and LinPac Plastics Limited (Pontefract, Yorkshire, UK) in association with several soft fruit suppliers have clearly demonstrated the shelf-life extending potential of this active packaging device (McGrath, 2000). This O_2 emitter/CO_2 scavenger sachet enables high O_2 levels to be maintained within high O_2 MA packs of respiring fresh prepared produce whilst simultaneously controlling CO_2 below levels that may cause physiological damage to produce. Also, the inclusion of this sachet within high O_2 MA packs of fresh prepared produce that have a high intrinsic respiration rate and/or produce volume/gas volume ratio will prevent excessive depletion of in-pack O_2 levels and build-up of in-pack CO_2 levels. In addition, this sachet could also be utilised in low O_2 MA packs of fresh prepared produce to prevent the development of undesirable anaerobic conditions during chilled storage.
- Thoroughly investigate the potential synergy of high O_2 MAP and other active packaging devices (e.g. moisture absorbers, ethylene scavengers and antimicrobial films) and suitable edible coatings and films (Day, 1994; Baldwin *et al.*, 1995; Nussinovitch and Lurie, 1995; Rooney, 1999). Selection criteria of promising active packaging devices and edible coatings and films should be based on their technical efficacy, cost, regulatory status and consumer acceptability (Day, 2000).
- Carry out further underpinning research investigations on the effects of high O_2 MAP on the various spoilage and pathogenic microorganisms associated with fresh prepared produce items. Also, further research is merited on the effects of high O_2 MAP on the beneficial nutritional components present in fresh produce and on the complex biochemical reactions and physiological processes that occur during storage.
- Establish optimal high O_2 MAP applications for extending the quality shelf-life and assuring the microbial safety of further fresh prepared produce items and combination food products which consist of respiring produce and non-respiring food items (e.g. ready meals, pizzas, kebabs, etc.). Initial trials carried out by CCFRA have already clearly demonstrated that high O_2 MAP is capable of extending the achievable shelf-life of several chilled ready meals, in comparison with CO_2/N_2 MAP and industry-standard air packing (Day, 2001b).

With regard to more general aspects of fresh prepared produce, the following knowledge gaps and suggested research directions are highlighted, in order to assist researchers in the future.

- Provision of packaging film permeability data on commercial laminations and coextrusions at realistic chilled temperatures (0–10ºC) and relative humidities (85–95%). At the present time, virtually all gas permeability data is quoted for single films at unrealistic storage temperatures and relative humidities (e.g. 23ºC and 0% RH).
- Provision of extensive respiration rate data on a wide variety of fresh prepared produce items at different chilled temperatures and under various gaseous storage conditions. At the present time, most respiration rate data available is for whole produce items stored in air.
- Provision of data on the physiological tolerance of fresh prepared produce items to low (and possibly high) O_2 levels and elevated CO_2 levels. Currently, extensive data is available on the tolerance of whole produce items to low O_2 and high CO_2 levels (Kader *et al.*, 1989) but there is a dearth of information on the tolerance of fresh prepared produce items to varying gaseous levels.
- Provision of information on the residual effects of MAP on individual fresh prepared produce items after subsequent pack opening and storage in air.
- Thoroughly investigate an integrated approach to minimal processing techniques, which cover the entire chain 'from farm to fork', so as to maintain the quality and assure the microbial safety of fresh prepared produce (Ahvenainen, 1996).
- Carry out further investigations on new and innovative natural preservatives, such as those produced by lactic acid bacteria and those derived from herbs and spices (Kets, 1999).
- Devise improved washing and decontamination procedures for fresh prepared produce that are based on safe non-chlorine alternatives.
- Develop peeling and cutting machinery that can process fresh produce more gently and hence extend the quality shelf-life of fresh prepared produce.
- Devote more resources into refrigeration equipment, design and logistics so that optimal storage temperatures for fresh prepared produce can be maintained throughout the entire chill chain.

10.6 References

AHVENAINEN, R. (1996) New approaches in improving the shelf-life of minimally processed fruit and vegetables. *Trends in Food Science and Technology* 7 (6), 179–87.

AIR PRODUCTS (1995) *The Freshline® guide to modified atmosphere packaging (MAP)*. Air Products Plc, Basingstoke, Hants., UK, pp 1–66.

AMANATIDOU, A. (2001) High oxygen as an additional factor in food preservation. Ph.D. Thesis, Wageningen University, The Netherlands.

BALDWIN, E.A., NISPEROS-CARRIEDO, M.O. and BAKER, R.A. (1995) Use of edible coatings to preserve quality of lightly (and slightly) processed products. *Critical Reviews in Food Science and Nutrition* 35, 509–24.

BCGA (1998) The safe application of oxygen enriched atmospheres when packaging food. *British Compressed Gases Association Guidance Note GD5*, BCGA, Eastleigh, Hants., UK.

BETTS, G.D. (1996) *A code of practice for the manufacture of vacuum and modified atmosphere packaged chilled foods*. Guideline No. 11, CCFRA, Chipping Campden, Glos., UK.

BRODY, A.L. and THALER, M.C. (1996) Argon and other noble gases to enhance modified atmosphere food processing and packaging. *Proceedings of IoPP conference on 'Advanced technology of packaging'*, Chicago, Illinois, USA, 17th November.

DAY, B.P.F. (1992) Guidelines for the good manufacturing and handling of modified atmosphere packed food products. *Technical Manual No. 34*, CCFRA, Chipping Campden, Glos., UK.

DAY, B.P.F. (1994) Modified atmosphere packaging and active packaging of fruits and vegetables. In: *Minimal Processing of Foods, VTT Symposium Series 142*, VTT, Espoo, Finland, pp 173–207.

DAY, B.P.F. (1998) Novel MAP – a brand new approach. *Food Manufacture* 73 (11), 22–24.

DAY, B.P.F. (2000) Consumer acceptability of active and intelligent packaging. *Proceedings of the Conference on 'Active and Intelligent Packaging: ideas for tomorrow or solutions for today'*, TNO Nutrition and Food Research, Zeist, The Netherlands.

DAY, B.P.F. (2001a) *Fresh prepared produce: GMP for high oxygen MAP and non-sulphite dipping. Guideline No. 31*, CCFRA, Chipping Campden, Glos., UK.

DAY, B.P.F. (2001b) *Novel high oxygen MAP for chilled combination food products. R&D Report No. 125*, CCFRA, Chipping Campden, Glos., UK.

DAY, B.P.F. (2002) Industry guidelines for high oxygen MAP of fresh prepared produce. *Proceedings of the Postharvest Unlimited Conference*, Universiteit Leuven, Belgium, 12–14 June.

DAY, B.P.F. and WIKTOROWICZ, R. (1999) MAP goes on-line. *Food Manufacture* 74 (6), 40–1.

FATH, D. and SOUDAIN, P. (1992) Method for the preservation of fresh vegetables. US Patent No. 5128160.

GONZALEZ RONCERO, M.I. and DAY, B.P.F. (1998) The effects of novel MAP on fresh prepared produce microbial growth. *Proceedings of the Cost 915 Conference*, Ciudad Universitaria, Madrid, Spain, 15–16th October.

GÖZÜKARA, Y. (2000). The effect of oxygen and carbon dioxide atmospheres on the quality of packaged fresh-cut lettuce. MSc. thesis, University of Melbourne, Werribee, Victoria, Australia.

HARTLEY, D.R. (2000) The product design perspective on fresh produce packaging. *Postharvest News and Information* 11 (3), 35N–38N.

JACXSENS, L, DEVLIEGHERE, F., VAN DER STEEN, C. and DEBEVERE, J. (2002). Effect of high oxygen modified atmosphere packaging on microbial growth and sensorial qualities of fresh-cut produce. *International J. Food Micro.* 71 (2), 197–210.

KADER, A.A. and BEN-YEHOSHUA, S. (2001). Effects of superatmospheric oxygen levels on postharvest physiology and quality of fresh fruits and vegetables. *Postharvest Biology and Technology* 20 (1), 1–13.

KADER, A.A., ZAGORY, D. and KERBEL, E.L. (1989) Modified atmosphere packaging of fruits and vegetables. *Critical Reviews in Food Science and Nutrition* 28 (1), 1–30.

KETS, E.P.W. (1999) Applications of natural anti-microbial compounds. In: *Proceedings of the International Conference on 'Fresh-cut Produce'*, Campden & Chorleywood Food Research Association, Chipping Campden, Glos., UK.

MCEVILEY, A.J., IYENGAR, R. and OTWELL, W.S.(1992) Inhibition of enzymatic brewing in foods and beverages. *Critical Reviews in Food Science and Nutrition* 32 (3), 253–73.

MCGRATH, P. (2000) Smart fruit packaging. *Grower* 133 (22), 15–16.

NUSSINOVITCH, A. and LURIE, S. (1995) Edible coatings for fruits and vegetables. *Postharvest News and Information* 6(4), 53N–57N.

O'BEIRNE, D. (1999) Modified atmosphere packed vegetables and fruit – an overview. In: *Proceedings of the International Conference on 'Fresh-cut Produce'*, Campden & Chorleywood Food Research Association, Chipping Campden, Glos., UK.

PEREZ, A.G. and SANZ, C. (2001). Effect of high oxygen and high carbon dioxide atmospheres on strawberry flavor and other quality traits. *J. Agric. Fd. Chem.* 49 (5), 2370–5.

ROONEY, M. (1999) Active and intelligent packaging of fruit and vegetables. In: *Proceedings of the International Conference on 'Fresh-cut Produce'*, Campden & Chorleywood Food Research Association, Chipping Campden, Glos., UK.

SAPERS, G.M. (1993) Browning of foods: control by sulfites, oxidants and other means. *Food Technology* 47 (10), 75–84.

SPENCER, K. (1999) Fresh-cut produce applications of noble gases. In: *Proceedings of the International Conference on 'Fresh-cut Produce'*, Campden & Chorleywood Food Research Association, Chipping Campden, Glos., UK.

THOM, S.R. and MARQUIS, R.E. (1984) Microbial growth modification by compressed gases and hydrostatic pressure. *Applied Environmental Microbiology* 47 (4), 780.

WSZELAKI, A.L. and MITCHAM, E.J. (2001). Effects of superatmospheric oxygen on strawberry fruit quality and decay. *Postharvest Biology and Technology* 20 (2), 125–33.

10.7 Acknowledgements

CCFRA gratefully acknowledges the financial support of the EU FAIR Programme and industrial Club Members for the work described in this chapter. The research contributions of CCFRA's EU FAIR partners (ATO-DLO, The Netherlands; SIK, Sweden; VTT, Finland; University of Limerick, Ireland; and INN, Italy) are also gratefully acknowledged.

11

MAP, product safety and nutritional quality

F. Devlieghere and J. Debevere, Ghent University, Belgium and M I Gil, CEBAS-CSIC, Spain

11.1 Introduction

Modified atmosphere packaging (MAP) may be defined as 'the enclosure of food products in gas-barrier materials, in which the gaseous environment has been changed' (Young *et al*., 1988). Because of its substantial shelf-life extending effect, MAP has been one of the most significant and innovative growth areas in retail food packaging. The potential advantages and disadvantages of MAP have been presented by Farber (1991), Parry (1993) and Davies (1995).

Whilst there is considerable information available regarding suitable gas mixtures for different food products, there is still a lack of scientific detail regarding many aspects relating to MAP. These include:

- mechanism of action of carbon dioxide (CO_2) on microorganisms
- safety of MAP packaged food products
- effect of MAP on the nutritional quality of packaged food products.

Current research and gaps in knowledge are discussed in the following sections.

11.2 Carbon dioxide as an antimicrobial gas

The gases that are applied in MAP today are basically O_2, CO_2 and N_2. The last has no specific preservative effect but functions mainly as a filler gas to avoid the collapse that takes place when CO_2 dissolves in the food product. CO_2, because of its antimicrobial activity, is the most important component in applied gas mixtures. When CO_2 is introduced into the package, it is partly dissolved in the water phase and the fat phase of the food. This results, after equilibrium, in a certain concentration of dissolved CO_2 ($[CO_2]_{diss}$) in the water phase of the product. Devlieghere *et al.* (1998) have demonstrated that the growth inhibition of microorganisms in modified atmospheres is determined by the concentration of dissolved CO_2 in the water phase.

The effect of the gaseous environment on microorganisms in foods is not as well understood by microbiologists and food technologists as are other external factors, such as pH and a_w. Despite numerous reports of the effects of CO_2 on microbial growth and metabolism, the 'mechanism' of CO_2 inhibition still remains unclear (Dixon and Kell, 1989; Day, 2000). The question of whether any specific metabolic pathway or cellular activity is critically sensitive to CO_2 inhibition has been examined in several studies. The different proposed mechanisms of action are:

1. Lowering the pH of the food.
2. Cellular penetration followed by a decrease in the cytoplasmic pH of the cell.
3. Specific actions on cytoplasmic enzymes.
4. Specific actions on biological membranes.

When gaseous CO_2 is applied to a biological tissue, it first dissolves in the liquid phase, where hydration and dissociation lead to a rapid pH decrease in the tissue. This drop in pH, which depends on the buffering capacity of the medium (Dixon and Kell, 1989), is not large in food products. In fact, the pH drop in cooked meat products only amounted to 0.3 pH units when 80% of CO_2 was applied in the gas phase with a gas/product volume ration of 4:1 (Devlieghere *et al.*, 2000b). Several studies have proved that the observed inhibitory effects of CO_2 could not solely be explained by the acidification of the substrate (Becker, 1933; Coyne, 1933).

Many researchers have documented the rapidity with which CO_2 in solution penetrates into the cell. Krogh (1919) discovered that this rate is 30 times faster than for oxygen (O_2) under most circumstances. Wolfe (1980) suggested the inhibitory effects of CO_2 are the result of internal acidification of the cytoplasm. Eklund (1984) supported this idea by pointing out that the growth inhibition of four bacteria obtained with CO_2 had the same general form as that obtained with weak organic acids (chemical preservatives), such as sorbic and benzoic acid. Tan and Gill (1982) also found that the intracellular pH of *Pseudomonas fluorescens* fell by approximately 0.03 units for each 1 mM rise in extracellular CO_2 concentration.

CO_2 may also exert its influence upon a cell by affecting the rate at which particular enzymatic reactions proceed. One way this may be brought about is to cause an alteration in the production of a specific enzyme, or enzymes, via induction or repression of enzyme synthesis (Dixon, 1988; Dixon and Kell, 1989; Jones, 1989). It was also suggested (Jones and Greenfield, 1982; Dixon and Kell, 1989) that the primary sites where CO_2 exerts its effects are the enzymatic carboxylation and decarboxylation reactions, although inhibition of other enzymes has also been reported (Jones and Greenfield, 1982).

Another possible factor contributing to the growth-inhibitory effect of CO_2 could be an alteration of the membrane properties (Daniels *et al.*, 1985; Dixon and Kell, 1989). It was suggested that CO_2 interacts with lipids in the cell membrane, decreasing the ability of the cell wall to uptake various ions. Moreover, perturbations in membrane fluidity, caused by the disordering of the lipid bilayer, are postulated to alter the function of membrane proteins (Chin *et al.*, 1976; Roth, 1980).

Studies examining the effect of a CO_2 enriched atmosphere on the growth of microorganisms are often difficult to compare because of the lack of information regarding the packaging configurations applied. The gas/product volume ratio and the permeability of the applied film for O_2 and CO_2 will influence the amount of dissolved CO_2 and thus the microbial inhibition of the atmosphere. For this reason, the concentration of dissolved CO_2 in the aqueous phase of the food should always be measured and mentioned in publications concerning MAP (Devlieghere *et al.*, 1998).

Only a few publications deal with the effect of MAP on specific spoilage microorganisms. Gill and Tan (1980) compared the effect of CO_2 on the growth of some fresh meat spoilage bacteria at 30ºC. Molin (1983) determined the resistance to CO_2 of several food spoilage bacteria. Boskou and Debevere (1997; 1998) investigated the effect of CO_2 on the growth and trimethylamine production of *Shewanella putrefaciens* in marine fish, and Devlieghere and Debevere (2000) compared the sensitivity for dissolved CO_2 of different spoilage bacteria at 7ºC. In general, Gram-negative microorganisms such as *Pseudomonas, Shewanella* and *Aeromonas* are very sensitive to CO_2. Gram-positive bacteria show less sensitivity and lactic acid bacteria are the most resistant. Most yeasts and moulds are also sensitive to CO_2. The effect of CO_2 on psychrotrophic food pathogens is discussed in Section 11.3.

11.3 The microbial safety of MAP: *Clostridium botulinum* and *Listeria monocytogenes*

Modified atmospheres containing CO_2 are effective in extending the shelf-life of many food products. However, one major concern is the inhibition of normal aerobic spoilage bacteria and the possible growth of psychrotrophic food pathogens, which may result in the food becoming unsafe for consumption before it appears to be organoleptically unacceptable. Most of the pathogenic bacteria can be inhibited by low temperatures (<7ºC). At these conditions, only psychrotrophic pathogens can proliferate. The effect of CO_2 on the different psychrotrophic foodborne pathogens is described below.

A particular concern is the possibility that psychrotrophic, non-proteolytic strains of *C. botulinum* types B, E and F are able to grow and produce toxins under MAP conditions. Little is known about the effects of modified atmosphere storage conditions on toxin production by *C. botulinum*. The possibility of inhibiting *C. botulinum* by incorporating low levels of O_2 in the package does not appear to be feasible. Miller (1988, cited by Connor *et al.*, 1989) reported that psychrotrophic strains of *C. botulinum* are able to produce toxins in an environment with up to 10% O_2. Toxin production by *C. botulinum* type E, prior to spoilage, has been described in three types of fish, at O_2 levels of 2% and 4% (O'Connor-Shaw and Reyes, 2000). Dufresne *et al.* (2000) also proposed that additional barriers, other than headspace O_2 and film, need to be considered to ensure the safety of MAP trout fillets, particularly at moderate temperature abuse conditions.

The probability of one spore of non-proteolytic *C. botulinum* (types B, E and F) being toxicogenic in rock fish was outlined in a report by Ikawa and Genigeorgis (1987). The results showed that the toxigenicity was significantly affected ($P<0.005$) by temperature and storage time, but not by the used modified atmosphere (vacuum, 100% CO_2, or 70% CO_2/30% air). In Tilapia fillets, a modified atmosphere (75% CO_2/25% N_2), at 8°C, delayed toxin formation by *C. botulinum* type E, from 17 to 40 days, when compared to vacuum packaged fillets (Reddy *et al.*, 1996). Similar inhibiting effects were recorded for salmon fillets and catfish fillets, at 4°C (Reddy *et al.*, 1997a and 1997b). Toxin production from non-proteolytic *C. botulinum* type B spores was also retarded by a CO_2 enriched atmosphere (30% CO_2/70% N_2) in cooked turkey at 4°C but not at 10°C nor at 15°C (Lawlor *et al.*, 2000). Recent results in a study by Gibson *et al.* (2000) also indicated that 100% CO_2 slows the growth rate of *C. botulinum*, and that this inhibitory effect is further enhanced with appropriate NaC1 concentrations and chilled temperatures.

Listeria monocytogenes is considered a psychrotrophic foodborne pathogen. Growth is possible at 1°C (Varnam and Evans, 1991) and has even been reported at temperatures as low as −1.5C (Hudson *et al.*, 1994). The growth of *L. monocytogenes* in food products, packaged under modified atmospheres, has been the focus of several, although in some cases contradicting, studies (Garcia de Fernando *et al.*, 1995). In general, *L. monocytogenes* is not greatly inhibited by CO_2 enriched atmospheres (Zhao *et al.*, 1992) although when combined with other factors such as low temperature, decreased water activity and the addition of Na lactate the inhibiting effect of CO_2 is significant (Devlieghere *et al.*, 2001). *Listeria* growth in anaerobic CO_2 enriched atmosphere has been demonstrated in lamb in an atmosphere of 50:50 CO_2/N_2, at 5°C (Nychas, 1994); in frankfurter type sausages in atmospheres of distinct proportions of CO_2/N_2, at 4, 7 and 10°C (Krämer and Baumgart, 1992) and in pork in an atmosphere of 40:60 CO_2/N_2, at 4°C (Manu-Tawiah *et al.*, 1993). However, other authors have not detected growth in chicken anaerobically packaged in 30:70 CO_2/N_2, at 6°C (Hart *et al.*, 1991); in 75:25 CO_2/N_2 at 4°C (Wimpfheimer *et al.*, 1990) and at 4°C in 100% CO_2 in raw minced meat (Franco-Abuin *et al.*, 1997) or in buffered tryptose broth (Szabo and Cahill, 1998). Several investigations demonstrated possible growth of *L. monocytogenes* on modified atmosphere packaged fresh-cut vegetables, although the results depended very much on the type of vegetables and the storage temperature (Berrang *et al.*, 1989a; Beuchat and Brackett, 1990; Omary *et al.*, 1993; Carlin *et al.*, 1995;

Carlin *et al.*, 1996a and 1996b; Zhang and Farber, 1996; Juneja *et al.*, 1998; Bennick *et al.*, 1999; Jacxsens *et al.*, 1999; Liao and Sapers, 1999; Thomas *et al.*, 1999; Castillejo-Rodriguez *et al.*, 2000).

There is no agreement about the effect of incorporating O_2 in the atmosphere on the antimicrobial activity of CO_2 on *L. monocytogenes* (Garcia de Fernando *et al.*, 1995). However, this effect could be very important in practice, as the existence of residual O_2 levels after packaging, and the diffusion of O_2 through the packaging film, can result in substantial O_2 levels during the storage of industrially 'anaerobically' modified atmosphere packaged food products. Most publications suggest there is a decrease in the inhibitory effect of CO_2 on *L. monocytogenes* when O_2 is incorporated into the atmosphere. Experiments on raw chicken showed *L. monocytogenes* failed to grow at 4, 10 and 27°C, in an anaerobic atmosphere containing 75% CO_2 and 25% N_2 (Wimpfheimer *et al.*, 1990). However, an aerobic atmosphere containing 72.5% CO_2, 22.5% N_2, and 5% O_2 did not inhibit the growth of *L. monocytogenes*, even at 4°C. *L. monocytogenes* was also only minimally inhibited on chicken legs, in an atmosphere containing 10% O_2 and 90% CO_2 (Zeitoun and Debevere, 1991). There was no significant difference in the inhibitory effect of CO_2, between the range of 0% and 50%, when 1.5% O_2, or 21% O_2 was present in the atmosphere of gas packaged brain heart infusion agar plates (Bennik *et al.* 1995). When *L. monocytogenes* was cultured in buffered nutrient broth, at 7.5°C, in atmospheres containing 30% CO_2, with four different O_2 concentrations (0, 10, 20 and 40%), the results showed that bacterial growth increased with the increasing O_2 concentrations (Hendricks and Hotchkiss, 1997).

11.4 The microbial safety of MAP: *Yersinia enterocolitica* and *Aeromonas* spp.

Yersinia enterocolitica is generally regarded as one of the most psychrotrophic foodborne pathogens. Growth of *Y. enterocolitica* was reported in vacuum packaged lamb at 0°C (Doherty *et al.*, 1995; Sheridan and Doherty, 1994; Sheridan *et al.*, 1992), beef at −2°C (Gill and Reichel, 1989), pork at 4°C (Bodnaruk and Draughon, 1998; Manu-Tawiah *et al.*, 1993), fresh chicken breasts (Özbas *et al.*, 1997) and roast beef at 3°C but not at −1.5°C (Hudson *et al.*, 1994).

CO_2 retards the growth of *Y. enterocolitica* at refrigerated temperatures. The effect of CO_2 on the growth of *Y. enterocolitica* has been described by several authors. Some of the results are shown in Table 11.1. Oxygen also seems to play an inhibiting role on the growth of *Y. enterocolitica* (Garcia de Fernando *et al.*, 1995). To ensure total inhibition of *Y. enterocolitica* in O_2 poor atmospheres and at realistic temperatures throughout the cooling chain, high CO_2 concentrations in the headspace are necessary.

Aeromonas species are able to multiply in food products stored in refrigerated conditions. Growth of *A. hydrophila* has been detected at low temperatures in a

Table 11.1 Growth of *Yersina enterocolitica* in different atmospheres

Product type	pH	Temp. (°C)	Storage time (days)	Atmosphere (%O_2/ CO_2/N_2)	Increase (log cfu/g)	Reference
Beef	>6.0	−2	126	0/100/0	0	Gill and Reichel (1989)
			63	vacuum	2.4	
		0	98	0/100/0	0	
			49	vacuum	4.1	
		2	42	0/100/0	0	
			35	vacuum	5.1	
		5	35	0/100/0	1.9	
			17	vacuum	5.5	
		10	10	0/100/0	3.4	
			5	vacuum	4.0	
Sliced roast beef	6.1	−1.5	112	0/100/0	0	Hudson *et al.* (1994)
			56	vacuum	4.2	
		3	70	0/100/0	3.8	
			21	vacuum	4.7	
Pork	5.57 (normal)	4	30	0/100/0	0	Bodnaruk and Draughon (1998)
			25	vacuum	1.7	
	6.21 (high)		30	0/100/0	1.7	
			25	vacuum	2.6	
Pork chops	6.0	4	35	0/20/80	4.1	Manu-Tawiah *et al.* (1993)
			35	0/40/60	4.0	
			35	10/40/50	4.0	
			35	vacuum	4.1	
Lamb	5.4–5.8	0	28	80/20/0	1.2	Doherty *et al.* (1995)
			28	0/50/50	3.9	
			28	0/100/0	1.6	
			28	vacuum	5.9	
			28	80/20/0	6.8	
			28	0/50/50	8.5	
			28	0/100/0	5.6	
			28	vacuum	8.1	

variety of vacuum packaged fresh products, such as chicken breasts at 3°C (Özbas *et al.*, 1996), lamb at 0°C under high pH conditions (Doherty *et al.*, 1996), and at −2°C (Gill and Reichel, 1989), and in sliced roast beef at 1.5°C (Hudson *et al.*, 1994). Devlieghere *et al.* (2000a) developed a model, predicting the influence of temperature and CO_2 on the growth of *A. hydrophila*. Proliferation of *A. hydrophila* is greatly affected by CO_2 enriched atmospheres. Some reports regarding the effect of CO_2 on the growth of *A. hydrophila* on meat are summarised in Table 11.2.

In a study by Berrang *et al.* (1989b), regarding controlled atmosphere storage of broccoli, cauliflower and asparagus stored at 4°C and 15°C, fast proliferation of *A. hydrophila* was observed at both temperatures, but growth was not significantly affected by gas atmosphere. Garcia-Gimeno *et al.* (1996) published the survival of *A. hydrophila* on mixed vegetable salads (lettuce, red cabbage and carrots)

Table 11.2 Growth of *Aeromonas hydrophila* in different atmospheres

Product type	pH	Temp. (°C)	Storage time (days)	Atmosphere ($\%O_2/CO_2/N_2$)	Increase (log cfu/g)	Reference
Beef	>6.0	−2	126	0/100/0	0	Gill and Reichel (1989)
			63	vacuum	1.0	
		0	98	0/100/0	0	
			49	vacuum	3.1	
		2	42	0/100/0	0	
			35	vacuum	3.0	
		5	35	0/100/0	0	
			17	vacuum	3.0	
		10	10	0/100/0	3.8	
			5	vacuum	5.8	
Sliced roast beef	6.1	−1.5	112	0/100/0	0	Hudson *et al.* (1994)
			56	vacuum	4.3	
		3	70	0/100/0	3.1	
			21	vacuum	4.6	
Lamb	5.4–5.8	0	45	80/20/0	0	Doherty *et al.* (1996)
			45	0/50/50	0	
			45	0/100/0	0	
			45	vacuum	0	
		5	45	80/20/0	0	
			45	0/50/50	0	
			45	0/100/0	0	
			45	vacuum	0	
Lamb	>6.0	0	42	80/20/0	0	Doherty *et al.* (1996)
			42	0/50/50	0	
			42	0/100/0	0	
			42	vacuum	4.1	
		5	42	80/20/0	4.2	
			42	0/50/50	1.7	
			42	0/100/0	0	
			42	vacuum	4.0	

packaged under MA (initial 10% of O_2-10% CO_2, after 48h 0% O_2-18% CO_2) and stored at 4°C while at 15°C a fast growth was noticed (5 log units in 24h). The combination of high CO_2 concentration and low temperature was revealed as responsible for the inhibition of growth. Bennik *et al.* (1995) concluded from their solid-surface model that at MA-conditions, generally applied for minimally processed vegetables (1–5% O_2 and 5–10% CO_2), growth of *A. hydrophila* is possible. Growth was virtually the same under 1.5% and 21% O_2. The behaviour of a cocktail of *A. caviae* (HG4) and *A. bestiarum* (HG2) in air or in low O_2-low CO_2 atmosphere was investigated in fresh-cut vegetables: no difference between both atmospheres was observed on grated carrots, a decreased growth on shredded Belgian endive and Brussels sprouts in MA but an increased growth on shredded iceberg lettuce in MA storage (Jacxsens *et al.*, 1999).

11.5 The effect of MAP on the nutritional quality of non-respiring food products

By using modified atmosphere packaging, the shelf-life of the packaged products can be extended by 50–200%, however, questions could arise regarding the nutritional consequences of MAP on the packaged food products. This section will discuss the effect of MAP on the nutritional quality of non-respiring food products while the effect of MAP on the nutritional value of respiring products, such as fresh fruits and vegetables, will be discussed in detail in the following sections.

Very little information is available about the influence of MAP on the nutritional quality of non-respiring food products. In most cases, for packaging non-respiring food products, oxygen is excluded from the atmosphere and therefore one should expect a retardation of oxidative degradation reactions. Moreover, modified atmosphere packaged food products should be stored under refrigeration to allow CO_2 to dissolve and perform its antimicrobial action. At these chilled conditions, chemical degradation reactions have only a limited importance.

No information is available regarding the nutritional consequences of enriched oxygen concentrations in modified atmospheres which can be applied for packaging fresh meat and marine fish. Some oxidative reactions can occur with nutritionally important compounds such as vitamins and polyunsaturated fatty acids. However, no quantitative information is available about these degradation reactions in products packaged in O_2 enriched atmospheres.

11.6 The effect of MAP on the nutritional quality of fresh fruits and vegetables: vitamin C and carotenoids

During the last few years many studies have demonstrated that fruit and vegetables are rich sources of micronutrients and dietary fibre. They also contain an immense variety of biologically active secondary metabolites that provide the plant with colour, flavour and sometimes antinutritional or toxic properties (Johnson *et al.*, 1994). Among the most important classes of such substances are vitamin C, carotenoids, folates, flavonoids and more complex phenolics, saponins, phytosterols, glycoalkaloids and the glucosinolates.

The nutrient content of fruit and vegetables can be influenced by various factors such as genetic and agronomic factors, maturity and harvesting methods, and postharvest handling procedures. There are some postharvest treatments which undoubtedly improve food quality by inhibiting the action of oxidative enzymes and slowing down deleterious processes. Storage of fresh fruits and vegetables within the optimum range of low O_2 and/or elevated CO_2 atmospheres for each commodity reduces their respiration and C_2H_4 production rates (Kader, 1986; Kader, 1997). Optimum CA retards loss of chlorophyll, biosynthesis of carotenoids and anthocyanins, and biosyntheses and oxidation of phenolic compounds.

In general, CA influences flavour quality by reducing loss of acidity, starch to sugar conversion, and biosynthesis of aroma volatiles, especially esters. Retention of ascorbic acid and other vitamins results in better nutritional quality, including antioxidant activity, of fruits and vegetables when kept in their optimum CA (Kader, 2001). However, little information is available on the effectiveness of controlled atmospheres or modified atmosphere packaging (CA/MAP) on nutrient retention during storage. The influence of CA/MAP on the antioxidant constituents related to nutritional quality of fruits and vegetables, including vitamin C, carotenoids, phenolic compounds, as well as glucosinolates will be reviewed here.

11.6.1 Vitamin C

Vitamin C is one of the most important vitamins in fruits and vegetables for human nutrition. More than 90% of the vitamin C in human diets is supplied by the intake of fresh fruits and vegetables. Vitamin C is required for the prevention of scurvy and maintenance of healthy skin, gums and blood vessels. Vitamin C, as an antioxidant, reduces the risk of arteriosclerosis, cardiovascular diseases and some forms of cancer (Simon, 1992). Ascorbic oxidase has been proposed as the major enzyme responsible for enzymatic degradation of L-ascorbic acid (AA). The oxidation of AA, the active form of vitamin C, to dehydroascorbic acid (DHA) does not result in loss of biological activity since DHA is readily re-converted to L-AA *in vivo*. However, DHA is less stable than AA and may be hydrolysed to 2,3-diketogulonic acid, which does not have physiological activity (Klein, 1987) and it has therefore been suggested that measurements of vitamin C in fruits and vegetables in relation to their nutritional value should include both AA and DHA.

The vulnerability of different fruits and vegetables to oxidative loss of AA varies greatly, as indeed do general quality changes. Low pH fruits (citrus fruits) are relatively stable, whereas soft fruits (strawberries, raspberries) undergo more rapid changes. Leafy vegetables (e.g. spinach) are very vulnerable to spoilage and AA loss, whereas root vegetables (e.g. potatoes) retain quality and AA for many months (Davey *et al.*, 2000). Fruits and vegetables undergo changes from the moment of harvest and since L-AA is one of the more reactive compounds it is particularly vulnerable to treatment and storage conditions. In broad terms, the milder the treatment and the lower the temperature the better the retention of vitamin C, but there are several interacting factors that affect AA retention (Davey *et al.*, 2000). The rate of postharvest oxidation of AA in plant tissues has been reported to depend upon several factors such as temperature, water content, storage atmosphere and storage time (Lee and Kader, 2002).

The effect of controlled atmospheres on the ascorbate content of intact fruit has not been extensively studied. The results vary among fruit species and cultivars, but the tendency is for reduced O_2 and/or elevated CO_2 levels to enhance the retention of ascorbate (Weichmann, 1986; Kader *et al.*, 1989). A reduction in temperature and of O_2 concentration in the storage atmosphere have been described as the two treatments which contribute to preserve vitamin C in fruits and vegetables (Watada, 1987). Delaporte (1971) and others observed that loss of AA can be reduced by storing apples in a reduced oxygen atmosphere. However, Haffner *et al.* (1997) have shown that AA levels in various apple cultivars decreased more under ultra low oxygen (ULO) compared to air storage. On the other hand, increasing CO_2 concentration above a certain threshold seems to have an adverse effect on vitamin C content in some fruits and vegetables. It has been reported that the effect of elevated CO_2 level and storage temperature and duration (Weichmann, 1986). Bangerth (1977) observed accelerated AA losses in apples and red currants stored in elevated CO_2 atmospheres. Vitamin C content was reduced by high CO_2 concentrations (10–30% CO_2) in strawberries and blackberries and only a moderate to negligible effect was found for black currants, red currants and raspberries (Agar *et al.*, 1997).

Storage of sweet pepper for six days at 13°C in CO_2 enriched atmospheres resulted in a reduction in AA content (Wang, 1977). Wang (1983) noted that 1% O_2 retarded AA degradation in Chinese cabbage stored for three months at 0°C. He observed that treatments with 10 or 20% CO_2 for five or ten days produced no effect, and 30 or 40% CO_2 increased AA decomposition. Veltman *et al.* (1999) have observed a 60% loss in AA content of 'Conference' pears after storage in 2% O_2 + 10% CO_2. There were no data available to show whether a parallel reduction in O_2 concentration alleviates the negative CO_2 effect. Agar *et al.* (1997) proposed that reducing O_2 concentration in the storage atmosphere in the present of high CO_2 had little effect on the vitamin C preservation. The only beneficial effect of low O_2 alleviating the CO_2 effect could be observed when applying CO_2 concentrations lower than 10%.

In fresh-cut products, high CO_2 concentration in the storage atmosphere has also been described to cause degradation of vitamin C. Thus, concentrations of 5, 10 or 20% CO_2 caused degradation of vitamin C in fresh-cut kiwifruit slices (Agar *et al.*, 1999). Enhanced losses of vitamin C in response to CO_2 higher than 10% may be due to the stimulating effects on oxidation of AA and/or inhibition of DHA reduction to AA (Agar *et al.*, 1999). In addition, vitamin C content decreased in MAP-stored Swiss chard (Gil *et al.*, 1998a) as well as in potato strips (Tudela *et al.*, 2002). In contrast, MAP retarded the conversion of AA to DHA that occurred in air-stored jalapeno pepper rings (Howard *et al.*, 1994; Howard and Hernandez-Brenes 1998). Wright and Kader (1997a) found no significant losses of vitamin C occurred during the post-cutting life of fresh-cut strawberries and persimmons for eight days in CA (2% O_2, air + 12% CO_2, or 2% O_2 + 12% CO_2) at 0°C.

In studies of cut broccoli florets and intact heads of broccoli CA/MAP resulted in greater AA retention and shelf-life extension in contrast to air-stored samples (Barth *et al.*, 1993; Paradis *et al.*, 1996). Retention of AA was found in fresh-cut lettuce packaged with nitrogen (Barry-Ryan and O'Beirne, 1999). They suggest that high levels of CO_2 (30–40%) increased AA losses by conversion into DHA due to availability of oxygen in lettuce (Barry-Ryan and

O'Beirne, 1999). This fact has also been shown in sweet green peppers (Petersen and Berends, 1993). The reduction of AA and the relative increase in DHA could be an indication that high CO_2 stimulates the oxidation of AA, probably by ascorbate peroxidase as in the case of strawberries (Agar *et al.*, 1997) and of spinach (Gil *et al.*, 1999). Mehlhorn (1990) demonstrated an increase in ascorbate peroxidase activity in response to ethylene.

High CO_2 at injurious concentrations for the commodity may reduce AA by increasing ethylene production and therefore the activity of ascorbate peroxidase. Ascorbate oxidase from green zucchini fruit, which catalyses the oxidation of AA to DHA, has been found to be unstable and to lose activity below pH 4 (Maccarrone *et al.*, 1993). This could partially explain the lower DHA content of the strawberries (pH 3.4–3.7) and the higher DHA content of the persimmons (pH 5.4–6.0) (Wright and Kader, 1997a) as well as the tendency of some vegetables at pH near to neutral to lose AA during storage (Gil *et al.*, 1998b).

In conclusion, the loss of vitamin C after harvest can be reduced by storing fruits and vegetables in atmosphere of reduced O_2 and/or up to 10% CO_2 as Lee and Kader (2002) have reported. CA conditions do not have a beneficial effect on vitamin C if high CO_2 concentrations are involved, although the concentrations above which CO_2 affects the loss of AA must be estimated for each commodity (Kader, 2001).

11.6.2 Carotenoids

Carotenoids form one of the more important classes of plant pigments and play a crucial role in defining the quality parameters of fruit and vegetables. Their role in the plant is to act as accessory pigments for light harvesting and in the prevention of photo-oxidative damage, as well as acting as attractants for pollinators. The best documented and established function of some of the carotenoids is their provitamin A activity, especially of β-carotene. A-Carotene and β-crytozanthin also possess provitamin A activity, but to a lesser extent than does β-carotene. Many yellow, orange or red fruit and root vegetables contain large amounts of carotenoids, which accumulate in the chloroplast during ripening or maturation. In some cases, the carotenoids present are simple, e.g. β-carotene in carrot or lycopene in tomato, but in other cases complex mixtures of unusual structures are found, e.g. in *Capsicum*.

Carotenoids are found in membranes, as microcrystals, in association with protcins or in oil droplets. *In vivo*, carotenoids are stabilised by these molecular interactions, that are also important in determining the bioavailability of the carotenoids. Plant materials do not contain vitamin A, but provide carotenoids that are converted to vitamin A after ingestion. Provitamin A carotenoids found in significant quantitites in fruits may have a role in cancer prevention by acting as free radical scavengers (Britton and Hornero-Mendez, 1997). Lycopene, although it has no provitamin A activity, has been identified as a particularly effective quencher of singlet oxygen *in vitro* (Di Mascio *et al.*, 1989) and as an anticarcinogenic (Giovannucci, 1999). Carotenoids are unstable when exposed to acidic pH, oxygen or light (Klein, 1987). The effect of controlled and modified atmospheres on the carotenoid content of intact fruits has not been well studied. Modified atmospheres including either reduced O_2 or elevated CO_2 are generally considered to reduce the loss of provitamin A, but also to inhibit the biosynthesis of carotenoids (Kader *et al.*, 1989). Reducing O_2 to lower concentrations enhanced the retention of carotene in carrots (Weichmann, 1986). The carotene content of leeks was found to be higher after storage in 1% O_2 + 10% CO_2 than after storage in air (Weichmann, 1986).

Few studies on the effect of CA storage on the provitamin A carotenoid content of fresh-cut products have been published. Wright and Kader (1997b) found for sliced peaches and persimmons, that the limit of shelf-life was reached before major losses of carotenoids occurred. Low changes in carotenoids have been observed in minimally processed pumpkin stored for 25 days at 5ºC in MAP (Baskaran *et al.*, 2001). Petrel *et al.* (1998) found no changes on the carotenoid content of ready to eat oranges after 11 days at 4ºC in MAP (19% O_2 + 5% CO_2 and 3% O_2 + 25% CO_2). In addition, the content of β-carotene in broccoli florets increased at the end of CA storage (2% O_2 + 6% CO_2) and remained stable after returning the samples to ambient conditions for 24 h (Paradis *et al*, 1996). Lutein, the major carotenoid in green bean tissue, also showed an accumulation after 13 days of CA storage (1% O_2 + 3% CO_2) and in these conditions retained carotenoids up to 22 days at 8ºC (Cano *et al.*, 1998). However, Sozzi *et al.* (1999) have observed that CA of 3% O_2 and 20% CO_2 both alone and together with ethylene prevented total carotenoid and lycopene biosynthesis on tomato. After exposing the fruits to air, total carotenoids and lycopene increased but were in all cases significantly lower than those which were held in air.

11.7 The effect of MAP on the nutritional quality of fresh fruits and vegetables: phenolic compounds and glucosinolates

11.7.1 Phenolic compounds

There is considerable evidence for the role of antioxidant constituents of fruits and vegetables in the maintenance of health and disease prevention (Ames *et al.*, 1993). Epidemiological studies show that consumption of fruits and vegetables with high phenolic content correlates with reduced cardio- and cerebrovascular diseases and cancer mortality (Hertog *et al.*, 1997). Recent work is also beginning to highlight the relation of flavonoids and other dietary phenolic constituents to these protective effects. They act as antioxidants by virtue of the free radical scavenging properties of their constituent hydroxyl groups (Kanner *et al.*, 1994; Vinson *et al.*, 1995). The biological properties of phenolic compounds are very variable and include anti-platelet action, antioxidant, antiinflamatory, antiumoral and oestrogenic activities, which might suggest their potential in the prevention of coronary heart diseases and cancer (Hertog *et al.*, 1993; Arai *et al.*, 2000).

In the last few years there has been an increasing interest in determining relevant dietary sources of antioxidant phenolics and red fruits such as strawberries, cherries, grapes and pomegranates have received considerable attention due to their antioxidant activity. However, storage under CA/MAP conditions has been focused on keeping the visual properties and few studies have been made on the effect on the nutritional quality. Generally an increase in phenolics is considered a positive attribute and enhances the nutritional value of plant product. However, many secondary metabolites typical of wild species of fruits or vegetables have toxic effects although they are not considered here. In addition, the organoleptic and nutritional characteristics of fruit and vegetables are strongly modified by the appearance of brown pigments. Oxidative browning is mainly due to the enzyme polyphenol oxidase (PPO) which catalyses the hydroxylation of monophenols to *o*-diphenols and, in a second step, the oxidation of colourless *o*-diphenols to highly coloured *o*-quinones (Vámos-Vigyázó, 1981). The *o*-quinones non-enzymatically polymerise and give rise to heterogeneous black, brown or red pigments called melanins decreasing the organoleptic and nutritional qualities (Tomás-Barberán *et al.*, 1997; Tomás-Barberán and Espin, 2002).

Controlled atmospheres and modified atmosphere packaging (MAP) can directly influence the phenolic composition as reflected in the changes observed in anthocyanins. Carbon dioxide-enriched atmospheres (>20%) used to reduce decay and extend the postharvest life of strawberries induced a remarkable decrease in anthocyanin content of internal tissues compared with the external ones (Gil *et al.*, 1997). Holcroft and Kader (1999) related the decrease in strawberry colour under CO_2 atmosphere, with a decrease of important enzyme activity involved in the biosynthesis of anthocyanins, phenylalanine ammonialyase (PAL; EC 4.3.1.5) and glucosyltransferase (GT; EC 2.4.1.9.1). A moderated CO_2 atmosphere (10%) prolongs the storage life and maintains quality and adequate red colour intensity of pomegranate arils (Holcroft *et al.*, 1998). However, the arils of pomegranates stored in air were deeper red than were those of the initial controls and of those stored in a CO_2 enriched atmosphere.

Modified atmospheres can also have a positive effect on phenolic-related quality, as in the case of the prevention of browning of minimally processed lettuce (Saltveit, 1997; Gil *et al.*, 1998b). In addition, modified atmosphere packaging of minimally processed red lettuce (2–3% O_2 + 12–14% CO_2) decreased the content of flavonol and anthocyanins of pigmented lettuce tissues when compared to air storage (Gil *et al.*, 1998b). The increase of soluble phenylpropanoids observed in the midribs of minimally processed red lettuce after storage in air was avoided under MAP. When minimally processed Swiss chard was stored in MAP (7% O_2 + 10% CO_2), no effect was observed on flavonoid content after eight days cold storage when compared to that stored in air (Gil *et al.*, 1998b). In addition, the total flavonoid content of fresh-cut spinach remained quite constant during storage in both air and MAP atmosphere (Gil *et al.*, 1999).

Abnormal browning frequently occurs when fruits are stored in very low oxygen atmospheres. Extended treatment in pure nitrogen enhances the appearance of brown surfaces in fruits, which then rot rapidly when they are returned to air (Macheix *et al.*, 1990). These observations are probably the result of cell disorganisation under anaerobiosis, but may also be related to variations in phenolic metabolism.

There is a decrease in all phenolic compounds (e.g. anthocyanins, flavonols, and caffeoyl tartaric and *p*-coumaroyl tartaric acids) in both skin and pulp of grape berries rapidly brought under anaerobiosis in CO_2 enriched atmosphere (Macheix *et al.*, 1990). Anaerobiosis generally appears to be harmful for the fruit products formed, with the frequent appearance of unwanted browning or loss of anthocyanins. In contrast, this treatment becomes necessary in the case of removal of astringency from persimmon fruit by means of an atmosphere of CO_2 or N_2. These treatments result in the production of acetaldehyde, and deastringency is due to the insolubilisation of kaki-tannin by reaction with the acetaldehyde (Haslam *et al.*, 1992).

11.7.2 Glucosinolates

Brassica vegetables, such as cabbage, Brussels sprouts, broccoli and cauliflower are an important dietary source for a group of secondary plant metabolites known as glucosinolates. The sulphur-containing glucosinolates are present as glucosides and can be hydrolysed by the endogenous plant enzyme myrosinase (thioglucoside glucohydrolase EC 3.2.3.1). Myrosinase and the flucosinolates are physically separated from each other in the plant cell and therefore hydrolysis can only take place when cells are damaged, e.g. by cutting or chewing (Verkerk *et al.*, 2001). The hydrolysis generally results in further breakdown of glucosinolates into isothiocyanates, nitriles, thiocyanates, indoles and oxazolidinethiones. Glucosinolate degradation products contribute to the characteristic flavour and taste of *Brassica* vegetables.

Glucosinolates and their biological effects have been reviewed in detail (Rosa *et al.*, 1997). Indol-3-ylmethylglucosinolates, which occur in appreciable amounts in several *Brassica* vegetables, are of interest for their potential contribution of anticarcinogenic compounds to the diet (Loft *et al.*, 1992) and so broccoli has been associated with a decreased risk of cancer based on several beneficial properties such as the level of vitamin C, fibre and glucosinolates. The glucosinolate content in *Brassica* vegetables can vary depending on the variety, cultivation conditions, harvest time and climate. Storage and processing of the vegetables can also greatly affect the glucosinolate content. Processes such as chopping, cooking and freezing influence the extent of hydrolysis of glucosinolates and the composition of the final products (Verkert *et al.*, 2001).

There are a few reports describing the effects of storage on the glucosinolate content; for instance the storage of white and red cabbage for up to five months at 4°C which does not seem to affect the levels of glucosinolates (Berard and Chong, 1985). However, there is still little information about the influence of

CA/MAP on total or individual glucosinolate content of *Brassica* vegetables but an increase in total glucosinolate content was reported in broccoli florets when stored in air or CA while the absence of O_2 with a 20% CO_2 resulted in total loss (Hansen *et al.*, 1995).

11.8 References

AGAR I T, STREIF J and BANGERTH F (1997), 'Effect of high CO_2 and controlled atmosphere on the ascorbic and dehydroascorbic acid content of some berry fruits', *Postharvest Biol Technol,* **11**, 47–55.

AGAR I T, MASSANTINI R, HESS-PIERCE B and KADER A A (1999), 'Postharvest CO_2 and ethylene production and quality maintenance of fresh-cut kiwifruit slices', *J Food Sci,* **64**, 433–40.

AMES B M, SHIGENA M K and HAGEN T M (1993), 'Oxidants, antioxidants and the degenerative diseases of ageing', *Proc Natl Acad Sci USA,* **90**, 7915–22.

ARAI Y, WATANABE S, KIMIRA M, SHIMOI K, MOCHIZUKI R and KINAE N (2000), 'Dietary intakes of flavonols, flavones and isoflavones by Japanese women and the inverse correlation between quercetin intake and plasma LDL cholesterol', *J Nutr,* **130**, 2378–83.

BANGERTH F (1977), 'The effect of different partial pressures of CO_2, C_2H_4, and O_2 in the storage atmosphere on the ascorbic acid content of fruits and vegetables', *Qual Plant,* **27**, 125–33.

BARRY-RYAN C and O'BEIRNE D (1999), 'Ascorbic acid retention in shredded iceberg lettuce as affected by minimal processing', *J Food Sci,* **64**, 498–500.

BARTH M M, KERBEL E L, PERRY A K and SCHMIDT S J (1993), 'Modified atmosphere packaging affects ascorbic acid, enzyme activity and market quality of broccoli', *J Food Sci,* **57**, 954–7.

BASKARAN R H, PRASAD R and SHIVAIAH K M (2001), 'Storage behaviour of minimally processed pumpkin (*Cucurbiat maxima*) under modified atmosphere packaging conditions', *Eur Food Res Technol,* **212**, 165–9.

BECKER Z E (1933), 'A comparison between the action of carbonic acid and other acids upon the living cell', *Protoplasma,* **25**, 161–75.

BENNICK M H J, SMID E J, ROMBOUTS F M and GORRIS L G M (1995), 'Growth of psychrotrophic foodborne pathogens in a solid surface model system under the influence of carbon dioxide and oxygen', *Food Microbiol,* **12**, 509–19.

BENNICK M, VAN OVERBEEK W, SMID E and GORRIS L (1999), 'Biopreservation in modified atmosphere stored mungbean sprouts: the use of vegetable-associated bacteriogenic lactic acid bacteria to control the growth of *Listeria monocytogenes*', *Letters in Applied Microbiology,* **28**, 226–32.

BERARD L and CHONG C (1985), 'Influences of storage on glucosinolate fluctuations in cabbage', *Acta Hort,* **157**, 29–44.

BERRANG M, BRACKETT R and BEUCHAT L (1989a), 'Growth of *Listeria monocytogenes* on fresh vegetables stored under a controlled atmosphere',

J Food Prot, **52**(10), 702–5.

BERRANG M, BRACKETT R and BEUCHAT L (1989b), 'Growth of *Aeromonas hydrophila* on fresh vegetables stored under a controlled atmosphere', *Applied and Environmental Microbiology,* **55**, 2167–71.

BEUCHAT L and BRACKETT R (1990), 'Survival and growth of *L. monocytogenes* on lettuce as influenced by shredding, chlorine treatment, modified atmosphere packaging and temperature', *Journal of Food Science,* **55**(3), 755–8, 870.

BODNARUK P W and DRAUGHON F A (1998), 'Effect of packaging atmosphere and pH on the virulence and growth of *Yersinia enterocolitica* on pork stored at 4 degrees', *Food Microbio,* **15**(2), 129–36.

BOSKOU G and DEBEVERE J (1997), 'Reduction of trimethylamine oxide by *Shewanella* spp. under modified atmospheres *in vitro*', *Food Microbiol,* **14**, 543–53.

BOSKOU G and DEBEVERE J (1998), '*In vitro* study of TMAO reduction by *Shewanella putrefaciens* isolated from cod fillets packed in modified atmosphere', *Food Additives and Contaminants,* **15**(2), 229–36.

BRITTON G and HORNERO-MÉNDEZ D (1997), 'Carotenoids and colour in fruit and vegetables', in *Phytochemistry of Fruit and Vegetables,* F A Tomás-Barberán, R J Robins (eds), Oxford, Oxford University Press.

CANO P, MONREAL M, DE ANCOS B and ALIQUE R (1998), 'Effects of oxygen levels on pigment concentrations in cold-stored green beans (*Phaseolus vulgaris* L. Cv Perona)', *J Agric Food Chem,* **46**, 4164–70.

CARLIN F, NGUYEN-THE C AND ABREU DA SILVA A (1995), 'Factors affecting the growth of *L. monocytogenes* on minimally processed fresh endive', *Journal of Applied Bacteriology,* **78**, 636–46.

CARLIN F, NGUYEN-THE C, ABREU DA SILVA A and COCHET C (1996a), 'Effects of carbon dioxide on the fate of *L. monocytogenes,* of aerobic bacteria and on the development of spoilage in minimally processed fresh endive', *International Journal of Food Microbiology,* **32**, 159–72.

CARLIN F, NGUYEN-THE C and MORRIS C (1996b), 'The influence of the background microflora on the fate of *Listeria monocytogenes* on minimally processed fresh broad leaved endive', *Journal of Food Protection,* **59**(7), 698–703.

CASTILLO-RODRIGUEZ A, BARCO-ALCALA E, GARCIA-GIMENO R and ZURERA-COSANO G (2000), 'Growth modelling of *Listeria monocytogenes* in packaged fresh green asparagus', *Food Microbiology,* **17**, 421–7.

CHIN J H, TRUDELL J R and COHEN E N (1976), 'The compression-ordering and solubility-disordering effects of high pressure gases on phospholipid bilayers', *Life Sciences,* **18**, 489–98.

CONNOR D E, SCOTT V N and BERNARD D T (1989), 'Potential *Clostridium botulinum* hazards associated with extended shelf-life refrigerated foods: a review', *J Food Safety,* **10**, 131–53.

COYNE F P (1933), 'The effect of carbon dioxide on bacterial growth with special reference to the preservation of fish. Part II', *J Soc Chem Ind* (London),

52, 19–24.

DANIELS J A, KRISHNAMURTHI R and RIZVI S S H (1985), 'A review of effects of carbon dioxide on microbial growth and food quality', *J Food Prot,* **48**, 532–7.

DAVEY M W, VAN MONTAGU M, INZE D, SANMARTIN M, KANELLIS A, SMIRNOFF N, BENZIE I J J, STRAIN J J, FAVELL D and FLETCHER J (2000), 'Plant $_{L}$ascorbic acid: chemistry, function, metabolism, bioavailability and effects of processing', *J Sci Food Agri,* **80**, 825–60.

DAVIES A R (1995), 'Advances in modified-atmosphere packaging', in *New Methods in Food Preservation,* G W Gould (ed.), London, Blackie Academic and Professional, 304–20.

DAY B P F (2000), 'Chilled storage of foods, principles', in *Encyclopaedia of Food Microbiology,* R K Robinson, C A Batt and P D Patel (eds), San Diego, Academic Press, 403–10.

DELAPORTE N (1971), 'Effect of oxygen content of atmosphere on ascorbic acid content of apple during controlled atmosphere storage', *Lebens Wiss Technol,* **4**, 106–12.

DEVLIEGHERE F and DEBEVERE J (2000), 'Influence of dissolved carbon dioxide on the growth of spoilage bacteria', *Lebens Wiss Technol,* **33**, 531–7.

DEVLIEGHERE F, DEBEVERE J and VAN IMPE J (1998), 'Concentration of carbon dioxide in the water-phase as a parameter to model the effect of a modified atmosphere on microorganisms', *Int J Food Microbiol,* **43**, 105–13.

DEVLIEGHERE F, LEFEVERE I, MAGNIN A and DEBEVERE J (2000a), 'Growth of *Aeromonas hydrophila* on modified atmosphere packaged cooked meat products, *Food Microbiol,* **17**, 185–96.

DEVLIEGHERE F, GEERAERD A H, VERSYCK K J, BERNAERT H, VAN IMPE J H and DEBEVERE J (2000b), 'Shelf life of modified atmosphere packaged cooked meat products: addition of Na-lactate as a fourth shelf-life determinative factor in a model and product validation', *Int J Food Microbiol,* **58**, 93–106.

DEVLIEGHERE F, GEERAERD A H, VERSYCK K J, VANDEWAETERE B, VAN IMPE J and DEBEVERE J (2001), 'Growth of *Listeria monocytogenes* in modified atmosphere paced cooked meat products: a predictive model', *Int J Food Microbiol,* **18**, 53–66.

DI MASCIO P, KAISER S and SIES H (1989), 'Lycopene as the most efficient biological carotenoid singlet oxygen quencher', *Arch Biochem Biophys,* **274**, 532–8.

DIXON N M (1988), *Effects of CO_2 on Anaerobic Bacterial Growth and Metabolism,* PhD thesis, University College of Wales, Aberystwyth.

DIXON N M and KELL D B (1989), 'The inhibition by CO_2 of the growth and metabolism of micro-organisms', *J Appl Bacteriol,* **67**, 109–36.

DOHERTY A, SHERIDAN J J, ALLEN P, MCDOWELL D A, BLAIR I S and HARRINGTON D (1995), 'Growth of *Yersinia enterocolitica* O:3 on modified atmosphere packaged lamb', *Food Microbiol,* **12**(3), 251–7.

DOHERTY A, SHERIDAN J J, ALLEN P, MCDOWELL D A, BLAIR I S and HARRINGTON D

(1996), 'Survival and growth of *Aeromonas hydrophilia* on modified atmosphere packaged normal and high pH lamb', *International Journal of Food Microbiol,* **28**(3), 379–92.

DUFRESNE I, SMITH J P, JIUN-NI-LIU and TARTE I (2000), 'Effect of headspace oxygen and films of different oxygen transmission rate on toxin production by *Clostridium botulinum* type E in rainbow trout fillets stored under modified atmospheres', *J Food Safety,* **20**(3), 157–75.

EKLUND T (1984), 'The effect of carbon dioxide on bacterial growth and on uptake processes in bacterial membrane vesicles', *Int J Food Microbiol,* **1**, 179–85.

FARBER J M (1991), 'Microbiological aspects of modified atmosphere packaging technology – a review', *J Food Prot,* **54**, 58–70.

FRANCO-ABUIN C M, ROZAS-BARRERO J, ROMERO-RODRIGUEZ M A, CEPEDA-SAEZ A and FENTE-SAMPAYO C (1997), 'Effects of modified atmosphere packaging on the growth an survival of *Listeria* in raw minced beef', *Food Sci and Technol Int,* **3**, 285–90.

GARCIA DE FERNANDO G D, NYCHAS G J E, PECK M W and ORDONEZ J A (1995), 'Growth/survival of psychrotrophic pathogens on meat packaged under modified atmospheres', *Int J Food Microbiol,* **28**, 221–31.

GARCIA-GIMENO R, SANCHEZ-POZO M, AMARO-LOPEZ M and ZURERA-COSANO G (1996), 'Behaviour of *Aeromonas hydrophila* in vegetables salads stored under modified atmosphere at 4 and 15C', *Food Microbiol,* **13**, 369–74.

GIBSON A M, ELLIS-BROWNLEE R C L, CAHILL M E, SZABO E A, FLETCHER G C and BREMER P J (2000), 'The effect of 100% CO_2 on the growth of non-proteolytic *Clostridium botulinum* at chill temperatures', *Int J Food Microbiol,* **54**, 39–48.

GIL M I, HOLCROFT D M and KADER A A (1997), 'Changes in strawberry anthocyanins in response to carbon dioxide treatments', *J Agric Food Chem,* **45**, 1662–7.

GIL M I, FERRERES F and TOMÁS-BARBERÁN F A (1998a), 'Effect of modified atmosphere packaging on the flavonoids and vitamin C content of minimally processed Swiss chard (*Beta vulgaris* subsp. Cycla)', *J Agric Food Chem,* **46**, 2007–12.

GIL M I, CASTAÑER M, FERRERES F, ARTÉS F and TOMÁS-BARBERÁN F A (1998b), 'Modified-atmosphere packaging of minimally processed Lollo Rosso (*Lactuca sativa*)', *Z Lebensm Unters Forsch,* **206**, 350–4.

GIL M I, FERRERES F and TOMÁS-BARBERÁN F A (1999), 'Effect of postharvest storage and processing on the antioxidant constituents (flavonoids and vitamin C) of fresh-cut spinach', *J Agric Food Chem,* **47**, 2213–17.

GILL C O and REICHEL M P (1989), 'Growth of the cold-tolerant pathogens *Yersinia enterocolitica, Aeromonas hydrophila* and *Listeria monocytogenes* on high-pH beef packaged under vacuum or carbon dioxide', *Food Microbiol,* **6**, 223–30.

GILL C O and TAN K H (1980), 'Effect of carbon dioxide on growth of meat spoilage bacteria', *Appl Environ Microbiol,* **39**, 317–19.

GIOVANNUCCI E (1999), 'Tomatoes, tomato-based products, lycopene and cancer: review of the epidemiologic literature', *J Natl Cancer Inst,* **91**, 317–31.

HAFFNER K, JEKSRUD W K and TENGESDAL G (1997), 'L-ascorbic acid contents and other quality criteria in apples (*Malus domestica* Borkh) after storage in cold store and controlled atmosphere', *Postharvest Horticultural Series* No **16**, University of California, E J Micham (ed.).

HANSEN M, MOLLER P, SORENSEN H and CANTWELL M (1995), 'Glucosinolates in broccoli stored under controlled atmosphere', *J Amer Soc Hort Sci,* **120**, 1069–74.

HART C D, MEAD G C and NORRIS A P (1991), 'Effects of gaseous environment and temperature on the storage behaviour of *Listeria monocytogenes* on chicken breast meat', *J Appl Bacteriol,* **70**, 40–6.

HASLAM E, LILLEY T H, WARMINSKI E, LIAO H, CAI Y, MARTIN R, GAFFNEY S H, GOULDING P N and LUCK G (1992), 'Polyphenol complexation', in *Phenolic Compounds in Food and Their Effect on Health I,* C T Ho, C Y Lee, M T Huang (eds), Washington, DC, American Chemical Society.

HENDRICKS M T and HOTCHKISS J H (1997), 'Effect of carbon dioxide on the growth of *Pseudomonas fluorescens* and *Listeria monocytogenes* in aerobic atmospheres', *J Food Prot,* **60**, 1548–52.

HERTOG M G L, FESKENS E J M, HOLLMAN P C H, KATAN M B and KROMHOUT D (1993), 'Dietary antioxidant flavonoids and risk of coronary heart disease: the Zutphen elderly study', *The Lancet,* **342**, 1007–11.

HERTOG M G L, SWEETNAM P M, FEHILY A M, ELWOOD P C and KROMHOUT D (1997), 'Antioxidant flavonols and ischaemic heart disease in a Welsh population of men. The Caerphilly study', *Am J Clin Nutr,* **65**, 1489–94.

HOLCROFT D M and KADER A A (1999), 'Carbon dioxide-induced changes in color and anthocyanin synthesis of stored strawberry fruit', *HortSci,* **34**, 1244–8.

HOLCROFT D M, GIL M I and KADER A A (1998), 'Effect of carbon dioxide on anthocyanins, phenylalanine ammonia lyase and glucosyltransferase in the arils of stored pomegranates', *J Amer Soc Hort Sci,* **123**, 136–40.

HOWARD L R and HERNANDEZ-BRENES C (1998), 'Antioxidant content and market quality of jalapeno pepper rings as affected by minimal processing and modified atmosphere packaging', *J Food Quality,* **21**, 317–27.

HOWARD L R, SMITH R T, WAGNER A B, VILLALON B and BURNS E E (1994), 'Provitamin A and ascorbic acid content of fresh pepper cultivars (*Capiscum annuum*) and processed jalapenos', *J Food Sci,* **59**, 362–5.

HUDSON J A, MOTT S J and PENNEY N (1994), 'Growth of *Listeria monocytogenes, Aeromonas hydrophila* and *Yersinia enterocolitica* on vacuum and saturated carbon dioxide controlled atmosphere packaged sliced roast beef', *J Food Prot,* **57**(3), 204–8.

IKAWA J T and GENIGEORGIS C (1987), 'Probability of growth and toxin production by non-proteolitic *Clostridium botulinum* in rockfish fillets stored under modified atmospheres', *Int J Food Microbiol,* **4**, 167–81.

JACXSENS L, DEVLIEGHERE F, FALCATO P and DEBEVERE J (1999), 'Behaviour of

Listeria monocytogenes and *Aeromonas* spp. on fresh cut produce packaged under equilibrium modified atmosphere', *J Food Prot,* **62,** 1128–35.

JOHNSON I T, WILLIAMSON G M and MUSK S R R (1994), 'Anticarcinogenic factors in plant foods: a new class of nutrients', *Nutr Res Rev,* **7**, 175–204.

JONES M V (1989), 'Modified atmospheres', in *Mechanisms of Action of Food Preservation Procedures,* G W Gould (ed.), Elsevier Science, 247–84.

JONES R P and GREENFIELD P F (1982), 'Effect of carbon dioxide on yeast growth and fermentation', *Enzyme & Microbial Technol,* **4**, 210–84.

JUNEJA V, MARTIN S and SAPERS G (1998), 'Control of *L. monocytogenes* in vacuum-packaged pre-peeled potatoes', *J Food Sci,* **63**, 911–14.

KADER A A (1986), 'Biochemical and physiological basis for effects of controlled and modified atmospheres on fruits and vegetables', *Food Technol,* **40**, 99–100, 102–4.

KADER A A (1997), 'Biological bases of O_2 and CO_2 effects on postharvest-life of horticultural perishables', *Postharvest Horticultural Series* No **18**, University of California, M E Saltveit (ed.).

KADER A A (2001), 'Physiology of CA treated produce', *8th International Controlled Atmosphere Research Conference,* Rotterdam, Oosterhaven.

KADER A A, ZAGORY D and KERBEL E L (1989), 'Modified atmosphere packaging of fruit and vegetables', *Crit Rev Food Sci Nutr,* **28**, 1–30.

KANNER J, FRANKEL E, GRANIT R, GERMAN B and KINSELLA J E (1994), 'Natural antioxidants in grapes and wines', *J Agric Food Chem,* **42**, 64–9.

KLEIN B P (1987), 'Nutritional consequences of minimal processing fruits and vegetables', *J Food Qual,* **10**, 179–83.

KRÄMER K H and BAUMGART J (1992), 'Brühwurstaufschnitt hemmung von *Listeria monocytogenes* durch eine modifizierte atmosphäre', *Fleischwirtschaft,* **72**, 666–8.

KROGH A (1919), 'The rate of diffusion of gases through animal tissues with some remarks on the coefficient of invasion', *J Physiol,* **52**, 391–408.

LAWLOR K A, PIERSON M D, HACKNEY C R, CLAUS J R and MARCY J E (2000), 'Non-proteolytic *Clostridium botulinum* toxigenesis in cooked turkey stored under modified atmospheres', *J Food Prot,* **63**, 1511–16.

LEE S K and KADER A A (2002), 'Preharvest and postharvest factors influencing vitamin C content of horticultural crops', *Postharvest Biol Technol,* **20**, 207–20.

LIAO C and SAPERS G (1999), 'Influence of soft rot bacteria on growth of *Listeria monocytogenes* on potato slices', *J Food Prot,* **62**, 343–8.

LOFT S, OTTE J, POULSEN H E and SORENSEN H (1992), 'Influence of intact and myrosinase-treated indolyl glucosinolates on the metabolism *in vivo* of metronidazole and antipyrine in rat', *Food Chem Toxicology,* **30**, 927–35.

MACCARRONE M, D'ANDREA G, SALUCCI M L, AVIGLIANO L and FINAZZI-AGRO A (1993), 'Temperature, pH, and UV irradation effects on ascorbate oxidase', *Phytochemistry,* **35**, 795–8.

MACHEIX J J, FLEURIET A and BILLOT J (1990), *Fruit Phenolics*, Boca Raton,

Florida, CRC Press.

MANU-TAWIAH W, MYERS D J, OLSON D G and MOLINS R A (1993), 'Survival and growth of *Listeria monocytogenes* and *Yersinia enterocolitica* in pork chops packaged under modified gas atmospheres', *J Food Sci,* **58**, 475–9.

MEHLHORN H (1990), 'Ethylene-promoted ascorbate peroxidase activity protects plants against hydrogen peroxide, ozone and paraquat', *Plant Cell Envrion,* **13**, 971–6.

MOLIN G (1983), 'The resistance to carbon dioxide of some food related bacteria', *European J Appl Microbiol Biotechnol,* **18**, 214–17.

NYCHAS G J E (1994), 'Modified atmosphere packaging of meats', in *Minimal Processing of Foods and Process Optimization, an Interface,* R P Singh, F A R Oliveira (eds), London CRC Press, 417–36.

O'CONNOR-SHAW R E and REYES V G (2000), 'Use of modified-atmosphere packaging', in *Encyclopedia of Food Microbiology,* R K Robinson, C A Batt, P D Patel (eds), San Diego, Academic Press, 410–15.

OMARY M, TESTIN R, BAREFOOT S and RUSHING J (1993), 'Packaging effects on growth of *Listeria innocua* in shredded cabbage', *Journal of Food Science,* **58**, 623–6.

ÖZBAS Z Y, VURAL H and AYTAC S A (1996), 'Effect of modified atmosphere and vacuum packaging on the growth of spoilage and inoculated pathogenic bacteria on fresh poultry', *Z Lebensm Unters Forsch,* **203**, 326–32.

ÖZBAS Z Y, VURAL H and AYTAC S A (1997), 'Effects of modified atmosphere and vacuum packaging on the growth of spoilage and inoculated pathogenic bacteria on fresh poultry', *Fleischwirtschaft,* **77**, 1111–16.

PARADIS C, CASTAIGNE F, DESROSIERS T, FORTIN J, RODRIGUE N and WILLEMOT C (1996), 'Sensory, nutrient and chlorophyll changes in broccoli florets during controlled atmosphere storage', *J Food Qual,* **19**, 303–16.

PARRY R T (1993), 'Introduction', in *Principles and Applications of Modified Atmosphere Packaging of Food,* R T Parry (ed.), Glasgow, Blackie Academic and Professonal, 1–18.

PETERSEN M A and BERENDS H (1993), 'Ascorbic acid and dehydroascorbic acid content of blanched sweet green pepper during chilled storage in modified atmospheres', *Z Lebens Unters Forsch,* **197**, 546–9.

PETREL M T, FERNáNDEZ P S, ROMOJARO F and MARTíNEZ A (1998), 'The effect of modified atmosphere packaging on ready-to-eat oranges', *Lebensm Wiss Technol,* **31**, 322–8.

REDDY N R, PARADIS A, ROMAN M G, SOLOMON H M and RHODEHAMEL E J (1996), 'Toxin development by *Clostridium botulinum* in modified atmosphere-packaged fresh Tilapia fillets during storage', *J Food Sci,* **61**, 632–5.

REDDY N R, ROMAN M G, VILLANUEVA M, SOLOMON H M, KAUTTER D A and RHODEHAMEL E J (1997a), 'Shelf life and *Clostridium botulinum* toxin development during storage of modified atmosphere-packaged fresh catfish fillets', *J Food Sci,* **62**, 878–84.

REDDY N R, SOLOMON H M, YEP H, ROMAN M G and RHODEHAMEL E J (1997b), 'Shelf life and toxin development by *Clostridium botulinum* during storage of modified-atmosphere-packaged fresh aquacultured salmon fillets', *J Food Prot,* **60**, 1055–63.

ROSA E A S, HEANEY R K, FENWICK G R and PORTAS C A (1997), 'Glucosinolate in crop plants', *Hort Rev,* **19**, 99–215.

ROTH S H (1980), 'Membrane and cellular actions for anesthetic agents', *Federation Proceedings,* **39**, 1595–9.

SALTVEIT M E (1997), 'Physical and physiological changes in minimally processed fruits and vegetables', in *Phytochemistry of Fruit and Vegetables,* F A Tomás-Barberán, R J Robins (eds), Oxford, Oxford University Press.

SHERIDAN J J and DOHERTY A (1994), 'Growth of *Yersinia enterocolitica* on modified atmosphere packaged lamb', *Proceedings 40th International S.Iia.13. Congress on Meat Science Technology,* The Hague, Holland.

SHERIDAN J J, DOHERTY A and ALLEN P (1992), *Improving the Safety and Quality of Meat and Meat Products by Modified Atmosphere and Assessment by Novel Methods,* FLAIR 89055 Interim. 2nd year report, EEC SGXII, Brussels.

SIMON J A (1992), 'Vitamin C and cardiovascular disease: a review', *J Am Coll Nutr,* **11**, 107–25.

SOZZI G, TRINCHERO G D and FRASCHINA A A (1999), 'Controlled-atmosphere storage of tomato fruit: low oxygen or elevated carbon dioxide levels alter galactosidase activity and inhibit exogenous ethylene action', *J Sci Food Agric,* **79**, 1065–70.

SZABO E A and CAHILL M E (1998), 'The combined affects of modified atmosphere, temperature, nisin and ALSO-T-M 2341 on the growth of *Listeria monocytogenes*', *Int J Food Microbiol,* **43**(1/2), 21–31.

TAN K H and GILL C O (1982), 'Physiological basis of CO_2 inhibition of a meat spoilage bacterium, *Pseudomonas fluorescens*', *Meat Sci,* **7**, 9–17.

THOMAS C, PRIOR O and O'BEIRNE D (1999), 'Survival and growth of *Listeria* species in a model ready-to-use vegetable product containing raw and cooked ingredients as affected by storage temperature and acidification', *International Journal of Food Science and Technology,* **34**, 317–24.

TOMÁS-BARBERÁN F A and ESPÍN J C (2002), 'Phenolic compounds and related enzymes as determinants of quality in fruit and vegetables', *J Sci Food Agric,* **81**, 853–76.

TOMÁS-BARBERÁN F A, GIL M I, CASTAÑER M, ARTÁS F and SALTVEIT M E (1997), 'Effect of selected browning inhibitors on harvested lettuce stem phenolic metabolism', *J Agric Food Chem,* **45**, 583–9.

TUDELA J A, ESPÍN J C and GIL M I (2002), 'Vitamin C retention in fresh-cut potatoes', *Postharvest Biol Techno,* in press.

VÁMOS-VIGYÁZÓ L (1981), 'Polyphenol oxidase and peroxidase in fruits and vegetables', *CRC Crit Rev Food Sci Nutr,* **15**, 49–127.

VARNAM A H and EVANS M G (1991), *Foodborne pathogens: an Illustrated Text,* London, Wolfe Publishing.

VELTMAN R H, SANDERS M G, PERSIJN S T, PEPPELENBOS H W and OOSTERHAVEN J

(1999), 'Decreased ascorbic acid levels and brown core development in pears (*Pyrus communis* L. cv. Conference)', *Physiol Plant,* **107**, 39–45.

VERKERT R, DEKKER M and JONGEN W M F (2001), 'Postharvest increase of indolyl glucosinolates in response to chopping and storage of *Brassica* vegetables', *J Sci Food Agric,* **81**, 953–8.

VINSON J A and HONTZ B A (1995), 'Phenol antioxidant index: comparative antioxidant effectiveness of red and white wines', *J Agric Food Chem,* **43**, 401–3.

WANG C Y (1977), 'Effects of CO_2 treatment on storage and shelf life of sweet pepper', *J Amer Soc Hortc Sci,* **102**, 808–12.

WANG C Y (1983), 'Postharvest responses of Chinese cabbage to high CO_2 treatments or low O_2 storage', *J Amer Soc Hortc Sci,* **108**, 125–9.

WATADA A E (1987), *Vitamins,* New York, Marcel Dekker.

WEICHMANN J (1986), 'The effect of controlled-atmosphere storage on the sensory and nutritonal quality of fruits and vegetables', *Hort Rev,* **8**, 101–27.

WIMPFHEIMER L, ALTMAN N S and HOTCHKISS J H (1990), 'Growth of *Listeria monocytogenes* Scott A, serotype 4 and competitive spoilage organisms in raw chicken packaged under modified atmospheres and in air', *Int J Food Microbiol,* **11**, 205–14.

WOLFE S K (1980), 'Use of CO and CO_2 enriched atmospheres for meats, fish and produce', *Food Technol,* **34**, 55–9.

WRIGHT K P and KADER A A (1997a), 'Effect of slicing and controlled-atmosphere storage on the ascorbate content and quality of strawberries and persimmons', *Postharvest Biol Techno,* **10**, 39–48.

WRIGHT K P AND KADER A A (1997b), 'Effect of controlled-atmosphere storage on the quality and carotenoid content of sliced persimmons and peaches', *Postharvest Biol Technol,* **10**, 89–97.

YOUNG L L, REVERIE R D and COLE A B (1988), 'Fresh red meats: a place to apply modified atmospheres', *Food Technol,* **42**(9), 64–6, 68–9.

ZEITOUN A A M and DEBEVERE J M (1991), 'Inhibition, survival and growth of *Listeria monocytogenes* on poultry as influenced by buffered lactic acid treatment and modified atmosphere packaging', *Int J Food Microbiol,* **14**, 161–70.

ZHANG S and FARBER J (1996), 'The effects of various disinfectants against *L. monocytogenes* on fresh-cut vegetables', *Food Microbiol,* **13**, 311–21.

ZHAO Y, WELLS J H and MARSHALL D L (1992), 'Description of log phase growth for selected microorganisms during modified atmosphere storage', *J Food Proc Engin,* **15**, 299–317.

12

Reducing pathogen risks in MAP-prepared produce

D. O'Beirne and G. A. Francis, University of Limerick, Ireland

12.1 Introduction

Modified atmosphere packaged (MAP) prepared fresh produce provides substrates and environmental conditions conducive to the survival and growth of microorganisms. Minimal processing treatments such as peeling and slicing disrupt surface tissues, expose cytoplasm and provide a potentially richer source of nutrients than intact produce (Brackett, 1994; Barry-Ryan and O'Beirne, 1998, 2000). This, combined with high Aw and either close to neutral (vegetables) or low acid (many fruits) tissue pH, facilitate microbial growth (Beuchat, 1996).

These products can harbour large and diverse populations of microorganisms, and counts of 10^5–10^7 CFU/g are frequently present. Most bacteria present are Gram-negative rods, predominantly *Pseudomonas*, *Enterobacter* or *Erwinia* species (Brocklehurst *et al.*, 1987; Garg *et al.*, 1990; Magnuson *et al.*, 1990; Manvell and Ackland, 1986; Marchetti *et al.*, 1992; Nguyen-the and Prunier, 1989). The organisms present and counts are affected by product type and storage conditions. Lactic acid bacteria have been detected in mixed salads and grated carrots, and may predominate in salads when held at abuse (30ºC) temperatures (Manvell and Ackland, 1986). Yeasts commonly isolated include *Cryptococcus*, *Rhodotorula*, and *Candida* (Brackett, 1994). Webb and Mundt (1978) surveyed 14 different vegetables for moulds. The most commonly isolated genera were *Aureobasidium*, *Fusarium*, *Mucor*, *Phoma*, *Rhizopus*, and *Penicillium*.

A number of important human pathogens can also be found in MAP prepared produce. Their presence is a consequence of contamination during agricultural production (mainly from contaminated seed, soil, irrigation water, and air), during harvesting and manual preparation (human contact) or during machine processing and packaging (contaminated work surfaces/packaging materials/ equipment). Cross-contamination by end-users after pack opening can also occur.

By extending shelf-life and protecting product quality, MAP prepared produce systems can provide sufficient time for pathogens to grow to significant numbers on otherwise acceptable fresh foods (Berrang *et al.*, 1989b). The risk of food poisoning is greatest in products eaten raw without any further preparation. While the food safety record of these products is good, a comprehensive understanding of the implications of this technology for pathogen survival and growth is required in order to optimise production systems and to inform HACCP protocols. The effects of MAP technology on the survival and growth of non-pathogens and on the interaction between pathogens and non-pathogens is also important (Francis and O'Beirne, 1998b). Non-pathogens are both potential competitors of pathogens and important indicators of product spoilage. While considerable progress has been made in the past decade in our understanding of the safety of these novel and complex food systems there are still significant gaps in knowledge requiring further research.

12.2 Measuring pathogen risks

A range of pathogens have been isolated from raw produce (Brackett, 1999; Francis *et al.*, 1999) and foodborne infections have been linked to the consumption of raw vegetables and fruits (Tables 12.1 and 12.2). While pathogens have also been isolated from MAP prepared produce (see Table 12.1) relatively few foodborne infections have been directly linked with this range of products. Those that have been linked include an outbreak of botulism ultimately linked to an MAP dry coleslaw product (Solomon *et al.*, 1990) and a *Salmonella* Newport outbreak linked to ready-to-eat salad vegetables (PHLS, 2001). There was also an outbreak of shigellosis linked to shredded lettuce (Davis *et al.*, 1988) though exactly how this product was packaged is unclear. Increasing consumption of fresh produce in the United States has been paralleled by an increase in produce-linked food poisoning outbreaks (NACMCF, 1999). Contributory factors include the increased range and diversity of products available to consumers and the elimination of seasonality by almost year-round availability of many commodities. This diversity and availability has been achieved by increased globalisation of the produce trade, and has brought with it new food safety risks and challenges. While the main pathogens of concern are still non-proteolytic *Clostridium botulinum, Listeria monocytogenes, Yersinia entercolitica* and *Aeromonas hydrophila*, there are important emerging threats from viral and protozoan pathogens.

There are a number of difficulties in estimating the magnitude of the true microbial risk from fresh produce and MAP fresh produce. Studies where samples of produce are examined for the presence of pathogens are, of necessity, limited in size and may not accurately reflect global contamination levels. In addition, surveys showing the absence of pathogens may receive less attention than those showing their presence, and this may distort the true picture. A recent examination of 127 fresh produce items from the Washington DC area (Thunberg *et al.*, 2002) showed low levels of contamination, no *Salmonella* or *Campylobacter* contamination, and seven samples positive for *L. monocytogenes*. On the other hand, food poisoning incidents related to fresh produce may be under-reported. By comparison with those linked to meat and poultry, outbreaks related to produce do not have the same pathogen and product characteristics which assist in recognition, investigation, and reporting (NACMCF, 1999). For example, the short shelf-lives, complex distribution and universal consumption of fresh produce make produce-implicated outbreaks more difficult to pin down. Even when produce is almost certainly implicated, the exact point of contamination is difficult to prove beyond doubt. Of 27 examples of produce-linked food poisoning outbreaks considered by NACMCF, investigators had definitively identified the point of contamination in only two. The main pathogens of concern in MAP produce are discussed below, focusing on sources and levels of contamination, and their likely health risk to consumers.

12.2.1 *Listeria monocytogenes*

L. monocytogenes is a Gram-positive rod which causes several diseases in man including meningitis, septicaemia, still-births and abortions (ICMSF, 1996). It is considered ubiquitous in the environment, being isolated from soil, faeces, sewage, silage, manure, water, mud, hay, animal feeds, dust, birds, animals and man (Al-Ghazali and Al-Azawi, 1990; Gunasena *et al.*, 1995; Gray and Killinger, 1966; Nguyen-the and Carlin, 1994; Welshimer, 1968). Contamination of vegetables by *L. monocytogenes* may occur through agricultural practices, such as irrigation with polluted water or use of contaminated manure (Nguyen-the and Carlin, 1994; Geldreich and Bordner, 1971). It may also occur during processing (see Section 12.3.3). *L. monocytogenes* has been isolated from minimally processed vegetables at rates ranging from 0% (Farber *et al.*, 1989; Fenlon *et al.*, 1996; Gohil *et al.*, 1995; Petran *et al.*, 1988) to 44% (Arumugaswamy *et al.*, 1994; Beckers *et al.*, 1989; Doris and Seah, 1995; Harvey and Gilmour, 1993; MacGowan *et al.*, 1994; McLauchlin and Gilbert, 1990; Sizmur and Walker, 1988; Velani and Roberts, 1991). In France (Nguyen-the and Carlin, 1994) and Germany (Lund, 1993) levels of $>10^2$ CFU/g are unacceptable, while in the UK and USA the organism must be absent in 25g.

Of particular concern is the organism's ability to grow at refrigeration temperatures; the minimum temperature for growth is reported to be −0.4°C (Walker and Stringer, 1987). It is also facultatively anaerobic, capable of survival/growth under the low O_2 concentrations within MA packages of prepared vegetables. While counts generally remain constant at 4°C (Farber *et al.,* 1998), they can increase to high numbers at mild abuse temperatures (8°C),

Table 12.1 Occurrence of pathogens on minimally processed produce

Vegetable	Number (and %) of positive samples	Country and comments	Reference
Listeria monocytogenes			
Cucumber slices	4/5 (80%)	Malaysia	Arumugaswamy *et al.*, 1994
Bean-sprouts	6/7 (85%)	Malaysia	Arumugaswamy *et al.*, 1994
Coleslaw	2/92 (2.2%)	Canada	Schlech *et al.*, 1983
	2/50 (4%)	Singapore	Doris and Seah, 1995
	3/39 (7.7%)	United Kingdom	MacGowan *et al.*, 1994
Pre-packed mixed salads	3/21 (14.3%)	Northern Ireland	Harvey and Gilmour, 1993
	4/60 (6.7%)	United Kingdom	Sizmur and Walker, 1988
Chopped lettuce	5/39 (13%)	Canada	Odumeru *et al.*, 1997
Cut and packaged lettuce	3/120 (2.5%)	Australia	Szabo *et al.*, 2000
Prepared mixed vegetables	8/42 (19%)	United Kingdom (contamination during processing suspected; <200/g present)	Velani and Roberts, 1991
Fresh cut salad vegetables	11/25 (44%)	The Netherlands ($<10^2$/g present)	Beckers *et al.*, 1989
Chicory salads	(8.8%)	France (<1/g present)	Nguyen-the and Carlin, 1994
Prepared vegetables	1/26 (3.8%)	United Kingdom	MacGowan *et al.*, 1994

Processed vegetables and salads	(13%)	United Kingdom	McLaughlin and Gilbert, 1990
Aeromonas **spp.**			
Cut lettuce	66/120 (55%)	Australia	Szabo *et al.*, 2000
Salad mix	12/12 (100%)	Italy	Marchetti *et al.*, 1992
Prepared salads	(21.6%)	UK	Fricker and Tompsett, 1989
E. coli **O157:H7**			
Salad mix	0/63 (0%)	US	Lin *et al.*, 1996
Clostridium botulinum			
MAP Salad mix	2/350 (0.6%)	US	Lilly *et al.*, 1996
MAP cabbage	1/337 (0.3)	US	Lilly *et al.*, 1996
MAP green pepper	1/201 (0.5%)	US	Lilly *et al.*, 1996
Salmonella **spp.**			
Salad mix	1/159 (0.6%)	Egypt	Saddik *et al.*, 1985
Endive	2/26 (7.7%)	Netherlands	Tamminga *et al.*, 1978
Yersinia **spp.**			
Cut and packaged lettuce	71/120 (59%)	Australia	Szabo *et al.*, 2000
Prepacked salads	3/3 (100%)	UK	Brocklehurst *et al.*, 1987
Campylobacter jejuni			
Mushrooms	3/200 (1.5%)	United States	Doyle and Schoeni, 1986

particularly after anti-microbial dipping treatments or within nitrogen flushed packages (Francis and O'Beirne, 1997). However, evidence is emerging that levels of virulence may vary greatly among *L. monocytogenes* strains, and that some serotypes found in MAP produce may be different (and less virulent) than those isolated in food poisoning outbreaks (Beuchat and Ryu, 1997). Further research is required to determine the significance of different *L. monocytogenes* strains for human health.

12.2.2 *Clostridium botulinum*

Cl. botulinum is a member of the genus *Clostridium*, characterised as Gram-positive, rod-shaped, endospore forming, obligate anaerobes (Varnum and Evans, 1991). The foodborne Clostridia have been comprehensively reviewed by McClane (1997) and Dodds and Austin (1997). *Cl. botulinum* is divided into numerous sub-divisions, based on the serological specificity of the neurotoxin produced, and physiological differences between strains. Human botulism is normally attributed to sub-species antigenic types A, B, E and occasionally type F. Endospores of *Cl. botulinum* are ubiquitous, being distributed in soils, aquatic sediments and the digestive tract of animals and birds.

Vegetables are potentially contaminated during growth, harvesting and processing (Rhodehamel, 1992). Despite their ubiquity, a recent study identified only 0.36% of pre-cut MAP vegetables to be contaminated with *Cl. botulinum* spores (Lilly *et al.*, 1996). In the case of mushrooms, a much lower incidence of *Cl. botulinum* was reported (Notermans *et al.*, 1989) than had been reported previously (Hauschild *et al.*, 1978), a change attributed to hygienic improvements in growing techniques.

The possibility of growth and toxin production by *Cl. botulinum* before obvious spoilage has long been of concern in over-wrapped mushrooms (Sugiyama and Yang, 1975) and in vacuum packaged prepared potatoes (O'Beirne and Ballantyne,1987). In addition, sufficiently anoxic conditions are frequently observed in MA packages where the respiration rate of the product is not matched by the permeability of the packaging used. Anoxic conditions may also develop within MAP produce where edible coatings are used (Guilbert *et al.*, 1996). Highly permeable or perforated over-wrapping films have been used for fresh mushrooms and low storage temperatures and short shelf-lives have been requirements in prepared potato products (IFST, 1990). In the case of other items of vacuum packaged/MAP prepared produce, the data suggest that spoilage is likely to preceed toxin production (Larson *et al.*, 1997; Petran *et al.*, 1995), with a probability of 1 in 10^5 for toxin production to occur prior to obvious spoilage (Larson *et al.*, 1997). However, there is a report linking a botulism outbreak with coleslaw prepared from a MAP dry coleslaw mix (Solomon *et al.*, 1990). The short shelf-lives of retail packs and the good control of temperature/modest storage lives of catering packs are likely to minimise such risks, but there is need for vigilance and further research.

12.2.3 *Escherichia coli* O157:H7

E. coli, type species of the type Enterobacteriaceae genus, *Escherichia*, is a common inhabitant of the gastrointestinal tract of mammals. Despite the commensal status of the majority of strains, pathogenic strains, particularly enterohaemorrhagic *E. coli* O157:H7, have emerged as highly significant foodborne pathogens. Gastroenteritis and haemorrhagic colitis are classical symptoms, while complications including thrombocytopenic purpura and haemolytic uraemic syndrome have been documented (Martin *et al.*, 1986), the latter potentially leading to renal failure and death in 3–5% of juvenile cases (Karmali *et al.*, 1983; Griffin and Tauxe, 1991).

The principal reservoir of *E. coli* O157:H7 is believed to be the bovine gastrointestinal tract (Wells *et al.*, 1991; Doyle *et al.*, 1997). Hence, contamination of meat and other food products with faeces is a significant risk factor. Contamination of, and survival of the organism in natural water sources make these also potential sources in the distribution of infection, particularly if untreated water is used to wash produce. The potential for cross-contamination during distribution and domestic storage are also of concern. Information regarding contamination rates of MAP prepared vegetables is limited. Recent surveys in the UK and US failed to find this pathogen (FDA, 2001).

12.2.4 *Aeromonas hydrophila*

Aeromonas hydrophila is a motile, Gram-negative, rod-shaped bacterium in the family Vibrionaceae. It causes a broad spectrum of infections (septicaemia, meningitis, endocarditis) in humans, often in immunocompromised hosts, and *Aeromonas* spp. have been associated epidemiologically with travellers diarrhoea. Its significance as a human pathogen has been reviewed by Altwegg and Geiss (1989).

A. hydrophila is considered to be ubiquitous and has been isolated from many sources. The best known sources are treated and untreated water, and animals associated with water, such as fish and shellfish (ICMSF, 1996). Hazen *et al.* (1978) isolated *A. hydrophila* from the vast majority of aquatic environments. *A. hydrophila* is also associated with soil (Brandi *et al.*, 1996) and with a range of foods including fresh vegetables. Foods in which *A. hydrophila* was isolated were most likely contaminated by water, soil or animal faeces.

A. hydrophila possesses a number of characteristics of concern in relation to MAP prepared vegetables. It is a psychrotroph; it grows slowly at 0ºC, but temperatures of 4–5ºC will support growth in foods. It is also a facultative anaerobe, capable of growing in atmospheres containing low concentrations of oxygen. Marchetti *et al.* (1992) isolated high counts (10^3–10^6/g) of *A. hydrophila* in commercial MAP prepared vegetable salads. *Aeromonas* spp. were also recovered from green salad, coleslaw (Hudson and De Lacy, 1991), pre-made salad samples (Fricker and Tompsett, 1989), mayonnaise salad samples (Knøchel and Jeppesen, 1990) and commercial mixed vegetable salads

(García-Gimeno *et al.*, 1996). By contrast, none of the vegetable samples from shops in Sweden was positive for *Aeromonas* spp. (Krovacek *et al.*, 1992).

12.2.5 *Salmonella*

Salmonella, a genus of the family Enterobacteriaceae, are characterised as Gram-negative, rod-shaped bacteria. Pathogenic species include *S.* Typhimurium, *S.* Enteritidis, *S.* Heidelberg, *S.* Saint-paul, and *S.* Montevideo. Salmonellae are mesophiles, with optimum temperatures for growth of 35–43°C. The growth rate is substantially reduced at <15°C, while the growth of most salmonellae is prevented at <7°C. *Salmonella* are facultatively anaerobic, capable of survival in low O_2 atmospheres.

These organisms are abundant in faecal material, sewage and sewage-polluted water; consequently they may contaminate soil and crops with which they come into contact. Sewage sludge may contain high numbers of salmonellae and, if used for agricultural purposes, will disseminate the bacterium. Once introduced into the environment, salmonellae remain viable for months (ICMSF, 1996). Potential contamination from workers who handle produce in the field or in processing plants is of great concern (see Section 12.4). Salmonellae have not generally been found in MAP produce, though they have been isolated from bean-sprouts (20%) in Malaysia (Arumugaswamy *et al.*, 1994).

12.2.6 *Yersinia enterocolitica*

Y. enterocolitica is currently considered to be the most significant genus member with respect to foodborne disease (Varnum and Evans, 1991). Traditional gastrointestinal symptoms, potentially mediated through the activity of a heat-stable enterotoxin, may develop into suppurative and autoimmune complications (Robins-Browne, 1997). The psychrotrophic status of *Y. enterocolitica* is potentially of great significance with regard to refrigerated MAP prepared produce.

Y. enterocolitica occupies a broad range of ecosystems including the intestinal tract, birds, flies, fish and a variety of terrestrial and aquatic ecosystems. However, most environmental isolates lack virulence markers and are of doubtful significance for human or animal health (Delmas and Vidon, 1985). Isolation of *Yersinia* spp. from raw vegetables has been reported at rates ranging from 3.3% (Tassinari *et al.*, 1994) to 46.1% (Delmas and Vidon, 1985), although specific isolation rates of pathogenic *Y. enterocolitica* strains are likely to be significantly lower.

12.2.7 *Campylobacter jejuni*

Since their principal identification as human gastrointestinal pathogens in the 1970s (Butzler *et al.*, 1973; Skirrow, 1977) members of the thermophilic campylobacters, e.g. *C. jejuni*, have emerged as major human gastrointestinal pathogens (Ketley, 1997). Despite fastidious growth requirements, members of the genus survive at refrigeration temperatures for extended periods within nutrient limited environments. This property, combined with the low infective dose (Robinson, 1981) and their microaerophilic nature, indicates the potential significance of the genus with respect to refrigerated MAP prepared produce.

Campylobacter are zoonotic pathogens, being primarily associated with the intestinal tracts of wild and domestic animals (Thomas *et al.*, 1995) and are distributed throughout the environment through vehicles including birds, surface water and flies. Inappropriate food preparation and handling procedures may lead to the cross-contamination of fresh produce with *Campylobacter* from uncooked meats, and such errors could have resulted in the identification of MAP prepared vegetable products as sources of infection (Bean and Griffin, 1990; Altrkruse *et al.*, 1994). A Canadian study of 296 fresh-cut MAP vegetable products detected no *Campylobacter* contamination (Odomeru *et al.*, 1997).

12.2.8 *Shigella* species

The genus *Shigella* is composed of four species, *S. dysenteriae*, *S. sonnei*, *S. boydii* and *S. flexneri,* all of which are pathogenic to humans at a low dose of infection. Fruits and vegetables may become contaminated with *Shigella* via infected food handlers or through the use of contaminated manure and irrigation water (FDA, 2001; Saddik *et al.*, 1985). Several outbreaks of shigellosis have been attributed to contaminated produce (Freudland *et al.*, 1987; see Table 12.1) and a 1986 outbreak of shigellosis was traced back to commercially distributed shredded packaged lettuce (Davis *et al.*, 1988). Despite their mesophilic status, *Shigella* can survive on lettuce stored at 5°C for seven days (Davis *et al.*, 1988) and on coleslaw at 4°C for 16 days with numbers decreasing slightly during storage (Rafii and Lundsford, 1997).

12.2.9 Viral and protozoan pathogens

The significance of viruses with respect to foodborne disease is clear with the inclusion of Norwalk virus, Hepatitis A virus and 'other viruses' within the top ten causes of foodborne disease outbreaks in the USA (1983–1987; Cliver, 1997). Outbreaks caused by hepatitis A virus, calicivirus and Norwalk-like viruses have been associated with the consumption of frozen raspberries and strawberries, melons, lettuce, watercress and diced tomatoes (Beuchat, 1996; Hedberg and Osterholm, 1993; Hutin *et al.*, 1999; Lund and Snowdon, 2000; Rosenblum *et al.*, 1990). Viruses can be transmitted by infected food handlers, through the fecal-oral route, and have been isolated from sewage and untreated water used for crop irrigation. Despite their significance, data regarding the effects of food preparation and storage conditions on the survival and infectivity of viruses is extremely limited, partly through the complexity of viral detection assays. Nonetheless, the potential of several viruses to survive on vegetables for periods exceeding their normal shelf-life has been identified (Badawy *et al.*,

Table 12.2 Foodborne infections linked to the consumption of raw fruits and vegetables

Pathogen	Product suspected	No. of cases	Location	Reference
Bacteria				
L. monocytogenes	Shredded cabbage in coleslaw	41	Canada	Schlech *et al.*, 1983
	Raw tomatoes, lettuce and celery	20	Boston, US	Ho *et al.*, 1986
Cl. botulinum	Shredded cabbage in coleslaw	4	Florida, US	Solomon *et al.*, 1990
	Chopped garlic in oil	37	British Columbia	Solomon and Kautter, 1988
Salmonella spp.	Sliced watermelon	39	Michigan, US	Blostein, 1993
	Cantaloupe melon	22	Canada	Deeks *et al.*, 1998
	Cress sprouts	31	UK	Feng, 1997
	Mung sprouts	143	UK	O'Mahony *et al.*, 1990
	Tomatoes	85	Multi-state US	Susman, 1999
	Tomatoes	174	Multi-state US	Tauxe, 1997
E. coli O157:H7	Cantaloupe melon	9	Oregon, US	Del Rosario and Beuchat, 1995
	Radish sprouts	6561	Japan	WHO, 1996
	Alfalfa sprouts	108	US	CDC, 1997a
	Lettuce	70	Montana, US	Ackers *et al.*, 1998
	Lettuce	23	Canada	Preston *et al.*, 1997
Shigella sonnei	Watermelon	15	Sweden	Freudlund *et al.*, 1987
	Shredded lettuce	347	Texas	Davis *et al.*, 1988
	Lettuce	140	Texas	Martin *et al.*, 1986
	Lettuce	118	Norway, UK, Sweden, Spain	Kapperud *et al.*, 1995
	Parsley	310	Multi-state US	CDC, 1999

Bacillus cereus	Soy, mustard and cress sprouts	4	Texas	Portnoy *et al.*, 1976
Yersinia enterocolitica	Beansprouts	16	US	Cover and Aber, 1989
Camylobacter jejuni	Salad	330	Canada	Allen, 1985
	Lettuce	14	Oklahoma, US	CDC, 1998a
Viruses				
Hepatitis A virus	Raspberries (frozen)	24	Scotland	Reid and Robinson, 1987
	Strawberries (frozen)	242 + 14 suspect	Multistate, US	Hutin *et al.*, 1999
	Lettuce	103	Florida, US	Lowry *et al.*, 1989
	Watercress	129	Tennessee	CDC, 1971
	Diced tomatoes	92	Arkansas, US	Lund and Snowdon, 2000
Norwalk virus	Melon	206	UK	Lund and Snowdon, 2000
	Fresh-cut fruit	>217	Hawaii	Herwaldt *et al.*, 1994
	Raspberries (frozen)	>500	Finland	Lund and Snowdon, 2000
Parasites				
Cyclospora cayetanensis	Raspberries	1465	20 US states & Canada	Herwaldt and Ackers, 1997
	Raspberries	1012	Multi-state US & Canada	Herwaldt and Beach, 1999
	Blackberries	104	Canada	Herwaldt, 2000
	Baby lettuce leaves	>91	Florida, US	Herwaldt and Beach, 1999
	Basil	>308	Multi-state US	CDC, 1997b
Cryptosporidium parvum	Green onions	54	Washington	CDC, 1998b
Giardia	Lettuce and onions	21	New Mexico	CDC, 1989

1985; Konowalchuk and Speirs, 1975; Sattar *et al.*, 1994). Survival appears to be dependent upon temperature and moisture content (Bidawid *et al.*, 2001; Konowalchuk and Speirs, 1975); however, little information is available on the effects of MAP on virus survival.

The protozoan parasites *Giardia lamblia*, *Cyclospora cayetanensis* and *Cryptosporidium parvum* have been the cause of serious foodborne outbreaks involving berries (Herwaldt, 2000; Herwaldt and Ackers, 1997), lettuce and onions (CDC, 1989) and raw sliced vegetables (Mintz *et al.*, 1993). These organisms normally gain access to produce before harvest, usually as a result of contaminated manure or irrigation water and poor hygiene practices by food handlers (Beuchat, 1996). The lack of sensitive methods for determining the survival or inactivation of oocysts has hampered incidence studies and studies focused on the effects of minimal processing and packaging. However, the increase in produce-linked outbreaks due to these organisms (see Table 12.2) indicates that research is needed to examine the behaviour of foodborne protozoan parasites on MAP produce.

12.3 Factors affecting pathogen survival

Pathogen survival on produce is influenced by a number of interdependent factors, principally storage temperature, product type/product combinations (e.g. vegetables combined with cooked ingredients), minimal processing operations (e.g. slicing, washing/disinfection), package atmosphere and competition from the natural microflora present on produce.

12.3.1 Storage temperature

Storage temperature is the single most important factor affecting survival/growth of pathogens on MAP produce. Storage of produce at adequate refrigeration temperatures, will limit pathogen growth to those that are psychrotrophic; *L. monocytogenes*, *Y. enterocolitica*, non-proteolytic *Cl. botulinum* and *A. hydrophila* being amongst the most notable. Although psychrotrophic organisms, such as *L. monocytogenes*, are capable of growth at low temperatures, reducing the storage temperature (≤4°C) will significantly reduce the rate of growth (Beuchat and Brackett, 1990a; Carlin *et al.*, 1995). *L. monocytogenes* populations remained constant or decreased on packaged vegetables stored at 4°C, while at 8°C, growth of *L. monocytogenes* was supported on all vegetables, with the exception of coleslaw mix (Francis and O'Beirne, 2001a). Thus even mild temperature abuse during storage permits more rapid growth of psychrotrophic pathogens (Berrang *et al.*, 1989a; Carlin and Peck, 1996; Conway *et al.*, 2000; Farber *et al.*, 1998; García-Gimeno *et al.*, 1996; Rodriguez *et al.*, 2000).

Mesophilic pathogens, such as *Salmonella* and *E. coli* O157:H7, are unable to grow where temperature control is adequate (i.e. ≤4°C). However, if temperature abuse occurs, they may then grow. Survival of *Salmonella* in produce stored for extended periods in chilled conditions may be of concern (Piagentini *et al.*, 1997; Zhuang *et al.*, 1995); *Salmonella* survived on a range of vegetables for more than 28 days at 2–4°C (ICMSF, 1996). *E. coli* O157:H7 populations survived on produce stored at 4°C and proliferated rapidly when stored at 15°C (Richert *et al.*, 2000). Reducing the storage temperature from 8 to 4°C significantly reduced growth of *E. coli* O157:H7 on MAP vegetables; however, viable populations remained at the end of the storage period at 4°C (Francis and O'Beirne, 2001a).

The survival of viruses on produce also depends upon temperature. Survival of Hepatitis A virus on lettuce was significantly lower at room temperature than at 4°C (Bidawid *et al.*, 2001). These results are consistent with those of Bagdasaryan (1964), as well as with those of Badawy *et al.* (1985), who found the greatest survival rates of viruses were at refrigeration temperatures. The behaviour of protozoan parasites on refrigerated produce is not known. However, the increase in incidence of produce-linked outbreaks due to these organisms indicates that research in this area is necessary.

Besides its direct effect on pathogen survival/growth, temperature may indirectly affect pathogen growth. Temperature determines the respiration rate of produce, and therefore changes in gas atmospheres within packages, which may influence pathogen growth. Reducing the storage temperature also reduces the growth of the mesophilic spoilage microflora. In the absence of spoilage microflora, high populations of pathogens may be achieved and the item consumed because it is not perceived as spoiled. The elimination or significant inhibition of spoilage microorganisms should not be practised, as their interactions with pathogens may play an integral role in product safety.

Guidelines for handling chilled foods, published by the UK Institute of Food Science and Technology (IFST, 1990), recommend a storage temperature range of 0–5°C for prepared salad vegetables, noting that some vegetables may suffer damage if kept at the lower end of this temperature range. Strict control of refrigeration temperature throughout the chill-chain is crucial for maintaining microbiological safety.

12.3.2 Product type/product combinations

Produce may include whole or sliced/diced fruits, leaves, stems, roots, tubers or flowers (Burnett and Beuchat, 2001). While all produce items have factors in common, each product has a unique combination of compositional and physical characteristics and will have specific growing, harvesting and processing practices, and storage conditions.

Survival/growth of pathogens on produce varies significantly with the type of product (Austin *et al.*, 1998; Carlin and Nguyen-the, 1994; Jacxsens *et al.*, 1999). Dry coleslaw mix was largely unsuitable for *L. monocytogenes* and *E. coli* O157:H7 growth while significant growth of the pathogens occurred on shredded lettuce (Francis and O'Beirne, 2001a, b). Product factors that may affect pathogen survival and/or growth include: pH, presence of competitive microflora and/or naturally occurring antimicrobials and respiration rate/packaging interactions.

Product pH strongly influences the survival/growth of pathogens. Most vegetables have a pH of ≥ 5.0, and consequently support the growth of most foodborne bacteria. Many fruits have acidic pH; however, a number of melons/soft fruits have pH values ≥ 5.0 which will support growth of pathogens (Beuchat, 1996; NACMCF, 1999; Escartin *et al.*, 1989; Lund 1992; Nguyen-the and Carlin, 1994). *L. monocytogenes* survived and grew on apple slices and cantaloupe melon (Conway *et al.*, 2000; Ukuku and Fett, 2002), and whole tomatoes (Beuchat and Brackett, 1991). Acid tolerance is common in *E. coli* O157:H7 and *Salmonella* serotypes and these organisms can survive/grow in acidic produce (Dingman, 2000; Liao and Sapers, 2000; Ukuku and Sapers, 2001; Wei *et al.*, 1995; Zhuang *et al.*, 1995).

Some plant tissues have naturally occurring antimicrobials that provide varying levels of protection against pathogens (Lund, 1992; Sofos *et al.*, 1998). The inhibitory effects of raw carrots and carrot juice on growth of *L. monocytogenes* have been reported (Beuchat *et al.*, 1994; Beuchat and Brackett, 1990b; Jacxsens *et al.*, 1999; Nguyen-the and Lund, 1991). Garlic and onion extracts exhibited antimicrobial properties, red chicory was antagonistic to certain *Pseudomonas* spp. as well as to *A. hydrophila*, and cooked cabbage and Brussels sprouts were inhibitory towards *Listeria* (Beuchat *et al.* 1986; Beuchat and Brackett, 1990b; Jacxsens *et al.*, 1999; Nguyen-the and Carlin, 1994).

MAP produce harbours a large and diverse microflora. Effects of competition between the indigenous microflora and pathogens on MAP produce may play an important role in product safety (see Section 12.3.5). Beansprouts did not support good growth of *L. monocytogenes* or *E. coli* O157:H7, due presumably to competition from high populations of background microflora, inhibition from the relatively high in-pack CO_2 levels (25–30%) and the more limited nutrient availability of intact vegetables (Francis and O'Beirne, 2001a).

Minimally processed produce may be combined with cooked ingredients. Growth of *L. monocytogenes* on raw endive was probably limited by nutrient availability, but reached higher numbers when sweetcorn was added (Carlin *et al.*, 1996b; Nguyen-the *et al.*, 1996). The addition of cooked products to raw vegetables supplied a source of nutrients and permitted rapid growth of both spoilage and pathogenic populations on such products (Thomas and O'Beirne, 2000).

12.3.3 Minimal processing operations

The unit operations employed during the production of minimally processed produce (handling, peeling, slicing, washing, packaging) cause the destruction of surface cells, affect product respiration rate and pH, and release nutrients and possibly antimicrobial substances from the plant cells (Brackett, 1994), which will in turn affect the behaviour of pathogens.

In general, pathogens will not grow on uninjured surfaces of fresh intact produce; however, cutting or slicing facilitates contamination by pathogens and subsequent survival and/or growth. Injuries to the wax layer, cuticle and underlying tissues increased bacterial adhesion and growth (Han *et al.*, 2000a, 2001; Seo and Frank, 1999; Takeuchi and Frank, 2001; Takeuchi *et al.*, 2000). Consequently, minimising damage throughout harvesting and processing reduces the chances of pathogen contamination, penetration and growth (Liao and Cooke, 2001).

Pathogens can become attached to processing equipment (slicers, shredders) and once attached (biofilms) are very difficult to remove by chemical sanitisers (Bremer *et al.*, 2001; Frank and Koffi, 1990; Garg *et al.*, 1990; Jöckel and Otto, 1990; Nguyen-the and Carlin, 1994). Indeed, *L. monocytogenes* has been recovered from the environment of processing operations used to prepare minimally processed vegetables (Zhang and Farber, 1996), highlighting the importance of strict hygiene during processing. Recommendations implemented to ensure quality and safety of produce relate to good manufacturing practices (see Section 12.4; Koek *et al.* (1983), microbial specifications for the processed product, and proper storage conditions (Nguyen-the and Carlin, 1994).

Washing/antimicrobial dipping

Washing in tap water removes soil and other debris, some of the surface microflora, and cell contents and nutrients released during slicing that help support growth of microorganisms (Bolin *et al.*, 1977). However, water washing had minimal effects on microorganisms on fresh produce (Beuchat, 1992; Nguyen-the and Carlin, 1994; Brackett, 1987; Adams *et al.*, 1989; Izumi, 1999) and due to the re-use of wash water in industry may result in cross-contamination of food products and food-preparation surfaces (Beuchat and Ryu, 1997; Brackett, 1992; Beuchat, 1996; Garg *et al.*, 1990).

A variety of antimicrobial wash solutions have been used to reduce populations of microorganisms on fresh produce. The effectiveness of disinfection depends on a number of factors including: (i) type of treatment, (ii) type, numbers and physiology of the target microorganism(s), (iii) product type, (iv) disinfectant concentration, (v) pH of the disinfectant solution, (vi) exposure time, (vii) temperature of washing water and (viii) general sanitation of plant and equipment (Adams *et al.*, 1989; Best *et al.*, 1990; El-Kest and Marth, 1988a,b).

Chlorine (50–300ppm) is the most frequently used disinfectant for fresh fruits and vegetables; added to water as a solid, liquid or gas (Adams *et al.*, 1989; Anon., 1973; Beuchat and Ryu, 1997; Lund, 1983). Total microbial populations were reduced about 1000-fold when lettuce was dipped in water containing 300ppm total chlorine, but no effect was seen against microbial populations on red cabbage or carrots (Garg *et al.*, 1990). Generally, no more than 2- to 3-$\log_{10}$ reductions of bacteria on produce after chlorine treatment have been reported (Adams *et al.*, 1989; Beuchat, 1992; 1999).

The effects of chlorine in removing pathogens from produce have been studied. *L. monocytogenes* counts on Brussels sprouts were reduced approximately 100-fold by chlorine treatment (200mg/l), 10-fold more than those treated with water (Brackett, 1987). The maximum $\log_{10}$ reductions of *L. monocytogenes*, after treatment with chlorine (200ppm), were 1.7 for lettuce and 1.2 for cabbage (Zhang and Farber, 1996). Dipping coleslaw and lettuce in a chlorine solution (100ppm) reduced initial *L. innocua* and *E. coli* populations, but resulted in enhanced survival during extended storage at 8ºC (Francis and O'Beirne, 2002). Chlorine (100–200ppm) was only marginally effective at reducing *E. coli* levels on lettuce tissue surfaces (Beuchat, 1999), apple surfaces (Wisniewsky *et al.*, 2000; Wright *et al.*, 2000) and broccoli florets (Behrsing *et al.*, 2000). *Salmonella* populations on alfalfa sprouts were reduced by about 2 $\log_{10}$ CFU/g after treatment with 500ppm chlorine, and to undetectable levels after treatment with 2,000ppm chlorine (Beuchat and Ryu, 1997). Ten-minute exposures of *Y. enterocolitica* on shredded lettuce to 100 and 300ppm chlorine resulted in population reductions of 2–3 $\log_{10}$ cycles (Escudero *et al.*, 1999). In the same study, a combination of 100ppm chlorine and 0.5% lactic acid inactivated *Y. enterocolitica* by >6 log cycles, suggesting that *Y. enterocolitica* may be more sensitive to chlorine than other pathogens.

Chlorine, used at concentrations currently permitted in the industry to wash fresh produce, cannot be relied upon to eliminate pathogens (see Chapter 23). The ineffectiveness of chlorine treatment may be due to a number of factors. The hydrophobic nature of the waxy cuticle on produce protects surface contaminants from exposure to chlorine which does not penetrate or dissolve these waxes/oils (Adams *et al.*, 1989). In addition, microbial cells may become embedded in crevices, creases or injured tissues and are inaccessible to chlorine treatments (Adams *et al.*, 1989; Lund, 1983; Koseki *et al.*, 2001; Seo and Frank, 1999; Takeuchi and Frank, 2000, 2001). Organic matter (fruit and vegetable components) neutralises chlorine, rendering it inactive against microorganisms (Beuchat, 1996; Beuchat *et al.*, 1998; Lund, 1983). It is important to sanitise injured surfaces before cutting as once cut or injured surfaces are contaminated by pathogens, it is very difficult to remove these attached and growing microorganisms. The most useful effect of chlorine may be in inactivating vegetative cells in washing water and on equipment during processing as part of a HACCP system, thus avoiding build-up of bacteria and cross-contamination (Wilcox *et al.*, 1994).

A concern regarding the use of chlorine dips is that pathogens may not be fully eliminated by commercial treatments, while at the same time natural competitive organisms may be removed. *L. monocytogenes* inoculated onto disinfected (10% hydrogen peroxide) endive leaves grew better than on water-rinsed produce (Carlin *et al.*, 1996b) and dipping lettuce in a chlorine (100ppm) solution followed by storage at 8ºC, significantly enhanced *Listeria* growth compared with undipped samples (Francis and O'Beirne, 1997). Disinfection before contamination with the pathogen occurs may increase growth of the pathogen because populations of competing microflora have been removed (Bennik *et al.*, 1996). Therefore, temperature management (i.e. ≤4ºC) after reduction of microbial populations is crucial for microbial safety.

Due to the ineffectiveness of chlorine in removing pathogens from produce and increasing concern over the production of chlorinated organic compounds and their impact on human and environmental safety, a variety of other disinfectants, including acidic electrolysed water (Park *et al.*, 2001), peroxyacetic acid (Park and Beuchat, 1999), chlorine dioxide (Zhang and Farber, 1996), hydrogen peroxide (Sapers and Simmons, 1998), organic acids (Karapinar and Gonul, 1992), trisodium phosphate (Zhang and Farber, 1996) and ozone (Burrows *et al.*, 1999) have been evaluated (Beuchat, 1999) (see Chapter 23). However, none of the sanitiser treatments tested is likely to be totally effective against all pathogens, and behaviour of pathogens during subsequent storage remains unpredictable (Beuchat and Ryu, 1997; Escudero *et al.*, 1999; Park and Beuchat, 1999; Zhang and Farber, 1996).

Viruses and protozoan cysts on fruits and vegetables generally exhibit higher resistance to disinfectants than do bacteria or fungi (Beuchat, 1998). Feline caliciviruses were very resistant to commercial disinfectants; however, peroxyacetic acid and H_2O_2 were effective at decontaminating strawberries and lettuce when used at four-fold higher concentrations than generally recommended (Gulati *et al.*, 2001). Treatment of *Cryptosporidium parvum* oocysts with 1ppm ozone for five minutes resulted in <1 $\log_{10}$ inactivation (Korich *et al.*, 1990).

12.3.4 Package atmosphere

When a sliced product is packaged, it continues respiring thereby modifying the gas atmosphere inside the package. Ideally, O_2 levels will fall from the 21% found in air to 2–5%, and CO_2 levels will increase to the 3–10% range. Atmospheres within MA packages might be cause for public health concern in at least three ways. The atmospheres and refrigeration temperatures employed may inhibit the development of some spoilage aerobic microorganisms (Daniels *et al.*, 1985; Farber, 1991). Consequently, their suppression may facilitate pathogen survival/growth, without the product showing obvious signs of spoilage. Secondly, MAP increases the shelf-life of products, thus increasing the time available for pathogens to grow. Over-extending the shelf-life may allow development of significant populations, particularly if combined with exposure to even modest abuse temperatures. Thirdly, although the low levels of O_2 (2–5%) within packages (e.g. at 4°C) should inhibit growth of obligate anaerobes such as *Cl. botulinum*, if packages are subjected to temperature abuse, they may become anaerobic as a result of increased product respiration. This could enable growth and toxin production by *Cl. botulinum* to occur (see Section 12.2.2). In addition to the target atmospheres described above, there is evidence of significant incidence of 'unintended' atmospheres in commercial practice. Where the gas permeability of packaging films is insufficient, produce with high respiration rates may generate MAs which are anoxic and/or contain high levels (>20%) of CO_2.

Of particular concern with refrigerated MAP produce is the growth of psychrotrophic, facultatively anaerobic and microaerophilic microorganisms, which can tolerate refrigeration temperatures and low O_2 atmospheres (Bennik *et al.*, 1995), and a number of studies have indicated that MAP may select for such pathogens (Hintlian and Hotchkiss, 1986; Brackett, 1994; Beuchat and Brackett, 1990a; Kallender *et al.*, 1991). There are inconsistencies in the literature regarding the effects of MAP on the growth of *L. monocytogenes*. Numerous researchers have shown that survival of *L. monocytogenes* on produce remains largely unaffected by MAP (Amatanidou *et al.*, 1999; Berrang *et al.*, 1989b; Beuchat and Brackett; 1990a, 1991; Conway *et al.*, 1998, Jacxsens *et al.*, 1999). However, in other work, nitrogen flushing combined with storage at 8°C enhanced the growth of *L. monocytogenes* on shredded lettuce (Francis and O'Beirne, 1997) and shredded chicory salads (Ringlé *et al.*, 1991).

Several studies have demonstrated that *A. hydrophila* can grow rapidly on vegetables stored at 4–5°C and MAP does not significantly affect its growth (Berrang *et al.*, 1989a). Austin *et al.* (1998) found that some samples of MAP vegetables appeared organoleptically acceptable when *Cl. botulinum* toxin was detected (see Section 12.2.2). *Salmonella* and *E. coli* O157:H7 can grow under MAP conditions; however, there is insufficient information available on whether atmospheres inhibit or enhance their growth. CO_2 had little or no inhibitory effect on growth of *E. coli* O157:H7 on shredded lettuce stored at 13 or 22°C, and growth potential was increased in an atmosphere of $O_2/CO_2/N_2$: 5/30/65, compared with growth in air (Abdul-Raouf *et al.*, 1993; Diaz and Hotchkiss, 1996). A recent study, investigating *C. jejuni* survival on MAP vegetables found that refrigeration temperatures in combination with a MA (2% O_2, 18% CO_2 and 80% N_2) were favourable for the organism (Tran *et al.*, 2000). The highest rates of Hepatitis A virus survival on lettuce stored at 4°C (12 days) was observed under 70% CO_2/30% N_2 and 100% CO_2 (Bidawid *et al.*, 2001).

In 'unintended' atmospheres, high CO_2 levels may develop within packages. Carlin *et al.* (1996a) examined the survival of *L. monocytogenes* on chicory leaves stored at 10°C in air, or under 10%, 30% or 50% CO_2, with 10% O_2 and found that *L. monocytogenes* grew better as the concentration of CO_2 increased. The growth rate of *A. hydrophila* decreased with increasing CO_2 concentrations, but maximum population densities were not affected by CO_2 concentrations of up to 50% (Bennik *et al.*, 1995). Novel, alternative techniques to low O_2 MAP are the use of high O_2 (i.e. >70% O_2) atmospheres and noble gases (Day, 1996; see Chapter 10). In an agar-based study to investigate the effects of high O_2 (90%) and moderate CO_2 (10–20%) concentrations on foodborne pathogens at 8°C, Amanatidou *et al.* (1999) noted inhibitory action against *L. monocytogenes*, *A. hydrophila*, *S.* Typhimurium, *S.* Enteritidis and *E. coli*.

Studies to determine the behaviour of *Y. enterocolitica* on MAP produce have not been published and information describing the survival of *Campylobacter* spp. on MAP produce is extremely limited. In addition, the behaviour of protozoan parasites under MAP is not known. Therefore, more research to determine the survival of these and other pathogens on MAP produce is warranted.

12.3.5 Competition between the indigenous microflora and pathogen

MAP produce harbours large populations of native microorganisms including pseudomonads, lactic acid bacteria (LAB) and *Enterobacteriaceae* (Francis *et al.*, 1999; Nguyen-the and Carlin, 1994). The background microflora provide indicators of temperature abuse largely by causing detectable spoilage, and can vary significantly for each product and during storage.

LAB can exert antibacterial effects due to one or more of the following mechanisms: lowering the pH (Raccach and Baker, 1979); generating H_2O_2 (Price and Lee, 1970); competing for nutrients (Iandolo *et al.*, 1965); and possibly by producing antimicrobial compounds, such as bacteriocins (Arihara *et al.*, 1993; Harris *et al.*, 1989; Klaenhammer, 1988). Cai *et al.* (1997) reported that a large portion of LAB isolates from bean sprouts inhibited the growth of *L. monocytogenes*. Strains of LAB were reported to inhibit *A. hydrophila*, *L. monocytogenes*, *S.* Typhimurium, and *Staphylococcus aureus* on vegetable salads (Vescovo *et al.*, 1996). Competition from LAB may limit pathogen growth on produce, but there is insufficient data available to prove this conclusively.

Various researchers have reported antagonism by the native microflora of vegetables against *Listeria* (Francis and O'Beirne, 1998a,b; Liao and Sapers, 1999). Reducing the background microflora of endive leaves (Carlin *et al.*, 1996b) and shredded lettuce (Francis and O'Beirne, 1997) resulted in enhanced growth of *Listeria*. A mixed bacterial population isolated from endive or lettuce reduced *Listeria* growth in vegetable media (Carlin *et al.*, 1996b; Francis and O'Beirne, 1998a, b). However, the inhibitory effects were dependent on gas atmosphere; in 3% O_2 (balance N_2) growth of the mixed population was inhibited while *L. monocytogenes* proliferated (Francis and O'Beirne, 1998a). Fluorescent pseudomonads have previously been shown to stimulate growth of *L. monocytogenes* in various foods, due to the release of potential nutrients by pseudomonads (Nguyen-the and Carlin, 1994; Liao and Sapers, 1999; Marshall and Schmidt, 1991). Bennik *et al.* (1996) found that strains of fluorescent pseudomonads slightly reduced final population densities of *L. monocytogenes* in an endive leaf medium. *P. fluorescens* and *P. viridiflava* inhibited growth of *L. monocytogenes* on potato slices while *Erwinia carotovora* and *Xanthomonas campestris* did not affect its growth (Liao and Sapers, 1999). *Enterobacter* isolates (*Enterobacter cloacae*, *Enterobacter agglomerans*) significantly reduced *L. monocytogenes* growth during storage on a model medium; however, the inhibitory activities of *Enterobacter* spp. decreased as the concentration of CO_2 increased (Francis and O'Beirne, 1998a). Del Campo *et al.* (2001) also found that *Enterobacteriaceae* (*Enterobacter agglomerans*, *Rhanella aquatilis*) reduced maximum population densities of *L. monocytogenes* in minimal media, presumably due to competition for glucose and/or amino acids. Ukuku and Fett (2002) reported that the native microflora of cantaloupe melon, especially the yeast and mould populations, might have out-competed *L. monocytogenes* for colonisable space and available nutrients, thus resulting in the decline of populations of *L. monocytogenes*.

Growth of the background microflora also significantly affected the growth and toxigenesis of *Cl. botulinum* in refrigerated foods (Hutton *et al.*, 1991; Hauschild, 1989) and Larson and Johnson (1999) demonstrated the ability of spoilage microflora to protect against *Cl. botulinum* outgrowth. *A. hydrophila* has been reported to be a poor competitor with LAB and other spoilage organisms (Palumbo and Buchanan, 1988). MAP and chill temperatures, combined with the use of a *Lactobacillus casei* inoculum, reduced growth of *A. hydrophila* on vegetables such as lettuce (Vescovo *et al.*, 1997). Competitive microflora had a significant effect on the growth of *E. coli* O157:H7 in broth media; *Hafnia alvei* significantly inhibited the growth of *E. coli* O157:H7 at 37ºC, whereas *Pseudomonas fragi* inhibited growth of the pathogen at 15ºC (Duffy *et al.*, 1999). Little is known about the mechanism by which *Salmonella* manages to compete with natural microflora and survive on plant products (Liao and Cooke, 2001). Wells and Butterfield (1997) demonstrated that salmonellae grew better on vegetables when co-cultured with *Erwinia carotovora* or *P. viridiflava*, two major causes of bacterial soft-rot.

Complex interactions with the indigenous microflora may have significant effects on survival/ growth of pathogens. More research needs to be done to examine the influence of gas atmospheres, background microflora and storage temperatures on the survival/growth of pathogens, including foodborne viruses and protozoan parasites on MAP produce in order to ensure that novel mild preservation technology can continue to be applied safely.

12.3.6 Minimal processing and stress responses

Pathogenic bacteria can respond or adapt to sub-lethal stresses encountered in minimal processing in ways that increase their resistance to more severe treatments and enable better survival in foods (Buncic and Avery, 1998; Abee and Wouters, 1999; Gahan and Hill, 1999). Apart from the enhanced survival in foods and increased resistance to subsequent food processing/preservation treatments, adapted or hardened pathogens may also have enhanced virulence (Abee and Wouters, 1999; Gahan and Hill, 1999; Rouquette *et al.*, 1998).

Two of the best studied adaptive tolerance responses are to heat (heat stress response) and to acid (acid tolerance response, ATR). Acid adapted *L. monocytogenes*, *Salmonella* and *E. coli* O157:H7 survived significantly better in acidic foods such as salad dressing and fruit juices, when compared to non-adapted cells (Gahan *et al.*, 1996; Leyer and Johnson, 1992). Acid adaptation induces acid tolerance to more severe or normally lethal acid, but it can also induce cross-protection against other environmental stresses such as thermal and osmotic stress (Leyer and Johnson, 1993; Lou and Yousef, 1997; O'Driscoll *et al.*, 1996). Equally other stresses can induce acid tolerance. Acid adaptation enhanced survival of *L. monocytogenes* during storage in packages of vegetables which had relatively high in-pack CO_2 levels (25–30% in MAP coleslaw and bean sprouts; Francis and O'Beirne, 2001b). *E. coli* O157:H7 survived in an acidic environment better at 4ºC than at 10ºC, which implies that induction of acid tolerance may enhance resistance to low temperature (Conner and Kotrola, 1995).

12.3.7 Implications of strain variation among pathogens

The selection of strain(s) of a particular pathogen to be used in survival studies is extremely important as different strains may behave differently on MAP produce. Unpublished work carried out by the authors has shown that strains of *L. monocytogenes* differ significantly in their inherent ability to survive/grow on MAP vegetables. In addition, there was significant variation among strains in their inherent stress resistance characteristics; some strains may be more resistant to the stressful conditions encountered in foods and during food processing.

Although the response of *L. monocytogenes* to food related growth factors (e.g. temperature, pH, gas atmosphere) has been studied extensively, in most studies only one strain has been tested. In studies where multiple isolates were examined, significant strain variation in resistance existed among *L. monocytogenes* isolates (Begot *et al.*, 1997; Barbosa *et al.*, 1994; Buncic *et al.*, 2001; Dykes and Moorhead, 2000; Mackey *et al.*, 1990; Palumbo *et al.*, 1995) and there were some differences between serotypes (Davies and Adams, 1994; Embarek and Huss, 1993; Sörqvist, 1994). Junttila *et al.* (1988) reported that there were differences in ability of *L. monocytogenes* strains to grow at low temperatures, with strains in the serotype 1/2 capable of growth at colder temperatures than strains of serotype 4b. Evidence also suggests that *L. monocytogenes* strains differ at the molecular level; however, little is known of the attributes that contribute to the ability of certain strains to cause disease, an ability that can vary significantly between individual strains (Barbour *et al.*, 1996; Del Corral *et al.*, 1990; Tabouret *et al.*, 1991; Brosch *et al.*, 1993; Farber and Peterkin, 1991; Rocourt, 1994).

The diversity of the genus *Salmonella* has been observed in many different forms, from genetic to physiological observations. The ability of *E. coli* O157:H7 to tolerate heat was strain dependent (Clavero *et al.*, 1998; Duffy *et al.*, 1999) and survival of *E. coli* O157:H7 on vegetables depended on bacterial strain and product type (Francis and O'Beirne, 2001a). Different strains of pathogens may respond differently to treatments including mild acid, low temperature and gas atmosphere, which may result in variations in the ability of surviving populations to cause human disease (Buncic *et al.*, 2001).

12.4 Improving MAP to reduce pathogen risks

The pathogen risks from MAP produce cannot be totally eliminated, but they can be minimised by applying best practice at every stage – agricultural production, pre-processing, processing, distribution and final use. At all stages, strategies to minimise contamination by pathogens, product storage at ≤4°C and education/training of workers and consumers are important recurring themes. Clearly, Good Agricultural Practices (GAP) and Good Manufacturing Practices (GMP) need to be put in place to minimise hazards, and many Codes of Practice have been published by national agencies (e.g. FSAI, 2001) and industry sectors. However, there may be insufficient data available on which to base a comprehensive validated Hazard Analysis and Critical Control Points (HACCP) Programme for most produce items (NACMCF, 1999).

12.4.1 Production of raw materials

Agricultural production practices can have major implications for contamination of raw produce with pathogens (Gorny and Zagory, 2000), and producer awareness of their role in assuring food safety is vitally important. This is an extremely complex arena with great diversity in crop production methods, scale, environmental factors, etc. (FDA, 1998). Land subject to flooding or on which animals have grazed should be avoided (Brackett, 1999). Improperly composted sewage or animal manure should not be applied to land where vegetables for processing are grown. However, persistence of *L monocytogenes* has been demonstrated even in treated sewage sludge (Al Ghazali and Al Azawi, 1986). Proximity to animal production facilities may also be a significant cross-contamination hazard. Irrigation should be carried out with clean water, pretreated if necessary (Robinson and Adams, 1978) and applied as trickle irrigation at ground level rather than as an overall spray (NACMCF, 1999). However, serious deficits in water quality and availability exist globally, with water pollution from sewage and animal production facilities posing serious problems.

Workers involved in harvesting and handling should be trained in the principles of good sanitation and provided with adequate washing/toilet facilities in fields and packhouses (Brackett, 1999). Harvesting equipment should be thoroughly cleaned and sanitised. Birds such as gulls and pigeons, wild animals, domestic animals and insects should be excluded from packhouses and processing areas (NACMCF, 1999). Wild birds are known to disseminate *Campylobacter, Salmonella, Vibrio cholera, Listeria* species and *E.coli* O157, apparently picked up from feeding on garbage, sewage, etc; control of preharvest contamination of produce by wild birds is particularly difficult (Beuchat and Ryu, 1997).

Increasing globalisation of produce supplies poses serious new challenges (Tauxe, 1997) and knowledge of contamination levels in imported produce is minimal (Beuchat and Ryu, 1997). The only rational solution is the extension of the requirement for GAPs to wherever primary production takes place.

12.4.2 Minimal processing

Based on the data discussed in Section 12.3, processing can be geared to minimise opportunities for pathogen contamination and growth. Starting at harvest, bruising and cutting should be minimised prior to processing (Liao and

Cooke, 2001). Immediately prior to processing, preliminary decontamination should be carried out by removing outer leaves, soil, etc., from produce using sharp sanitised knives for any cutting. Peeling, cutting, shredding, etc., should be carried out with equipment designed to cause the minimum of tissue disruption, as severe processing may facilitate more effective contamination and subsequent growth by pathogens (Gleeson *et al.*, 2002). Severe processing may also reduce the effectiveness of subsequent anti-microbial treatments (Han *et al.*, 2000a; Han *et al.*, 2000b; Liao and Cooke, 2001; Liao and Sapers, 2000; Takeuchi and Frank, 2000, 2001). GMP should include effective surface and machine sanitising to eliminate the risk of pathogen contamination from the processing environment or from machines used in processing (Zhang and Farber, 1996; Nguyen-the and Carlin, 1994). Food safety experts should be consulted by engineers designing processing equipment to ensure ease of sanitisation (Beuchat and Ryu, 1997). Human contact should be eliminated or minimised to reliable trained staff.

Although its benefits are questioned (Brackett, 1999), anti-microbial dipping is probably a valuable tool for reducing numbers of potential pathogens (Beuchat and Ryu, 1997). State-of-the-art effective systems are available and should be used. Some of these greatly reduce the levels of chlorine needed (Varoquaux, 2001). Special care should be exercised to avoid contamination after dipping. Post-processing risks introduced by anti-microbial dips (Francis and O'Beirne, 1997; Carlin *et al.*, 1996b; Bennik *et al.*, 1996) should be addressed in HACCP protocols: the most important of these are measures to ensure that products are stored at $\leq$4°C at all times and the use of conservative use-by dates. Where alternatives to chlorine are being introduced, any differences in their anti-microbial effects should be understood and taken into account.

12.4.3 Modified atmosphere packaging

Packaging materials must be carefully selected to ensure that their gas permeability properties match the respiration rates of the products being packaged. This is necessary in order to achieve package atmospheres within the technically useful range of 2–5% O_2 and 3–10% CO_2 (Cliffe-Byrnes *et al.*, 2003; Barry-Ryan *et al.*, 2000). Technical advice from researchers and packaging suppliers is essential (see Section 12.6), though user-friendly software may be developed to assist industry in the future. Poor 'package-product compatibility' will result in the creation of unintended atmospheres with uncertain microbiological implications (Bennik *et al.*, 1998).

The use of coatings with gas barrier properties can be a feature of MAP produce (Guilbert *et al.*, 1996), but more information is needed on their effects on internal atmospheres. Other novel elements of MAP include the use of high oxygen and noble gas enriched atmospheres (see Chapter 10). While atmospheres with 80% oxygen have been used in MAP of fresh meat for a few decades and appear safe, less is known about the effects of noble gases such as argon on microbial ecology. The microbial quality of the final packaged product should be monitored to ensure that it complies with international guidelines (see Francis *et al.*, 1999).

12.4.4 Distribution and final use

Ensuring that temperatures are kept at or below 4ºC throughout the cold chain is essential for microbial safety and requires considerable attention to detail. Refrigerated distribution requires suitably designed vehicles, properly loaded to allow for air movement (Brackett, 1999). At supermarket level, LeBlanc *et al.* (1996) found 90% of produce items above 4^0C in supermarket chill cabinets. Problems can also arise at consumer level where products are held for extended periods in cars or experience elevated temperatures in (poorly operating) domestic refrigerators. Time temperature indicators embedded in the packaging may have a significant role in ensuring that safe storage temperatures are used (see Chapter 6). Distributors, retailers and consumers must be educated on the importance of low storage temperatures. Consumers also need to be educated on the nature of minimal processing technologies for fresh foods, in particular that consumption of apparently fresh food beyond its use-by date is potentially hazardous.

General principles of good hygiene must apply throughout the distribution chain, particularly avoiding cross-contamination. For example, refrigerated trucks carrying vegetables on an outward journey may be used to 'backhaul' animals or raw meats (Brackett, 1999). Truck use needs to be monitored and vehicles appropriately sanitised. In the food service sector training of operatives is important since many of these are teenagers and may require food hygiene training within the school system, as they receive little food preparation experience in the modern home (Beuchat and Ryu, 1997).

12.4.5 HACCP strategies

While HACCP principles are being applied by many growers, manufacturers, and distributors based on current knowledge, the US National Advisory Committee on Microbiological Criteria for Foods claim that there is insufficient evidence to put in place a comprehensive validated system for fresh produce (NACMCF, 1999). Model farm-to-table HACCP protocols have been developed for only a few commodities (sprouted seeds, shredded lettuce and tomatoes) but even these have not been completely validated. According to NACMCF, further research is needed to provide data and technology for the validated control measures needed for GAP and GMP.

12.5 Future trends

The recent rapid growth in the volume of produce consumption and in the globalisation of sourcing can be expected to continue. There will be improved information on emerging and existing pathogens and their interaction with production and processing technologies. There will be greater application of new and existing technology and of best practice.

12.5.1 Production of raw materials

Greater emphasis can be expected on the development and application of GAP protocols, particularly for use of water and manure, for worker hygiene and for transportation of produce. Serious efforts will be made to apply these protocols to production in developing as well as industrialised countries. These initiatives should result in a safer, more reliable raw material stream.

12.5.2 Processing, packaging, and distribution

Current interest in alternatives to chlorine dipping are likely to result in novel chemical treatments and the application of physical treatments such as UV radiation (see Chapter 23). Ionising radiation may also be used either alone or in combination with other treatments, as a means of extending the shelf-life of produce (Diehl, 1995; Langerak, 1978). Doses in the range of <1 to 3 kGy have been shown to reduce or eliminate populations of pathogens and postharvest spoilage organisms on produce (Farkas, 1997) and salmonellae were not recovered from alfalfa sprouts irradiated with 0.5 kGy (Rajkowski and Thayer, 2000). However, despite the efficiency, safety and suitability to products with surface contamination (O'Beirne, 1989) the use of irradiation will depend on its acceptance by consumers.

Greater use of edible coatings (e.g. sucrose polyesters of fatty acids, cellulose derivatives, etc.) to food surfaces can be expected (Krochta and De Mulder-Johnston, 1997; Baldwin *et al.*, 1995). Edible coatings (e.g. hydroxypropyl methylcellulose) can extend shelf-life, and with the inclusion of anti-microbials, reduce the potential growth of pathogens (Zhuang *et al.*, 1996).

Other additional novel processing steps may be introduced such as inoculation of MAP produce with organisms inhibitory to one or more pathogens. For example, strains of LAB inhibited *A. hydrophila*, *L. monocytogenes*, *S. typhimurium*, and *Staphylococcus aureus* on vegetable salads (Vescovo *et al.*, 1996) and use of a *Lactobacillus casei* inoculum, reduced growth of *A. hydrophila* on MAP vegetables such as lettuce (Vescovo *et al.*, 1997).

More reliable package atmospheres can be expected as a result of improved materials, temperature-responsive smart packaging and better software to define gas permeability requirements for individual products. There will be widespread commercial application of active packaging with anti-microbial and other properties.

Stress responses of pathogens and cross-protection must be considered when current food processing technologies are being modified or new ones developed. These responses are particularly significant in minimal processing/packaging technology where the imposition of one sub-lethal stress may lead to the induction of multiple stress responses that may reduce the efficacy of later treatments (Hill *et al.*, 1995). Strategies to prevent such stress responses would facilitate the development of improved procedures for prevention of pathogen survival and growth. For example, new decontamination technologies will be developed which provoke minimal levels of stress response. More generally, micro-array technology will be used to assess the response of both plant materials and pathogens to processing and storage regimes.

In relation to temperature control, greater use of IT can be expected in wireless and internet based data collection/operator alerting systems such as those developed by Freshloc Technologies Inc. for monitoring product temperatures during transportation.

12.5.3 Research

Research trends driven by the needs of this sector include greater understanding of emerging pathogens, particularly viruses and protozoan parasites; greater understanding of processes of produce contamination generally, and of how to prevent them; the development of new effective decontamination technologies; development and application of active and intelligent packaging. In order to improve surveillance for food-borne illness, there is a need for greater use of molecular techniques for sub-typing of pathogens (serotyping/molecular typing). This technology can help establish sources/points of contamination, links between geographically isolated outbreaks of food poisoning with a common source (NACMCF, 1999), and provide other types of data which will help develop HACCP protocols which can be validated.

12.6 Sources of further information and advice

Campden and Chorleywood Food Research Association (1996), Code of practice for the manufacture of vacuum and modified atmosphere packaged chilled foods with particular regards to the risks of botulism, Guideline No. 11.

Campden and Chorleywood Food Research Association (1992), Guidelines for the good manufacture and handling of modified atmosphere packed food products, Technical Manual No. 34.

Codex Alimentarius http://www.fao.org/es*/esn/codex/

Food and Drug Administration (1998), Guide to minimize microbial food safety hazards for fresh fruits and vegetables. http:www.foodsafety.gov/~dms/prodguid.html

Food and Drug Administration (2000), Kinetics of microbial inactivation for alternative food processing technologies. http:vm.cfsan.fda.gov/~comm/ift-toc.html

Food Safety Authority of Ireland (FSAI, 2001), Code of practice for food safety in the fresh produce supply chain in Ireland, Code of practice No. 4. ISBN 0953918343.

Institute of Food Science and Technology (1990), Guidelines for the handling of chilled foods. ISBN 0905367073.

Institute of Food Science and Technology (1992), Guidelines to good catering practice. ISBN 090536709X.

International Fresh-cut Produce Association (1996), in *Food Safety Guidelines for the Fresh-cut Product Industry*, 3rd edn, Alexandria (VA), IFPA. 125p.

12.7 References

ABDUL-RAOUF, U.M., BEUCHAT, L.R. and AMMAR, M.S., (1993), 'Survival and growth of *E. coli* O157:H7 on salad vegetables', *Appl. Environ. Microbiol.*, 59, 1999–2006.

ABEE, T. and WOUTERS, J.A., (1999), 'Microbial stress response in minimal processing', *Int. J. Food Microbiol.*, 50, 65–91.

ACKERS, M., MAHON, B.E., LEAHLY, E., GOODE, B., DAMROW, T., HAYES, P.S., BIBB, W.F., RICE, D.H., BARRETT, T.J., HUTWAGNER, L. *et al.*, (1998), 'An outbreak of *Escherichia coli* O157:H7 infections associated with leaf lettuce consumption', *J. Infect. Dis.*, 177, 1588–93.

ADAMS, M.R., HARTLEY, A.D. and COX, L.J., (1989), 'Factors affecting the efficacy of washing procedures used in the production of prepared salads', *Food Microbiol.*, 6, 69–77.

ALLEN, A.B., (1985), 'Outbreak of campylobacteriosis in a large educational institution – British Columbia.', *Can Dis. Weekly Rep.*, 2, 28–30.

AL-GHAZALI, M.R. and AL-AZAWI, S.K., (1986), 'Detection and enumeration of *Listeria monocytogenes* in a sewage treatment plant in Iraq', *J. Appl Bacteriol*, 60, 251–4.

AL-GHAZALI, M.R. and AL-AZAWI, S.K., (1990), *Listeria monocytogenes* contamination of crops grown on soil treated with sewage sludge cake. *J. Appl Bacteriol.*, 69, 642–7.

ALTRKRUSE S.F., HUNT, J.M., TOLLEFSON, L.K. and MADDEN, J.M., (1994), 'Food and animal sources of human *Campylobacter jejuni*', *J. Am. Vet. Med. Assoc.*, 204, 57–61.

ALTWEGG, M. and GEISS, H.K., (1989), '*Aeromonas* as a human pathogen', *Crit. Rev. Microbiol.*, 16, 253–86.

AMANATIDOU, A., SMID, E.J. and GORRIS, L.G.M., (1999), 'Effect of elevated oxygen and carbon dioxide on the surface growth of vegetable-associated microorganisms', *J. Appl. Microbiol.*, 86, 429–38.

ANONYMOUS, (1973), 'Determination of chlorine in water', *Chem. Proc.*, 19, 13–15.

ARIHARA, K., CASSENS, R.G. and LUCHANSKY, J.B., (1993), 'Characterization of bacteriocins from *Enterococcus faecium* with activity against *Listeria monocytogenes*', *Int. J. Food Microbiol.*, 19, 123–34.

ARUMUGASWAMY, R.K., RUSUL RAHAMAT ALI, G. and NADZRIAH BTE ABD. HAMID, S., (1994), 'Prevalence of *Listeria monocytogenes* in foods in Malaysia',

Int. J. Food Microbiol., 23, 117–21.

AUSTIN, J.W., DODDS, K.L., BLANCHFIELD, B. and FARBER, J.M., (1998), 'Growth and toxin production by *Clostridium botulinum* on inoculated fresh-cut packaged vegetables', *J. Food Prot.*, 61, 324–8.

BADAWY, A.S., GERBA, C.P. and KELLEY, L.M., (1985), 'Survival of rotavirus SA-11 on vegetables', *Food Microbiol.*, 2, 199–205.

BAGDASARYAN, G.A., (1964), 'Survival of viruses of the enterovirus group (poliomyelitis, echo, coxsackie) in soil and on vegetables', *J. Hyg. Epidemiol. Microbiol. Immunol.*, 8, 497–505.

BALDWIN, E.A., NISPEROS-CARRIEDO, M.O. and BAKER, R.A., (1995), 'Use of edible coatings to preserve quality of lightly (and slightly) processed products', *Crit. Rev. Food Sci. Nutr.*, 35, 509–24.

BARBOUR, A.H., RAMPLING, A. and HORMAECHE, C.E., (1996), 'Comparison of the infectivity of isolation of *Listeria monocytogenes* following intragastric and intravenous inoculation in mice', *Microb. Pathogenesis*, 20, 247–53.

BARBOSA, W.B., CABEDO, L., WEDERQUIST, H.J., SOFOS, J.N. and SCHMIDT, G.R., (1994), 'Growth variation among species and strains of *Listeria* in culture broth', *J. Food Prot.*, 57, 765–9, 775.

BARRY-RYAN, C. and O'BEIRNE, D., (1998), 'Effects of slicing method on the quality and storage-life of modified atmosphere packaged carrot discs', *J. Food Sci.*, 63, 851–6.

BARRY-RYAN, C. and O'BEIRNE, D., (2000), 'Effects of peeling method on the quality of ready-to-use carrots', *Int. J. Food Sci. Technol.*, 35, 243–54.

BARRY-RYAN, C., J.M.PACUSSI AND O'BEIRNE, D., (2000), 'Quality of shredded carrots as affected by packaging film and storage temperature', *J. Food Sci.*, 65, 726–30.

BEAN, N.H. and GRIFFIN, P.M., (1990), 'Foodborne disease outbreaks in the United States, 1973–1987: pathogens, vehicles and trends', *J. Food Prot.*, 53, 807–14.

BECKERS, H.J., IN'T VELD, P.H., SOENTORO, P.S.S. and DELFGOU-VAN ASH, E.H.M., (1989), 'The occurrence of *Listeria* in foods', symposium proceedings *Foodborne Listeriosis,* Wiesbaden, Hamburg, September 7, 1988, 83–97.

BEGOT, C., LEBERT, I. and LEBERT, A., (1997), 'Variability of the response of 66 *Listeria monocytogenes* and *Listeria innocua* strains to different growth conditions', *Food Microbiol.*, 14, 403–12.

BEHRSING, J., WINKLER, S., FRANZ, P. and PREMIER, R., (2000), 'Efficacy of chlorine for inactivation of *Escherichia coli* on vegetables', *Postharvest Biol. Technol.*, 19, 187–92.

BENNIK, M.H.J., SMID, E.J., ROMBOUTS, F.M. and GORRIS, L.G.M., (1995), 'Growth of psychrotrophic foodborne pathogens in a solid surface model system under the influence of carbon dioxide and oxygen', *Food Microbiol.*, 12, 509–19.

BENNIK, M.H.J., PEPPELENBOS, H.W., NGUYEN-THE, C., CARLIN, F., SMID, E.J. and GORRIS, L.G.M., (1996), 'Microbiology of minimally processed, modified atmosphere packaged chicory endive', *Postharvest Biol. Technol.*, 9, 209–21.

BENNIK, M.H.J., VORSTMAN,W., SMID, E.J. and GORRIS, L.G.M., (1998), 'The influence of oxygen and carbon dioxide on the growth of prevalent *Enterobacteriaceae* and *Pseudomonas* species isolated from fresh and controlled-atmosphere-stored vegetables', *Food Microbiol.*, 15, 459–69.

BERRANG, M.E., BRACKETT, R.E. and BEUCHAT, L.R., (1989a), 'Growth of *Aeromonas hydrophila* on fresh vegetables stored under a controlled atmosphere, *Appl. Environ. Microbiol.*, 55, 2167–71.

BERRANG, M.E., BRACKETT, R.E. and BEUCHAT, L.R., (1989b), 'Growth of *Listeria monocytogenes* on fresh vegetables stored under controlled atmosphere', *J. Food Prot.*, 52, 702–5.

BEST, M., KENNEDY, M.E. and COATES, F., (1990), 'Efficacy of a variety of disinfectants against *Listeria* spp.' *Appl. Environ. Microbiol.*, 56, 377–80.

BEUCHAT, L.R., (1992), 'Surface disinfection of raw produce', *Dairy Food Environ. Sanitat.*, 12, 6–9.

BEUCHAT, L.R., (1996), 'Pathogenic microorganisms associated with fresh produce', *J. Food Prot.*, 59, 204–16.

BEUCHAT, L.R., (1998), 'Surface decontamination of fruits and vegetables eaten raw: a review', World Health Organization, Food Safety Unit, WHO/FSF/FOS/98.2, *www.who.int/fsf/fos982~1*.

BEUCHAT, L.R., (1999), 'Survival of enterohemorrhagic *Escherichia coli* O157:H7 in bovine feces applied to lettuce and the effectiveness of chlorinated water as a disinfectant', *J. Food Prot.*, 62, 845–9.

BEUCHAT, L.R. and BRACKETT, R.E., (1990a), 'Survival and growth of *Listeria monocytogenes* on lettuce as influenced by shredding, chlorine treatment, modified atmosphere packaging and temperature', *J. Food Sci.*, 55, 755–8, 870.

BEUCHAT, L.R. and BRACKETT, R.E., (1990b), 'Inhibitory effect of raw carrots on *Listeria monocytogenes*', *Appl. Environ. Microbiol.*, 56, 1734–42.

BEUCHAT, L.R. and BRACKETT, R.E., (1991), 'Behaviour of *Listeria monocytogenes* inoculated into raw tomatoes and processed tomato products', *Appl. Environ. Microbiol.*, 57, 1367–71.

BEUCHAT, L.R., BRACKETT, R.E., HAO, D.Y.-Y. and CONNER, D.E., (1986), 'Growth and thermal inactivation of *Listeria monocytogenes* in cabbage and cabbage juice', *Can. J. Microbiol.*, 32, 791–5.

BEUCHAT, L.R. and RYU, J.-H., (1997), 'Produce handling and processing practices', *Emerg. Infect. Dis.*, 3, 459–65.

BEUCHAT, L.R., NAIL, B.V., ADLER, B.B. and CLAVERO, M.R.S., (1998), 'Efficacy of spray application of chlorinated water in killing pathogenic bacteria on raw apples, tomatoes and lettuce', *J. Food Prot.*, 61, 1305–11.

BEUCHAT, L.R., BRACKETT, R.E. and DOYLE, M.P., (1994), 'Lethality of carrot juice to *Listeria monocytogenes* as affected by pH, sodium chloride and temperature', *J. Food Prot.*, 57, 470–4.

BIDAWID, S., FARBER, J.M. and SATTAR, S.A., (2001), 'Survival of hepatitis A virus on modified atmosphere packaged (MAP) lettuce', *Food Microbiol.*, 18, 95–102.

BLOSTEIN, J., (1993), 'An outbreak of *Salmonella javiana* associated with consumption of watermelon', *J. Environ. Health*, 56, 29–31.

BOLIN, H.R., STAFFORD, A.E., KING, A.D. JR. and HUXSOLL, C.C., (1977), 'Factors affecting the storage stability of shredded lettuce', *J. Food Sci.*, 42, 1319–21.

BRACKETT, R.E., (1987), 'Antimicrobial effect of chlorine on *Listeria monocytogenes*', *J. Food Prot.*, 50, 999–1003.

BRACKETT, R.E., (1992), 'Shelf stability and safety of fresh produce as influenced by sanitation and disinfection', *J. Food Prot.*, 55, 804–14.

BRACKETT, R.E., (1994), 'Microbiological spoilage and pathogens in minimally processed refrigerated fruits and vegetables, in Wiley, R.C., *Minimally processed refrigerated fruits and vegetables*, New York, Chapman and Hall, 269–312.

BRACKETT, R.E., (1999), 'Incidence, contributing factors, and control of bacterial pathogens in produce', *Postharvest Biol. Technol.*, 15, 305–11.

BRANDI, G., SISTI, M., SCHIAVANO, G.F., SALVAGGIO, L. and ALBANO, A., (1996), 'Survival of *Aeromonas hydrophila*, *Aeromonas caviae* and *Aeromonas sobria* in soil', *J. Appl. Bacteriol.*, 81, 439–44.

BREMER, P.J., MONK, I. and OSBORNE, C.M., (2001), 'Survival of *Listeria monocytogenes* attached to stainless steel surfaces in the presence or absence of *Flavobacterium* spp.', *J. Food Prot.*, 64, 1369–76.

BROCKLEHURST, T.F., ZAMAN-WONG, C.M. and LUND, B.M., (1987), 'A note on the microbiology of retail packs of prepared salad vegetables', *J. Appl. Bacteriol.*, 63, 409–15.

BROSCH, R., CATIMEL, B., MILON, G., BUCHRIESER, C., VINDEL, E. and ROCOURT, J., (1993), 'Virulence heterogeneity of *Listeria monocytogenes* strains from various sources (food, human, animal) in immunocompetent mice and its association with typing characteristics', *J. Food Prot.*, 56, 296–301, 312.

BUNCIC, S. and AVERY, S.M., (1998), 'Effects of cold storage and heat-acid shocks on growth and verotoxin 2 production of *Escherichia coli* O157:H7', *Food Microbiol.*, 15, 319–28.

BUNCIC, S., AVERY, S.M., ROCOURT, J. and DIMITRIJEVIC, M., (2001), 'Can food-related environmental factors induce different behaviour in two key serovars, 4b and 1/2a, of *Listeria monocytogenes*?' *Int. J. Food Microbiol.*, 65, 210–12.

BURNETT, S.L. and BEUCHAT, L.R., (2001), 'Human pathogens associated with raw produce and unpasteurized juices, and difficulties in de-contamination', *J. Ind. Microbiol. Biot.*, 27, 104–10.

BURROWS, J., YUAN, J., NOVAK, J., BOISROBERT, C. and HAMPSON, B., (1999), 'Ozone application in food processing', *Fresh Cut*, April, 50–4.

BUTZLER, J. P., DEKEYSER, P., DETRAIN, M. and DEHAEN, F., (1973), 'Related *Vibrios* in stools', *J. Paediatr*, 82, 493–5.

CAI, Y., NG, L.K. and FARBER, J.M., (1997), 'Isolation and characterization of nisin-producing *Lactococcus lactis* subsp. *lactis* from bean-sprouts', *J. Appl. Microbiol.*, 83, 499–507.

CAMPDEN AND CHORLEY FOOD RESEARCH ASSOCIATION, (1992), Guidelines for the good manufacture and handling of modified atmosphere packed food products, Technical Manual No. 34.

CAMPDEN AND CHORLEY FOOD RESEARCH ASSOCIATION, (1996), Code of practice for the manufacture of vacuum and modified atmosphere packaged chilled foods with particular regard to the risks of botulism, guideline No. 11.

CARLIN, F. and NGUYEN-THE, C., (1994), 'Fate of *Listeria monocytogenes* on four types of minimally processed green salads', *Lett. Appl., Microbiol.*, 18, 222–6.

CARLIN, F., NGUYEN-THE, C., ABREU DA SILVA, A. and COCHET, C., (1996a), 'Effects of carbon dioxide on the fate of *Listeria monocytogenes*, of aerobic bacteria and on the development of spoilage in minimally processed fresh endive', *Int. J. Food Microbiol.*, 32, 159–72.

CARLIN, F., NGUYEN-THE, C. and MORRIS, C.E., (1996b), 'The influence of the background microflora on the fate of *Listeria monocytogenes* on minimally processed fresh broad leaved endive (*Cichorium endivia* var. *latifolia*)', *J. Food Prot.*, 59, 698–703.

CARLIN, F., NGUYEN-THE, C. and ABREU DA SILVA, A. (1995), 'Factors affecting the growth of *Listeria monocytogenes* on minimally processed fresh endive', *J. Appl. Bacteriol.*, 78, 636–46.

CARLIN, F. and PECK, M.W., (1996), 'Growth of and toxin production by non-proteolytic *Cl. botulinum* in cooked pureed vegetables at refrigeration temperatures', *Appl. Environ. Microbiol.*, 62, 3069–72.

CDC, (1971), 'Infectious Hepatitis Tennessee', *MMWR*, 20, 357.

CDC, (1989), 'Epidemiological notes and reports common source outbreak of giardiasis New Mexico', *MMWR*, 38, 405–7.

CDC, (1997a), 'Outbreaks of *Escherichia coli* infection associated with eating alfalfa sprouts – Michigan and Virginia, June-July, 1997', *MMWR*, 46, 741–4.

CDC, (1997b), 'Outbreak of cyclosporiasis Northern Virginia, Washington D.C., Baltimore, Maryland metropolitan area, 1997', *MMWR*, 46, 689–91.

CDC, (1998a), 'Outbreak of *Campylobacter enteritis* associated with cross-contamination of food – Oklahoma, 1996', *MMWR*, 47, 129–31.

CDC, (1998b), 'Foodborne outbreak of cryptosporidiosis Spokane, Washington, 1997, *MMWR*, 47, 565–567.

CDC, (1999), 'Outbreaks of *Shigella sonnei* infection associated with eating fresh parsley – United States and Canada, July–August, 1998', *MMWR*, 48, 285–9.

CLAVERO, M.R.S., BEUCHAT, L.R. and DOYLE, M.P., (1998), 'Thermal inactivation of *Escherichia coli* O157:H7 isolated from ground beef and bovine feces, and suitability of media for enumeration', *J. Food Prot.*, 61, 285–9.

CLIFFE-BYRNES,V., MCLAUGHLIN, C.P. and O'BEIRNE, D., (2003), 'Effects of packaging film and storage temperature on the quality of modified atmosphere packaged dry coleslaw mix'. *Int. J. Food Science and Technology*, 38, 187–99.

CLIVER, D.O., (1997), 'Foodborne viruses', in Doyle M.P, Beuchat, L.R. and Montville, T.J., *Food Microbiology – Fundamentals and Frontiers*, Washington DC, ASM Press, 437–46.

CONNER, D.E. and KOTROLA, J.S., (1995), 'Growth and survival of *Escherichia coli* O257:H7 under acidic conditions', *Appl. Environ. Microbiol.*, 382–5.

CONWAY, W.S., JANISIEWICZ, W.J., WATADA, A.E. and SAMS C.E., (1998), 'Survival and growth of *Listeria monocytogenes* on fresh-cut apple slices', *Phytopathol.*, 88, S18.

CONWAY, W.S., LEVERENTZ, B., SAFTNER, R.A., JANISIEWICZ, W.J., SAMS, C.E. and LEBLANC, E., (2000), 'Survival and growth of *Listeria monocytogenes* on fresh-cut apple slices and its interaction with *Glomerella cingulata* and *Penicillium expansum*', *Plant Dis.*, 84, 177–81.

COVER, T.L. and ABER, R.C., (1989), '*Yersinia enterocolitica'*, *New Engl. J. Med.*, 321, 16–24.

DANIELS, J.A., KRISHNAMURTHI, R. and RIZVI, S.S.H., (1985), 'A review of the effect of carbon dioxide on microbial growth and food quality', *J. Food Prot.*, 48, 532–7.

DAVIES, E.A. and ADAMS, M.R., (1994), 'Resistance of *Listeria monocytogenes* to the bacteriocin nisin', *Int. J. Food Microbiol.*, 21, 341–7.

DAVIS, H., TAYLOR, J.P., PERDUE, J.N., STELMA, J.G.N., HUMPHREYS, J.J.M., ROWNTREE, R. and GREENE, K.D., (1988), 'A shigellosis outbreak traced to commercially distributed shredded lettuce', *Am. J. Epidemiol.*, 128, 1312–21.

DAY, B.P.F., (1996), 'High oxygen modified atmosphere packaging for fresh prepared produce', *Postharvest News Info.*, 7, 1N-34N.

DEEKS, S., ELLIS, A., CIEBIN, B., KHAKHRIA, R., NAUS, M. and HOCKIN, J., (1998), '*Salmonella oranienburg*, Ontario', *Can. Comm. Dis. Rep.*, 24, 177–9.

DEL CAMPO, J., CARLIN, F. and NGUYEN-THE, C., (2001), Effects of epiphytic Enterobacteriaceae and pseudomonads on the growth of *Listeria monocytogenes* in model media', *J. Food Prot.*, 64, 721–4.

DEL CORRAL, F., BUCHANAN, R.L., BENCIVENGO, M.M. and COOKE, P.H., (1990), 'Quantitative comparison of selected virulence associated characteristics in food and clinical isolates of *Listeria*', *J. Food Prot.*, 53, 1003–9.

DELMAS, C.L. and VIDON, D.J.M., (1985), 'Isolation of *Y. enterocolitica* and related species from food in France', *Appl. Environ. Microbiol.*, 50, 767–71.

DEL ROSARIO, B.A. and BEUCHAT, L.R., (1995), 'Survival and growth of enterohemorrhagic *Escherichia coli* O157:H7 in cantaloupe and watermelon', *J. Food Prot.*, 58, 105–7.

DIAZ, C. and HOTCHKISS, J. H., (1996), 'Comparative growth of *E. coli* O157:H7, spoilage organisms and shelf life of shredded iceberg lettuce stored under modified atmospheres', *J. Sci. Food Agricul.*, 70, 433–8.

DIEHL, J.F., (1995), *Safety of irradiated foods*, 2nd revised edition, New York, Marcel Dekker Inc.

DINGMAN, D.W., (2000), 'Growth of *Escherichia coli* O157:H7 in bruised apple (*Malus domestica*) tissue as influenced by cultivar, date of harvest and source', *Appl. Environ. Microbiol.*, 66, 1077–83.

DODDS, K.L. and AUSTIN, J.W., (1997), *Clostridium botulinum*. In: *Food Microbiology – Fundamentals and Frontiers* (edited by M.P. Doyle, L.R. Beuchat and T.J. Montville). pp. 288–304. Washington DC: American Society of Microbiology Press.

DORIS, L.K.NG and SEAH, H.L., (1995), 'Isolation and identification of *Listeria monocytogenes* from a range of foods in Singapore', *Food Control*, 6, 171–3.

DOYLE, M.P. and SCHOENI, J.L., (1986), 'Isolation of *Campylobacter jejuni* from retail mushrooms', *Appl. Environ. Microbiol.*, 51, 449–50.

DOYLE, M.P., ZHAO, T., MENG, J. and ZHAO, S., (1997), *E. coli* 0157:H7. In: *Food Microbiology – Fundamentals and Frontiers* (edited by M.P. Doyle, L.R. Beuchat and T.J. Montville). pp. 171–91. Washington DC: American Society of Microbiology Press.

DUFFY, G., WHITING, R.C. and SHERIDAN, J.J., (1999), 'The effects of a competitive microflora, pH and temperature on the growth kenetics of *Escherichia coli* O157:H7', *Food Microbiol.*, 16, 299–307.

DYKES, G.A. and MOORHEAD, S.M., (2000), 'Survival of osmotic and acid stress by *Listeria monocytogenes* strains of clinical or meat origin', *Int. J. Food Microbiol.*, 56, 161–6.

EL-KEST, S.E. and MARTH, E.H., (1988a), '*Listeria monocytogenes* and its inactivation by chlorine: A review', *Lebensm.-Wiss. u. Technol.*, 21, 346–51.

EL-KEST, S.E. and MARTH, E.H., (1988b), 'Temperature, pH, and strain of pathogen as factors affecting inactivation of *Listeria monocytogenes* by chlorine', *J. Food Prot.*, 51, 622–5.

EMBAREK, P.K.B. and HUSS, H.H., (1993), 'Heat resistance of *Listeria monocytogenes* in vacuum packaged pasteurised fish fillets', *Int. J. Food Microbiol.*, 20, 85–90.

ESCARTIN, E.F., CASTILLO AYALA A. and LOZANO, J.S., (1989), 'Survival and growth of *Salmonella* and *Shigella* on sliced fresh fruit', *J. Food Prot.*, 52, 471–2, 483.

ESCUDERO, M.E., VELAZQUEZ, L., DI GENARO, M.S. and DE GUZMAN, A.M.S., (1999), 'Effectiveness of various disinfectants in the elimination of *Yersinia enterocolitica* on fresh lettuce', *J. Food Prot.*, 62, 665–9.

FARBER, J.M., (1991), 'Microbiological aspects of modified atmosphere packaging technology: A review', *J. Food Prot.*, 54, 58–70.

FARBER, J.M. and PETERKIN, P.I., (1991), '*Listeria monocytogenes*: a food-borne pathogen', *Microbiol. Rev.*, 55, 476–511.

FARBER, J.M., SANDERS, G.W. and JOHNSON, M.A., (1989), 'A survey of various foods for the presence of *Listeria* species', *J. Food Protect.*, 52, 456–8.

FARBER, J.M., WANG, S.L., CAI, Y. and ZHANG, S., (1998), 'Changes in populations of *Listeria monocytogenes* inoculated on packaged fresh-cut vegetables', *J. Food Prot.*, 61, 192–5.

FARKAS, J., (1997), 'Physical methods of food preservation', in Doyle, M.P.,

Beuchat, L.R., and Monteville, T.J., *Food Microbiology: Fundamentals and Frontiers*, Washington DC, American Society for Microbiology, 497–519.

FDA, (2000), Kinetics of microbial inactivation for alternative processing technologies. http:vm.cfsan.fda.gov/~comm/ift-foc.html

FDA, (2001), 'FDA survey of imported fresh produce', FY 1999 field assignment, FDA-CFSAN, Office of Plant and Dairy Foods and Beverages. *http://www.cfsan.fda.gov/~dms/prodsur6.html*

FDA/CFSAN, (1998), 'Guide to minimize microbial food safety hazards for fresh fruits and vegetables', Washington D.C., 36p.

FENG, P., (1997), 'A summary of background information and foodborne illness associated with the consumption of sprouts', Food and Drug Administration, Center for Food Safety and Applied Nutrition, http://www.cfsan.fda.gov/~mow/sprouts.html.

FENLON, D.R., WILSON, J. and DONACHIE, W., (1996), 'The incidence and level of *Listeria monocytogenes* contamination of food sources at primary production and initial processing', *J. Appl. Bacteriol.*, 81, 641–50.

FRANCIS, G.A. and O'BEIRNE, D., (1997), 'Effects of gas atmosphere, antimicrobial dip and temperature on the fate of *Listeria innocua* and *Listeria monocytogenes* on minimally processed lettuce', *Int. J. Food Sci. Technol.*, 32, 141–51.

FRANCIS, G.A. and O'BEIRNE, D., (1998a), 'Effects of storage atmosphere on *Listeria monocytogenes* and competing microflora using a surface model system', *Int. J. Food Sci. Technol.*, 33, 465–76.

FRANCIS, G.A. and O'BEIRNE, D. (1998b), 'Effects of the indigenous microflora of minimally processed lettuce on the survival and growth of *L. monocytogenes*,' *Int. J. Food Sci. Technol.*, 33, 477–88.

FRANCIS, G.A., THOMAS, C. and O'BEIRNE, D., (1999), 'Review paper: The microbiological safety of minimally processed vegetables', *Int. J. Food Sci. Technol.*, 34, 1–22.

FRANCIS, G.A. and O'BEIRNE, D., (2001a), 'Effects of vegetable type, package atmosphere and storage temperature on growth and survival of *Escherichia coli* O157:H7 and *Listeria monocytogenes*', *J. Ind. Microbiol. Biot.*, 27, 111–16.

FRANCIS, G.A. and O'BEIRNE, D., (2001b), 'Effects of acid adaptation on the survival of *Listeria monocytogenes* on modified atmosphere packaged vegetables', *Int. J. Food Sci. Technol.*, 36, 477–87.

FRANCIS, G.A. and O'BEIRNE, D., (2002), 'Effects of vegetable type and antimicrobial dipping on survival and growth of *Listeria innocua* and *E. coli*', *Int. J. Food Sci. Technol.*, 37, 711–18.

FRANK, J.F. and KOFFI, R.A., (1990), 'Surface-adherent growth of *Listeria monocytogenes* is associated with increased resistance to surfactant sanitisers and heat', *J. Food Prot.*, 53, 550–4.

FREUDLUND, H., BACK, E., SJOBERG, L. and TORNQUIST, E., (1987), 'Watermelon as a vehicle of transmission of Shigellosis', *Scand. J. Infect. Dis.*, 19, 219–21.

FRICKER, C.R. and TOMPSETT, S., (1989), '*Aeromonas* spp. in foods: a significant cause of food poisoning', *Int. J. Food Microbiol.*, 9, 17–23.

FSAI, (2001), *Code of practice for food safety in the fresh produce supply chain in Ireland*, Food safety Authority of Ireland, Dublin, 74p.

GAHAN, C.G.M. and HILL, C., (1999), 'The relationship between acid stress responses and virulence in *Salmonella typhimurium* and *Listeria monocytogenes'*, *Int. J. Food Microbiol.*, 50, 93–100.

GAHAN, C.G.M., O'DRISCOLL, B. and HILL, C., (1996), 'Acid adaptation of *Listeria monocytogenes* can enhance survival in acidic foods and during milk fermentation', *Appl. Environ. Microbiol.*, 62, 3128–32.

GARCÍA-GIMENO, R.M., SANCHEZ-POZO, M.D., AMARO-LÓPEZ, M.A. and ZURERA-COSANO, G. (1996), 'Behaviour of *Aeromonas hydrophila* in vegetable salads stored under modified atmosphere at 4 and 15ºC', *Food Microbiol.*, 13, 369–74.

GARG, N., CHUREY, J.J. and SPLITTSTOESSER, D.F., (1990), 'Effect of processing conditions on the microflora of fresh-cut vegetables', *J. Food Prot.*, 53, 701–3.

GELDREICH, E.E. and BORDNER, R.H., (1971), 'Fecal contamination of fruits and vegetables during cultivation and processing for market: A review', *J. Food Technol.*, 34, 184–95.

GLEESON, E., FRANCIS, G.A. and O'BEIRNE, D., (2002), 'Survival and growth of *Escherichia coli* and *Listeria innocua* on fresh-cut vegetables as affected by process severity'. Proc. 32nd Ann. Food Science and Technology Conference, University College Cork, p.119.

GOHIL, V.S., AHMED, M.A., DAVIES, R. and ROBINSON, R.K., (1995), 'Incidence of *Listeria* spp. in retail foods in the United Arab Emirates', *J. Food Prot.*, 58, 102–4.

GORNY, J.R. and ZAGORY, D., (2000), 'Produce food safety', in Gross, K.C., Saltveit, M.E.and Wang, C.Y., *The commercial storage of fruits vegetables and florist and nursery stocks*, Washington, DC, USDA Handbook 66, 35p.

GRAY, M.L. and KILLINGER, A.H., (1966), '*Listeria monocytogenes* and listeric infections', *Bacteriol. Rev.*, 30, 309–82.

GRIFFIN, P.M. and TAUXE, R.V., (1991), 'The epidemiology of infections caused by *E. coli* O157:H7, other enterohaemorrhagic *E. coli*, and the associated haemolytic syndrome', *Epidemiol. Rev.*, 13, 60–98.

GUILBERT, S.,GONTARD, N. and GORRIS, L., (1996), 'Prolongation of shelf-life of perishable food products using biodegradable films and coatings', *Lebensm Wiss Technol.*, 29, 10–17.

GULATI, B.R., ALLWOOD, P.B., HEDBERG, C.W. and GOYAL, S.M., (2001), 'Efficacy of commonly used disinfectants for the inactivation of calicivirus on strawberry, lettuce and a food-contact surface', *J. Food Prot.*, 64, 1430–4.

GUNASENA, D.K., KODIKARA,C.P., GANEPOLA, K. and WIDANAPATHIRANA, S., (1995), 'Occurrence of *Listeria monocytogenes* in food in Sri Lanka', *J. Nat. Sci. Council Sri Lanka*, 23, 107–14.

HAN, Y., LINTON, R.H., NIELSEN, S.S. and NELSON, P.E., (2000a), 'Inactivation of *Escherichia coli* O157:H7 on surface-uninjured and -injured green pepper (Capsicum annuum L.) by chlorine dioxide gas as demonstrated by confocal laser scanning microscopy', *Food Microbiol.,* 17, 643–55.

HAN, Y., SHERMAN, D.M., LINTON, R.H., NIELSEN, S.S. and NELSON, P.E., (2000b), 'The effects of washing and chlorine dioxide gas on survival and attachment of *Escherichia coli* O157:H7 to green pepper surfaces', *Food Microbiol.*, 17, 521–33.

HAN, Y., LINTON, R.H., NIELSEN, S.S. and NELSON, P.E., (2001), 'Reduction of *Listeria monocytogenes* on green peppers (*Capsicum annuum* L.) by gaseous and aqueous chlorine dioxide and water washing and its growth at 7ºC', *J. Food Prot.*, 64, 1730–8.

HARRIS, L.J., DAESCHEL, M.A., STILES, M.E. and KLAENHAMMER, T.R., (1989), 'Antimicrobial activity of lactic acid bacteria against *Listeria monocytogenes*', *J. Food Prot.*, 52, 384–7.

HARVEY, J. and GILMOUR, A., (1993), 'Occurrence and characteristics of *Listeria* in foods produced in Northern Ireland', *Int. J. Food Microbiol.*, 19, 193–205.

HAUSCHILD, A.H.W., (1989), '*Cl. botulinum*', in Hauschild, A.H.W. and Dodds K.L., *Cl. botulinum*: *Ecology and Control in Foods*, New York, Marcel Dekker, 111–89.

HAUSCHILD, A.H.W., ARIS, B. and HILSHEIMER, R., (1978), '*Cl. botulinum* in marinated products', *Can. Instit. Food Sci. Technology J.*, 8, 84–87.

HAZEN, T.C., FLIERMANS, C.B., HIRSCH, R.P. and ESCH, G.W., (1978), 'Prevalence and distribution of *Aeromonas hydrophila* in the United States', *Appl. Environ. Microbiol.*, 36, 731–8.

HEDBERG, C.W. and OSTERHOLM, M.T., (1993), 'Outbreaks of foodborne and waterborne viral gastroenteritis', *Clin. Microbiol. Rev*., 6, 199–210.

HERWALDT, B.L., (2000), '*Cyclospora cayetanensis*: a review focusing on the outbreaks of cyclosporiasis in the 1990s', *Clin. Infect. Dis*., 31, 1040–57.

HERWALDT, B.L. and ACKERS, M.L., (1997), 'An outbreak in 1996 of cyclosporiasis associated with imported raspberries', *New Engl. J. Med.*, 336, 1548–56.

HERWALDT, B.L. and BEACH, M.J., (1999), 'The return of *Cyclospora* in 1997: another outbreak of cyclosporiasis in North America associated with imported berries', *Ann of Intern Med.*, 130, 210–19.

HERWALDT, B.L., LEW, J.F., MOE, C.L., LEWIS, D.C., HUMPHREY, C.D., MONROE, S.S., PON, E.W. and GLASS, R.I., (1994), 'Characterization of a variant strain of norwalk virus from a foodborne outbreak of gastroenteritis on a cruise ship in Hawaii', *J. Clin. Microbiol*., 32, 861–6.

HILL, C., O'DRISCOLL, B. and BOOTH, I.R., (1995), 'Acid adaptation and food poisoning microorganisms', *Int. J. Food Microbiol*., 28, 245–54.

HINTLIAN, C.B. and HOTCHKISS, J.H., (1986), 'The safety of modified atmosphere packaging: A review', *Food Technol.*, 40, 70–76.

HO, J.L., SHANDS, K.N., FREIDLAND, G., ECKIND, P. and FRASER, D.W., (1986), 'An outbreak of type 4b *Listeria monocytogenes* infection involving patients from eight Boston hospitals', *Arch. Internal Med.*, 146, 520–24.

HUDSON, J.A. and DE LACY, K.M., (1991), 'Incidence of motile aeromonads in New Zealand retail foods', *J. Food Prot.*, 54, 696–9.

HUTIN, Y.J.F., POOL, V., CRAMER, E.H., NAINAN, O.V., WETH, J., WILLIAMS, I.T., GOLDSTEIN, S.T., GENSHEIMER, K.F., BELL, B.P., SHAPIRO, C.N., *et al.*, (1999), 'A multistate foodborne outbreak of hepatitis A', *New Engl. J. Med.*, 340, 595–602.

HUTTON, M.T., CHEHAK, P.A. and HANLIN, J.H., (1991), 'Inhibition of botulinum toxin production by *Pedicoccus acidilactici* in temperature abused refrigerated foods', *J. Food Safety*, 11, 255–67.

IANDOLO, J.J., CLARK, C.W., BLUHM, L. and ORDAL, Z.J., (1965), 'Repression of *Staphylococcus aureus* in associative culture', *Appl. Microbiol.*, 13, 646–9.

INTERNATIONAL COMMISSION ON MICROBIOLOGICAL SPECIFICATIONS FOR FOODS (ICMSF), (1996), *Microorganisms in Foods. 5. Microbiological Specifications of Food Pathogens*, London, Blackie Academic and Professional.

INTERNATIONAL FRESH-CUT PRODUCE ASSOCIATION, 1996), in *Food Safety Guidelines for the Fresh-cut Product Industry*, 3rd edn, Alexandria (VA), IFPA, 125p

INSTITUTE OF FOOD SCIENCE AND TECHNOLOGY (IFST), (1990), *Guidelines for the handling of chilled foods*, 2nd edition', London, UK, ISBN 0905367073.

INSTITUTE OF FOOD SCIENCE AND TECHNOLOGY (IFST), (1992), *Guidelines to Good Catering Practice*, London, UK, ISBN 090536709X.

IZUMI, H., (1999), 'Electrolyzed water as a disinfectant for fresh-cut vegetables', *J. Food Sci.*, 64, 536–9.

JACXSENS, L., DEVLIEGHERE, F., FALCATO, P. and DEBEVERE, J., (1999), 'Behaviour of *Listeria monocytogenes* and *Aeromonas* spp. on fresh-cut produce packaged under equilibrium-modified atmosphere', *J. Food Prot.*, 62, 1128–35.

JÖCKEL, VON J. and OTTO, W., (1990), 'Technological and hygienic aspects of production and distribution of pre-cut vegetable salads', *Arch. Lebensmittelhyg.*, 41, 149–52.

JUNTTILA, J.R., NIEMELA, S.I. and HIRN, J., (1988), 'Minimum growth temperatures of *Listeria monocytogenes* and non-haemolytic *Listeria*', *J. Appl. Bacteriol.*, 65, 321–7.

KALLENDER, K.D., HITCHINS, A.D., LANCETTE, G.A., SCHMIEG, J.A., GARCIA, G.R., SOLOMON, H.M. and SOFOS, J.N., (1991), 'Fate of *Listeria monocytogenes* in shredded cabbage stored at 5 and 25ºC under a modified atmosphere', *J. Food Prot.*, 54, 302–4.

KAPPERUD, G., RORVIK, L.M., HASSELTVEDT, V., HOIBY, E.A., IVERSEN, B.G., STAVELAND, K., JOHNSEN, G., LEITAO, J., HERIKSTAD, H., ANDERSSON, Y. *et al.*, (1995), Outbreak of *Shigella sonnei* infection traced to imported iceberg lettuce', *J. Clin. Microbiol.*, 33, 609–14.

KARAPINAR, M. and GONUL, S.A., (1992), 'Effects of sodium bicarbonate, vinegar, acetic and citric acids on growth and survival of *Yersinia enterocolitica*', *Int. J. Food Microbiol.*, 16, 343–7.

KARMALI, M.A., STEELE, B.T., PETRIC, M. and LIM, C., (1983), 'Sporadic cases of haemolytic uremic syndrome associated with faecal cytotoxin and cytotoxin-producing *E. coli*', *Lancet*, i, 619–20.

KETLEY, J. M., (1997), 'Pathogenesis of enteric infection by *Campylobacter*', *Microbiol.*, 143, 5–21.

KLAENHAMMER, T.R., (1988), 'Bacteriocins of lactic acid bacteria', *Biochimie*, 70, 337–49.

KNØCHEL, S. and JEPPESEN, C., (1990), 'Distribution and characteristics of *Aeromonas* in food and drinking water in Denmark', *Int. J. Food Microbiol.*, 10, 317–22.

KONOWALCHUK, J. and SPEIRS, J.I., (1975), 'Survival of enteric viruses on fresh vegetables', *J. Milk Food Technol.*, 37, 132–4.

KOEK, P.C., DE WITTE, Y. and DE MAAKER, J., (1983), 'The microbial ecology of prepared raw vegetables', in Roberts, T.A. and Skinner F.A., *Food Microbiology: Advances and Prospects*, London, Academic Press, 231–40.

KORICH, D.G., MEAD, J.R., MADORE, M.S., SINCLAIR, N.A. and STERLING, C.R., (1990), 'Effects of ozone, chlorine dioxide, chlorine and monochloroamine on *Cryptosporidium parvum* oocyst viability', *Appl. Environ. Microbiol.*, 56, 1423–8.

KOSEKI, S., YOSHIDA, K., ISOBE, S. and ITOH, K., (2001), 'Decontamination of lettuce using acidic electrolysed water', *J. Food Prot.*, 64, 652–8.

KROCHTA, J.M. and DE MULDER-JOHNSTON, C., (1997), 'Scientific status summary, edible and biodegradable polymer films: challenges and opportunities', *Food Technol.*, 51, 61–70.

KROVACEK, K., FARIS, A. and MANSSON, I., (1992), 'Growth of and toxin production by *Aeromonas hydrophila* and *Aeromonas sobria* at low temperatures', *Int. J. Food Microbiol.*, 13, 165–76.

LANGERAK, D.I., (1978), 'The influence of irradiation and packaging on the quality of prepacked vegetables', *Ann. Nutr. Aliment*, 32, 569–86.

LARSON, A.E. and JOHNSON,. E.A., (1999), 'Evaluation of botulinal toxin production in packaged fresh-cut cantaloupe and honeydew melons', *J. Food Prot.*, 62, 948–52.

LARSON, A.E., JOHNSON, E.A., BARMORE, C.R. and HUGHES, M.D., (1997), 'Evaluation of the botulism hazard from vegetables in modified atmosphere packaging', *J. Food Prot.*, 60, 1208–14.

LEBLANC, D.I., START, R., MACNEIL, B., GOGUEN, B. and BEAULIEU, C., (1996), 'Perishable food temperatures in retail stores', *Refrig. Sci. Technol. Proc.*, 6, 42–51.

LEYER, G.L. and JOHNSON, E.A., (1992), 'Acid adaptation promotes survival of *Salmonella* spp. in cheese', *Appl. Environ. Microbiol.*, 58, 2075–80.

LEYER, G.L. and JOHNSON, E.A., (1993), 'Acid adaptation induces cross-protection against environmental stresses in *Salmonella typhimurium*', *Appl.*

Environ. Microbiol., 59, 1842–7.

LIAO, C.-H. and COOKE, P.H., (2001), 'Response to trisodium phosphate treatment of *Salmonella* Chester attached to fresh-cut green pepper slices', *Can. J. Microbiol.*, 47, 25–32.

LIAO, C.-H. and SAPERS, G.M., (1999), 'Influence of soft rot bacteria on growth of *Listeria monocytogenes* on potato tuber slices', *J. Food Prot.*, 62, 343–8.

LIAO, C.-H. and SAPERS, G.M., (2000), 'Attachment and growth of *Salmonella* Chester on apple fruits and in vivo response of attached bacteria to sanitiser treatments', *J. Food Prot.*, 63, 876–83.

LILLY, T., SOLOMON, H.M. and RHODEHAMEL, E.J., (1996), 'Incidence of *Clostridium botulinum* in vegetables packaged under vacuum or modified atmosphere', *J. Food Prot.*, 59, 59–61.

LIN, C.M., FERNANDO, S.Y. and WEI, C.I., (1996), 'Occurrence of *Listeria monocytogenes*, *Salmonella* spp., *Escherichia coli* and *E. coli* O157:H7 in vegetable salads', *Food Control*, 7, 135–40.

LOU, Y. and YOUSEF, A.E., (1997), 'Adaptation to sublethal environmental stresses protects *Listeria monocytogenes* against lethal preservation factors', *Appl. Environ. Microbiol.*, 63, 1252–5.

LOWRY, P.W., LEVINE, R., STROUP, D.F., GUNN, R.A., WILDER, M.H. and KONIGSBERG, C., (1989), 'Hepatitis A outbreak on a floating restaurant in Florida, 1986', *Am. J. Epidemiol.*, 129, 155–64.

LUND, B.M., (1983), 'Bacterial spoilage', in Dennis, C., *Post-harvest Pathology of Fruits and Vegetables*, London, Academic Press, 219–257.

LUND, B.M., (1992), 'Ecosystems in vegetable foods,' *J. Appl. Bacteriol. Symposium supplement*, 73, 115S–126S.

LUND, B.M., (1993), 'The microbiological safety of prepared salad vegetables', *Food Technol. Int. Eur.*, 196–200.

LUND, B.M. and SNOWDON, A.L., (2000), 'Fresh and processed fruits, Chapter 27, in Lund, B.M., Baird-Parker, T.C. and Gould, G.W., *The microbiological safety and quality of food, Volume 1*, Gaithersburg, Aspen, 738–58.

MCCLANE, B.A., (1997), *Clostridium perfringens*. In: *Food Microbiology – Fundamentals and Frontiers* (edited by M.P. Doyle, L.R. Beuchat and T.J. Montville). pp. 305–26. Washington DC: American Society of Microbiology Press.

MACGOWAN, A.P., BOWKER, K., MCLAUCHLIN, J., BENNETT, P.M. and REEVES, D.S., (1994), 'The occurrence and seasonal changes in the isolation of *Listeria* spp. in shop bought food stuffs, human faeces, sewage and soil from urban sources', *Int. J. Food Microbiol.*, 21, 325–34.

MACKEY, B.M., PRITCHET, C., NORRIS, A. and MEAD, G.C., (1990), 'Heat resistance of *Listeria*: strain differences and effects of meat type and curing salts' *Lett. Appl. Microbiol.*, 10, 251–5.

MCLAUCHLIN, J. and GILBERT, R.J., (1990), '*Listeria* in foods, Report from the PHLS Committee on *Listeria* and listeriosis', *PHLS Microbiol. Dig.*, 7, 54–5.

MAGNUSON, J.A., KING, A.D. JR. and TÖRÖK, T., (1990), 'Microflora of partially processed lettuce', *Appl. Environ. Microbiol.*, 56, 3851–4.

MANVELL, P.M. and ACKLAND, M.R., (1986), 'Rapid detection of microbial growth in vegetable salads at chill and abuse temperatures', *Food Microbiol.*, 3, 59–65.

MARCHETTI, R., CASADEI, M.A. and GUERZONI, M.E., (1992), 'Microbial population dynamics in ready-to-use vegetable salads', *Ital. J. Food Sci.*, 4, 97–108.

MARSHALL, D.L. and SCHMIDT, R.H., (1991), 'Physiological evaluation of stimulated growth of *Listeria monocytogenes* by *Pseudomonas* species in milk', *Can. J. Microbiol.*, 37, 594–9.

MARTIN, D.L., GUSTAFSON, T.L., PELOSI, J.W., SUAREZ, L. and PIERCE, G.V., (1986), 'Contaminated produce – a common source for two outbreaks of *Shigella* gastroenteritis', *Am. J. Epidemiol.*, 124, 229–305.

MINTZ, E.D., HUDSON-WRAGG, M., MSHAR, P., CARTTER, M.L. and HADLER, J.L., (1993), 'Foodborne giardiasis in a corporate office setting', *J. Infect. Dis.*, 167, 250–3.

NATIONAL ADVISORY COMMITTEE ON MICROBIOLOGICAL CRITERIA FOR FOODS (NACMCF), (1999), 'Microbiological safety evaluations and recommendations on fresh produce', *Food Control*, 10, 117–43.

NGUYEN-THE, C. and CARLIN, F., (1994), 'The microbiology of minimally processed fresh fruits and vegetables', *Crit. Rev. Food Sci.*, 34, 371–401.

NGUYEN-THE, C. and LUND, B.M., (1991), 'The lethal effect of carrot on *Listeria* species', *J. Appl. Bacteriol.*, 70, 479–88.

NGUYEN-THE, C. and PRUNIER, J.P., (1989), 'Involvement of pseudomonads in the deterioration of 'ready-to-use' salads', *Int. J. Food Sci. Technol.*, 24, 47–58.

NGUYEN-THE, C., HALNA-DU-FRETAY, B. and ABREU DA SILVA, A., (1996), 'The microbiology of mixed salad containing raw and cooked ingredients without dressing', *Int. J. Food Sci. Technol.*, 31, 481–7.

NOTERMANS, S., DUFRENNE, J. and GERRITS, J.P.G., (1989), 'Natural occurrence of *Cl. botulinum* on fresh mushrooms (*Agaricus bisporus*)', *J. Food Prot.*, 52, 733–6.

O'BEIRNE, D., (1989), Irradiation of fruits and vegetables – applications and issues. *Prof. Hort.*, 3, 12–19.

O'BEIRNE, D. and BALLANTYNE, A., (1987), 'Some effects of modified atmosphere packaging and vacuum packaging in combination with antioxidants on quality and storage life of chilled potato strips', *Int. J. Food Sci. Technol.*, 22, 515–23.

O'DRISCOLL, B., GAHAN, C.G.M. and HILL, C., (1996), 'Adaptive acid tolerance response in *Listeria monocytogenes*: isolation of an acid-tolerant mutant which demonstrates increased virulence', *Appl. Environ. Microbiol.*, 62, 1693–8.

ODUMERU J.A., MITCHELL S.J., ALVES D.M., LYNCH J.A., YEE A.J., WANG S.L., STYLIADIS S., and FARBER J.M., (1997), 'Assessment of the microbiological quality of ready-to-use vegetables for health-care food services', *J. Food Prot.*, 60, 954–60.

O' MAHONY, M., COWDEN, J., SMYTH, B., LYNCH, D., HALL, M., ROWE, B., TEARE, E.L., TETTMAR, R.E., RAMPLING, A.M., COLES, M., GILBERT, R.J., KINGCOTT, E. and BARTLETT, C.L.R., (1990), 'An outbreak of *Salmonella saintpaul* infection associated with beansprouts', *Epidemiol. Infect.*, 104, 229–35.

PALUMBO, M. S., BEERS, S.M., BHADURI, S. and PALUMBO, S.A., (1995), 'Thermal resistance of *Salmonella* spp. *Listeria monocytogenes* in liquid egg yolk and egg yolk products' *J. Food Prot.*, 58, 960–6.

PALUMBO, S.A. and BUCHANAN, R.L., (1988), 'Factors affecting the growth or survival of *Aeromonas hydrophila* in foods', *J. Food Safety*, 9, 37–51.

PARK, C.-M. and BEUCHAT, L.R., (1999), 'Evaluation of sanitisers for killing *Escherichia coli* O157:H7, *Salmonella* and naturally occurring microorganisms on cantaloupes, honeydew melons and asparagus', *Dairy Food Environ. Sanit.*, 19, 842–7.

PARK, C.-M., HUNG, Y.-C., DOYLE, M.P., EZEIKE, G.O.I. and KIM, C., (2001), 'Pathogen reduction and quality of lettuce treated with electrolysed oxidizing and acidified chlorinated water', *J. Food Sci.*, 66, 1368–72.

PETRAN, R.L., ZOTTOLA, E.A. and GRAVANI, R.B., (1988), 'Incidence of *Listeria monocytogenes* in market samples of fresh and frozen vegetables', *J. Food Sci.*, 53, 1238–40.

PETRAN, R.L., SPERBER, W.H. and DAVIS, A.B., (1995), '*Cl. botulinum* toxin formation in romaine lettuce and shredded cabbage: Effects of storage and packaging conditions', *J. Food Prot.*, 58, 624–7.

PHLS (PUBLIC HEALTH LABORATORY SYSTEM), (2001), '*Salmonella* Newport infection in England associated with the consumption of ready to eat salad', *Eurosurveillance Weekly*, June, 26.

PIAGENTINI, A.M., PIROVANI, M.E., GÜEMES, D.R., DI PENTIMA, J.H. and TESSI, M.A., (1997), 'Survival and growth of *Salmonella hadar* on minimally processed cabbage as influenced by storage abuse conditions', *J. Food Sci.*, 62, 616–18, 631.

PORTNOY, B.L., GOEPFERT, J.M. and HARMON, S.M., (1976), 'An outbreak of *Bacillus cereus* food poisoning resulting from contaminated vegetable sprouts', *Am. J. Epidemiol.*, 103, 589–94.

PRESTON, M., DAVIDSON, R., HARRIS, S., THUSUSKA, J., GOLDMAN, C., GREEN, K., LOW, D., PROCTOR, P., JOHNSON, W. and KHAKHRIA, R., (1997), 'Hospital outbreak of *Escherichia coli* O157:H7 associated with a rare phage type-Ontario', *Can. Comm. Dis. Rep.*, 23, 33–37.

PRICE, R.J. and LEE, J.S., (1970), 'Inhibition of *Pseudomonas* species by hydrogen peroxide producing lactobacilli', *J. Milk Food Technol.*, 33, 13–18.

RACCACH, M. and BAKER, R.C., (1979), 'The effect of lactic acid bacteria on some properties of mechanically deboned poultry meat', *Poultry Sci.*, 58, 144–7.

RAFII, F. and LUNSFORD, P., (1997), 'Survival and detection of *Shigella flexneri* in vegetables and commercially prepared salads', *J. AOAC Int.*, 80, 1191–7.

RAJKOWSKI, K.T. and THAYER, D.W., (2000), 'Reduction of *Salmonella* spp. and strains of *Escherichia coli* O157:H7 by gamma radiation of inoculated sprouts', *J. Food Prot.*, 63, 871–5.

REID, T.M.S. and ROBINSON, H.G., (1987), 'Frozen raspberries and hepatitis A', *Epidem. Inf.*, 98, 109–12.

RHODEHAMEL, E.J., (1992), 'FDA's concerns with sous vide processing', *Food Technol.*, 46, 73–6.

RICHERT, K.J., ALBRECHT, J.A., BULLERMAN, L.B. and SUMNER, S.S., (2000), 'Survival and growth of *Escherichia coli* O157:H7 on broccoli, cucumber and green pepper', *Dairy Food Environ. Sanit.*, 20, 24–8.

RINGLÉ, P., VINCENT, J.P. and CATTEAU, M., (1991), 'Evolution de *Listeria* dans les produits de 4ème gamme', in Lahellec, C., *Les Microorganismes Contaminants dans les Industries Agroalimentaires: Colonisation, Détection, Maitrise*, Paris: 7ème Colloque de la Section Microbiologie Alimentaire, 13–14 March 1991, Société Française de Microbiologie, pp. 324–8.

ROBINS-BROWNE, R.M., (1997), *Y. enterocolitica.* In: *Food Microbiology – Fundamentals and Frontiers* (edited by M.P. Doyle, L.R. Beuchat and T.J. Montville), pp. 192–215. Washington DC: American Society of Microbiology Press.

ROBINSON, D.A., (1981), 'Infective dose of *Campylobacter jejuni* in milk', *British Med. J.*, 282, 1584.

ROBINSON, I. and ADAMS, R.P., (1978), 'Ultra-violet treatment of contaminated irrigation water and its effect on the bacteriological quality of celery at harvest', *J. Appl Bacteriol*, 45, 83–90.

ROCOURT, J., (1994), '*Listeria monocytogenes*: the state of the science', *Dairy, Food Environ. Sanit.*, 14, 70–82.

RODRIGUEZ, A.M.C., ALCALA, E.B., GIMENO, R.M.G. and COSANO, G.Z., (2000), 'Growth modelling of *Listeria monocytogenes* in packaged fresh green asparagus', *Food Microbiol.*, 17, 421–7.

ROSENBLUM, L.S., MIRKIN, I.R., ALLEN, D.T., SAFFORD, S. and HADLER, S.C., (1990), 'A multifocal outbreak of hepatitis A traced to commercially distributed lettuce', *AJPH*, 80, 1075–9.

ROUQUETTE, C., DE CHASTELLIER, C., NAIR, S. and BERCHE, P., (1998), 'The ClpC ATPase of *Listeria monocytogenes* is a general stress protein required for virulence and promoting early bacterial escape from the phagosome of macrophages', *Mol. Microbiol.*, 27, 1235–45.

SADDIK, M.F., EL-SHERBEENY, M.R. and BRYAN, F.L., (1985), 'Microbiological profiles of Egyptian raw vegetables and salads', *J. Food Prot.*, 48, 883–6.

SAPERS, G.M. and SIMMONS, G.F., (1998), 'Hydrogen peroxide disinfection of minimally processed fruits and vegetables', *Food Technol.*, 52, 48–52.

SATTAR, S.A., SPRINGTHORPE, V.S. and ANSARI, S.A., (1994), 'Rotavirus', in Hui, Y.H., Gorham, J.R., and Cliver, D.O. *Foodborne disease Handbook Volume 2*, New York, Marcel Dekker 81–111.

SCHLECH, W.F., LAVIGNE, P. M., BORTOLUSSI, R.A., ALLEN, A.C., HALDANE, E.V., WORT, A.J., HIGHTOWER, A.W., JOHNSON, S.E., KING, S.H., NICHOLLS, E.S. and BROOME, C.V., (1983), 'Epidemic listeriosis-evidence for transmission by food', *New England J. Med.*, 308, 203–6.

SEO, K.H. and FRANK, J.F., (1999), 'Attachment of *Escherichia coli* O157:H7 to lettuce leaf surface and bacterial viability in response to chlorine treatment as demonstrated by using confocal scanning laser microscopy', *J. Food Prot.*, 62, 3–9.

SIZMUR, K. and WALKER, C.W., (1988), '*Listeria* in prepacked salads', *Lancet*, i, 1167.

SKIRROW, M.B., (1977), '*Campylobacter* enteritis; a new disease', *British Med. J.*, 2, 9–11.

SOFOS, J.N., BEUCHAT, L.R., DAVIDSON, P.M. and JOHNSON, E.A., (1998), 'Naturally occurring antimicrobials in food', Report 132, Council for Agric. Sci. Technol. Task Force.

SOLOMON, H.M. and KAUTTER, D.A., (1988), 'Outgrowth and toxin production by *Clostridium botulinum* in bottled chopped garlic', *J. Food Prot.*, 51, 862–5.

SOLOMON, H.M., KAUTTER, D.A., LILLY, T. and RHODEHAMEL, E.J., (1990), 'Outgrowth of *Cl. botulinum* in shredded cabbage at room temperature under modified atmosphere', *J. Food Prot.*, 53, 831–3.

SÖRQVIST, S., (1994), 'Heat resistance of different serovars of *Listeria monocytogenes*', *J. Appl. Bacteriol.,* 76, 383–8.

SUGIYAMA, H. AND YANG, K.H., (1975), 'Growth potential of *Cl. botulinum* in fresh mushrooms packaged in semi-permeable plastic film', *Appl. Microbiol.*, 30, 964–9.

SUSMAN, E., (1999), 'Nationwide outbreak of Salmonella linked to tomatoes: experts call for stricter measures to decrease spread of disease causing bacteria', WebMD Health, *http://my.webmd.com/content/article/1728.50099*.

SZABO, E.A., SCURRAH, K.J. and BURROWS, J.M., (2000), 'Survey for psychrotrophic bacterial pathogens in minimally processed lettuce', *Lett. Appl. Microbiol.*, 30, 456–60.

TABOURET, M., DERYCKE, J., AUDURIER, A. and POUTREL, B., (1991), 'Pathogenicity of *Listeria monocytogenes* isolates in immunocompromised mice in relation to listeriolysin production', *J. Med. Microbiol.,* 34, 13–18.

TAKEUCHI, K. and FRANK, J.F., (2000), 'Penetration of *Escherichia coli* O157:H7 into lettuce tissues as affected by inoculum size and temperature and the effect of chlorine treatment on cell viability', *J. Food Prot.*, 63, 434–40.

TAKEUCHI, K. and FRANK, J.F., (2001), 'Quantitative determination of the role of lettuce leaf structures in protecting *Escherichia coli* O157:H7 from chlorine disinfection', *J. Food Prot.*, 64, 147–51.

TAKEUCHI, K., MATUTE, C.M., HASSAN, A.N. and FRANK, J.F., (2000), 'Comparison of the attachment of *Escherichia coli* O157:H7, *Listeria monocytogenes*, *Salmonella* Tyhimurium and *Pseudomonas fluorescens* to lettuce leaves', *J. Food Prot.*, 63, 1433–7.

TAMMINGA, S.K., BEUMER, R.R. and KAMPELMACHER, E.H., (1978), 'The hygienic quality of vegetables grown in or imported into the Netherlands: a tentative study', *J. Hyg. Camb*, 80, 143–54.

TASSINARI, A., FRANCO, B.D.G. and LANDGRAF, M., (1994), 'Incidence of *Yersinia* spp. in food in Sao Paulo, Brazil', *Int. J. Food Microbiol.*, 21, 263–70.

TAUXE, R.V., (1997), 'Emerging foodborne diseases: an evolving public health challenge', *Dairy Food Environ. Sanit.*, 17, 788–95.

THOMAS, C. and O'BEIRNE, D., (2000), 'Evaluation of the impact of short-term temperature abuse on the microbiology and shelf-life of a model ready-to-use vegetable combination product', *Int. J. Food Microbiol.*, 59, 47–57.

THOMAS, C., MABEY, M., HILL, D.J. and KENWARD, M.A., (1995), '*Campylobacter* - The next challenge', *Int. Food Hyg.*, 6, 13–15.

THUNBERG, R.L., TRAN, T.T., BENNETT, R.W., MATTHEWS, R.N. and BELAY, N., (2002), 'Microbial evaluation of selected fresh produce obtained at retail markets'. *J. Food Prot.*, 4, 677–82.

TRAN, T.T., THUNBERG, R.L., BENNETT, R.W. and MATTHEWS, R.N., (2000), 'Fate of *Campylobacter jejuni* in normal air, vacuum and modified atmosphere packaged fresh-cut vegetables (abstract)', in *Proceedings of 114th Annual Meeting, Association of Official Analytical Chemists, Sepember 10–14, Philadelphia*, 86.

UKUKU D.O. and FETT, W., (2002), 'Behaviour of *Listeria monocytogenes* inoculated on cantaloupe surfaces and efficacy of washing treatments to reduce transfer from rind to fresh-cut pieces', *J. Food Prot.*, 65, 924–30.

UKUKU D.O. and SAPERS, G.M., (2001), 'Effect of sanitiser treatments on *Salmonella stanley* attached to the surface of cantaloupe and cell transfer to fresh-cut tissues during cutting practices', *J. Food Prot.*, 64, 1286–91.

VARNUM, A.H. and EVANS, M.G., (1991), *Foodborne pathogens – an illustrated text*. England, Wolfe Publishing.

VAROQUAUX, P., (2001), 'Unit operations for fresh-cut produce', *Proc. Second International Conference on Fresh-Cut Produce*, Campden and Chorleywood Food and Drink Research Association, Chipping Campden, UK.

VELANI, S. and ROBERTS, D., (1991), '*Listeria monocytogenes* and other *Listeria* spp.in prepacked mixed salads and individual salad ingredients', *PHLS Microbiol. Dig.*, 8, 21–2.

VESCOVO, M., TORRIANI, S., ORSI, C., MACCHIAROLO, F. and SCOLARI, G., (1996), 'Application of antimicrobial-producing lactic acid bacteria to control pathogens in ready-to-use vegetables', *J. Appl. Bacteriol.*, 81, 113–19.

VESCOVO, M., SCOLARI, G., ORSI, C., SINIGAGLIA, M. and TORRIANI, S., (1997), 'Combined effects of *Lactobacillus casei* inoculum, modified atmosphere packaging and storage temperature in controlling *Aeromonas hydrophila* in ready-to-use vegetables', *Int. J. Food Sci. Technol*, 32, 411–19.

WALKER, S.J. and STRINGER, M.F., (1987), 'Growth of *Listeria monocytogenes* and *Aeromonas hydrophila* at chill temperatures', *Campden Food Preservation Research Association Technical Memorandum* No. 462., CFPRA: Chipping Campden, U.K.

WEBB, T.A. and MUNDT, J.O., (1978), 'Molds on vegetables at the time of harvest', *Appl. Environ. Microbiol.*, 35, 655–8.

WEI, C.I., HUANG, T.S., KIM, J.M., LIN, W.F., TAMPLIN, M.L. and BARTZ, J.A., (1995), 'Growth and survival of *Salmonella montevideo* on tomatoes and disinfection with chlorinated water', *J. Food Prot.*, 58, 829–36.

WELLS, J.M. and BUTTERFIELD, J.E., (1997), '*Salmonella* contamination associated with bacterial soft rot of fresh fruits and vegetables in the marketplace', *Plant Dis.*, 81, 867–72.

WELLS, J.G., SHIPMAN, L.D., GREENE, K.D., SOWERS, E.G., GREEN, J.H., CAMERON, D.N., DOWNES, F.P., MARTIN, M.L., GRIFFIN, P.M., OSTROFF, S.M., POTTER, M.E., TAUXE, R.V. and WACHSMUTH, I.K., (1991), 'Isolation of *E. coli* O157:H7 and other Shiga-like toxin producing *E. coli* from dairy cattle', *J. Clin. Microbiol.*, 29, 985–9.

WELSHIMER, H.J., (1968), 'Isolation of *Listeria monocytogenes* from vegetation', *J. Bacteriol.*, 95, 300–3.

WHO, (1996), 'Enterohaemorrhagic *Escherichia coli* infection', *Weekly Epidemiol. Rec.*, 30, 229–30.

WILCOX, F., TOBBACK P. and HENDRICKX, M., (1994), 'Microbial safety assurance of minimally processed vegetables by implementation of the Hazard Analysis Critical Control Point system', *Acta. Aliment.*, 23, 221–38.

WISNIEWSKY, M.A., GLATZ, B.A., GLEASON, M.L. and REITMEIER, C.A., (2000), 'Reduction of *Escherichia coli* O157:H7 counts on whole fresh apples by treatment with sanitisers', *J. Food Prot.*, 63, 703–8.

WRIGHT, J.R., SUMNER, S.S., HACKNEY, C.R., PIERSON, M.D. and ZOECKLEIN, B.W., (2000), 'Reduction of *Escherichia coli* O157:H7 on apples using wash and chemical sanitiser treatments', *Dairy Food Environ. Sanitat.*, 20, 120–6.

ZHANG, S. and FARBER, J.M., (1996), 'The effects of various disinfectants against *Listeria monocytogenes* on fresh-cut vegetables', *Food Microbiol.*, 13, 311–21.

ZHUANG, R.-Y., BEUCHAT, L.R. and ANGULO, F.J., (1995), 'Fate of *Salmonella montevideo* on and in raw tomatoes as affected by temperature and treatment with chlorine', *Appl. Environ. Microbiol.*, 61, 2127–31.

ZHUANG, R., BEUCHAT, L.R., CHINNAN, M.S., SHEWFELT, R.L. and HUANG, Y-W., (1996), 'Inactivation of *Salmonella* Montevideo on tomatoes by applying cellulose-based edible films', *J. Food Prot.* 59, 808–12.

13

Detecting leaks in modified atmosphere packaging

E. Hurme, VTT Biotechnology, Finland

13.1 Introduction

Package integrity is an essential requirement for maintaining the high quality of, for example, sterilised foods and modified atmosphere packaged foods. The increasing focus on quality assurance is putting demands on verification of food package integrity. The foremost noticeable package integrity problem is probably leaking seals, particularly with flexible plastic packages which are more prone to mechanical damage than traditional rigid metal packages. A non-destructive leak test device allowing evaluation of every container produced is, therefore, of interest to food manufacturers.

Non-destructive package leak testing equipment detects defective packages immediately in the packaging line. This can be considered as an integral part of packaging process control. The most effective way to detect a package leakage, non-destructively, throughout the whole distribution chain from the manufacturer to the consumer is a leak indicator permanently attached to the package. One key element in selecting a proper leak test device and leak indicator is knowledge of the leakages, which are critical to the product shelf-life.

This chapter reviews the integrity requirements of flexible food packages, non-destructive package leak test methods, and intelligent leak indicators for modified atmosphere packages.

13.2 Leakage, product safety and quality

Before the selection of leak-testing methods for different packages can be made, it is essential to have information concerning the required integrity of different

Table 13.1 Deterioration factors related to critical channel leakages in different packaged foods

	Aseptic and sterilised foods	Ready-to-eat-meals	Baked goods	Dried goods
Most important deteriorative factors	microbial changes, oxidation	microbial changes, oxidation	oxidation, mould growth, moisture changes	oxidation, moisture changes
Critical leakage diameter in package	5–25 μm[1,3,5,9–12]	30-50 μm [6,12–13]	50 μm[13]	> 130 μm[13]

package types and products. That is, how big a leakage can there be without the packed product deteriorating microbiologically or chemically before the use-by date, and how small a leakage should the leak testing method detect (Table 13.1).

In many studies leakages of around 10 μm in diameter have been demonstrated, under strict conditions, to cause microbial contamination in model packages[1,2] and in commercially processed and packaged aseptic packages.[3,4,5] The critical leakage size causing accelerated quality deterioration in gas-flushed modified atmosphere packages (MAP) may, however, vary considerably between different products and packaging methods. Small leakages (hole diameter < 169μm, hole length 3mm) in gas packages have even been reported to retain the quality of packed minced meat steaks better than in intact packages.[6] Other recent studies have, on the other hand, revealed accelerated quality deterioration of raw marinated chicken breast and raw rainbow trout[7] and pizza[8] in gas packages with leakages as small as 30μm and 55μm (hole length 3mm), respectively. Table 13.1 summarises the most important deterioration factors of different packaged foods and studies concerning critical leakages.

13.3 Package leak detection during processing

13.3.1 Methods in use

Food package and seal integrity is widely verified using destructive manual methods, such as a biotest, electrolytic test, dye penetration test and bubble test. The major drawbacks of destructive test methods are that it is not possible to check every package produced, and the tests are often laborious. An automated, reliable, 100% in-line non-destructive leak test machine allowing testing of every container produced would, therefore, be of interest to companies. This kind of package testing would serve as an immediate process control tool, resulting in an overall cost reduction in terms of a reduced number of packages

Table 13.2 Commercial methods for non-destructive food package leakage detection

Test stimulus	Test response	Application
External pressure	Package movement* Pressure decay*	Packages with headspace
External vacuum	Package movement* Pressure decay* Tracer gas (H_2, CO_2, He, SF_6)	Packages with headspace
Internal pressure	Squeezer movement* Package movement*	Packages with headspace
Internal vacuum	Package movement*	All packages
Machine vision	Image change*	Foil packages

* On-line application available.

lost both in production and in destructive testing. Also, complaints and returns from retailers and consumers relating to leaky packages and deteriorated products would be diminished.

Much interest has concentrated on plastic food packages to define their integrity requirements,[1,2,4,8,14,15] to research and develop new non-destructive test methods,[16-19] and to evaluate the reliability of commercial non-destructive test methods.[15,20] In-line non-destructive test equipment should meet demands such as: reliable identification and rejection of all the defective packages produced; fast leak detection; non-damaging to the product; easy to use and maintain; and reasonable supply and operating costs.

Most non-destructive leak inspection systems for flexible and semi-rigid packages are based on a stimulus response technique: the stimulus to the package can be, for example, ultrasound,[18] pressure,[22] tracer gas like helium,[23] carbon dioxide[16] or hydrogen[21] and the response can be, for example, sound/beam reflection, package movement, pressure change, or tracer gas detection (Table 13.2). In recent years, numerous new patents suitable for non-destructive food or medical package integrity testing have been published.

Although tracer gas detection is a very sensitive method, detection of pressure differential is perhaps currently the most popular method employed for flexible and semi-rigid packages with a headspace. Commercial pressure differential methods are typically based either (i) on detection of an external rise or fall in pressure in a test chamber created outside the package with compressed air or a vacuum pump, respectively, or (ii) on detection of an internal fall in pressure created inside the package either mechanically or by heat. Evaluation studies of commercial automated non-destructive leak detectors based on detection pressure differentials revealed that these test methods – although much used in industry – were not capable of reliably detecting leakages that were proven to be penetrable by harmful microbes.[20,22,23]

13.3.2 Novel tracer gas system for in-line application

Tracer gas leak detection methods are very sensitive, and the most commonly used gas has been helium. Another possibility is to use the more economical carbon dioxide as a tracer gas. Carbon dioxide is often routinely used as a protective packaging gas in food packages, which eliminates the need for the special addition of tracer gas in the package. However, introduction of automatic in-line leak detectors based on helium or carbon dioxide tracer gases has not been successful. The reasons for this have possibly been the relatively high operating and supply costs of the helium method, or the unfavourable physical characteristics of the carbon dioxide method.

A novel leak-detection system has recently developed at VTT using hydrogen (H_2) as a tracer gas.[21,24] The leak tester utilising H_2 and a very sensitive hydrogen detector is very effective and fast and is especially suitable for MAP. For example, at least 30μm diameter holes in a gas-flushed package have been demonstrated to be reliably detected within one second.[21] Using this method, a package containing H_2 tracer gas is positioned in a specially designed test chamber. A vacuum pressure is then drawn into the test chamber and the package expands due to the increased pressure differential between the package walls. Trace amounts of H_2 are then forced out of leaking packages through a pipe in which a H_2 sensor is positioned towards the gas flow. The sensor connected to the H_2 detector reacts to the H_2, and immediately gives an electrical signal to the H_2 detector.

H_2 has many characteristics advantageous to its use as a tracer gas in leak detection. First of all, it is a colourless, odourless, tasteless and non-toxic gas at atmospheric temperatures and pressures. A non-flammable concentration (<5% in air) of hydrogen is sufficient for sensitive leak detection. Nevertheless, the tracer gas concentration in the package headspace is proportional to the leak detection sensitivity and speed; even concentrations as low as 0.5% can be used to detect relatively small leakages. The low background concentration of H_2 in air, only 0.5ppm, enables sensitive leak detection. That is, the minimum detection limit of H_2 escaping from a defective package is very low. In comparison, the carbon dioxide and helium concentrations in air are 300 and 5 ppm, respectively. Hydrogen is also the lightest of all gases (molecular weight: H_2 2.0, He 4.0, CO_2 44.0, air 29.0g/mol) thus reducing the risk of background gas contamination in the leak test area. For example, carbon dioxide as a heavier gas than air may accumulate in the leak test area creating a risk of false readings.

13.4 Package leak indicators during distribution

The modified atmosphere package for non-respiring food typically has a low (0–2%) oxygen (O_2) concentration and a high (20–80%) carbon dioxide (CO_2) concentration. Hence, a leak means a considerable increase in O_2 concentration and a decrease in CO_2 concentration. If the package leaks, microbial growth is likely to take place. This means that CO_2 may in some cases accumulate in package. In the worst case, the CO_2 concentration will remain high despite leakage and microbial growth. Thus, the leak indicators for modified atmosphere packages should rely on the detection of oxygen rather than on the detection of CO_2.

13.4.1 Visual oxygen indicators

At present, the main application of commercially available O_2-sensitive package indicators is to ensure the proper functioning of oxygen absorption; companies that also deal with O_2 absorbers have developed the indicators. For example, Mitsubishi Gas Chemical Company in Japan has greatly contributed to the development of O_2 absorbers and was the first to commercialise O_2-absorbing sachets under the trade name 'Ageless'.[25] The 'Ageless-Eye' sachets containing an O_2 indicator tablet have been designed to confirm that the 'Ageless' absorbers are functioning properly. The manufacturer claims that indicator tablet turns from blue into pink within 2–3 hours after O_2 has reached a zero concentration at 25°C and into blue again in about five minutes when it is in contact with O_2. Also some other Japanese companies like Toppan Printing have been active in developing oxygen indicators.

A typical visual O_2 indicator consists of a redox dye, i.e., a reducing compound and an alkaline compound. In addition to these main components, compounds such as a solvent (typically water and/or alcohol) and bulking agent (e.g. zeolite, silica gel, cellulose materials, polymers) are added to the indicator. The indicator can be formulated as a tablet[26–27] or a printed layer[28–30] or it can be laminated in a polymer film.[31] The redox dyes of the indicators are oxidised by O_2 and a colour change can be observed. The most common dye used in the indicators is methylene blue, which is typically white in the reduced state and blue in the oxidised state. Other redox dyes used in O_2 indicators are 2,6-dichloroindophenol[32] and *N,N,N′,N′*-tetramethyl-*p*-phenylenediamine.[33] A reducing compound is added to the O_2 indicator to reduce the dye and to keep it in the reduced state during the packaging process. Common reducing compounds for O_2 indicators are reducing sugars, but inorganic salts as well as reduction by irradiation have also been used. An alkaline compound is added to the indicator to maintain the pH on the alkaline side and thus prevent too rapid an oxidation reaction of the dye.[34–35] Inorganic compounds, such as sodium hydroxide, potassium hydroxide, calcium hydroxide and magnesium hydroxide, have typically been used for this purpose.[27,30]

A different approach to constructing a visual O_2 indicator was introduced by Krumhar & Karel[29] who developed a two-step colour reaction. In the first reaction step O_2-sensitive material is oxidised and the formation of an acid or peroxide occurs. These components will cause a colour change in the specific colorant included in the system. Oxidative enzyme-based oxygen indicators have been described by Ahvenainen *et al.*[36] and Gardiol *et al.*[37–38]

13.4.2 Invisible oxygen indicators

In addition to the purely visual O_2 indicators, some other systems can also be considered as indicators even if external equipment is needed. These systems possess, however, an internal indicator attached to the package and can be interpreted non-destructively. The concept of luminescent dyes quenched by O_2 as indicators for food packages was preliminarily introduced by Reininger *et al.*[39] This optical method can be used for quantitative measurement of O_2 concentration in a non-destructive manner. Maurer[40] suggests a system using the conversion of O_2 to ozone with the aid of UV radiation or an electric field. The presence of ozone is shown with a potassium iodide/starch indicator strip.

A more recent approach is an optical oxygen-sensing method developed at TNO. The measurement principle is based on the fluorescence quenching of a metal-organic fluorescent dye, which is immobilised in a hydrofobic polymer. The dye is excited by an excitation pulse, after which the dye emits fluorescent light proportional to O_2 concentration. The dye is claimed to be very sensitive to O_2 and the measurement can take less than one second.[41] The system can be used for measuring O_2 in gas and dissolved in water. In principle, this method could be used also for in-line application. However, these measurements need time after packaging to allow oxygen to enter into the package through a leakage.[42]

13.4.3 The applicability and restrictions of oxygen indicators

In MAPs the high sensitivity of oxygen indicators is not advantageous as the sensitive indicator might also react with the residual O_2, which is often entrapped in the modified-atmosphere package during the packaging procedure (typically <1.0%). Extreme sensitivity also complicates handling of the indicator and anaerobic conditions are required during the preparation of the indicator and the packaging procedure. As the O_2 concentration required for the colour change of most indicators is around 0.1% they cannot be applied to the leak indication of MAPs as such. It has been claimed that the colour change of O_2 indicators used in MAPs containing acidic CO_2 gas is not definite enough.[34,43–44] Many of the patented O_2 indicators are reversible in their colour change and change colour according to the prevailing O_2 concentration[27,45] However, the reversibility is undesirable if the indicator is used for leakage control since the O_2 entering the package through the leak will be consumed during the microbial growth that is likely to follow the loss of the package integrity. In the worst case, the indicator colour will be the same as for intact packages, even if the product has been spoiled.

A visual O_2 indicator designed specifically for leak detection of MAPs has been developed at VTT.[34] This indicator, which is based on an oxygen-sensitive dye, is suitable for the quality control of modified-atmosphere-packed products[13] and it contains, in addition to the oxygen-sensitive component, an oxygen-absorbing component, and can hence prolong the product's shelf-life. This leak indicator does not react with the residual O_2 entrapped in the modified-atmosphere package because the O_2-absorbing component with adjusted capacity for the residual O_2 is included in the indicator and, moreover, the indicator is included in a film composition which protects against oxidation of the indicator during packaging.

13.4.4 Carbon dioxide indicators

CO_2 is widely used as a protective gas in modified-atmosphere packaging. During the first 12 days after the packaging procedure CO_2 is dissolved into the product and its concentration in the head-space is decreased, the final concentration being even as low as half of the original. After this period (1–2 days) a considerable decrease in CO_2 concentration is an evident sign of leakage in a package. However, CO_2 is also produced in microbial metabolism and its accumulation in a package headspace can be considered to be a sign of microbial growth. A leak in a package (decrease in the CO_2) is often followed by microbial growth (increase in the CO_2) and, in the worst case, the CO_2 will remain constant even in the case of leakage and microbial spoilage. For these two reasons, CO_2 indicators as leak indicators appear not to be as reliable as O_2 indicators.

In their patent Balderson & Whitwood[44–46] describe a reversible CO_2 indicator suitable for modified atmosphere packages. The indicator consists of, for example, five indicator strips. The strips contain CO_2-sensitive indicator material consisting, for example, of an indicator anion and a lipophilic organic quaternary cation.[47] The colour change of each strip has been designed to take place when the CO_2 concentration is below a certain limit (e.g. 25%, 20%, 15%, 10% or 5%). The concentration of CO_2 is indicated by a change of colour in one or more of the strips. Sealed Air Ltd has produced a visible CO_2 indicator for MAPs.

13.4.5 Safety aspects

A self-evident requirement for internal indicators placed in the package headspace is their absolute safety. The legislative aspects are discussed in Chapters 19 and 22.

13.5 Future trends

Package integrity is an essential requirement for maintaining the high quality of, e.g., sterilised and modified atmosphere packaged foods. The increasing focus on quality assurance is putting demands on verification of food package and seal integrity. On the other hand, much effort is and will also be put into the development of new materials and packaging systems with minimised risk of package failures.

Non-destructive package leak detection systems installed in-line are not yet very widely used, mainly because of high costs and lack of reliability/sensitivity to find all defective packages. New reliable and cost-effective systems are needed. One candidate for this could be the use of hydrogen as a tracer gas. Another possibility could be oxygen indicator labels or dyes printed onto packaging material and read automatically at a distance.

Today, application of intelligent package leak-indicating systems in Europe has been limited to some time-temperature indicators. However, some food producers are increasingly seeking extra merchandising and safety features. Intelligent leak indicators marketed, e.g., as 'premium quality labels' can be seen to give added value to the product/brand image. The visible indicators are ideal in many cases, however, in the future it can be expected that an intelligent package can contain more complex invisible messages that can be read at a distance. A label could be introduced as a chip but advances in ink technology might enable the use of printed circuits as well. The security tags and radio frequency identity/tracebility tags are the first examples of electronic labelling. Another approach for the future is the development of different optically read systems.

Development of these 'next generation' intelligent labels/printing systems is very challenging, e.g., in terms of cost demands, effectiveness and logistics. Standardisation will undoubtedly be one of the key issues when new systems are pushed onto the market. The basic requirement for success in making intelligent systems work in real life is collaboration between research institutes, authorities, and companies from product manufacturers and raw material supplier to retailer.

13.6 References

1. CHEN, C., HARTE, B., LAI, C., PETSKA, J. and HENYON, D. Assessment of package integrity using a spray cabinet technique. *J. Food Prot.* 1991, **54**: 643–7.
2. KELLER, S.W., MARCY, J.E., BLACKISTONE, B.A., LACY, G.H., HACKNEY, C.R. and CARTER, W.H. Bioaerosol exposure method for package integrity testing. *J. Food Prot.* 1996, **59**: 768–71.
3. AHVENAINEN, R., MATTILA-SANDHOLM, T., AXELSON, L. and WIRTANEN, G. The effect of microhole size and foodstuff on the microbial integrity of aseptic plastic cups. *Pack. Techn. Sci.* 1992, **5**: 101–7.
4. BLACKISTONE, B.A., KELLER, S.W., MARCY, J.E., LACY, G.H., HACKNEY, C.R. and CARTER., W.H. Contamination of flexible pouches challenged by immersion biotesting. *J. Food Prot.* 1996, **59**: 764–7.
5. HURME, E., WIRTANEN, G., AXELSON-LARSSON, L., PACHERO, A. and AHVENAINEN, R.. Penetration of bacteria through microholes in semirigid aseptic and retort packages. *J. Food Prot.* 1997, **60**: 520–5.
6. EILAMO, M., AHVENAINEN, R., HURME, E., HEINIÖ, R-L. and MATTILA-SANDHOLM, T. The effect of package leakage on the shelf-life of modified atmosphere packed minced meat steaks and its detection. *Lebensm.-Wiss. -Techn.* 1995, **28**: 62–71.

7. RANDELL, K., AHVENAINEN, R.,` LATVA-KALA, K., HURME, E., MATTILA-SANDHOLM, T. and HYVÖNEN, L. Modified atmosphere-packed marinated chicken breast and rainbow trout quality as affected by package leakage. *J. Food Sci.* 1995, **60**: 667–72, 684.
8. AHVENAINEN, R., EILAMO, M. and HURME, E. Detection of improper sealing and quality deterioration of MA-packed pizza by a colour indicator. *Food Control* 1997, **8**: 177–84.
9. GILCHRIST, J.E., SHAH, D.B., RADLE, D.C. and DICKERSON., R.W. Leak detection in flexible retort pouches. *J. Food Prot.* 1989, **52**: 412–15.
10. LAMPI, R.A. Retort pouch: the development of a basic packaging concept in today's high technology era. *J. Food Proc. Eng.* 1980, **4**: 1–18.
11. MARCY, J.E. Integrity testing and biotest procedures for heat-sealed containers. In: (Blackiestone, B.A. and Harper, C.L., eds), *Plastic Package Integrity Testing – Assuring Seal Quality*. Institute of Packaging Professionals, Herndon, Virginia. pp. 35–48. 1995.
12. ROSE, D. *Risk factors associated with post process contamination of heat sealed semi-rigid packaging*. Techn. Mem. No. 708. CCFRA, Chipping Campden, Gloucestershire, 1994.
13. AHVENAINEN, R., HURME, E., RANDELL, K. and EILAMO, M. *The effect of leakage on the quality of gas-packed foodstuffs and the leak detection*. VTT Research Notes 1683, Espoo, Finland, 1995.
14. AXELSON, L., CAVLIN, S. and NORDSTRÖM, J. Aseptic integrity and microhole determination of packages by electrolytic conductance measurement. *Pack. Techn. Sci.* 1990, **3**: 141–62.
15. HURME, E., WIRTANEN, G., AXELSON-LARSSON, L. and AHVENAINEN, R. Testing of reliability of non-destructive pressure differential package leakage testers with semirigid aseptic cups. *Food Control*. 1998, **9**: 49–55.
16. JENSEN, P. Testing of package integrity based on CO_2 as a trace gas. *Proc. IAPRI Symp*. Reims, 9–12 October 1994, 9 p.
17. SAFVI, A., MEERBAUM, M., MORRIS, S., HARPER, C. and O'BRIEN W. Acoustic imaging of defects in flexible food packages. *J. Food Prot*. 1997, **60**: 309–14.
18. SONG, Y., LEE, H. and YAM, K. Feasibility of using a non-destructive ultrasonic technique for detecting defective seals. *Pack. Techn. Sci.* 1993, **6**: 37–42.
19. YAM, K.L. Pressure differential techniques for package integrity inspection. In (Blakistone, B.A. and Harper, C.L. eds) *Plastic Package Integrity Testing – assuring Seal Quality,* Institute of Packaging Professionals, Herndon, VA., pp. 137–46, 1995.
20. HURME, E., WIRTANEN, G. AXELSON-LARSSON, L. and AHVENAINEN, R. Reliability of non-destructive pressure differential package leakage testers using semirigid retort trays. *Food Sci. Techn.* 1998, **31**: 461–6.
21. HURME, E. and AHVENAINEN, R. 1998. A nondestructive leak detection method for flexible food packages using hydrogen as a tracer gas. *J. Food Prot*. 1998, **61**: 1165–9.

22. STAUFFER, T. Non-destructive in-line detection of leaks in food and beverage packages – an analysis of methods. *J. Pack. Techn.* 1988, **2**: 147–9.
23. BOJKOW, E., RICHTER, C. and PÖTZL, G. Helium leak testing of rigid containers. *Proc. IAPRI Symp*. Vienna, 23–26 September, 16 p. 1984.
24. HEIKKINEN, E., HURME, E. and AHVENAINEN, R. *US Patent 6279384*. Method for treating a product and a leak-detection device, 2001.
25. ABE, Y. Active packaging with oxygen absorbers. In (Ahvenainen, R., Mattila-Sandholm, T. and Ohlsson, T. eds) *Minimal Processing of Foods*, VTT Symposium 142. , pp. 209–23, 1994.
26. GOTO, M. Japanese *Patent JP 62-259059*. Oxygen Indicator, Mitsubishi Gas Chemical Co., Inc., Tokyo, Japan, 1987.
27. YOSHIKAWA, Y., NAWATA, T., GOTO, M. and FUJII, Y. *US Patent 4169811*. Oxygen Indicator, Mitsubishi Gas Chemical Co., Inc., Tokyo, Japan, 1979.
28. DAVIES, E.S. and GARNER, C.D. *GB 2298273* Oxygen Indicating Composition. The Victoria University of Manchester, Manchester, UK, 1996.
29. KRUMHAR, K. C. and KAREL, M. *US Patent 5096813*. Visual Indicator System. Massachusetts Institute of Technology, Cambridge, MA, USA, 1992.
30. YOSHIKAWA, Y., NAWATA, T., GOTO, M. and KONDO, Y. *US Patent 4349509*. Oxygen Indicator Adapted for Printing or Coating and Oxygen-Indicating Device. Mitsubishi Gas Chemical Co., Inc., Tokyo, Japan, 1982.
31. NAKAMURA, H., NAKAZAWA, N. and KAWAMURA, Y. *Japanese Patent JP 62-183834*. Food Oxidation Indicating Material – Comprises Oxygen Absorption Agent Containing Indicator Composed of Methylene Blue, Reducing Agent and Resin Binder. Toppan Printing Co., Ltd., Tokyo, Japan, 1987.
32. SHIROZAKI, Y. *Japanese Patent JP 2-57975*. Oxygen Indicator. Nippon Kayaku KK, Tokyo, Japan, 1990.
33. LENARVOR, N., HAMON, J.-R. and LAPINTE, C. *French Patent FR 2710751*. Detecting the presence and disappearance of a gaseous target substance – using an indicator which forms a coloured reaction product with the substance, and an antagonist which modifies the colour of the reaction product. ATCO, Caen, France, 1993.
34. MATTILA-SANDHOLM, T., AHVENAINEN, R., HURME, E. and JÄRVI-KÄÄRIÄINEN, T. *EP 0666977*. Oxygen sensitive colour indicator for detecting leaks in gas-protected food packages. Technical Research Centre of Finland (VTT), Espoo, Finland, 1998.
35. PERLMAN, D. and LINSCHITZ, H. *US Patent 4526752*. Oxygen Indicator for Packaging, 1985.
36. AHVENAINEN, R., PULLINEN, T., HURME, E., SMOLANDER, M. and SIIKA-AHO, M. *WO 9821120*. Package for decayable foodstuffs. Technical Research Centre of Finland (VTT), Espoo, Finland, 1998.
37. GARDIOL, A.E., HERNANDEZ, R.J., REINHAMMAR, B. and HARTE, B.R.

Development of a gas-phase oxygen biosensor using a blue copper containing oxidase. *Enz. Microb. Technol.* **18**: 347–52, 1996.
38. GARDIOL, A.E., HERNANDEZ, R.J. and HARTE, B.R. *US patent 5654164.* Method and device for reducing oxygen with a reduced oxidase with colour formation, 1997.
39. REININGER, F., KOLLE, C., TRETTNAK, W. and GRUBER, W. Quality Control of Gas-Packed Food by an Optical Oxygen Sensor. *Proc. of Int. Symp. on Food Packaging – Ensuring the Safety and Quality of Foods,* 11–13 September Budapest, ILSI, Budapest, Hungary. 1996.
40. MAURER, H. *Swiss Patent CH 654109.* Verfahren zum Nachweis von Sauerstoff in einer Verpackung. Tecan AB, Hombrechtikon, Switzerland, 1986.
41. ATKINSON, S. Non-invasive method for determining oxygen in food packaging. *Food, Cosmetics and Drug Packaging*. pp. 118–19, June 2000.
42. SAINI, D., KONIG, H. and DRAAIJER, A. In-line non-invasive detection of in-pack oxygen in PET containers. *Nova-Pack Europe 2001.* September 18–19, Munich, Germany, 2001.
43. AHVENAINEN, R. and HURME, E. Active and Smart Packaging for Meeting Consumer Demands for Quality and Safety. *Food Addit. Contaminants,* **14:** 753–63. 1997.
44. BALDERSON, S.N. and WHITWOOD, R.J. *US Patent 5439648.* Gas Indicator for a Package. Trigon Industries Ltd, Auckland, New Zealand, 1995.
45. KATSURA, H., MIYAZAKI, S. and EBASHI, S. *Japanese Patent JP 60-71956.* Oxygen Presence Indicator Consists of Cobalt Complex of Bis(salicylaldehyde) Alkylenediimine and/or Derivatives; Salicylaldehyde. Toyo Ink MGF KK, Tokyo, Japan, 1989.
46. BALDERSON, S.N. and WHITWOOD, R.J. *EP 0625467.* Tamper evident system with gas sensitive element. Trigon Industries Limited, Auckland, New Zealand, 1994.
47. MILLS, A. and MCMURRAY, H.N. *WO 91/05252.* Carbon Dioxide Monitor Abbey Biosystems Limited, Dyfed, UK, 1991.

14

Combining MAP with other preservation techniques

J.T. Rosnes, M. Sivertsvik and T. Skåra, NORCONSERV, Norway

14.1 Introduction

Modified atmosphere packaging (MAP) is widely used for many food products and is now a commercial and economic reality. MAP is common in markets that have a well established and controlled cold chain and that can sustain a high-priced quality product. However, MAP is a mild preservation method and a major concern is that MAP storage may not provide a sufficient level of safety for the extended storage of fresh chilled food products with regard to pathogenic bacteria. Other preservation steps may be necessary, in addition to MA packaging and low temperatures, in order to delay outgrowth of pathogens or toxin production beyond their point of spoilage. One feasible solution can be to use a combination of different preservation factors or techniques. This approach provides reliable, yet mild, multi-targeted preservation of foods, thereby facilitating improvements in food safety, quality and economics. The topic of this chapter is to outline the significance of combining MAP with other preservative techniques.

Food environments are generally stressful for bacteria because most nutrients are in the form of complex substrates whereby the conditions for bacterial growth are not optimal. The level of free moisture may be restricted and the presence of acids and other chemicals may be at stressful levels. In addition, there is often competition from other microorganisms which are present. Replacing the normal atmosphere with a modified atmosphere, i.e., other concentrations of O_2, CO_2 and N_2, will add additional stress to microorganisms and change the composition of the initial microbial flora. From the early development of MAP (Coyne, 1932; Coyne, 1933), it has been shown that MA can, on its own, inhibit growth of microorganisms. Higher levels of CO_2 have a bacteriostatic effect on microorganisms and properly designed MAP can double a product's shelf-life (Davies, 1995). In spite of 70 years of knowledge about CO_2 inhibition it is only in the last two decades that MAP has become a widely commercially used technology for storage and distribution of foods. This trend is mainly driven by the demands of modern consumers for pre-processed products that have a fresh appearance and are convenient and easy to prepare. The main focus in this chapter will therefore be on chilled MA packaged products, where pronounced effects of modified atmosphere packaging combined with preservation factors can be seen.

The potential of MAP to extend shelf-life for many foods is well documented, e.g., fish (Dalgaard *et al.*, 1993), sandwiches (Farber, 1991), salads and vegetables (Day, 1990), and meat (Gill, 1996). Several review articles outline the different aspects of MA packaging (Farber, 1991; Church and Parsons, 1995; Davies, 1995; Phillips, 1996; Sivertsvik *et al.*, 2002). A major concern associated with the use of MAP is that of product safety. The desired suppression of spoilage microorganisms extends the shelf-life if compared to food products stored in a normal air environment, and this may create opportunities for slower growing pathogenic bacteria. In particular the growth of psychrotrophic pathogens in refrigerated ready-to-eat food may create a health risk before the product is overtly spoiled (Farber, 1991).

Since some preservation procedures (e.g. chemical additives) used in food products act by inhibiting growth, instead of inactivating microorganisms, their contribution may be most beneficial when used against pathogens that form toxins in foods, or those that need to reach high numbers to cause foodborne illness, especially in healthy consumers. However, in order to protect consumers at risk from foodborne illness or against microbes with low infectious doses, there is a need for complete inactivation of pathogens and avoidance of recontamination of foods during processing, distribution, and preparation for consumption. For each specific MA-packaged product this must be done either before packaging or later by adjusting to correct preservation intensity in the product.

14.2 Combining MAP with other preservative techniques

The preservation of almost all foods in industrialised and developing countries is based on combinations of several factors that secure microbial safety, stability and sensory quality. This is true not only for traditional foods, but also for more novel products. The most important preservative methods in common use for food preservation are high temperature (heat treatment), low temperature, water activity (a_w), acidity (pH), redox potential (Eh), some preservatives, and a competitive flora (Leistner, 1992). The application of some processes using the aforementioned preservation methods at low intensity or concentrations is still in the exploratory and developmental stages. Other methods have obtained regulatory approval and are being introduced in HACCP plans and in the marketplace for consumer evaluation and acceptance.

The principle of combined preservation has been well described by Leistner *et al.*, and is often referred to as hurdle technology (Leistner, 1992; Leistner, 1995b; Leistner, 2002). Whilst the hurdle concept is widely accepted as a food preservation strategy, its potential, using MAP, has still to be fully realised. The intelligent selection of hurdles in terms of the number required, the intensity of each and the sequence of applications to achieve a specified outcome are expected to have significant potential for the future (McMeekin and Ross, 2002).

Homeostatsis is the tendency towards uniformity and stability in the internal status of living organisms. For instance, the maintenance of a defined pH within narrow limits is a prerequisite and feature of all living cells, and this applies to higher organisms, as well as microorganisms. In food preservation the homeostasis of microorganisms is a key phenomenon because if homeostasis of these organisms is disturbed by some preservation methods in foods, they will not multiply, i.e. they will remain in the lag phase or may even die before their homeostasis is re-established. Therefore, in actual fact, the preservation of food is achieved by disturbing, temporarily or permanently, the homeostasis of microorganisms in the food. In most foods microorganisms are able to operate homeostatically in order to react to the environmental stresses imposed by the applied preservation procedures. Applying additional preservation will inhibit repair of disturbed homeostasis and this requires extra energy from the microorganisms concerned. In MA products energy depletion increases as the intensity or concentration of preservation is increased and the restriction of the energy supply will inhibit the repair mechanisms of the microbial cells' factors and leads to growth inhibition or death.

14.2.1 Preservation focused on specific groups of microorganisms

If the true potential of some of the emerging preservation technologies, combined with MAP is to be realised, it will be important to develop systematic, kinetic data describing their efficiency against key target microorganisms. The type and numbers of microorganisms in the raw material have a direct influence on the effectiveness of MAP in inhibiting both spoilage organisms and pathogens. When adding extra preservation to packaged food, it is therefore important to understand which part of the bacterial population is inhibited and which is not. The shelf-life extension obtained with MA does not always give the same extension in safety. Pathogenic bacteria may gain advantage when the competing flora is inhibited, e.g., *Listeria monocytogenes* increased in numbers on raw chicken in 72.5:22.5:5 (CO_2:N_2:O_2) atmosphere at 4°C, irrespective of a decrease in the aerobic spoilage flora (Wimpfheimer *et al.*, 1990). Many MA packaged products of meat, vegetable and sea-food origin have common key target organisms. For chilled products psychrotrophic pathogens are the target, while in heat-treated ready meals spore-forming *Clostridium* and *Bacillus* species are the target organisms. There are five food-borne pathogenic bacteria known to be capable of growth below 5°C: *Bacillus cereus*, non-proteolytic *Clostridium botulinum* type E, B and F (group II), *Listeria monocytogenes,*

Table 14.1 Preservatives used to inhibit specific psychrotropic pathogens in combination with MAP

Organism	Relevant food	Preservative	References
Bacillus cereus	Dairy food Ready-to-eat food		(Kośeki and Itoh, 2002)
Non-proteolytic *Clostridium botulinum*	Ready-to-eat food	Irradiation	(Lambert *et al.*, 1991)
	Dinner	Microbial inhibition by *Bacillus* species	(Lyver *et al.*, 1998)
		NaCl	(Gibson *et al.*, 2000)
Listeria monocytogenes	Fish, meat, vegetables, fresh produce	Competitive microbial flora	(Liserre *et al.*, 2002; Wimpfheimer *et al.*, 1990; Francis and O'Beirne, 1998; Bennik *et al.*, 1999)
		Nisin	(Szabo and Cahill, 1998) (Fang and Lin, 1994)
		Na-lactate	(Devlieghere *et al.*, 2001; Pothuri *et al.*, 1996)
		Irradiation	(Thayer and Boyd, 2000; Thayer and Boyd, 1999)
		pH	(Francis and O'Beirne, 2001)
		Oregano essential oils	(Tsigarida *et al.*, 2000)
		High O_2 level	(Jacxsens *et al.*, 2001)
Yersinia enterocolitica	Pork	Lactate	(Barakat and Harris, 1999)
	Poultry	Lactic acid	(Grau, 1981)
		Background flora	(Kleinlein and Untermann, 1990)
		Low temperature	(Gill and Reichel, 1989)
Aeromonas hydrophila	Fish, shellfish	Heat	(Devlieghere *et al.*, 2000b)
	Mussels Meat	pH	(Doherty *et al.*, 1996)
Salmonella	Poultry	Sorbate	(Elliott and Gray, 1981)

Yersinia enterocolotica, and *Aeromonas hydrophila*. Consequently the ability of modified atmospheres to inhibit the growth of these organisms in foods under refrigerated storage is of vital importance and additional preservation factors have therefore been combined with MAP (Table 14.1). The main cause of concern, however, is the possible growth of non-proteolytic *C.botulinum*, because it is both anaerobic and low-temperature tolerant. Of particular concern is the fact that it may grow and produce toxin on the product before spoilage is detectable to the consumer.

Few non-thermal treatments can currently be relied upon to inactivate bacterial spores. Hence low-temperature storage must be combined with an additional preservation hurdle such as acidic formulation or salt to prevent spore outgrowth. Most food spoilage moulds species have an absolute requirement for oxygen and appear to be sensitive to high levels of CO_2. Consequently foods with low a_w values, such as bakery products, that are susceptible to spoilage by moulds can have their shelf-lives extended by MAP. Many yeasts are capable of growing in the complete absence of oxygen and most are comparatively resistant to CO_2. Although MAP can inhibit the growth of bacterial and fungal spoilage microorganisms, its effect on the survival of enteric viruses, including hepatitis A viruses (HAV), has not been well investigated. Both mussels and lettuce that are packaged in MAP may be a vehicle in the transmission of HAV (due to contact with contaminated water) and therefore can contribute to hepatitis A outbreaks (Cliver, 1997). Experiments by Bidawid *et al.* (2001) indicated that MAP does not influence HAV survival when present on the surface of produce with high CO_2 levels. This may have been attributed to the inhibition of spoilage-causing enzymatic activities in the lettuce, which may have reduced exposure of viruses to potential toxic by-products.

14.2.2 Preventative techniques combined with MAP

The main preservation techniques currently used act in one of three ways: (i) preventing the access of microorganisms to foods, (ii) inactivating them when they have gained access, or (iii) preventing or slowing down their growth when they have gained access and not been inactivated. During the past few years there has been increasing interest in modifying these approaches or in developing new ones, with the objective of reducing the severity of the more extreme techniques. Many such developments have involved new uses of existing techniques in new combinations to inhibit the growth of micro-organisms. Approaches where preservation techniques are used at lower intensity or at lower concentration, causes inactivation and bacterial growth inhibition to overlap. It is the safety level, the quality level or the outcome of inactivation or growth inhibition of target organisms that determines the final use of the chosen preservation method(s) (Table 14.2).

14.2.3 Hygienic conditions

Hygienic production is not a preservation method, but ingredients or raw material used in MAP should always be of superior quality, i.e. low bacterial numbers and preferably without pathogenic bacteria. This is a prerequisite for fresh products with increased shelf-life, and preservation should never be used to compensate for inadequate hygiene or poor raw material quality. A strategy for the control of pathogens and, to a large extent, spoilage microorganisms is basically one of exclusion, which requires reducing or eliminating the initial microbial load or preventing or minimising further contamination. Since MA packaged products are hermetically sealed, recontamination is eliminated and the hygienic pre-packaging conditions are the most important steps. An appropriate design and construction of the pre-packaging premises is necessary

Table 14.2 Uses and limitations of preservation technologies combined with MAP (Adapted from Leistner, 2002)

Effect in MAP	Preventing assess		Inactivation or growth/activity inhibition							
	Heat treatment	Ionising irradiation	NaCl	pH	Bacteriocins	Low Temp[1]	Preservatives	Na-lactate	Essential oils	SGS[2]
Killing spores	+	+	–	–	–	–	–	–	–	–
Killing veg. cells	+	+	±	±	±	±	+	–	–	–
Preventing growth	–	–	+	+	–	–	+	+	+	+
Solids	+	+	+	+	+	+	+	+	+	–
Liquids	+	+	+	+	+	+	+	+	+	+
In-pack treatments	+	+	–	–	–	+	–	–	–	+
In-line treatment	+	+	–	–	+	–	–	–	–	+

[1] Low temperature (super chilling and freezing) may also kill bacteria
[2] SGS = soluble gas stabilisation

to limit entry, multiplication, and spread of microorganisms in the environment where MA packaged foods are being produced or manufactured, in order to prevent or minimise cross-contamination of the products. New and hygienic design of production facilities, with elements from clean room technology, are now more frequently adopted in the production of high-priced products. These techniques meet the requirements of freeing the products from microorganisms by cross-contamination, decontaminating the packaging material, and sterilising air in contact with the product.

14.3 Heat treatment and irradiation

Refrigerated ready-to-eat meals and entrées, prepared salads, sandwiches, pizza, fresh pasta, soups, whole meals, and sauces are commonly packaged in MA after heat treatment. These products have received some form of heat treatment, and are for the most part 'low acid'. They are marketed refrigerated (−1 to +4°C) and require little preparation before consumption. There has been a recent expansion in the use of the combination of mild heating of vacuum-packaged foods, e.g., sous vide, and cook-and-chill products with controlled chill storage, particularly for catering but also for retail. MA packaging of cook-and-chill foods is now commonly used for processed minimal heat-treated ready meals. Many nursing homes and canteens currently receive heat-treated MA packaged meals prepared in a central kitchen unit. With this method the risk of recontamination of microorganisms after cooking must be taken into account. These ready-to-eat meals have a shelf-life of 7–14 days, depending on the amount of heat used.

The success of heat-treated ready meals results primarily from the inactivation of the vegetative microbial flora by mild heating. Another fact is that the spores of psychrotrophic bacteria, which can grow at low chill temperatures, are generally more heat sensitive than those of mesophiles and thermopiles, which cannot grow at these temperatures. The mild heating therefore destroys the cold-growing fraction of the potential spoilage flora, whilst the minimal thermal damage and conditions of low oxygen tension ensure high product quality. Shelf-lives at temperatures below about 3°C can therefore be very long, i.e., in excess of three weeks, with eventual spoilage resulting from the slow growth of psychrothropic strains of *Bacillus* and *Clostridium*. In order to ensure safety, heat processes equivalent to 90°C for 10 min. (ACMSF-Advisory Committee on the Microbiological Safety of Food, 1992) are generally regarded as sufficient to ensure inactivation of spores in the coldest-growing pathogenic sporeformers such as psychrotrophic strains of *Clostridium botulinum* (Notermans *et al.*, 1990; Lund and Peck, 1994). For lower heat treatments, strict limitations of shelf-life, efficient control of storage temperatures below 3.0°C or some form of intrinsic preservation is necessary.

During a three-year period, 2168 heat-treated, commercially available ready-made meals with a shelf-life of 3–5 weeks were examined for sporeforming bacteria (Nissen *et al.*, 2003). Three-quarters of the samples had less than ten bacteria/g the day after production, and none had more than 1000. Similar numbers were found at the end of the shelf-life. At abuse temperatures (20°C), the number of bacteria increased to 10^6–10^7 cfu/g in seven days. Three hundred and fifty isolates of spore-forming bacteria (aerobic and anaerobic) were collected and characterised as *Bacillus licheniformis*, *B. thuringiensis, B. megatherium, B. pumilis, B. subtilis, B. sphaeicus,* and *B.cereus*, but no *Clostridium* strains were detected. Growth experiments of 113 strains from this work showed that only 11 strains were able to grow at 7°C. Furthermore, none of the psychotropic strains were able to produce substantial amounts of toxins. These experiments show that spore-formers, especially *Bacillus* strains, survive mild heat treatments and some of their members may be a health risk in products with long shelf-lives or if stored at high temperatures. Further research on germination, growth and toxin production at chilled temperatures in modified atmosphere is required.

14.3.1 Low temperature (freezing, partial freezing, super chilling)

A low and stable temperature is a general prerequisite for many MA products and has a particular importance in fresh storage. Both enzymatic and microbiological activity are greatly influenced by temperature. Many bacteria are unable to grow at temperatures below 10°C and even psychrotrophic organisms grow very slowly, and with extended lag phases, at temperatures that approach 0°C. Temperature can, however, be used to achieve special effects in MA products. Guldager *et al.* (1998) and Bøknæs *et al.* (2000) have found that frozen (−20°C) and thawed cod fillets in MA had longer shelf-life than raw cod in MA. This shelf-life extension was most likely due to the inactivation of the spoilage bacterium *Photobacterium phosphoreum* during frozen storage. The use of frozen fillets as a raw material not only provides a more stable MAP product but also allows much greater flexibility for production and distribution. A similar effect was found when frozen and thawed salmon was packaged in MA. Here also the freezing eliminated *P. phosphoreum* and extended the shelf-life of MAP salmon at 2°C by 1–2 weeks (Emborg *et al.*, 2002).

Earlier experiments with whole gutted salmon have shown that MAP can be combined with super-chilling to extend further the shelf life and safety of fresh fish (Rosnes *et al.*, 1998; Rosnes *et al.*, 2001; Sivertsvik *et al.*, 1999). In this technique, also known as partial freezing, the temperature of the fish is reduced to between 1 or 2°C below the initial freezing point and some ice is formed inside the product (Gould and Peters, 1971). Under normal conditions, the gas atmosphere surrounding a MA product will insulate the product, leading to a longer time until it is satisfactorily chilled. Partial freezing eliminates this problem by reducing the temperature of the fish before packaging. These experiments showed that super-chilling can decrease the temperature before packaging and increase stored refrigeration capacity during storage, and thereby significantly decrease microbial growth at temperatures of 2–6°C, which is often found in chilled retail counters. Sikorski and Sun (1994) found that super-chilling can store enough refrigeration capacity to keep a core temperature < 0ºC during the first three weeks of chilled storage. A shelf-life extension of seven days has been obtained for super-chilled fish when compared to traditional ice stored fish of the same type (Leblanc and Leblanc, 1992). Untreated salmon steaks in MA, and partial frozen salmon steaks in MA, had an acceptable microbiological quality of 22 days at 0ºC, but were rejected by odour after 17 days. Salmon steaks in air had and acceptable microbiological quality for only eight days (Rosnes *et al.*, 2001). MAP is also being used to package products for frozen storage. The reasoning behind the use of MAP for ready-to-eat products is that they can be distributed frozen, then thawed and sold as chilled products but with an extended shelf-life (Morris, 1989).

14.3.2 Irradiation

The attraction of combining irradiation with MAP is that the modified atmospheres are not lethal to spoilage organisms and pathogens. The possibility exists, therefore, of using irradiation below the 'threshold' dose, i.e., the level at which spoilage organisms and pathogens are killed and below the level where undesirable organoleptic changes are introduced, in order to enhance the attractiveness of MAP. The effects of MAP/irradiation on sensory properties, and its effect upon depletion of vitamin content during storage, compared to untreated items, have been examined in detail. Studies on the effects of MAP/irradiation methods on nutritional quality showed that the deleterious effects of irradiation on vitamins can be removed by modifying storage atmospheres (Robins, 1991). For a radiation dose of 0.25 kGy and in an air atmosphere, 60% of the thiamine content was lost over the storage period, compared to a minimal loss in the non-irradiated control over the same period. The loss of α-tocopherol, exposed to 1 kGy irradiation, was some 50% over this period, compared to a similar minimal loss in the non-irradiated control sample. In both cases there were much reduced loss rates in N_2 atmospheres, which demonstrated that the effects of irradiation on these vitamins could be removed by modifying storage atmospheres.

The growth rate of surviving microorganisms was measured as a function of atmospheric composition for the irradiated and non-irradiated food samples, and the optimum lethal atmospheres were found to range from CO_2/N_2 : 25/75 to CO_2/N_2 : 50/50. Tests at 10ºC showed a similar trend, although the effectiveness of high concentrations of CO_2 was reduced. The major surviving organisms even in the irradiated packs were lactobacilli, in accordance with general expectations on their resistance to radiation.

A series of experiments on MAP/irradiation combination, for use with chicken and pork products, with the goal of optimising sensory quality have shown that each particular food item requires careful evaluation and that generalisation can lead to incorrect and inappropriate specifications for optimum storage. However, as one of several different treatment combinations aimed at

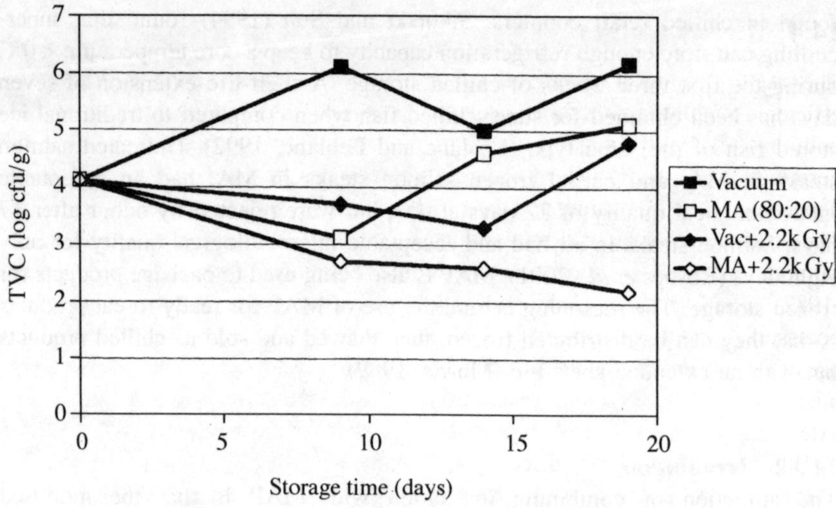

Fig. 14.1 Effect of irradiation on MAP and vacuum packaged cod fillets.

reducing mould in strawberries, the MAP/irradiation method gave the best results. Several studies have been carried out on the use of MAP/irradiation treatments in fish products, e.g., low dose irradiation extended the shelf-life of haddock fillets and cod fillets (Licciardello *et al.*, 1984) more than either process achieved on its own. Przybylski *et al.* (1989) examined fresh catfish fillets, processed with low dose irradiation in combination with MAP, and demonstrated that irradiation treatments with or without elevated carbon dioxide-modified atmosphere packaging significantly reduced the bacterial load and extended shelf-life from 5–7 days to between 20 and 30 days.

In an experiment, cod fillets were packaged in MA (80:20 CO_2:N_2) and under vacuum before irradiation with 2.2 kGy, and subsequent storage at 4ºC. The results (Fig. 14.1) showed a large inhibitory effect of irradiation on micro-organisms. The best results were observed when combining irradiation with MAP. The sensory shelf-life of irradiated MA cod and irradiated vacuum packaged cod was >24 days and 24 days accordingly. For non-irradiated MA cod the shelf-life was <14 days, and for vacuum packaged cod, <9 days. This should indicate a large potential for seafood product shelf-life extensions through the use of MAP combined with low-dose irradiation. However, before this method is widely accepted, several issues need to be resolved, such as legislative, scientific (food safety), and also consumer attitudes towards irradiated foods (Sivertsvik *et al.*, 2001). Nevertheless, all studies have shown that the advantages of MAP/irradiation treatment methods must be determined for specific applications with a fair degree of caution and this requires the ascertainment of exact conditions for every product in terms of microbiological safety.

14.4 Preservatives

The use of chemical preservatives (benzoic acid, sorbic acid) is often very efficient in inhibiting microbial growth. These molecules inhibit the outgrowth of both bacterial and fungal cells. Sorbic acid is also reported to inhibit the germination and outgrowth of bacterial spores. Their effect, however, is strongly dependent on the pH value, and their use is rarely recommended if pH exceeds 6. The effect of adding potassium sorbate to ice used for cooling of red hake and salmon, packaged in modified atmosphere was studied by Fey and Regenstein (1982). These authors found that a CO_2–O_2 atmosphere combined with 1% potassium sorbate ice was most satisfactory. Also other studies conducted on the use of sorbates in fish and fish products suggest that sorbates in combination with other compounds or techniques can be used as an effective preservative tool for extending the shelf-life of fish products (Thakur and Patel, 1994).

Elliott and Gray (1981) discovered growth inhibition of *Salmonella enteritidis* following exposure to a combination treatment of potassium sorbate (0.5, 1.5 or 2.5%) and modified atmospheres of 20, 60 and 100% CO_2 at pH 6.5, 6.0 or 5.5 at 10ºC. Dalgaard *et al.* (1998) found that including potassium sorbate was effective in reducing the growth of the specific spoilage organisms *P. phosphoreum* in model substrates. This may have a practical use in extending the shelf-life of MA packaged seafood. Cooked and brined shrimps, including benzoic, citric and sorbic acids, packaged in modified atmosphere were stored at 0, 5, 8, and 25ºC (Dalgaard and Jørgensen, 2000). The shrimps had a shelf-life of > 7 months at 0ºC, but spoiled in 4–6 days at 25ºC. This pronounced effect of temperature was explained by changes in spoilage at different storage temperatures.

14.4.1 Sodium chloride, a_w

Sodium chloride is an old preserving agent with antimicrobial importance that generally binds water and thus inhibits bacterial growth through reducing water activity. Recent focus, however, on the adverse effects of a high sodium intake on blood pressure, has led to a sharp decrease in salt consumption. Therefore, salt is less and less used in the preservation of food, and more only for taste purposes. Furthermore, addition of salt activated and stabilised proteolytic activity (alkaline proteases primarily) over a wider temperature and pH area in Atlantic salmon (Olsen *et al.*, 2002), although conclusive data on the effects remain to be presented.

A number of different fish species have been studied after being subjected to 5 minutes treatment in a brine solution (5% NaCl) and then packaged in different gas mixtures. Mitsuda *et al.* (1980) and Pastoriza *et al.* (1998) found good texture and greatly repressed colour change at 3ºC. The effects of an optimum gas mixture on hake slices when combined with a sodium chloride dip were studied. A delay in chemical, microbiological and sensorial alteration was found and total volatile bases (TVB) and microbiological levels were significantly lower when MAP-stored samples had been previously dipped in

NaCl. Additional effects which are important for MA packaged fish were reduced exudation, higher water binding capacity and increased time before MAP stored samples were rejected due to off-odours. The antimicrobial contribution of NaCl in a food system may also be influenced by the presence of other preservatives, e.g. benzoate, sorbate, phosphates, antioxidants, spices and liquid smoke.

14.4.2 Alteration of pH

pH influences spoilage due to its effect on the microorganism and enzyme activity (Ashie *et al.*, 1996). Daniels *et al.* (1985) claimed that the CO_2/bicarbonate ion has an observed effect on the permeability of the cell membranes, and that CO_2 is able to produce rapid acidification of the internal pH of the microbial cell, with possible ramifications relating to metabolic activities. Fey and Regenstein (1982) noted that CO_2 did not lower the pH of the fish. In most fish products, even though a reduced pH could be advantageous in order to reduce bacterial growth, with the iso-electric point of fish proteins being approx. 5.5, it will also lead to reduced water holding capacity as well as textural changes.

Devlieghere *et al.* (1998) modelled the effect of pH on the solubility of CO_2 and found that higher amounts of CO_2 can be dissolved in aqueous foods with high pH levels. In spite of the fact that higher concentrations of CO_2 are dissolved at higher pH, the preservative effect seems to be larger at low pH levels when in combination with a modified atmosphere. Beef with pH 6.3 in 100% N_2 at 5°C supported the growth of *E. cloacae* but not at pH 5.4 (Grau, 1981). Growth of *S. liquefaciens* was inhibited on beef with pH 5.4 in 100% N_2 at 5°C but grew to levels of 10^8 cfu/g in eight days on meat with pH 6.3. *Yersinia enterocolitica* failed to grow on beef ranging in pH from 5.4 to 5.9 under 100% N_2, but grew at pH 6.0–6.2. Under aerobic conditions pH had little effect on the growth of *Y. enterocolitica*.

In muscle foods, the initial decline in pH is reversed during later stages of post-mortem changes as a result of decomposition of nitrogenous compounds (Ashie *et al.*, 1996). This effect may be inhibited by addition of buffer-compounds like polyphosphates. These enhance the preservative role by (i) acting as metal ion chelators, (ii) acting as pH buffers, (iii) interacting with proteins to promote hydration and water binding capacity, and thus (iv) preventing lipid oxidation and microbial growth (Ellinger, 1972).

14.4.3 Organic acids

The antimicrobial properties of acetic, lactic, citric and malic acid have been utilised by the food industry for food preservation. It is generally accepted that the undissociated molecule of the organic acid or ester is responsible for the antimicrobial activity. Many weak acids, in their undissociated form, can penetrate the cell membrane and accumulate in the cytoplasm and acidify its interior. The activity of the lactate has also been attributed to its lowering of the water activity of the food product, but this can only partly explain its antimicrobial effects on meat products (Debevere, 1989; Houtsma *et al.*, 1993). Salts of organic acids, such as sodium and potassium lactate, are fully dissociated in aqueous solutions, and at the pH of an unfermented meat product, which is typically 6.0 to 6.5, the concentration of the undissociated form of the added lactate is low. The increased permeability of cellular membranes for lactic acid at higher pH values may be an important factor in understanding the antimicrobial activity of Na-lactate, as observed in neutral food media and food products. Cooked meat products packaged in oxygen-free atmospheres will spoil due to psychrotrophic lactic acid bacteria (Borch *et al.*, 1996) but with addition of Na-lactate the shelf-life will be prolonged (Debevere, 1989). Devlieghere *et al.* (2000a) examined the shelf-life of MA packaged cooked meat products after the addition of Na-lactate and found that a significant shelf-life extension was obtained through the use of Na-lactate, and this was more pronounced at low temperatures. A synergistic effect was reported between Na-lactate and CO_2, which could partly be explained by the pH lowering effect of CO_2. The use of buffered lactic acid systems on poultry enhanced the decontaminating effect and increased the shelf-life of poultry (Zeitoun and Debevere, 1990). Further studies on poultry showed that buffered lactic acid treatment and MAP had an inhibitory effect on *Listeria monocytogenes* and increased the shelf-life (Zeitoun and Debevere, 1991).

14.4.4 Essential oils

Essential oils are regarded as natural alternatives to chemical preservatives. Their practical application is limited due to flavour considerations, and their effectiveness is moderate due to their interaction with food ingredients and structures. The results obtained by Skandamis *et al.* (2002) showed that volatile compounds of oregano essential oil are capable of affecting both the growth and metabolic activity of the microbial association of meat stored at modified atmospheres. This inhibition was not as strong as that found in the contact of pure essential oil with microorganisms when added directly on the surface of meat (Skandamis and Nychas, 2001). These authors conclude that the volatile compounds of oregano essential oils improve the shelf-life of meat by (i) delaying the growth of specific spoilage organisms, (ii) inhibiting or restricting metabolic activities that cause spoilage through the production of spoilage microbial metabolites, and (iii) minimising the flavour concentration.

14.5 Other techniques

14.5.1 $Na_2CaEDTA$

Low levels of $Na_2CaEDTA$ (25 to 500 ppm) have been approved for use in some foods. This chelating agent has little effect on most of the microorganisms found in seafood (Dalgaard *et al.*, 1998), but more importantly does inhibit *P. phosphoreum* in MAP cod. In naturally contaminated MA packaged cod fillets, 500 $Na_2CaEDTA$ reduced the growth rate of *P. phosphoreum* by 40% and shelf-life was increased proportionally by 40%, from 15–17 days up to 21–23 days at 0°C. In aerobic stored cod fillets other microorganisms were responsible for spoilage and $Na_2CaEDTA$ had no influence on shelf-life.

14.5.2 Soluble gas stabilisation

The mode of action of the different gases used in MAP (including high O_2 concentrations) is discussed in other chapters. CO_2 gas, however, has special preservative effects in the package. Many bacteria are inhibited by very high CO_2 concentrations, and keeping a high CO_2 concentration in the product during shelf-life demands special techniques. One possible approach is to create a modified atmosphere for a product by either generating the CO_2 inside the package after packaging, or to dissolve the CO_2 into the product prior to packaging. Both methods can provide appropriate packages with smaller gas/product ratios, and thus decrease the package size. An example of the first method includes the use of either CO_2 generators or small amounts of dry ice (solid CO_2) inside the package. CO_2 generators are commercially available (Ageless, Tokyo, Japan) and could be used on their own in order to extend the shelf-life of foods (Sivertsvik, 1999). CO_2 could also be produced inside the packages by letting the exudates from the product react with a mixture of sodium carbonate and citric acid inside the drip pad, as described by Bjerkeng *et al.* (1995). The development of a 100% CO_2 atmosphere can be obtained by combining dry ice (approx. 1 g pr. kg product) and vacuum packaging. Care must be taken to avoid direct contact between the dry ice and the product, because of freeze burns. Whole salmon in plastic bags which have one-way valves to let excess CO_2 seep out, offered a superior quality product compared to ordinary MAP and traditional ice packaging methods using this dry ice.

When CO_2 is dissolved in the package prior to packaging (soluble gas stabilisation – SGS) the CO_2 is dissolved into the food product at low temperature (~0°C) and elevated pressures (> 2 atm). This is in contrast to ordinary MAP, where CO_2 is introduced into the package atmosphere at the time of packaging. The latter method can extend the shelf-life of different fish products either alone, combined with traditional MAP, or vacuum packaging (Sivertsvik, 1999). The additional benefit of this method compared to MAP, in addition to the inhibition of microorganisms that are obtained by dissolving CO_2, is that the possible degree of filling is significantly increased.

14.5.3 Protective microbes and their bacteriocins

The application of bacteriocins, i.e., antibacterial proteins produced by lactic acid bacteria (LAB), in combination with traditional methods of preservation and proper, hygienic processing can be effective in controlling spoilage and pathogenic bacteria. A wide range of bacteriocins is produced by LAB, and although these are found in fermented and non-fermented foods, nisin is currently the only bacteriocin widely used as a food preservative. Nisin is approved for use in over 40 countries and has been in use as a food preservative for over 50 years (Cleveland *et al.*, 2001). Since bacteriocins are isolated from foods such as meat and dairy products, which normally contain lactic acid bacteria, they have unknowingly been consumed for centuries. Today there are many examples of the effective use of nisin in food systems, e.g., cottage cheese (Ferreira and Lund, 1996), ricotta cheese (Davies *et al.*, 1997), skimmed milk (Wandling *et al.*, 1999), Bologna-type sausages (Davies *et al.*, 1999), lean beef (Cutter and Siragusa, 1998), and Kimchi (Choi and Parrish, 2000).

In principle, there are two common ways to use bacteriocins; by the addition of a starter culture which produces a bacteriocin which has the necessary inhibitory spectrum (Stiles, 1996), or the bacteriocin itself may be added as an ingredient at an early stage of the production process. A third way is by immobilising the bacteriocins on the packaging materials. Fang and Lin (1994) found that the numbers of *Pseudomonas fragi* on cooked tenderloin pork were reduced by MA storage, but were unaffected by nisin. In contrast to this, the growth of *L.monocytogenes* was prevented when samples were treated with 1×10^4 nisin IU/ml. In addition, the MAP (100% CO_2, 80% CO_2 + 20% air)/nisin (10^3, 10^4 IU/ml) combination system used in this study decreased the growth of both organisms, and the inhibition was more pronounced at 4ºC than at 20ºC. In a cocktail of seven *L. monocytogenes* isolates of food, human and environmental in origin, Szabo and Cahill (1998) found an increase in lag phase in all atmospheres when nisin was used. Increasing the concentration of nisin to 1250 IU/ml inhibited the growth of *L. monocytogenes* in all atmosphere combinations at 4 and 12ºC. The addition of nisin and/or a CO_2 atmosphere increased the shelf-life of cold smoked salmon from four weeks (5ºC) to five or six weeks (Paludan-Muller *et al.*, 1998).

Scannell *et al.* (2000) developed bioactive food packaging materials using immobilised bacteriocins lacticin 3147 and nisaplin. They found antimicrobial activity against the indicator strains *Lactococcus lactis*, *Listeria innocua* and *Staphylococcus aureus*. Adsorption of lacticin 3147 into plastic film was unsuccessful, but nisin bound well and the resulting film maintained its activity for a three-month period, both at room temperature and under refrigeration.

14.6 Consumer attitudes

The number of food types involved in carrying foodborne illness has increased, together with an increase in pathogenic micoorganisms documented as being transmitted through food. This makes it necessary to reconsider our approach to food preservation and pathogen control in order to meet these new challenges and to enhance food safety. However, MAP is regarded as a mild preservation method by most consumers, inducing minor changes of the inherent raw material qualities. A development towards using more preservatives in combination with MAP, e.g., additives or preservatives, and in some cases technologies with less well recognised effects (e.g. irradiation), may lead to a lowering of consumer acceptance for MA packaged foods. Consumer demands for both fresh and safe food, provides the producer with a dilemma; should he produce a product with a modest shelf-life or use preservatives to enhance product safety.

Most legislative authorities in Europe and the US aim at giving the consumer complete information about processes and packaging conditions. Therefore the producer must clearly state on the label which additives, preservatives or methods have been used and may therefore have an effect on the properties of the food. Some of the preservatives examined for use in combination with MAP, in products for daily use in households, may therefore be met with scepticism. Furthermore, the food additives benzoate and sorbate are often associated with a negative image. The food control authorities are also concerned because some preservation techniques may mask poor and improper raw material quality. Irradiation and to a certain degree preservatives, for example, used at harvest may decrease or delay the onset of microbial growth without delaying biochemical reactions. In the end this may provide for a long shelf-life in MA products measured by microbiological analysis without offering improvement in eating quality.

Irradiation treatments have been a matter of debate for a long time. Despite the advantages of irradiation both for the processor, retailer and consumer, irradiation is not widely used because of uncertainty regarding consumer acceptance, particularly given that there is a requirement to label all irradiated food in most countries. Research on consumer attitudes and a marked response to irradiated foods have shown that the public's knowledge is limited and that the acceptance of food in the fresh food category is limited.

14.7 Future trends

It is likely that the consumer demand for high quality, nutritious and ready-to-eat products will last for many years. Therefore MA packaged food products which use minimal preservation, contain few artificial additives and show little alteration from the raw product will be preferred. Preservation techniques which meet these requirements include the optimisation of gases (combinations and SGS), the use of low temperatures and the utilisation of protective microbes and their bacteriocines. Active packaging is an emerging technology in which the food, package and environment interact. This technology also includes different kinds of gas emitters and absorbers resulting in an extended shelf-life of the product. For many food products more relevant are oxygen absorbers and carbon dioxide producers used either alone to develop a modified atmosphere or in combination with a gas mixture.

A number of novel processes are now under development for microbial control of foods (Leistner and Gorris, 1997). Many of these processes can be used in the category with the aim of preventing microorganisms' access to foods, e.g., in improved heat processing like infra-red heating, electric volume heating, electric resistance/ohmic heating, high frequency (HF) or radio-frequency heating, microwave heating, inductive electric heating (Ohlsson, 2002b). There are also non-thermal methods like high pressure, pulsed electric fields, pulsed white light, ultra sound and ultraviolet radiation (Ohlsson, 2002a). These treatments are promising as pre-treatments to MA packaging and have been proposed for use as part of combinations in multiple hurdle systems. However, many of them are still at an experimental stage, with expensive and ineffective batch production.

An important question producers should address is the purpose or the need for using additional preservations to MA products. One obvious and sensible reason is that of increased safety, as previously described in this chapter. By using preservations together with MAP it is possible to get a long and safe shelf-life where target pathogens are under control. The definition of shelf-life is, however, not obvious. Most chilled raw or partly processed food products packaged in MA will have a limited period of good quality, then chemical and biochemical processes together with microbiological spoilage will decrease the sensory quality. After the period of good quality, a period with regular or even poor quality may follow, without producing safety hazards. Future use of preservation, next to safeguarding safety, should focus on prolonging the good quality lifespan of MA products. For heat-treated products, new methods that allow a faster and more even heat penetration may improve eating quality and survival of nutrients. Most processes or preservations used together with MAP do not prolong the high-quality period. An exception to this is low temperature and superchilling treatments which may inhibit both microbial spoilage and biochemical reactions.

14.8 Sources of further information and advice

Books on modified atmosphere packaging of foods

Farber, J.M. and Dodds, K.L. 1995. *Principles of modified atmosphere packaging and sous-vide packaging*. Technomic Publishing Co., Basel. pp. 464.

Brody, A.L. 1994. *Modified Atmosphere Food Packaging*. IoPPress. pp. 275.

Parry, R.T. 1993. *Principles and Applications of Modified Atmosphere Packaging of Foods*. Blackie Academic & Professional Publishing, London. pp. 305.

Ooraikul, B. and Stiles, M.E. 1991. *Modified Atmosphere Packaging of Food*. Ellis Horwood, New York, pp. 293.

Brody, A.L. 1989. *Controlled/Modified Atmosphere/Vacuum Packaging of Food*. Food & Nutrition Press, Trumbull. pp. 179.

Books on preservation and shelf-life of foods

Jujena, V.K. and Sofos, J.N. 2002. *Control of foodborne mircoorganisms.* Marcel Dekker Inc. pp. 535.

Man, D. and Jones, A. 2000. *Shelf-life evaluation of foods.* Aspen Publishers, Inc. Maryland. pp. 272.

Ohlsson, T. and Bengtsson, N. 2002. *Minimal processing technologies in the food industry.* CRC. Woodhead Publishing Limited, Cambridge, England.

Proceedings concerning modified atmosphere packaging of foods

Institute of Packaging Professionals. 1995. Proceedings from *MAPack'95 the leading edge conference on modified atmosphere packaging.* October 19–20, 1995, Anaheim, Ca.

Campden and Chorleywood Food Research Association. 1995. Proceedings from *Modified Atmosphere Packaging (MAP) and related technologies.* September 6–7, 1995, UK, Chipping Campden, UK.

Campden Food and Drink Research Association. 1990. Proceedings from *International Conference on Modified Atmosphere Packaging.* Parts 1 and 2, October 15–17, 1990, Stratford-upon-Avon, UK.

Guidelines on modified atmosphere packaging of foods

Day, B.P.F. 1992. *Guidelines for the good manufacturing and handling of modified atmosphere packed food products.* Technical Manual No. 34, Campden Food and Drink Research Association, Chipping Campden, Gloucestershire, UK.

14.9 References

ACMSF-Advisory commitee on the microbiological safety of food. (1992) Report on vacuum packaging and associated processes. London, UK, HMSO.

ASHIE, I. N. A., SMITH, J. P. and SIMPSON, B. K. (1996) Spoilage and shelf-life extension of fresh fish and shellfish. *Crit Rev Food Sci Nutr* **36**, 87–121.

BARAKAT, R. K. and HARRIS, L. J. (1999) Growth of *Listeria monocytogenes* and *Yersinia enterocolitica* on cooked modified-atmosphere-packaged poultry in the presence and absence of a naturally occurring microbiota. *Appl. Environ. Microbiol* **65**, 342–5.

BENNIK, M. H., VAN OVERBEEK, W., SMID, E. J. and GORRIS, L. G. (1999) Biopreservation in modified atmosphere stored mungbean sprouts: the use of vegetable-associated bacteriocinogenic lactic acid bacteria to control the growth of *Listeria monocytogenes*. *Lett. Appl. Microbiol* **28**, 226–32.

BIDAWID, S., FARBER, J. M. and SATTAR, S. A. (2001) Survival of hepatitis A viruses on modified atmosphere-packaged (MAP) lettuce. *Food Microbiology* **18**, 95–102.

BJERKENG, B., SIVERTSVIK, M., ROSNES, J. T. and BERGSLIEN, H. (1995) Reducing package deformation and increasing filling degree in packages of cod fillets in CO2-enriched atmospheres by adding sodium carbonate and citric acid to an exudate absorber. In *Foods and Packaging materials – Chemical Interactions* (Edited by Ackermann, P., Jägerstad, M., and Ohlsson, T.) pp. 222–7. The Royal Society of Chemistry, Cambridge, UK.

BØKNÆS, N., OSTERBERG, C., NIELSEN, J. and DALGAARD, P. (2000) Influence of freshness and frozen storage temperature on quality of thawed cod fillets stored in modified atmosphere packaging. *Lebensmittel Wissenschaft und Technologie* **33**, 244–8.

BORCH, E., KANT-MEUERMANS, M. L. and BLIXT, Y. (1996) Bacterial spoilage of meat and cured meat products. *Int. J Food Microbiol.* **33**, 103–20.

CHOI, Y. M. AND PARRISH, F. C. (2000) Selective control of lactobacilli in kimchi with nisin. *Lett. Appl. Microbiol* **30**, 173–7.

CHURCH, I. J. and PARSONS, A. L. (1995) Modified atmosphere packaging technology: A review. *J. Sci. Food Agric.* **67**, 143–52.

CLEVELAND, J., MONTVILLE, T. J., NES, I. F. and CHIKINDAS, M. L. (2001) Bacteriocins: safe, natural antimicrobials for food preservation. *Int. J Food Microbiol.* **71**, 1–20.

CLIVER, D. O. (1997) Virus transmission via food. *Food Technology* **51**, 71–8.

COYNE, F. P. (1932) The effect of carbon dioxide on bacterial growth with special reference to the preservation of fish. Part I. *J. Soc. Chem. Ind.* **51**, 119T–121T.

COYNE, F. P. (1933) The effect of carbon dioxide on bacterial growth with special reference to the preservation of fish. Part II. *J. Soc. Chem. Ind.* **52**, 19T–24T.

CUTTER, C. N. and SIRAGUSA, G. R. (1998) Incorporation of nisin into a meat binding system to inhibit bacteria on beef surfaces. *Lett. Appl. Microbiol* **27**, 19–23.

DALGAARD, P., GRAM, L. and HUSS, H. H. (1993) Spoilage and Shelf-Life of Cod Fillets Packed in Vacuum or Modified Atmospheres. *International Journal of Food Microbiology* **19**, 283–94.

DALGAARD, P., MUNOZ, L. G. and MEJLHOLM, O. (1998) Specific inhibition of *Photobacterium phosphoreum* extends the shelf-life of modified-atmosphere-packed cod fillets. *Journal of Food Protection* **61**, 1191–4.

DALGAARD, P. and JØRGENSEN, L. V. (2000) Cooked and brined shrimps packaged in a modified atmosphere have a shelf-life of >7 months at 0ºC but spoil in 4–6 days at 25ºC. *International Journal of Food Science and Technology* **35**, 431–42.

DANIELS, J. A., KRISHNAMURTHI, R. and RIZVI, S. S. H. (1985) A review of the effects of carbon dioxide on microbial growth and food quality. *J Food Prot.* **48**, 532–7.

DAVIES, A. R. (1995) Advances in Modified Atmosphere packaging. In *New methods of food preservation* (Edited by Gould, G. W.) pp. 304–20. Blackie Acedemic & Professional, Glasgow, UK.

DAVIES, E. A., BEVIS, H. E. and DELVES-BROUGHTON, J. (1997) The use of the bacteriocin, nisin, as a preservative in ricotta-type cheeses to control the food-borne pathogen *Listeria monocytogenes*. *Lett. Appl. Microbiol* **24**, 343–6.

DAVIES, E. A., MILNE, C. F., BEVIS, H. E., POTTER, R. W., HARRIS, J. M., WILLIAMS, G. C., THOMAS, L. V. and DELVES-BROUGHTON, J. (1999) Effective use of nisin to control lactic acid bacterial spoilage in vacuum-packaged Bologna-type sausage. *J. Food Prot.* **62**, 1004–10.

DAY, B. (1990) A Perspective of Modified Atmosphere Packaging of Fresh produce in Western Europe. *Food Science and Technology Today* **4**, 215.

DEBEVERE, J. M. (1989) The effect of sodium lactate on shelf-life of vacuum-packaged coarse liver pâté. *Fleischwirtschaft* **69**, 223–24.

DEVLIEGHERE, F., DEBEVERE, J. and VAN IMPE, J. (1998) Concentration of carbon dioxide in the water-phase as a parameter to model the effect of a modified atmosphere on microorganisms. *Int. J Food Microbiol* **43**, 105–13.

DEVLIEGHERE, F., GEERAERD, A. H., VERSYCK, K. J., BERNAERT, H., VAN IMPE, J. F. and DEBEVERE, J. (2000a) Shelf life of modified atmosphere packed cooked meat products: addition of Na-lactate as a fourth shelf-life determinative factor in a model and product validation. *International Journal of Food Microbiology* **58**, 93–106.

DEVLIEGHERE, F., LEFEVERE, I., MAGNIN, A. and DEBEVERE, J. (2000b) Growth of *Aeromonas hydrophila* in modified-atmosphere-packed cooked meat products. *Food Microbiology* **17**, 185–96.

DEVLIEGHERE, F., GEERAERD, A. H., VERSYCK, K. J., VANDEWAETERE, B., VAN IMPE, J. and DEBEVERE, J. (2001) Growth of *Listeria monocytogenes* in modified atmosphere packed cooked meat products: a predictive model. *Food Microbiology* **18**, 53–66.

DOHERTY, A., SHERIDAN, J. J., ALLEN, P., MCDOWELL, D. A., BLAIR, I. S. and HARRINGTON, D. (1996) Survival and growth of *Aeromonas hydrophila* on modified atmosphere packaged normal and high pH lamb. *International Journal of Food Microbiology* **28**, 379–92.

ELLINGER, R. H. (1972) Phosphates in food processing. In *Handbook of Food Additives* (Edited by Furia, T.) pp. 617–780. CRC Press, Boca Raton, FL.

ELLIOTT, P. and GRAY, R. J. H. (1981) Salmonella sensitivity in a sorbate/modified atmosphere combination system. *J. Food Protection* **44**, 903–8.

EMBORG, J., LAURSEN, B. G., RATHJEN, T. and DALGAARD, P. (2002) Microbial spoilage and formation of biogenic amines in fresh and thawed modified atmosphere-packed salmon (*Salmo salar*) at 2 degrees C. *Journal of Applied Microbiology* **92**, 790–9.

FANG, T. J. and LIN, L.-W. (1994) Growth of *Listeria monocytogenes* and *Pseudomonas fragi* on cooked pork in a modified atmosphere packaging/nisin combination. *J. Food Protection* **57**, 485.

FARBER, J. M. (1991) Microbiological Aspects of Modified-Atmosphere Packaging Technology – A review. *J. Food Prot.* **54**, 58–70.

FERREIRA, M. A. and LUND, B. M. (1996) The effect of nisin on *Listeria monocytogenes* in culture medium and long-life cottage cheese. *Lett. Appl. Microbiol* **22**, 433–8.

FEY, M. S. and REGENSTEIN, J. M. (1982) Extending Shelf-Life of Fresh Red Hake and Salmon Using CO_2–O_2 Modified Atmosphere and Potassium Sorbate Ice at 1°C. *J Food Sci* **47**, 1048–54.

FRANCIS, G. A. and O'BEIRNE, D. (1998) Effects of storage atmosphere on *Listeria monocytogenes* and competing microflora using a surface model system. *International Journal of Food Science and Technology* **33**, 465–76.

FRANCIS, G. A. and O'BEIRNE, D. (2001) Effects of acid adaptation on the survival of *Listeria monocytogenes* on modified atmosphere packaged vegetables. *International Journal of Food Science and Technology* **36**, 477–87.

GIBSON, A. M., ELLIS-BROWNLEE, R. C. L., CAHILL, M. E., SZABO, E. A., FLETCHER, G. C. and BREMER, P. J. (2000) The effect of 100% CO2 on the growth of nonproteolytic *Clostridium botulinum* at chill temperatures. *International Journal of Food Microbiology* **54**, 39–48.

GILL, C. O. (1996) Extending the storage life of raw chilled meats. *Meat Sci.* **43**, S99–S109.

GILL, C. O. and REICHEL, M. P. (1989) Growth of the cold-tolerant pathogens *Yersinia enterocolitica*, *Aeromonas hydrophila* and *Listera monocytogenes* on high-pH beef packaged under vacuum or carbon dioxide. *Food Microbiology* **6**, 223–30.

GOULD, E. and PETERS, J. A. (1971) On testing the freshness of frozen fish. In *Fishing News Books Ltd.* pp. 45–47. London.

GRAU, R. D. (1981) Role of pH, lactate, and anaerobiosis in controlling the growth of some fermentative, gram-negative bacteria on beef. *Appl. Environ. Microbiol* **42**, 1043–50.

GULDAGER, H. S., BØKNÆS, N., ØSTERBERG, C., NIELSEN, J. and DALGAARD, P. (1998) Thawed Cod Fillets Spoil Less Rapidly Than Unfrozen Fillets When Stored under Modified Atmosphere at 2°C. *J. Food Prot.* **61**, 1129–36.

HOUTSMA, P. C., DE WIT, J. C. and ROMBOUTS, F. M. (1993) Minimum inhibitory concentration (MIC) of sodium lactate for pathogens and spoilage organisms occurring in meat products. *Int. J Food Microbiol* **20**, 247–57.

JACXSENS, L., DEVLIEGHERE, F., VAN DER STEEN, C. and DEBEVERE, J. (2001) Effect of high oxygen modified atmosphere packaging on microbial growth and sensorial qualities of fresh-cut produce. *International Journal of Food Microbiology* **71**, 197–210.

KLEINLEIN, N. and UNTERMANN, F. (1990) Growth of pathogenic *Yersinia enterocolitica* strains in minced meat with and without protective gas with consideration of the competitive background flora. *Int. J Food Microbiol* **10**, 65–71.

KOSEKI, S. and ITOH, K. (2002) Effect of nitrogen gas packaging on the quality and microbial growth of fresh-cut vegetables under low temperatures. *J Food Prot.* **65**, 326–32.

LAMBERT, A. D., SMITH, J. P. and DODDS, K. L. (1991) Effect of initial O_2 and CO_2 and low-dose irradiation on toxin production by *Clostridium botulinum* in MAP fresh pork. *J. Food Prot.* **54**, 939–44.

LEBLANC, E. L. and LEBLANC, R. J. (1992) Determination of Hydrophobicity and

Reactive Groups in Proteins of Cod (*Gadus morhua*) Muscle During Frozen Storage. *Food Chemistry* **43**.

LEISTNER, L. (1992) Food preservation by combined methods. *Food Research International* **25**, 151–8.

LEISTNER, L. (1995a) Principles and applications of hurdle technology. In *New Methods of Food Preservation* (edited by Gould, G. W.) pp. 1–21. Blackie Acedemic & Professional, Glasgow.

LEISTNER, L. (1995b) Use of Hurdle Technology in Food Processing: Recent advances. In *Food Preservation by Moisture Control* (Edited by Barbosa-Cánovas, G. V. and Welti-Chanes, J.) pp. 377–96. Technomic Publishing Company, Inc., Basel, Switzerland.

LEISTNER, L. (2002) Hurdle technology. In *Control of foodborne microorganisms* (Edited by Juneja, V. K. and Sofos, J. N.) pp. 493–508. Marcel Dekker, Inc., Basel, Switzerland.

LEISTNER, L. and GORRIS, L. G. M. Food preservation by combining processes. Final report – FLAIR Concerted Action No. 7, Subgroup, 1–100. 1997. DG XII Flair, European Commission.

LICCIARDELLO, J., RAVESI, E., TUHKUNEN, B. and RACICOT, L. (1984) Effects of some potentially synergistic treatments in combination with 100 krad irradiation on the shelf-life of cod fillets. *J. Food Technol.* **49**, 1341.

LISERRE, A. M., LANDGRAF, M., DESTRO, M. T. and FRANCO, B. D. G. M. (2002) Inhibition of *Listeria monocytogenes* by a bacteriocinogenic *Lactobacillus sake* strain in modified atmosphere-packaged Brazilian sausage. *Meat Sci.* **61**, 449–55.

LUND, B. M. and PECK, M. W. (1994) Heat resistance and recovery of spores of non-proteolytic *Clostridium botulinum* in relation to refrigerated, processed foods with an extended shelf-life. *Soc. Appl. Bacteriol. Symp. Ser.* **23**, 115S–128S.

LYVER, A., SMITH, J. P., AUSTIN, J. and BLANCHFIELD, B. (1998) Competitive inhibition of *Clostridium botulinum* type E by Bacillus species in a value-added seafood product packaged under a modified atmosphere. *Food Research International* **31**, 311–19.

MCMEEKIN, T. A. and Ross, T. (2002) Predictive microbiology: providing a knowledge-based framework for change management. *Int. J. Food Microbiol.* **78**, 133–53.

MITSUDA, H., NAKAJIMA, K., MIZUNO, H. and KAWAI, F. (1980) Use of sodium chloride solution and carbon dioxide for extending shelf-life of fish fillets. *J. Food Sci.* **45**, 661–6.

MORRIS, C. E. (1989) Convenience for supermarket delis with CAP pizza. *Food Eng.* **61**, 53–54.

NISSEN, H., ROSNES, J. T., BRENDEHAUG, J. and KLEIBERG, G. H. (2003) Safety evaluations of sous vide ready-meals. *Letters Appl. Microbiol.* In press.

NOTERMANS, S., DUFRENNE, J. and LUND, B. M. (1990) Botulism risk of refrigerated, processed foods of extended durability. *J. Food Prot.* **53**, 1020.

OHLSSON, T. (2002a) Minimal processing of foods with non-thermal methods. In *Minimal processing technologies in the food industry* (edited by Ohlsson, T. and Bengtsson, N.) pp. 34–60. Woodhead Publishing Limited, Cambridge, England.

OHLSSON, T. (2002b) Minimal processing of foods with thermal methods. In *Minimal processing technologies in the food industry* (Edited by Ohlsson, T. and Bengtsson, N.) pp. 4–33. Woodhead Publishing Limited, Cambridge, England.

OLSEN, S. O., STEINSBØ, E. E., AND SKÅRA, T. (2002) The Influence of Potential Pre-Treatment and Processing Parameters on General Proteolytic Activity Characteristics in Atlantic salmon (*Salmo salar*), studied in a Model System. *Journal of Aquatic Food Product Technology* **11**, No. 3–4, 65–85.

PALUDAN-MULLER, C., DALGAARD, P., HUSS, H. H. and GRAM, L. (1998) Evaluation of the role of *Carnobacterium piscicola* in spoilage of vacuum- and modified-atmosphere-packed cold-smoked salmon stored at 5 degrees C. *Int. J Food Microbiol* **39**, 155–66.

PASTORIZA, L., SAMPEDRO, G., HERRERA, J. J. and CABO, M. L. (1998) Influence of sodium chloride and modified atmosphere packaging on microbiological, chemical and sensorial properties in ice storage of slices of hake (*Merluccius merluccius*). *Food Chem.* **61**, 23–28.

PHILLIPS, C. A. (1996) Review: Modified atmosphere packaging and its effects on the microbiological quality and safety of produce. *Int. J. Food Sci. & Technol.* **31**, 463–79.

POTHURI, P., MARSHALL, D. L. and MCMILLIN, K. W. (1996) Combined effects of packaging atmosphere and lactic acid on growth and survival of *Listeria monocytogenes* in crayfish tail meat at 4 degrees C. *J. Food Prot.* **59**, 253–6.

PRZYBYLSKI, L. A., FINERTY, M. W., GRODNER, R. M. and GERDES, D. L. (1989) Extension of shelf-life of iced fresh channel catfish fillets using modified atmosphere packaging and low dose irradiation. *J. Food Sci.* **54**, 269–73.

ROBINS, D. (1991) Combination treatments with food irradiation. In *The preservation of food by irradiation. A factual guide to the process and its effect on food* pp. 53–61. Dotesios Limited, Trowbridge, Wiltshire.

ROSNES, J. T., SIVERTSVIK, M., SKIPNES, D., NORDTVEDT, T. S., CORNELIUSSEM, C. and JAKOBSEN, Ø. (1998) *Transport of superchilled salmon in modified atmosphere*. International Insitute of Refrigeration – IIF/IIR, Nantes, France.

ROSNES, J. T., KLEIBERG, G. H. and FOLKVORD, L. (2001) Increased shelf-life of salmon (*Salmo salar*) steaks using partial freezing (liquid nitrogen) and modified atmosphere packaging (MAP). *Annales Societatis Scientiarum Færoensis* Supplementum XXVIII, 83–91.

SCANNELL, A. G., HILL, C., ROSS, R. P., MARX, S., HARTMEIER, W., ELKE and ARENDT, K. (2000) Development of bioactive food packaging materials using immobilised bacteriocins lacticin 3147 and nisaplin. *Int. J Food Microbiol* **60**, 241–9.

SIKORSKI, Z. E. and SUN, P. (1994) Preservation of seafood quality. In *Seafoods: Chemistry, Processing, Technology and Quality* (Edited by Shahidi, F. and Botta, J. R.) pp. 168. Blackie Academic and Professional, Glasgow, UK.

SIVERTSVIK, M. (1999) *Use of soluble gas stabilisation to extend shelf-life of fish.* Norconserv., WEFTA, 29th annual meeting, Oct. 10–14, 1999, Thessaloniki, Greece.

SIVERTSVIK, M., NORDTVEDT, T. S., AUNE, E. J. and ROSNES, J. T. (1999) *Storage quality of superchilled and modified atmosphere packaged whole salmon.* 20th International Congress of Refrigeration, International Institute of Refrigeration, Sydney, Australia.

SIVERTSVIK, M., BERGSLIEN, H. and ROSNES, J. T. (2001) Shelf-life extensions of seafood under modified atmospheres, Warsaw, Poland.

SIVERTSVIK, M., JEKSRUD, W. K. and ROSNES, J. T. (2002) A review of modified atmosphere packaging of fish and fishery products – significance of microbial growth, activities and safety. *Int. J. Food Science and Technol.* **37**, 107–27.

SKANDAMIS, P. N. and NYCHAS, G. J. E. (2001) Effect of oregano essential oil on microbiological and physico-chemical attributes of minced meat stored in air and modified atmospheres. *Journal of Applied Microbiology* **91**, 1011–22.

SKANDAMIS, P., TSIGARIDA, E. and NYCHAS, G. J. E. (2002) The effect of oregano essential oil on survival/death of *Salmonella typhimurium* in meat stored at 5 degrees C under aerobic, VP/MAP conditions. *Food Microbiology* **19**, 97–103.

STILES, M. E. (1996) Biopreservation by lactic acid bacteria. *Antonie Van Leeuwenhoek* **70**, 331–45.

SZABO, E. and CAHILL, M. E. (1998) The combined effect of modified atmosphere, temperature, nisin and ALTA 2341 on the growth of *Listeria monocytogenes. Int. J Food Microbiol* **43**, 21–31.

THAKUR, B. R. and PATEL, T. R. (1994) Sorbates in fish and fish products – A review. *Food Reviews International* **10**, 93–107.

THAYER, D. W. and BOYD, G. (1999) Irradiation and modified atmosphere packaging for the control of *Listeria monocytogenes* on turkey meat. *J. Food Prot.* **62**, 1136–42.

THAYER, D. W. and BOYD, G. (2000) Reduction of normal flora by irradiation and its effect on the ability of *Listeria monocytogenes* to multiply on ground turkey stored at PC when packaged under a modified atmosphere. *J. Food Prot.* **63**, 1702–6.

TSIGARIDA, E., SKANDAMIS, P. and NYCHAS, G. J. E. (2000) Behaviour of *Listeria monocytogenes* and autochthonous flora on meat stored under aerobic, vacuum and modified atmosphere packaging conditions with or without the presence of oregano essential oil at 5 degrees C. *Journal of Applied Microbiology* **89**, 901–9.

WANDLING, L. R., SHELDON, B. W. and FOEGEDING, P. M. (1999) Nisin in milk sensitizes Bacillus spores to heat and prevents recovery of survivors. *J*

Food Prot. **62**, 492–8.

WIMPFHEIMER, L., ALTMAN, N. S. and HOTCHKISS, J. H. (1990) Growth of *Listeria monocytogenes* Scott A, serotype 4 and competitive spoilage organisms in raw chicken packaged under modified atmospheres and in air. *Int. J Food Microbiol* **11**, 205–14.

ZEITOUN, A.A.M. and DEBEVERE, J. M. (1991) The effect of treatment with buffered lactic acid on microbial decontaminating on the shelf-life of poultry. *Int. J. Food Microbiol* **11**, 305–12.

ZEITOUN, A.A.M. and DEBEVERE, J. M. (1990) Inhibition, survival and growth of *Listeria monocytogens* on poultry as influenced by buffered lactic acid treatment and modified atmosphere packaging. *Int. J. Food Microbiol* **14**, 161–70.

15

Integrating MAP with new germicidal techniques

J. Lucas, University of Liverpool, UK

15.1 Introduction

Modified Atmospheric Packaging (MAP) is a precise description of this shelf-life extension technique (Bennett 1995). In the UK, MAP mainly involves the use of three gases – carbon dioxide, nitrogen and oxygen although other gases are used elsewhere. Products are packed in various combinations of these three gases depending on the physical and chemical properties of the food.

15.1.1 MAP and food preservation, food spoilage and shelf-life

Over time, food spoilage inevitably sets in and the rate at which it occurs depends on the physical structure and properties of the food itself, the type of microorganisms present and the environment the food is kept in. By carefully matching individual modified atmospheres to specific food products, adopting appropriate manufacturing, handling and packaging methods and observing recommended storage and display conditions, a retailer can successfully extend the shelf-life of most foodstuffs. Fine tuning this process can result in substantial benefits. Selecting the correct mixture of gases for the modified atmosphere is determined by looking at a combination of shelf-life and visual appearance. For the longest shelf-life red meat uses 100% carbon dioxide but the meat would not have the bright red colour desired by consumers. The redness of meat, an essential part of the consumer's decision to buy, can be maintained longer by using a MAP gas mixture between 60% and 80% oxygen. Once it has been accepted that it can, in certain cases, make economic sense to sacrifice some shelf-life to ensure visual appearance, then it has been established which mixture produces the best result for each product. The effect of the individual gases on

Table 15.1 MAP gas mixtures for food items

Food item	Retail gas mix O_2	CO_2	N_2	Storage temp. °C	Shelf-life days MAP gas	In air
Raw red meat	70	30		−1 to + 2	5–8 days	2–4 days
Raw offal	80	20		−1 to + 2	4–8 days	2–6 days
Raw poultry and game		30	70	−1 to + 2	10–21 days	4–7 days
Raw fish and seafood	30	40	30	−1 to + 2	4–6 days	2–3 days
Cooked, cured and processed meat products		30	70	0 to + 3	3–7 weeks	1–3 weeks
Cooked, cured and processed fish and seafood products		30	70	0 to + 3	7–21 days	5–10 days
Cooked, cured and processed poultry and game bird products		30	70	0 to + 3	7–21 days	5–10 days
Ready meals		30	70	0 to + 3	5–10 days	2–5 days
Fresh pasta products		50	50	0 to + 5	3–4 weeks	1–2 weeks
Bakery products		50	50	0 to + 5	4–12 weeks	4–14 days
Hard cheese		100		0 to + 5	2–12 weeks	1–4 weeks
Soft cheese		30	70	0 to + 5	2–12 weeks	1–4 weeks
Dried food products			100	Ambient	1–2 years	4–8 months
Cooked and dressed vegetable products		30	70	0 to + 3	7–21 days	3–14 days
Liquid food and beverage products			100	0 to + 3	2–3 weeks	1 week
Carbonated soft drinks		100		0 to + 3	1 year	6 months

both food and microorganisms will now be outlined. Table 15.1 gives summary advice on recommended gas mixtures, storage temperatures and achievable shelf-lives for 16 different foodstuffs.

There are sound commercial reasons why MA packed foods are in such demand in the UK. These are:

- extension of shelf-life by 50% to 500%
- minimisation of waste – restocking and ordering can become more flexible
- quality, presentation and visual appeal – all improved
- reduction of need for artificial preservatives
- increased distribution distances of products
- semi-centralised production is possible.

15.1.2 New germicidal techniques

No matter how effectively modified atmosphere technology is applied to food, no product can remain on the supermarket shelf indefinitely. For each food there is a recommended gas mixture, storage temperature and achievable shelf-life as given in Table 15.1. At the end of the shelf-life, a summary of the main sources of food spoilage and poisoning which have occurred under the MAP process is given in Table 15.2. In all cases the principal spoilage mechanism is microbial and the main microorganisms responsible for food poisoning for that particular product have been identified.

Over time, food spoilage inevitably sets in but the rate at which it occurs can be slowed down by combining germicidal and MAP techniques. Both UV and

Table 15.2 Sources of food spoilage and poisoning

Food item	Principal spoilage mechanisms	Some food poisoning hazards
Raw red meat	Colour change (red to brown). Microbial.	Clostridium species, Salmonella species, S. aureus, Bacillus species, Listeria monocytogenes, E. coli.
Raw poultry and game	Microbial.	Clostridium species, Salmonella species, S. aureus, Listeria monocytogenes, Campylobacter species.
Raw fish and seafood	Oxidative rancidity. Microbial.	Clostridium botulinum (non-proteolytic E, B and F) Vibris parahaemolyticus.
Ready meals	Microbial.	Clostridium species, Salmonella species, S. aureus, Bacillus species, Listeria monocytogenes, Yersinia enterocolitica
Bakery products	Microbial, staling. Physical separation. Moisture migration.	S. aureus, Bacillus species,
Cheese	Microbial, oxidative rancidity. Physical separation	Clostridium species, Salmonella species, S. aureus, Bacillus species, Listeria monocytogenes, E. coli.

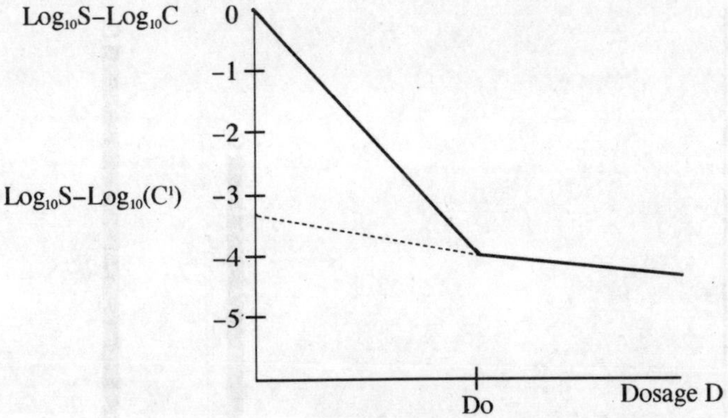

Fig. 15.1 Fraction of living microorganisms (S).

ozone are able to kill microorganisms therefore the combining of UV and ozone with modified atmospheric packaging (MAP) results in a safer product and an extended shelf-life. Compact germicidal systems can be incorporated within the MAP packaging process, resulting in a sustainable increase in shelf-life.

The survival (*S*) of microorganisms when exposed to either UV or ozone is represented by two rates of decay (Wekhof 2000) as follows

$$S = C \exp(-kD) \text{ for } D < D_o$$

$$S = C \exp(-mD) \text{ for } D < D_o$$

This relationship is illustrated in Fig. 15.1. The dosage *D* is the product of the UV or ozone intensity and duration (*t*) of exposure. There is an initial rapid rate of kill (*k*) to a level $(1 - C)$ and this is followed by a much slower kill rate (*m*). The value of *C* is of the order of 10^{-3}. Figure 15.2 shows a comparison of the dosages (D_o) required for UV, ozone and chlorine required to achieve a 99.9% kill level when compared with the dosage for *Escherichia coli* (*E. coli*) in water. They show comparative responses with a range of microorganisms.

The most likely explanation for the tailing off of the survival curves is the clumping effect suggested by various investigators – the tendency of micron-sized particles to clump together naturally. The clumping of bacteria cells protects a small percentage of bacteria and causes them to behave as if they had much higher resistance to both UV and ozone.

15.2 Ultraviolet radiation

Ultraviolet (UV) radiation is a form of energy that can be absorbed by and can bring about structural changes of systems (Koller 1965). The exposure of microbiological systems to UV radiation, within the wavelength range defined by Fig. 15.3, can dissociate the DNA, which are vital to metabolic and

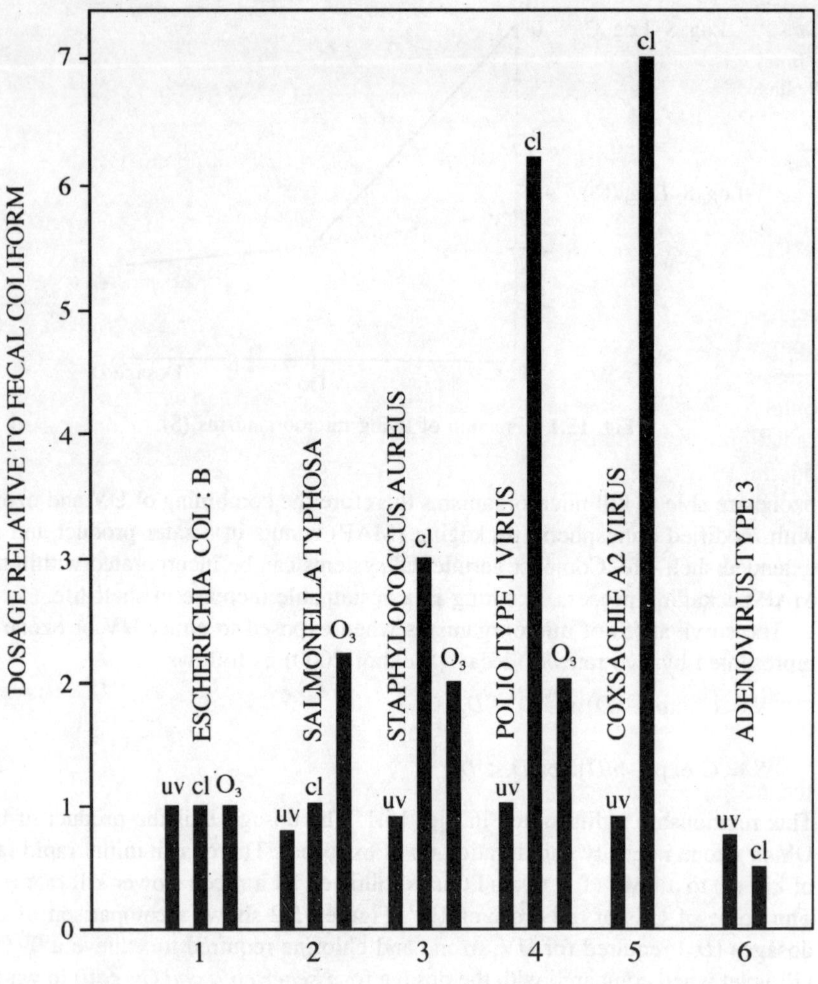

Fig. 15.2 Mortalities of bacteria and pathogens in sterilisation of water.

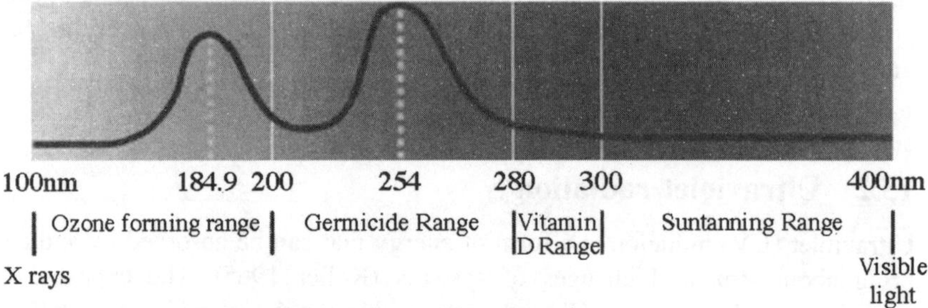

Fig. 15.3 Ultraviolet radiation spectrum.

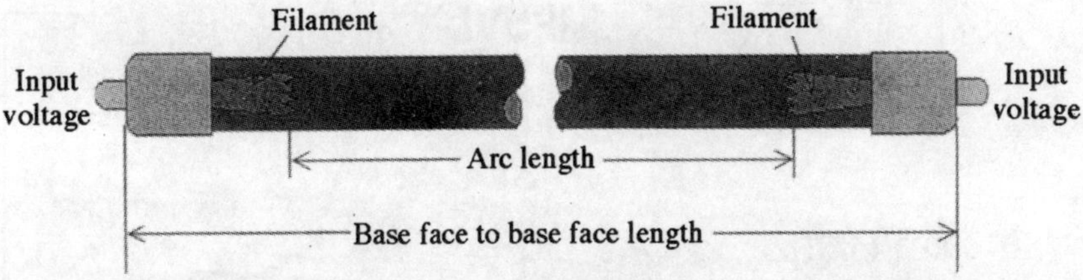

Fig. 15.4 Conventional low-pressure mercury lamp.

reproductive functions and thus inactivate the microorganisms. The most common source for producing light within a germicidal region is the low pressure mercury vapour lamp. At room temperature approximately 73% of the output radiation produces 254nm UV radiation, 19% produces 185nm UV radiation and 8% is output as a series of wavelengths 313, 365, 405, 436 and 546nm.

This is shown in Fig. 15.4. It operates with the same principle as a fluorescent lamp but without the phosphor coating. A voltage applied across the lamp generates an electric field E within the lamp which ionises the mercury vapour to produce UV light emission. The bulb is made of type 219 quartz which excludes light below 220nm. When operating at a temperature of 40°C this lamp emits 92% of its radiation at 254nm wavelength. The characteristics of this family of lamps are given in Table 15.3. They operate using a.c. (50Hz) mains power and produce an output of no more than 25W per metre lamp length.

Microwaves are high frequency electromagnetic waves generated by magnetrons, which can be stored in a resonance cavity made of metal or dielectric material (Wilson 1992). The principle is illustrated in Fig. 15.5 in which microwaves are launched into the lamp via a coupled metal cavity resonator. The electric field (E) ionises the mercury vapour in the lamp to produce the UV emission. The microwave frequency is 2.46GHz and is the same as that used in a microwave oven. This allows low cost magnetrons to be used (Kraszewski 1967). The lamps differ significantly from conventional UV lamps because they have no warm-up time, do not deteriorate with age, have adaptable shapes and can be used in pulsed mode. There is also the possibility of producing ozone and UV from the same lamp to produce a synergistic effect.

Table 15.3 Conventional UV lamp characteristics

Lamp and arc length (mm)	Lamp wattage W	Lamp current mA	UV output W	UV output @ 1000mm, μW/cm^2
212, 131	10	425	2.9	24
287, 206	14	425	3.9	35
436, 356	23	425	7.0	69
793, 711	37	425	12.8	131

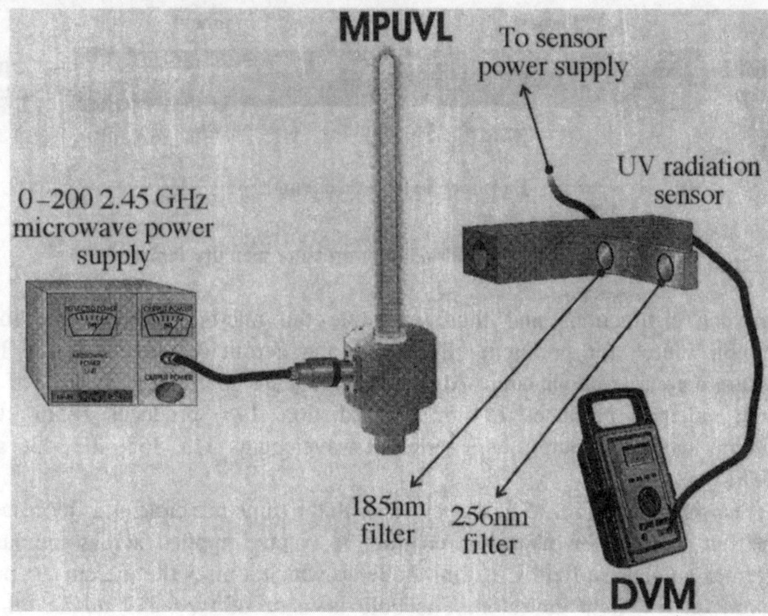

Fig. 15.5 The microwave UV lamp.

Two different lamp designs are shown in Fig. 15.6. The lamps are energised from one end and operate in free space to emit both 185nm and 254nm by using 214 quartz glass or can emit only 254nm by using 219 quartz glass (Al-Shamma'a *et al.* 2001). Because the microwaves produce a transverse electric field compared with the longitudinal electric field of the conventional lamp, the microwave lamp is able to emit UV of an order of magnitude higher in intensity, e.g., at least 250W/m.

UV light can be detected by silicon photodiodes having enhanced responses in the 190 to 400nm wavelength range. The 5.8mm^2 detector area is housed in a metal can package whilst the 33.6 and 100mm^2 devices are housed in ceramic packages (RS Components 1998). All packages incorporate a quartz window for enhanced spectral response. The device is illustrated in Fig. 15.7 with all

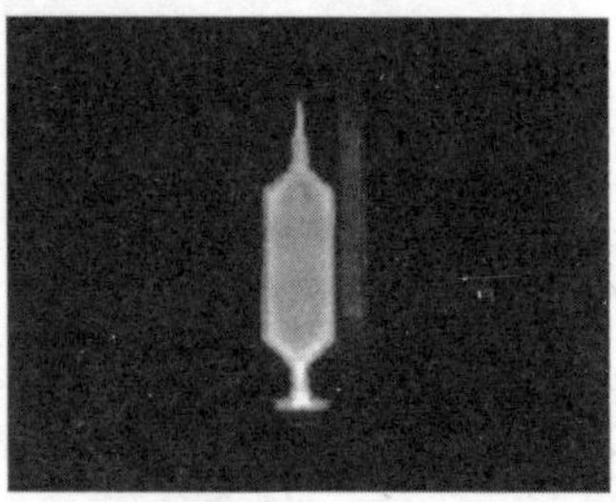

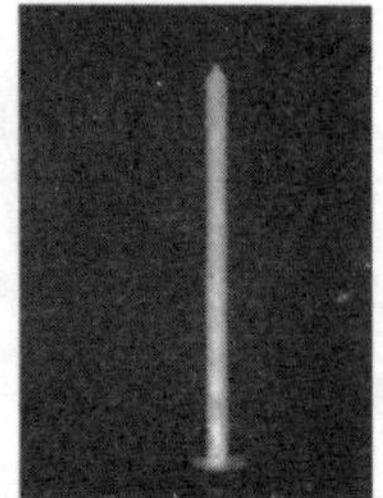

The UV flat lamp The UV cylindrical lamp

Fig. 15.6 Microwave UV lamp shapes.

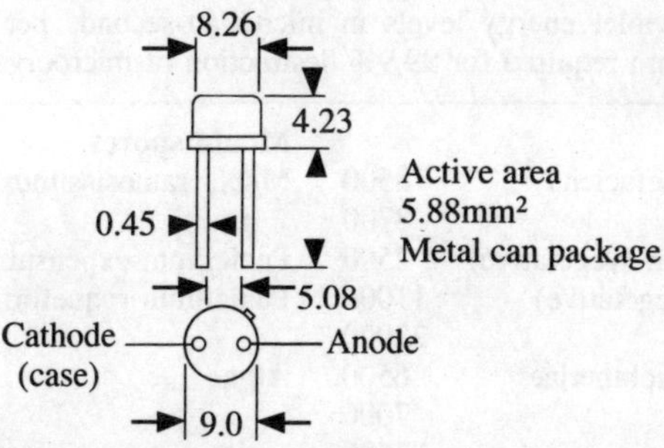

Fig. 15.7 The UV detector diode (mm units).

dimensions being given in mm. It operates with a voltage of 5V and a maximum current of 10mA. The electrical characteristics are given in Table 15.4 and the responsitivity in Fig. 15.8. The device produces a current output which is linear with input UV power.

UV light is able to kill microorganisms by using wavelengths within the germicidal region. The 254nm wavelength emitted from a mercury discharge is ideal for this action. The kill rate is usually represented by a logarithmic value of

Table 15.4 Diode characteristics

Active area mm^2	Active area mm	Responsivity amp/watt (typical) @ 190nm	@ 245nm	@ 340nm	Peak responsivity (typical)
5.8	2.4 × 2.4	0.12	0.14	0.19	950nm
33.6	5.8 × 5.8	0.12	0.14	0.19	950nm

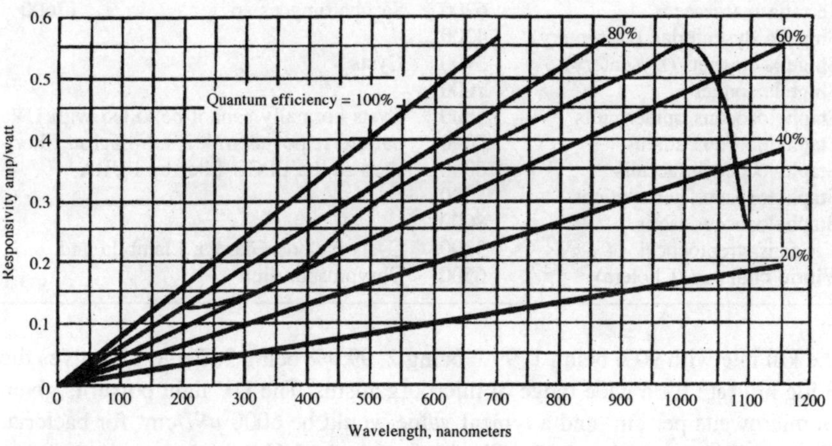

Fig. 15.8 Typical spectrum response and typical quantum efficiency curves.

Table 15.5 Ultraviolet energy levels in microwatt-seconds per square centimetre at wavelength of 254nm required for 99.9% destruction of microorganisms

Bacteria		**Mould spores**	
Agrobacterium tumefaciens	8500	Mucor ramosissimus (white gray)	35200
Bacillus anthraci	8700		
Bacillus megaterium (vegetative)	2500	Penicillum expensum	22000
Bacillus subtilis (vegetative)	11000	Penicillum roqueforti (green)	26400
Clostridium tetani	22000		
Corynebacterium diphtheriae	6500	**Algae**	
Escherichia coli	7000		
Legionella bozemanii	3500		
Legionella dumoffii	5500	Chlorella vulgaris	22000
Legionella gormonii	4900		
Legionella micdadei	3100		
Legionella longbeachae	2900		
Legionella pneumophila (Legionaires disease)	3800	**Viruses**	
Leptospira interrogans (Infectious Jaundice)	6000		
Mycobacterium tuberculosis	10000	Bacteriophage (e. coli)	6600
Neisseria cattarhalis	8500	Hepatitis virus	8000
Protius vulgaris	6600	Influenza virus	6600
Pseudomonas aeruginosa (laboratory strain)	3900	Poliovirus	21000
Pseudomonas aeruginosa (environmental strain)	10500	Rotavirus	24000
Rhodospirilium rubrum	6200	**Yeast**	
Salmonella enteritidis	7600		
Salmonella paratyphi (Enteric fever)	6100	Baker's yeast	8800
Salmonella typhimurium	15200	Brewer's yeast	6600
Salmonella typhosa (typhoid fever)	6000	Common yeast cake	13200
Sarcini lutea	26400	Saccharomyces var. ellipsoideus	13200
Serratia marcescens	6200	Saccharomyces sp	17600
Shigella dysenteriae (Dysentery)	4200		
Shigella flexneri (Dysentery)	3400	**Cysts**	
Shigella sonnei	7000		
Staphylococcus opidermidis	5800	Cysts normally cannot be killed with UV, but are removed with a sub-micron filter such as the EPCB filter by PURA	
Staphylococcus aureus	7000		
Staphylococcus faecalis	10000		
Staphylococcus hemolyticus	5500		
Staphylococcus lactis	8000		
Viridans streptococci	3800	Cysts include Giardia, Llambila and Chryptosporidiun	
Vibrio cholerae (Cholera)	6500		

the kill rate with 90% being 1, 99% being 2, 99.9% being 3. Table 15.5 gives the 3 log kill rate for a wide range of microorganisms. The UV light power is given in microwatts per cm^2 and a typical value would be 6000 $\mu W/cm^2$ for bacteria. Higher kill rates can be obtained by increasing the UV light dosage (intensity × time) but there is usually a limit attained for the kill rate.

15.3 Ozone

Ozone is toxic and concentrations in excess of 5ppm are required to produce a significant microbiocidal effect in a short exposure time consistent with modern high-speed production lines. Ozone is a compound in which three atoms of oxygen are combined to form the molecule O_3. It is a strong, naturally occurring oxidising and disinfecting agent. The weak bond holding ozone's third oxygen atom causes the molecule to be unstable. Because of this instability an oxidisation reaction occurs upon any collision between an ozone molecule and microorganisms (bacteria, viruses and cysts). Bacteria cells and viruses are split apart or inactivated through oxidisation of their DNA chains.

$$O_3 \quad + \quad X \quad = \quad O_2 \quad + XO$$

ozone microorganism = oxygen oxide

Ozone has a half life of 4 to 12 hours in air depending on the temperature and humidity of the ambient air. The half life in water ranges between seconds and hours depending on the temperature, pH and water quality.

Two commercial methods are used for generating ozone namely corona discharge and ultraviolet radiation. The corona discharge (CD) system is produced by passing air through a high voltage electric field which is close to the ignition voltage required for electrical breakdown. Typical operating conditions range from 5000 volts for high frequency voltages of 1000Hz to 16000 volts for low frequency voltages of 50Hz (mains frequency). Air (containing approximately 21% oxygen) or concentrated oxygen (up to 95% pure oxygen) dried to a minimum of −60ºC dew point passes through the corona which contains free electrons (e) which causes the oxygen (O_2) bond to split allowing two O atoms to collide with other O_2 molecules to create ozone

$$O_2 + e = 2O + e$$

$$O_2 + O = O_3$$

The ozone/gas mixture discharged from the CD ozone generator normally contains 1% to 3% when using dry air and 3% to 10% when using high purity oxygen.

As indicated in Fig. 15.9, the production of ozone with un-dried air (−10ºC) is less than half of that at the dew point of −60ºC. The figure alone shows the increase in the production of nitrogen oxides increases exponentially above −40ºC dew point. The nitrogen oxides dissolve in water creating nitric acid, which is corrosive to the CD system construction materials causing increased maintenance. Moisture can be removed by passing the air through molecular sieves, activated alumina, silica gel, membranes or by a combination of refrigeration and desiccation. Oxygen is concentrated in air by passing ambient air through molecular sieve material which absorbs moisture and nitrogen when pressurised to 2 bar. The production rates for commercial units are indicated in Table 15.6.

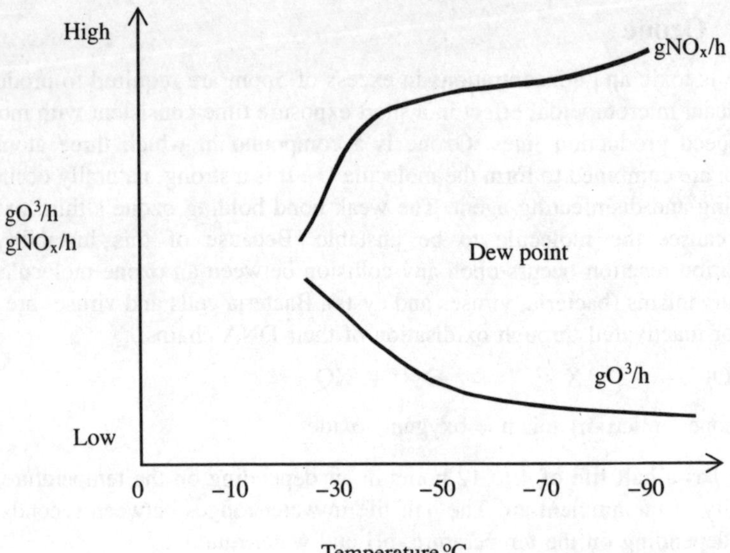

Fig. 15.9 Corona discharge ozone production.

Ozone is produced by irradiating ambient air with UV having wavelengths below 200nm. Longer wavelengths, around 250nm, are more efficient at destroying ozone rather than producing it. The energy of the UV splits some of the O_2 molecules into two O atoms which collide with other O_2 molecules to produce ozone (O_3). The system is shown in Fig. 15.10 for which air is flowed through a larger cylinder placed around the UV lamp. Because UV light sources are not monochromatic, both long and short wavelengths are generated therefore in UV systems ozone is simultaneously produced and destroyed. The concentration of ozone from the UV generator depends on the UV energy output of the lamp used, the enclosure surrounding the lamp, the temperature, humidity and oxygen content of the air and the volume of air flowing through the generator. Figure 15.11 shows the increase in production rate g/kWhr as the gas flow increases when using the microwave UV lamp. For a flow rate of 160lpm the production rate is 13gm/kWRr for a 45.9W lamp when using 214 quartz. By using refined quartz it is possible to transmit wavelengths as low as 160nm and for these conditions the ozone production rate is higher than 40g/kWhr.

Table 15.6 Ozone production rates for commercial ozone units

Ozone production g/h	Oxygen feed gas flow rate lpm	Power consumption with compressor W	Efficiency g/kWhr
16	4.5	835	19
30	9.0	1415	21
45	13.5	1930	23
60	18.0	2430	25

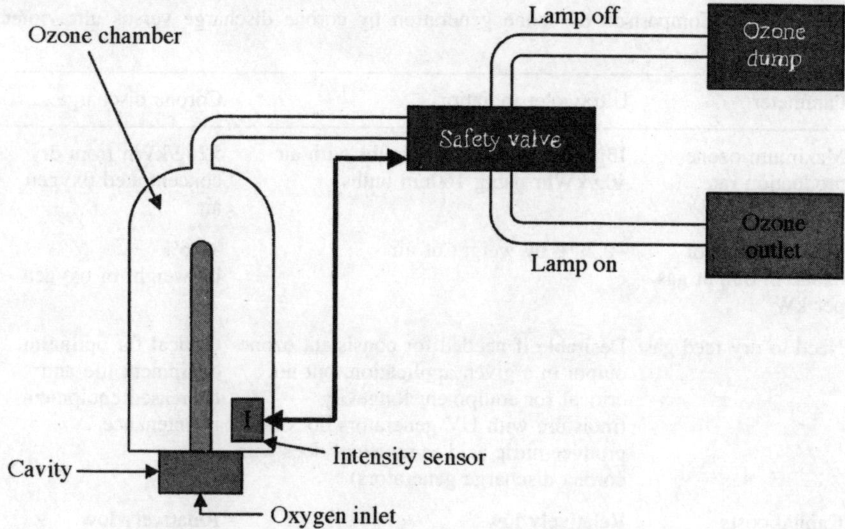

Fig. 15.10 Generation of ozone using 186nm UV.

Sensors for ozone employ electrochemical sensors. These sensors operate continually and require minimum maintenance. The rated ambient temperatures are between −25ºC to + 50ºC. The ranges can vary between 0 and 10ppm with a sensitivity of 0.1ppm up to a range between 0 and 100ppm. The output signal can be transmitted on a 4–20 mA current loop to remote displays or data logger. The device normally operates from mains power but hand-held sets operating from 12V d.c. batteries are available. The warm-up period is a few minutes and the response to ozone changes only takes a few seconds. Figure 15.12 shows the compact sensor system produced by ATI (Manchester UK).

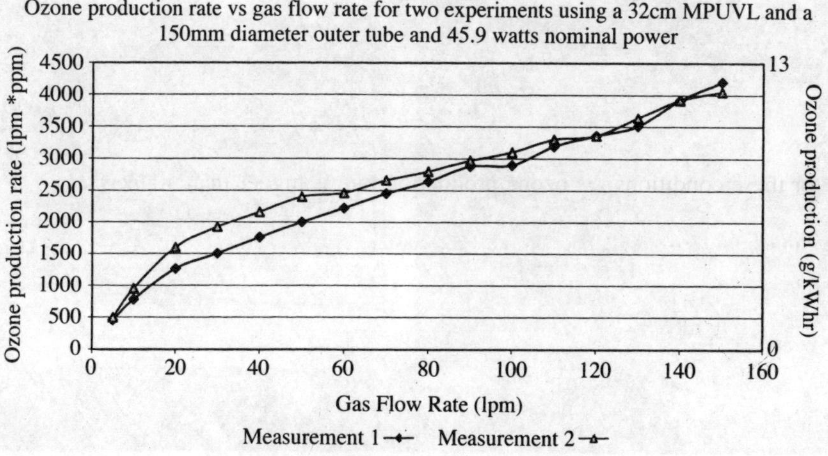

Fig. 15.11 Ozone results – compressed air.

Table 15.7 Comparison of ozone generation by corona discharge versus ultraviolet radiation

Parameter	Ultraviolet radiation	Corona discharge
Maximum ozone production rate	13g/kWh using185nm bulbs with air 40g/kWhr using 160nm bulbs	>25g/kWh from dry concentrated oxygen air
Concentration of ozone in output gas per kW	~0.29% by weight of air	~1.6% by weight of oxygen
Need to dry feed gas	Desirable if needed for consistent ozone output in a given application, but not critical for equipment longevity (moisture with UV generators does not produce nitric acid as moisture does with corona discharge generators)	Critical for optimum equipment life and decreased equipment maintenance
Capital costs	Relatively low	Relatively low
Operating costs (electrical energy)	High	High

Some results for ozone in water have been given in Fig. 15.2. The dosage of ozone for E coli bacteria is 0.5mg/litre of water with a kill rate of 99.9% being obtained. The results for the kill rates of ozone with contaminated air is given in Table 15.8.

The kill rate as a function of time is shown in Figs 15.13 and 15.14 for the *E. coli* and *S. aureus* bacteria. There is a two decay process with different rate

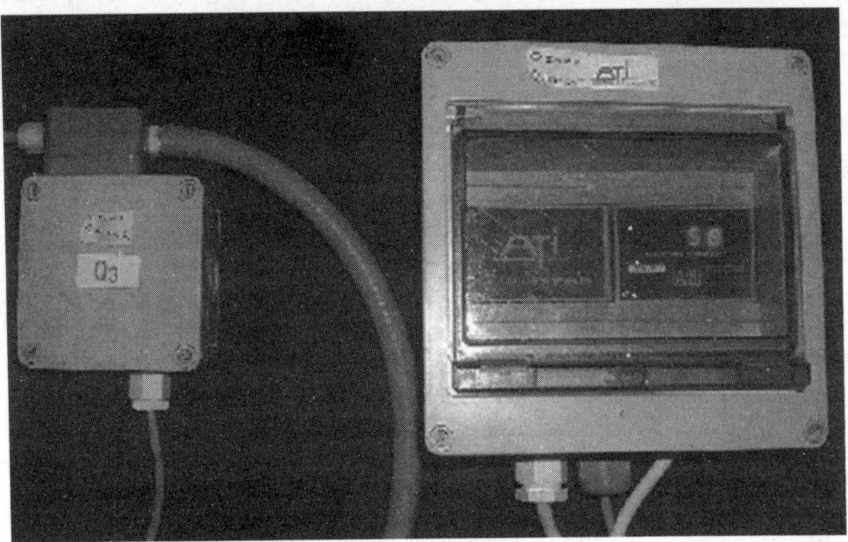

Fig. 15.12 ATI ozone sensor 1–100ppm.

Table 15.8 Bacteria kill rates and ozone dosages in contaminated air

Organism	Survival %	Ozone ppm	Time sec	Dosage (ppmxsec)
S. salivarius	2	0.6	600	360
S. epidermis	0.6	0.6	240	144
E. coli	0.056	300	15	4500
	0.007	631	15	9460
S. aureus	0.004	300	15	4500
	0.003	1500	15	22500

constants which occur simultaneously namely the rapid die-off of the individual bacterial cells and the slow death of resistant or protectively clumped bacteria. The response to ozone is as if two different species were present and the total effect of ozonation is simply the addition of the separate effects.

15.4 Integration with MAP

These are a range of machines such as horizontal form-fill-seal machines, thermoform-fill-seal machines, vacuum chambers and snorkels. Horizontal form-fill-seal (HFFS) or so-called flow pack machines are capable of making flexible pillow-pack pouches from only one reel of film. Horizontal form-fill-seal machines can also overwrap a pre-filled tray of product. Form-fill-seal

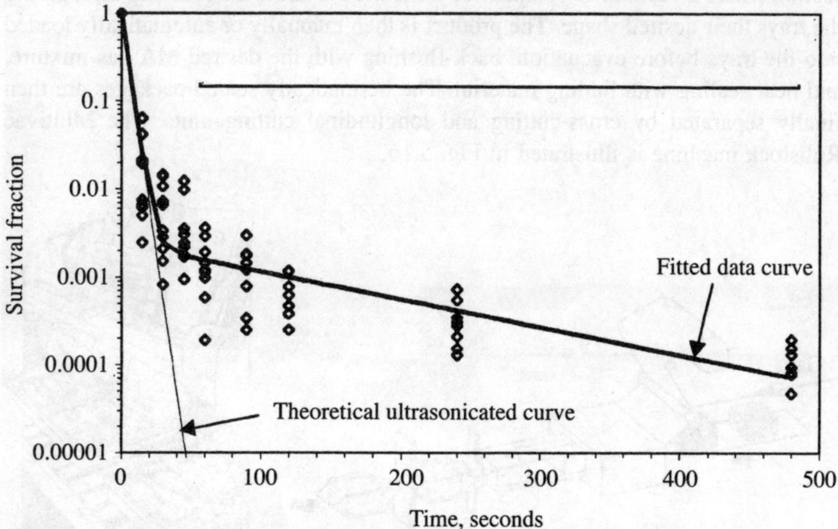

Fig. 15.13 Death curves for *E. coli* in ozonated air. (Ozone concentrations during this series of experiments varied from 300 to 1500ppm.)

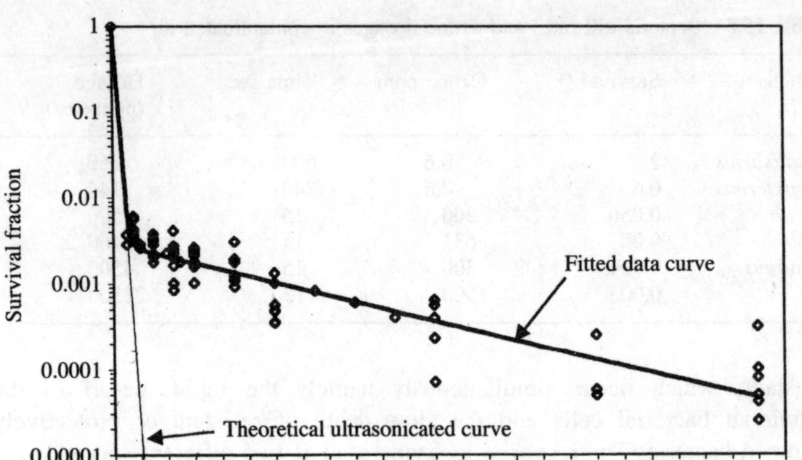

Fig. 15.14 Death curves for *S. aureus* in ozonated air. (Ozone concentrations during this series of experiments varied from 300 to 1500ppm.)

machines can be of two types – horizontal and vertical (VFFS) as illustrated in Fig. 15.15. VFFS machines are suitable for gravity-fed loose products such as coffee, snack products, grated cheese, salads, etc.

Thermoform-fill-seal machines produce packages consisting of a thermoformed semi-rigid tray which is hermetically sealed to a flexible lidding material. Rollstock film (typically PVC/PE) is automatically conveyed into a thermoforming section where a vacuum or compressed air is used to draw the film into dies, giving the trays their desired shape. The product is then manually or automatically loaded into the trays before evacuation, back-flushing with the desired MA gas mixture, and heat sealing with lidding material. The hermetically sealed packages are then finally separated by cross-cutting and longitudinal cutting units. The Multivac Rollstock machine is illustrated in Fig. 5.16.

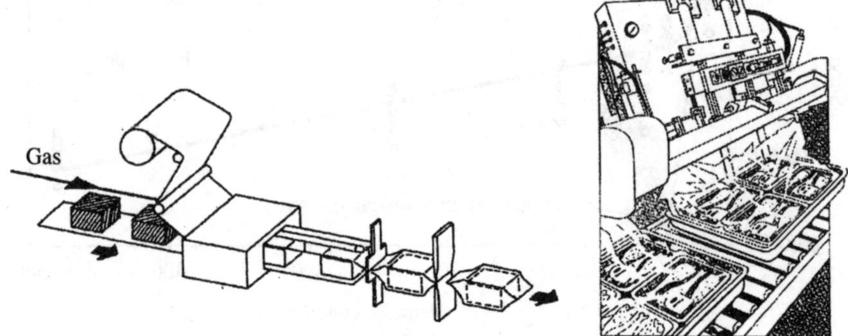

Fig. 15.15 Horizontal form-fill-seal machine.

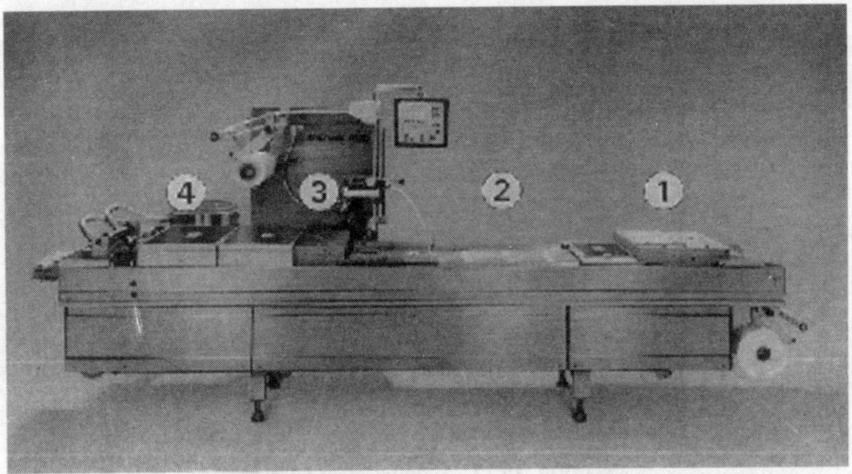

1 Reel of film 2 Automated filling station 3 Sealing station 4 Finished packs

Fig. 15.16 Multivac Rollstock machine.

The choice of films for MAP is largely determined by their gas and water vapour transmission rates. Materials such as polyester (PET), nylon, polyvinylidene chloride (PVdC) and ethylene vinyl alcohol copolymer (EVOH) provide good gas barriers but in many cases poor water vapour barriers. Polythene, polypropylene and ethylene vinyl acetate have gas transmission rates which are too high to maintain a chosen gas mixture or vacuum for long enough to provide an adequate shelf-life for most products. However, they are good barriers to water vapour and hence prevent products drying out or dry products becoming moist. Oxygen transmission rates of films for most applications lie in the range 10–125cm^3/m^2.day.atm.

15.4.1 Installation of the UV/ozone systems

The UV/ozone lamp shown in Fig. 15.5 has been incorporated into the Rollstock Machine shown in Fig. 15.16. The food is exposed to the ozone or UV before the lidding material is applied to the packaging, as shown in Fig. 15.17. The lamp is able to produce both UV and ozone either separately or as a combined output as illustrated in Fig. 15.18a, b, c. The UV germicidal effect is produced by the 254nm mercury line which is transmitted by both 214 and 219 quartz glass. The ozone is produced by the 185nm mercury line which is only transmitted by the 214 quartz glass as shown in Fig. 15.19. Hence the lamp may be arranged to give UV (Fig.15. 18a), ozone (Fig. 15.18b) or UV and ozone (Fig. 15.18c). Pyrex glass prevents the transmission of both 185nm and 254nm UV radiation.

The UV/ozone lamp has been mounted onto the Rollstock machine shown in Fig. 15.16. The arrangement is shown in Fig. 15.20a and b with the indication of the best location for the system. A gantry type arrangement to position the lamp

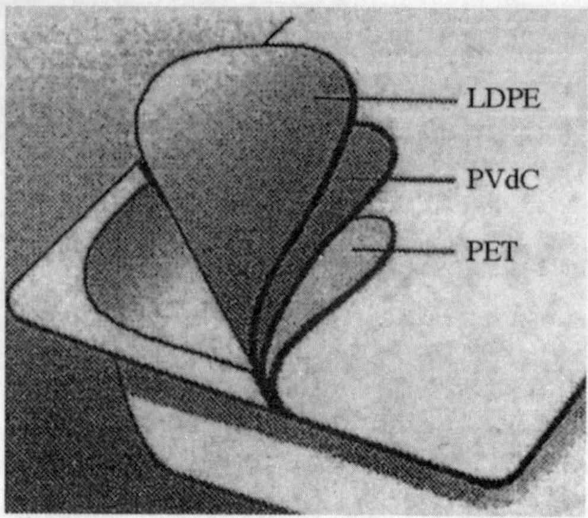

Fig. 15.17 Lidding material.

above the food trays is shown in Fig. 15.21a and b. This allows the lamp and reflector clearly to illuminate the food trays with UV, ozone or its combination. The solid construction of the device is shown in Fig. 15.22a and the absence of UV radiation close to the operating area is seen in Fig. 15.22b. The screening also maintains a trap for the ozone. The bacteria kill rate for the S. aureus microorganism is given in Fig. 15.23 and shows the killing actions of ozone and UV/ozone jointly producing 4 log performance.

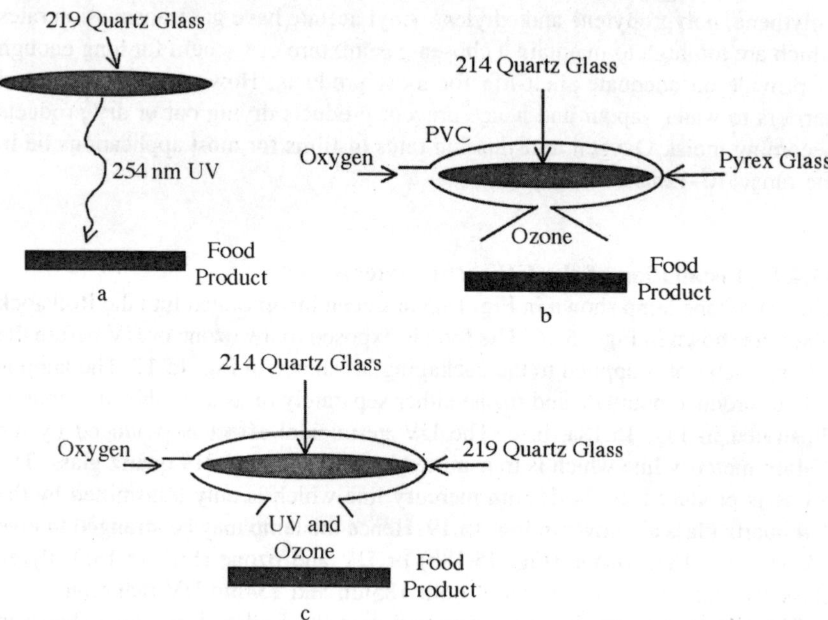

Fig. 15.18 UV/ozone germicidal arrangement. Type 214 (1mm thickness) Type 219 (1mm thickness).

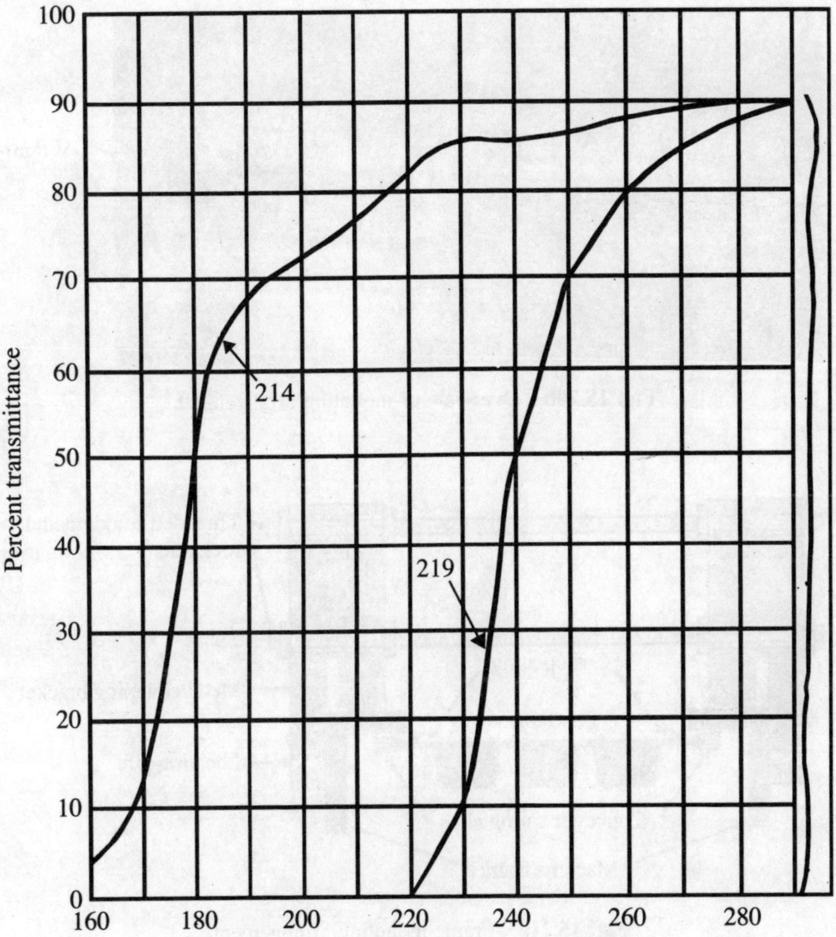

Fig. 15.19 Transmission of UV in quartz glass.

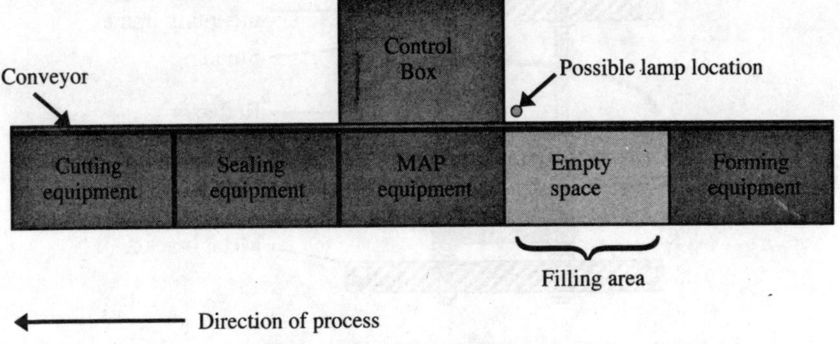

Fig. 15.20a Installation location.

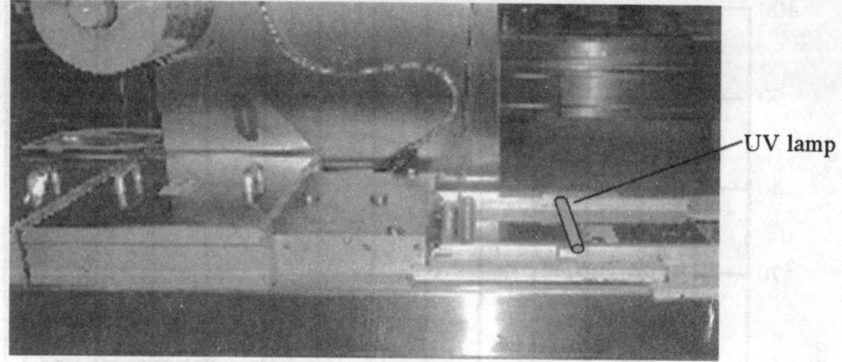

Fig. 15.20b Overview of mounting arrangement.

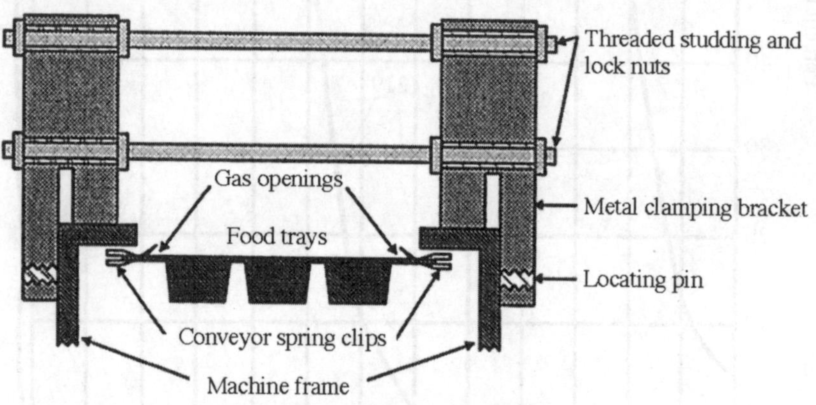

Fig. 15.21a Frame mounting arrangement.

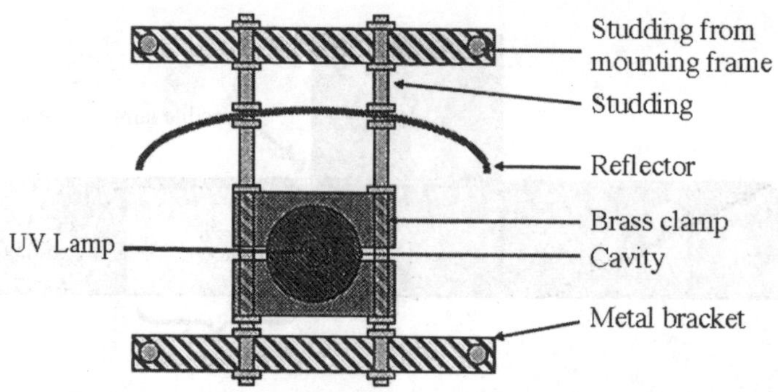

Fig. 15.21b Microwave cavity with UV lamp fixing bracket.

Fig. 15.22a Gantry mounting of the UV lamp.

Fig. 15.22b UV screening of UV radiation from operating personnel.

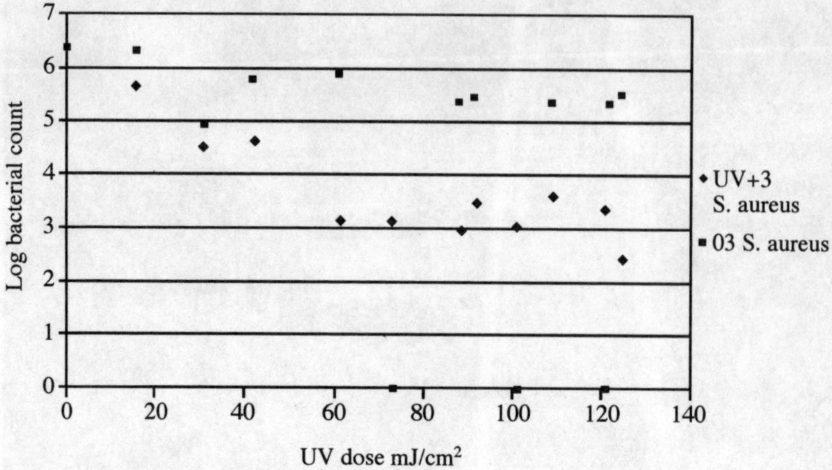

Fig. 15.23 Data for *S. aureus* and *Ps aeruginosa* after treatment with UV + ozone (44–67ppm).

15.5 Future trends

In order to obtain higher kill rates it is necessary to break up the microorganism clumps into single microorganisms. This is possible by using large doses of UV or ozone (Wekhof 2001). Figures 15.24 and 15.25 illustrate enhanced energy (J/cm^2) levels required when using white light. Such energy can be generated only by using a xenon flashlamp which produces a wide radiation spectrum from

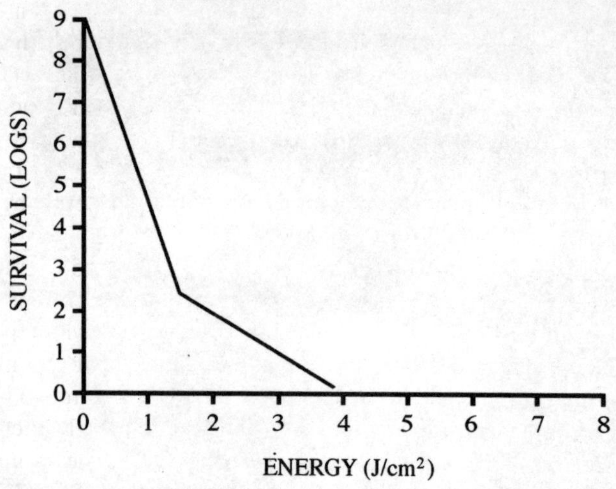

Fig. 15.24 Microbiological effects of pure white light on *E. coli* (petri dish test).

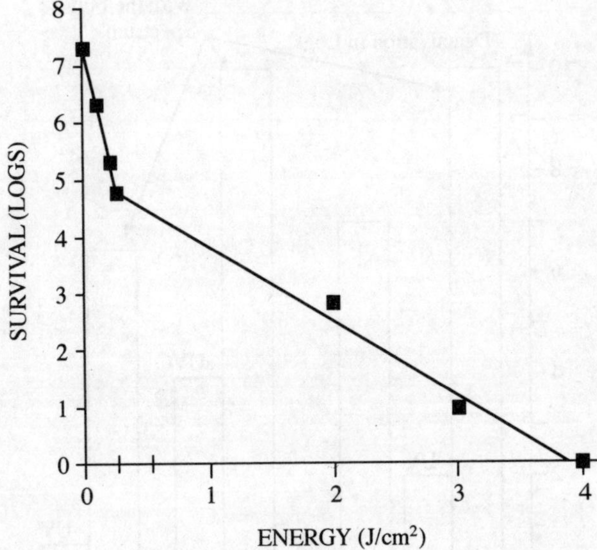

Fig. 15.25 Pure white light effect on *B. subtilis* spores (petri dish test).

1100nm to 200nm. The amount of germicidal radiation within the range 230 to 280nm is about 5% compared with 73% for the low-pressure lamp. The rate of kill for the single microorganism is therefore about 15 times less efficient for the flashlamp because of the reduced percentage of germicidal radiation. However, the breaking up of the clumps of microorganisms is mainly a thermal effect and can be achieved by all the intense emitted radiation and hence the observed faster kill rate for higher intensities.

The deactivation rate for *E. coli*, *B. subtilis* and *S. aureus* is given in Fig. 15.26. Filtering out the UV germicidal radiation from the spectrum produces a dramatic reduction in the obtained kill values. The kill rate with 254nm UV radiation for *E. coli* is 6mJ/cm^2 for a 3 log kill rate when using a conventional UV lamp and this has to be compared with 500mJ/cm^2 using a xenon flashlight. Likewise the kill rate for *B. subtilis* with 254nmUV radiation is 11mJ/cm^2 compared with 250mJ/cm^2. However, a large overall power has the ability to improve the overall kill rate by up to 4 logs of addition kill levels by using total radiation effects as shown in Fig. 15.25.

The advantage of the microwave system, shown in Fig. 15.5, is its ability to produce pulsed UV light at high pulse powers. Figure 15.27 shows the same average power of 6mJ/cm^2 being produced in 100μs pulses with repetition time of 70μs. The advantage of using UV over white light is that it is more readily absorbed by the substrate and hence less pulse power is required to break up the clumps of microorganisms. In addition the kill rate of the single microorganisms resulting from the breakup of clumps is also enhanced when compared with white light. Theory suggests that the surface temperature must rise to over 100°C during the pulse duration in order to detach the surface bacteria.

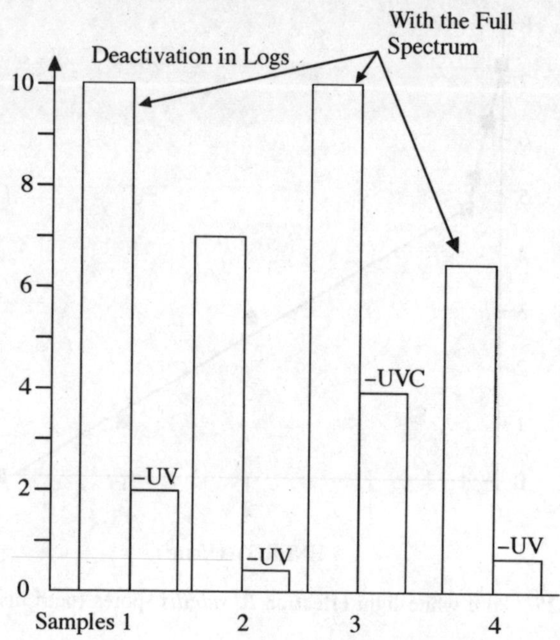

Fig. 15.26 Comparison of bacteria deactivation with a flashlamp for a full spectrum and for the UV filtered spectra. 1. *E. Coli* at 8 flashes of 12J/cm^2 of full spectra and 10 flashes each of 12J/cm^2 with UV filtering. 2. *B. subtilis* (vegetative form) at 1 flash (4 to 12) j/cm^2 of a full spectra and 15 flashes each of (8 to 10) j/cm^2 with the UV filtering. 3. *B. subtilis* (spores) at 1 flash of 8/cm^2 of a full spectra and 10 flashes of same energy with UVC filtering. 4. *S. aureus* at 1 flash of 2J/cm^2 of a full spectra and with 5 flashes at 4J/cm^2 each with UV filtering.

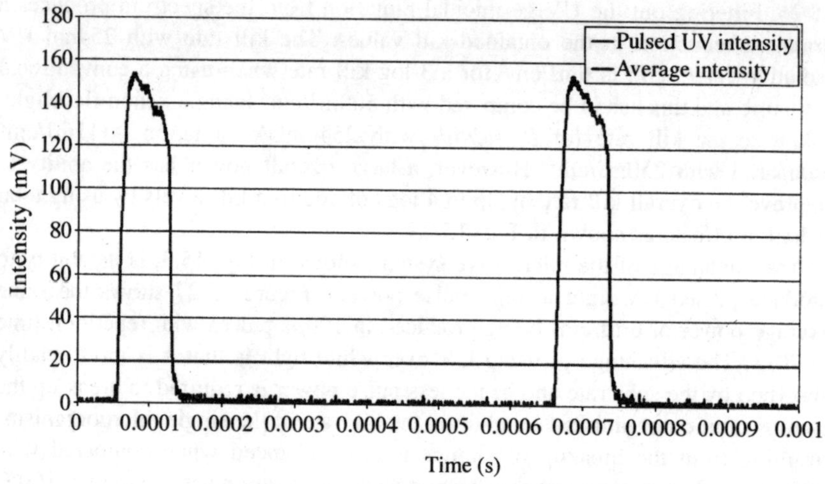

Fig. 15.27 Pulsed UV lamp waveform.

Equation 15.1 shows the surface temperature (T) as a function of time (t) after the surface has been irradiated by a plane wave light source of power P (W/m^2)

$$Pt = \rho\sigma T\sqrt{Dt} \qquad 15.1$$

with $D = K/\sigma\rho$.
where K = thermal conductivity W/(mk)
σ = specific heat kJ/(kgk)
ρ = density (kg/dm^3)

The diffusion distance (r) *is* $r = \sqrt{Dt}$. 15.2

For a specimen plate of thickness r then

$$Pr = KT. \qquad 15.3$$

For aluminium $K = 209$, $\sigma = 0.904$, $\rho = 2.7$,
whilst for glass $K = 0.81$, $\sigma = 0.84$, $\rho = 2.5$

In order to reach a surface temperature of 100°C then for a glass substrate $Pr =$ 81 and $t = 2.59r^2$. If $r =$ 10mm then $P =$ 8100W/m^2 (i.e. 800nmW/cm^2) and $t = 259\mu s$, alternatively if $r =$ 1mm then $P =$ 8W/cm^2 and $t = 2.59\mu s$. Thus a series of 8kW/m^2 pulses of 259μs duration will boil off the bacteria from the surface for destruction by the UV light.

The food industry is keen to adopt and exploit techniques that improve the safety and/or extend the shelf-life of food products without the use of preservatives. There is currently considerable interest in the use of UV and ozone particularly with the prospect of chlorine washing being discontinued for organic produce. One of the drawbacks of conventional UV lamps is that they do not work in shadowed areas. However, one of the advantages of microwave UV lamps is that their shape and size can be adapted to suit the product being irradiated. Some products such as sliced meat present a flat surface which lends itself readily to UV treatment. Other products such as bread have a porous crumb structure which is less easily sterilised by UV light but could be treated with ozone. Unfortunately ozone is toxic with maximum exposure levels of 0.2ppm. Doses in excess of this (2–5ppm minimum) are required to produce significant microbiocidal effects in a short exposure time consistent with modern high-speed production lines. Hence safety aspects for operating personnel need to be carefully considered.

Combining UV and ozone could provide sufficient sterilisation which when combined with MAP, results in a safer product and/or extended shelf-life. Some products containing fatty acids can unfortunately be oxidised by ozone leading to off flavours. It may be possible to counteract the oxidation during the sterilisation phase by use of an appropriate MAP gas system. Combined UV/ozone systems can provide more options for food (and packaging) sterilisation. They provide the option of 'flash' sterilisation. The challenge will be to determine the optimum frequency, intensity or waveform for the greatest biocidal effect. There is also the option of producing ozone and UV to produce a synergistic effect. This will also be combined with MAP gases. The combined UV/ozone system (Lucas and Al-Shamma'a 2001) has the following attributes.

- To kill bacterial growth and moulds by using recently invented, compact systems for producing UV radiation and ozone by microwaves.
- To enhance the shelf-life of food products by integrating the germicidal system into modified atmosphere packaging (MAP) machines (e.g., Multivac Chamber and Rollstock Machines).
- To destroy microbes in water washing systems for fruit and vegetable produce.
- To be applicable within a factory environment for a wide range of food products.

15.6 References

AL-SHAMMA'A, A.I., PANDITHAS, I. and LUCAS, J., (2001), 'Low Pressure Microwave Plasma UV Lamp for Water Purification and Ozone Production', *J. Phys D: Appl. Phys.* Special Issue, Vol 34, No. 14, 2775–81.

AL-SHAMMA'A, A.I., PANDITHAS, I., LUCAS J. and STUART R.A., (2001), 'Microwave Plasma UV Lamp for Water Purification', Accepted for Publication in the IEEE Transaction on Plasma Science, Ref: S016.

BENNETT, B. (1995), *The freshline Guide to Modified Atmospheric Packaging*, Crew, Air Products.

KOLLER, L. R. (1965) *Ultraviolet Radiation*, John Wiley and Sons, Inc., N.Y., Second Edition.

KRASZEWSKI, A. (1967) *Microwave Gas Discharge Devices*, Ilife Books Ltd., Second Edition.

LUCAS J. and AL-SHAMMA'A, A.I., (2001), Germicidal Modified Atmospheric Packaging, DEFRA, Food Link News, Newsletter for the Food Link programme, No. 36, 9.

RS COMPONENTS DATASHEET, *Photodiodes*, issued July 1998 (order code 298-4562).

WEKHOF, A. (2000), 'Disinfection with flash lamps', *PDA Journal of Pharmaceutical Science and Technology*, 54 (3), 264–76.

WEKHOF, A. (2001), 'Pulsed UV to Sterilise Packaging and to preserve foodstuffs', *AGIR Conference on New Sterilisation Technologies*, Talence, 1–6.

WILSON F.A. (1992) *An Introduction to Microwaves*, Bernard Babani (publishing) Ltd. ISBN 0 85934 257 3.

16

Improving MAP through conceptual models

M.L.A.T.M. Hertog, Katholieke Universiteit Leuven, Belgium and N.H. Banks, Zespri Innovation Ltd, New Zealand

16.1 Introduction

Conceptual models are descriptions of our understanding of a system that are used to shape the implementation of solutions to problems.[58] The quality and quantum of innovation that will occur in development of modified atmosphere packaging (MAP) strongly depends upon the insights gained from robust conceptual models of components of MAP. In this chapter, we outline a number of simple principles about modified atmosphere (MA) systems that we believe will assist industries that apply MA technology to move beyond the rather empirical 'pack-and-pray' approach that still predominates in commercial practice. This chapter will focus on the applications of MAP for the horticultural food industry, dealing with respiring plant produce, whole or minimally processed. However, most of the principles discussed will also hold for MAP of meat or processed food.

MA is generally used as a technique to prolong the keeping quality[64] of fresh and minimally processed fruits and vegetables.[75] In the widest sense of the term, MA technology includes controlled atmosphere storage, ultra low oxygen storage, gas packaging, vacuum packaging, passive modified atmosphere packaging and active packaging.[30,38,39,71] Each of these techniques is based on the principle that manipulating or controlling the composition of the surrounding atmospheres affects the metabolism of the packaged product, such that the ability to retain quality of the product can be optimised. The different techniques come with different levels of control to realise and/or maintain the composition of the atmosphere around the product. While controlled atmosphere storage can rely on a whole arsenal of machinery for this purpose, active packages rely on simple scavengers and/or emitters of gases such as oxygen, carbon dioxide, water or ethylene either integrated in the packing material or added in separate sachets. Passive MA packaging, as an extreme, relies solely on the metabolic activity of the packaged product to modify and subsequently maintain the gas composition surrounding the product.

Although much research has been done to define optimum MA conditions for a wide range of fresh food products,[37] the underlying mechanisms for the action of MA are still only superficially understood. The application of MA generally involves reducing oxygen levels (O_2) and elevating levels of carbon dioxide (CO_2) to reduce the respiratory metabolism.[38] Parallel to the effect on the respiratory metabolism, the energy produced to support other metabolic processes, and consequently these processes themselves, will be affected accordingly.[10] This still covers only part of the story of how MA can affect the metabolism of the packaged produce. The physiological effects of MA can be diverse and complex.[13] In MAP, the success of the package strongly depends on the interactions between the physiology of the packaged product and the physical aspects of the package; MAP is a conceptually demanding technology. Much of the work in the area of MAP has been, and still is, driven by practical needs of industry.[29] This has enabled commercial development based upon pragmatic solutions but has not always contributed substantially to advancing the conceptual basis upon which future innovation in MA technologies depends. As a result, there is a substantial potential for models to contribute to the field of MAP by making the complex and vast amount of, sometimes fragmental, expert knowledge available to packaging industries.

In this chapter, we bring together existing concepts, models and sub-models on MAP to build an overall conceptual model of the complex system of MAP. Starting from this overall model, dedicated models can be extracted for specific tasks or situations. The benefits and drawbacks of the modelling approach are discussed, together with an identification of the future developments needed to create advantage to MAP commercial operations.

16.2 Conceptual models

The ideal model integrating all critical aspects of MAP would inevitably have a multidisciplinary nature and a complexity that, at least in its mathematical form, is far beyond the scope of this chapter. Here we attempt to provide a sound conceptual model to assist understanding of the underlying mechanisms. Going in aggregation level from the macro (palletised packs) via the meso (individual packs) to the micro level (packaged product) the emphasis shifts from physics and engineering to include more and more biology; physiology and microbiology. In parallel to this shift, the level of complexity and uncertainty increases.

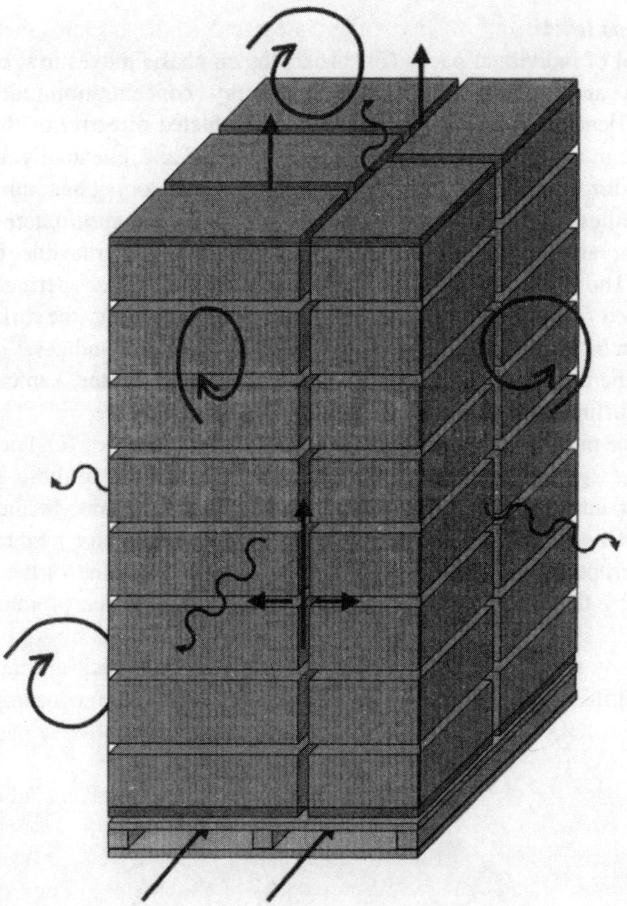

Fig. 16.1 A schematic outline at the macro level of MAP where forced airflow and turbulent convection are responsible for heat and mass transfer to and from the individual MA packs.

16.2.1 Macro level

The macro level is schematically presented in Fig. 16.1. Much research has been undertaken on heat and mass transfer, the effects of boundary layers and different flow patterns given different geometries, types of cooling and ventilation.[20,28] The same techniques have been applied to the storage of living and non-living food and non-food products all over the world. These techniques enable, in general, a good understanding of the storage environment of palletised or stacked packs, whether or not MA packs. Cooling is needed to remove heat from the packages and continuously to counteract the heat produced by the living product. Both forced airflow and turbulent convection are at this level major contributors to the transport of heat, water, gases and volatiles, to and from the packs.

16.2.2 Meso level

At the level of individual packs (Fig. 16.2) the emphasis moves towards natural convection and diffusion processes driven by concentration and thermal gradients. Heat produced by the product is conducted directly, or through the atmosphere in the package, to the packaging material and, eventually, is released to the air surrounding the pack. Water vapour, respiratory gases, ethylene and other volatiles are exchanged between the package atmosphere and the surrounding atmosphere by diffusion through (semi-) permeable packaging materials. Those packaging films can be either selective semi-permeable films or perforated films. In the case of perforated films especially, the diffusion rate of a gas can be influenced by a concurrent diffusion of a second gas.[57] A counter current generally hinders diffusion while a current in the same direction promotes diffusion of the first gas.

Inside the package, the metabolic gases are either consumed (O_2) or produced (H_2O, CO_2, C_2H_4 and other volatiles) by the product. Each of these gases may promote or inhibit certain parts of the product's metabolism. In the end, the overall metabolism of the packaged product is responsible for maintaining the product's properties. As long as the product properties relevant for the quality as perceived by the consumer stay above satisfactory levels the product remains acceptable.

The steady state gas conditions realised inside an MA pack are the result of both the influx and the efflux through diffusion and the consumption and

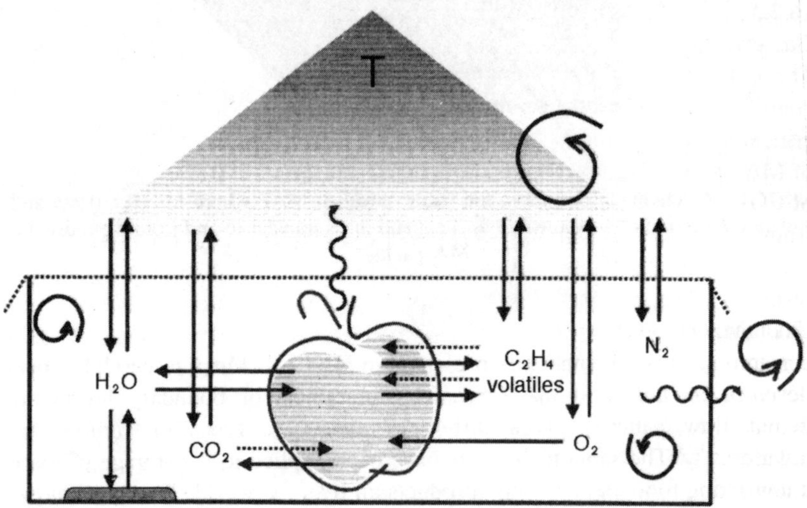

Fig. 16.2 A schematic outline at the meso level of MAP where heat and mass transfer from and to the packaged product are ruled by natural convection and diffusion processes. The packaging film acts like a selective semi-permeable barrier between the package and the surrounding atmosphere. Temperature has a marked effect on all processes going on at the meso level.

production by the product which are themselves strongly dependent on the composition of the package atmosphere.[35] For instance, water loss by the product is the main source for water accumulating in the pack atmosphere. The product elevates humidity levels within the pack to an extent that depends upon relative water vapour permeances of film and product. This elevated humidity inhibits further water loss to a progressively greater extent as relative humidity approaches saturation. This substantial benefit carries a risk of condensation that is exacerbated by temperature fluctuations. Condensation creates favourable conditions for microbial growth that will eventually spoil the product and, as water condenses on the film, will also depress the overall permeance of the package.

The time needed for a package to reach steady state is important as only from that moment on is the maximum benefit from MA being realised. In the extreme situation, the time to reach steady state could outlast the shelf-life of the packaged product. A typical example of how the atmospheric composition in an MA pack and gas exchange of the packaged product can change during time is illustrated in Fig. 16.3. The dynamics of reaching steady state depend upon the rates of gas exchange and diffusion and upon the dimensions of the package in relation to the amount of product contained. Packages with large void volumes take longer to reach steady state levels. Temperature has a major effect on the rates of all processes involved in establishing these steady state levels[4] and hence on the levels of the steady state gas conditions themselves.

16.2.3 Micro level

Gas exchange

The complexity of the biological system inherent in each fruit (Fig. 16.4) contributes significantly to the uncertainties in current knowledge on issues critical to the outcome of MA treatments. One of the central issues is the impact of MA upon the product's gas exchange, its consumption of O_2 and production of CO_2 (Fig. 16.5). Total CO_2 production consists of two parts, one part coming from the oxidative respiration in parallel to the O_2 consumption and the other part originating from the fermentative metabolism.[51] At high O_2 levels, aerobic respiration prevails. In this situation, the respiration quotient (RQ; ratio of CO_2 production to O_2 consumption), influenced by the type of substrate being consumed, remains close to unity. At lower oxygen levels, fermentation can develop, generally causing a substantial increase in RQ. This is due to an increased fermentative CO_2 production relative to an O_2 consumption declining towards zero. Besides the effect of O_2 on respiration and fermentation, CO_2 is known to inhibit gas exchange in some produce as well.

Although it would be convenient to consider gas exchange to be constant with time, there can be considerable ontogenetic drift in rates of gas exchange.[8] In so-called climacteric fruits especially, a respiration burst can be observed when the fruit starts to ripen. In addition, freshly harvested, mildly processed or handled fruit generally shows a temporary increased gas exchange rate.[10] Microbial infections can also stimulate gas exchange.[70]

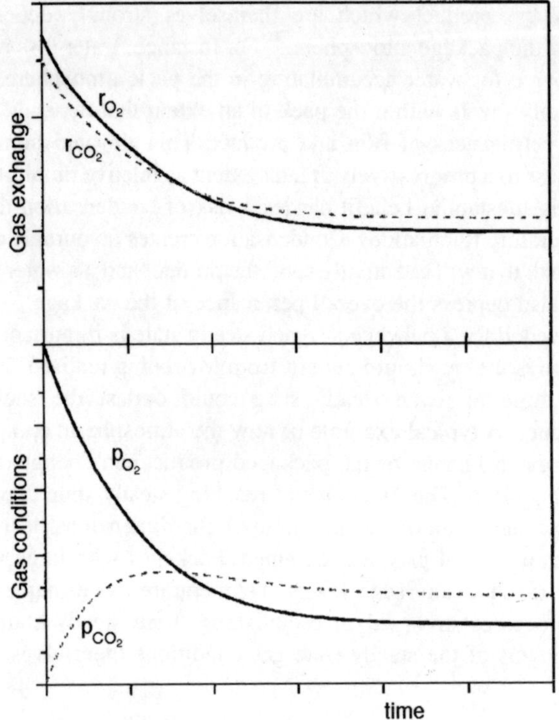

Fig. 16.3 A typical example of the dynamics of MA. Due to the gas exchange, CO_2 starts to accumulate while the O_2 level starts to decrease (bottom). In response to the changing gas conditions, gas exchange rates are inhibited (top). Driven by the increasing concentration gradients between package and surrounding atmosphere, O_2 and CO_2 start to diffuse through the packaging film. Combined, this slows down the change in gas conditions. Eventually, gas exchange by the product and diffusion through the film reach steady state levels at which the consumption and production of O_2 and CO_2 equals the influx and efflux by diffusion.

Gas diffusion

When one considers gas exchange as a function of O_2 and CO_2 levels, one is generally inclined to look at the atmospheric composition surrounding the product as the driving force. However, the actual place of action of gas exchange is inside the cells, in the mitochondria. Depending on the type of product, this means that an O_2 molecule has to diffuse through the boundary layer surrounding the product, through a wax layer, cracks, pores or stomata, through intercellular spaces, has to dissolve in water, and has to pass the cell membrane to get into the cell.[13] The CO_2 molecule produced by the gas exchange has to travel the same way in the opposite direction. The driving force for the diffusion comes from the partial pressure difference for O_2 and CO_2 between the fruit's internal and external atmospheres generated by the gas exchange. The intracellular, *in-situ*, O_2 and CO_2 concentrations are much more relevant for

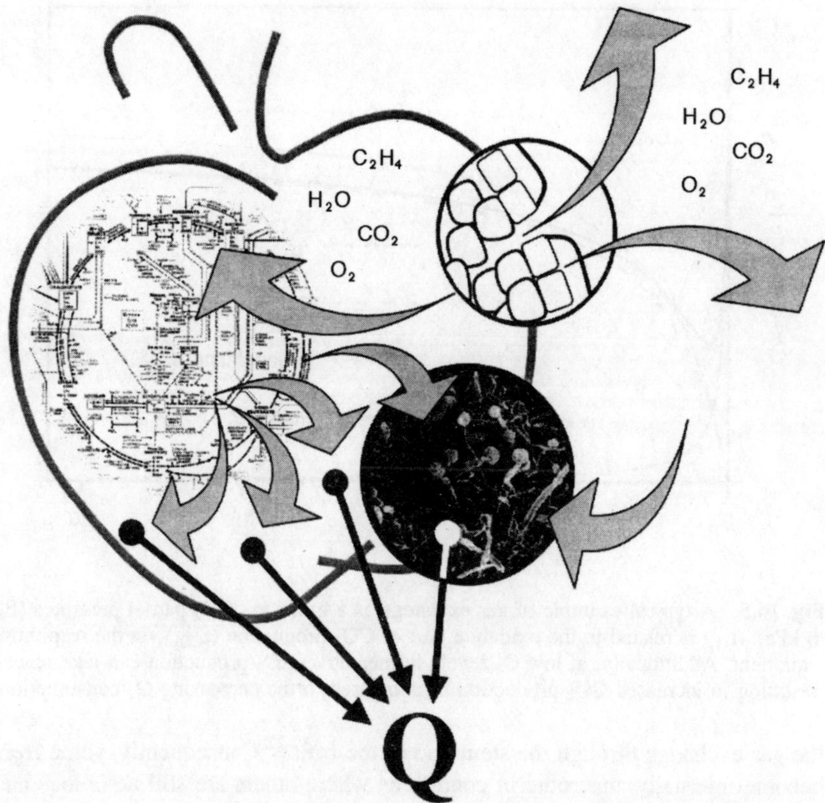

Fig. 16.4 A schematic outline at the micro level of MAP where the product is considered to generate its own MA conditions due to the resistance of the skin. The internal gas conditions are responsible for affecting large parts of the metabolism either directly or via the gas exchange. This will influence quality related product properties determining the quality (Q) as perceived by the consumer. Depending on the MA conditions, microbes can interact with the product's physiology influencing its final quality.

the gas exchange than the fruit external gas conditions. Generally, it is assumed however, that the largest resistance in the diffusion pathway from the surroundings into the fruit exists at the skin of the fruit.[12,14] Therefore the largest gradient in concentration occurs at the skin while the concentration differences within a fruit are small.

Even at identical external atmospheres, different species of fruit will have completely different internal gas compositions due to their different skin permeances. Fruit with a wax layer, like apples, have a much lower permeance than leafy vegetables like cabbages, which generally have a large amount of stomata present.[18] The skin permeance of different apple varieties will be strongly affected by thickness of their natural wax layers. Due to such a wax layer, the skin of tomato and bell pepper is relatively impermeable, forcing all

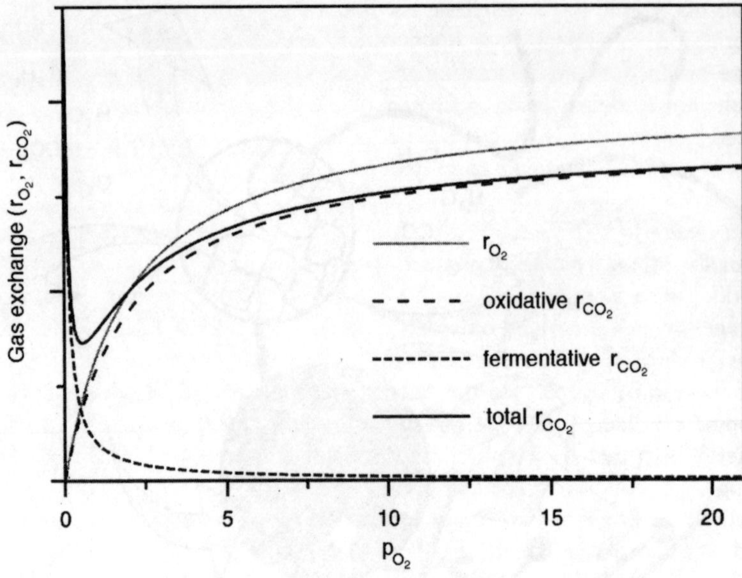

Fig. 16.5 A typical example of gas exchange as a function of O_2 partial pressures (P_{O_2} in kPa). (r_{O_2}) is related to the oxidative part of CO_2 production (r_{CO_2}) via the respiration quotient. Additionally, at low O_2 levels fermentative CO_2 production can take place resulting in increased CO_2 production as compared to the decreasing O_2 consumption.

the gas exchange through the stem end of the fruit.[22] Consequently, some fruits become internally anaerobic in conditions where others are still aerobic.

Water diffusion and water loss

The diffusion of water vapour is limited by skin permeance in the same way as the diffusion of O_2 and CO_2, the slight difference being that diffusion of O_2 and CO_2 takes place mainly through pores connected to intercellular spaces while water vapour is more easily released through the whole skin surface.[3,44] Water loss is driven by the partial pressure difference of water vapour between the fruit's internal (close to saturation) and external atmospheres. Water loss is an important issue in relation to the overall mass loss, firmness loss and shrivelling or wilting of the product. Inside an MA pack, water loss can also be responsible for generating conditions favourable for microbial growth (high RH).

Ethylene effects

Being a plant hormone, ethylene takes a special place among the gases and volatiles produced by the product because of its potential impact on the product's own metabolism. The pathways of biosynthesis and bio-action of ethylene are still subject to extensive study.[40] Most of the climacteric fruits show a peak of ethylene production at the onset of ripening. In most of these fruits, ripening can be triggered by exogenously supplied ethylene. This creates the situation that one ripening fruit in an MA pack will trigger the other fruit to ripen simultaneously, due to the ethylene accumulating in the pack. MA can inhibit the normal development and ripening of products postponing climacteric ethylene production thus extending the keeping quality of the product. With kiwifruit, however, advanced softening of the fruit occurs before ethylene is produced.[6] Although the fruit is not yet producing any ethylene, the softening process is extremely susceptible to exogenously applied ethylene.

Product quality
The quality of the packaged product is based on some subjective consumer evaluation of a complex of quality attributes (like taste, texture, colour, appearance) which are based on specific product properties (like sugar content, volatile production, cell wall structure).[61] These product properties generally change over time as part of the normal metabolism of the product. Those developmental changes that are directly influenced by O_2 or CO_2 or driven by the energy supplied by respiration or fermentation will all be affected by applying MA conditions, potentially extending the keeping quality of the product. Some processes are more affected than others due to the way they depend on atmospheric conditions. To understand the mode of action of MAP for a specific product, a good understanding of how the relevant product properties depend on gas conditions and temperature is required.

Spoilage and pathogenics
MA conditions can also provide conditions favourable to the growth of microbes potentially limiting the keeping quality of the packaged product due to rot. This is especially the case for soft fruits or minimally processed fruit and vegetable salads when high humidity levels are combined with a tasty substrate.[7] Some microbes are known to be opportunistic, waiting for their chance to invade the tissue when ripe, damaged or cut. In this case, MA conditions inhibiting the ripening of fruit in combination with proper handling and disinfection can prevent some of the problems. Other microbes more actively invade the tissue, causing soft patches on the fruit. More insight is needed on how MA can inhibit not only the metabolism of the product but also that of the microbes present on the products. High CO_2 levels are generally believed to suppress the growth of microbes, although sometimes the CO_2 levels needed to suppress microbial growth exceed the tolerance levels of the vegetable produce packaged.[7,17]

Variation
Although the general concept of MAP is now almost complete, there is one thing left that affects all other issues outlined so far; the effect of variation. Variation can occur on different levels, like time and spatial variation in temperature control in storage, irregularities in the stacking of cartons influencing ideal flow patterns, irregularities in the thickness or perforation of films or differences between batches of film used. However, the most important non-verifiable factor is biological variation. Besides the more obvious differences between cultivars, distinct differences exist between produce from different harvests, years, soils or

locations.[42] Even within one batch, considerable variation between individual items can occur.[67] The amount of biological variation that can be expected generally depends on the organisation level looked at. Within packages, the product generally comes from one grower resulting in a relatively homogenous batch with limited fruit-to-fruit variation. Comparing different pallets involves product potentially originating from different growers and different harvest dates result in a much larger variation.

When developing small consumer MA packages, variation in the rate of gas exchange is almost impossible to take into account. The larger the package, the more these differences tend to average out. However, in the case of fruit interactions, individual outliers can affect the other fruit in a pack, as with the spreading of rots, the onset of ripening through C_2H_4 production or with off-flavour development.

16.3 Mathematical models

Over the years, different elements of what has been discussed above have been subject to mathematical modelling. Other subjects are still to be explored. Models describing the physics of MAP are usually more fundamental than the ones describing the physiology of MAP. This is due to the increased complexity and the lack of knowledge of the underlying mechanisms. For this reason, empirical 'models' (arbitrary mathematical equations fitted to experimental data) still prevail in post-harvest physiology. This section gives an overview of the type of MAP-related models available in the literature with the emphasis on the physiological aspects of MAP.

16.3.1 Macro level

With the strong development of computers, rapidly increasing computational power becomes available to food and packaging engineers. Associated with this, engineers can add new numerical tools to their standard toolkit such as Computational Fluid Dynamics, infinite elements and finite differences. In general, when modelling heat and mass transfer, conservation laws are applied to formulate energy and mass balances.[20] The space under study is subdivided in a number of defined elements. Each of them is represented by one point within the three-dimensional space and is assumed to exchange mass and heat with its neighbouring elements according the heat and mass balances defined. The accuracy of such a model strongly depends on the number and size of elements defined and the knowledge of system input parameters. To improve both accuracy and computational time, smaller elements can be defined in areas with steep gradients and larger elements in the more homogeneous areas.

Theoretically, this approach is applicable at both the macro level to describe airflow in a cold room, at the meso level to describe diffusion within a pack, and at the micro level to describe gradients within the product. The main application is however at the macro level and to a lesser extent at the meso level when large bulk packages are involved.[69] For small consumer-size packages the simplification of treating the pack atmosphere as one homogeneous unit is generally acceptable. At the micro level, there are too many system inputs still undefined to enable formulation of such a model, not to mention parameterising and validating it.

16.3.2 Meso level

At the level of small consumer-size packages the physics simplifies to relatively easy diffusion equations based on Fick's law describing gradient-driven fluxes from point A to B through a medium with a certain resistance. Gas permeates into (or out of) the package faster with increased film area, with thinner films and with larger concentration differences.[23] The permeance of a film typically depends on the material used. With the current range of polymers available, a wide range in permeability can be realised. Most films are selective barriers with different permeances from the different gases.[24]

The standard industry test for determining permeance of a specific film is done at the single temperature of 23ºC using dry air conditions. The conditions at which a film is exposed in MAP of fresh produce, ranges however from zero to 25ºC and high humidity levels (>90%). Depending on the actual temperature, the permeance of the film changes accordingly. This temperature dependence is generally described using an Arrhenius equation. This is an exponential relationship originating from chemistry where it is used to describe the rate of chemical reactions as a function of temperature. The activation energy is the parameter quantifying temperature dependence. The higher the activation energy the faster permeance increases with increasing temperatures. An activation energy of zero means that the permeance does not change with temperature. The activation energy is characteristic for the film material used and is different for the different gases.

The effect of humidity and condensation on the permeance of films is widely recognised and still subject to study. At high humidities, water can be absorbed by the film changing the permeance for other gases as well. Furthermore, due to temperature changes, water can condense on the film forming an extra barrier for diffusion. Both aspects are still to be modelled. When perforated films are considered, diffusion through the film can be separated into two processes, diffusion through the film polymer and diffusion through the pores. Perforations are generally much less selective as this involves just diffusion through air. In addition, the effect of temperature on diffusion through pores (air) is much less as compared to its effect on diffusion through the polymer. As diffusion through the pores accounts for most of the total diffusion through a perforated film, the activation energies for perforated films are close to zero. Due to the effect of boundary layers, diffusion through pores is not linearly related to pore area and film thickness and some corrections have to be made depending on pore size and pore density. Models, originally developed to describe stomatal resistance in leaves, have been applied for this.[26]

The effect of concurrent diffusions has been modelled using Stefan-Maxwell equations.[57] These equations take into account the effect of collisions between counter currents of different species of molecules on their final diffusion rates and can explain some of the observed diversions from Fick's law of diffusion. The effect of pack volume on the dynamics of MAP is something that does not need to be modelled explicitly. As both diffusion and respiration are defined as a function of partial gas pressures, and as these partial pressures depend by definition on the number of molecules present per unit of volume, the volume is already incorporated implicitly. For instance, doubling the void volume of an MA pack means that twice the number of oxygen molecules are available. To reduce the oxygen concentration in the void volume to a certain level, twice the number of molecules have to be removed, which takes about twice as long.

16.3.3 Micro level

Several attempts have been made to model gas exchange either by empirical models or greatly simplified fundamental or kinetic models using, for instance, a single Arrhenius equation.[48,73] A more fundamental approach was used by Chevillotte[19] who introduced Michaelis Menten kinetics to describe respiration on the cell level. Lee[41] introduced and extended this approach in the post-harvest field to describe the respiration of whole fruit. After him, several other authors successfully applied this Michaelis Menten approach to a wide range of products[27,56,62] and extended the original Michaelis Menten equation to include different types of CO_2 inhibitions[51] and to account for the effect of temperature.[15,34,65] Traditionally, the effect of temperature was described using the Q_{10} system. More recently, the use of the Arrhenius equation has been favoured. The general applicability of the Michaelis Menten approach is probably due to the fact that it is simplified enough to enable parameterisation, and that it is detailed enough to account for the different phenomena observed.

Driven by dissatisfaction with the Michaelis Menten approach, as it may not describe the respiration of fresh produce because actual respiration is composed of many steps of metabolic reactions, Makino *et al.*[46,47] felt the need to develop an even more simplified model. Based on Langmuir absorption theory, an O_2 consumption model was developed which, in the end, appears to be an exact copy of the Michaelis Menten approach, with parameters meaning the same, only labelled differently. Instead of developing an alternative for the Michaelis Menten approach, Makino unintentionally reinvented it and validated its assumptions via an analogous mechanistic approach.

Although proven extremely applicable for practical use and indispensable for enhancing the understanding and interpretation of gas exchange data, the Michaelis Menten type of formulation is a considerable simplification of the biochemical reality. This stimulates ongoing research into generating models that are more detailed.[1] The developmental effect on gas exchange has not been modelled so far, except for some empirical corrections for an assumed drift of respiration during time.[10]

Burton[13] added a whole new dimension to MA research by stimulating research on internal atmosphere compositions of products as a key concept in the responses of fruits and vegetables to MA. They emphasised the concept of the skin being a barrier between external and internal atmospheres. In the same way that film permeance alters gas conditions inside the package, the skin alters internal gas atmospheres.[2] Basically, the fruit can be considered as the smallest possible MA package. The mathematics behind modelling internal atmospheres is the same as is applied in modelling pack atmospheres. Assuming the largest resistance in the diffusion pathway exists at the skin of the fruit, diffusion from the pack atmosphere to the fruit internal atmosphere can be described with a simple diffusion equation using the permeance and area of the skin. The relation between fruit internal and external atmosphere conditions can be completely understood from the combined effect of skin permeance and gas exchange characteristics. However, the gas exchange model now has to be parameterised as a function of fruit internal gas conditions instead of pack atmosphere conditions.

With regard to gas exchange, the combination of diffusion equations and Michaelis Menten type kinetics resulted in generally accepted and applicable models. As far as product specific issues are concerned, models are completely lacking or available only in a rudimentary form. However, to complete the overall MAP model we do need sub-models on how MAP is influencing the physiology of the packaged product beyond its gas exchange. How do the properties determining product quality depend on the gas conditions, either direct or via the changed gas exchange? The development of such models is severely hampered by the lack of physiological knowledge and complete sets of data for validation. Though empirical or statistical models can be useful to describe simple relationships found in a specific experiment, robust mechanistic models are needed to develop predictive models that can be applied under a wide range of conditions.

A relatively simple problem like shrivelling of apples due to water loss can be easily understood from the diffusion of water from the fruit's internal atmosphere into their external atmosphere.[45] The analysis of the results is hampered though, by the large biological variation in skin permeance.[43] However, due to its generic mechanistic approach, the model can easily be integrated within the larger MAP model for a wide range of products.

The colour change of some products (tomatoes,[63] cucumber[60]) has been successfully modelled. What remains to be investigated is how these colour changes are affected by the gas conditions. With the colour change of broccoli buds, MA conditions were shown to have an effect on the rate of colour change.[52] Whether this was directly related to the reduction in gas exchange was not tested. In the case of rot development in strawberries, Hertog *et al.*[32] assumed that the metabolic rate was the direct driving force for the process of ripening enabling microbes to develop rot. Reduction of spoilage under MA could be explained from this reduction of gas exchange.

An extremely complex and relevant issue of how ripening of (climacteric) fruits is affected by gas conditions has not been unravelled, let alone been modelled. However, based on some general concepts, Tijskens *et al.*[68] developed a simplified mechanistic model describing the softening of apples under MA including some of those climacteric developmental changes. Although this model is a strong simplification of the physiological reality, it shows the generic potential of well formulated mechanistic models.

Understanding the mode of action of MAP for a specific product requires knowledge of how the relevant product properties depend on the gas conditions (composition and temperature). This is what makes MAP a laborious exercise as each product can have different quality-determining product properties responding in slightly different ways to the MA conditions applied. One way to get around this is by developing generic models describing phenomena like shrivelling, softening, sweetening, mealiness, flesh browning or skin colour change that can be validated independently for a wide range of products.

Another option that has already proved itself successful is to stick to a more general level, describing keeping quality independent of the underlying product properties. This generic approach was originally developed by Tijskens and Polderdijk[64] to describe the effect of temperature on keeping quality for a wide range of commodities. This approach was extended to include the effect of MA, assuming all quality decay is driven by the metabolic rate.[31,52] Some further refinement would be needed to discriminate for instance between respiration- and fermentation-driven quality decay processes.[33] This approach can give an insight into how much MAP is able to extend keeping quality without the need to unravel the exact mechanism of how, for instance, firmness of apple is influenced by MA conditions on the biochemical level.

Modelling in microbiology has always been important. Usually growth curves are described as a function of temperature, pH, water activity and in response to the presence of competing microbes at well-defined growth media.[72] The relevance of microbes for MAP increased with the increasing demand for convenience foods stimulating the markets for MA packaged cut and slightly processed fruit and vegetable mixes. Low numbers of microbes in foods may already result in hazardous situations. However, the currently available models in predictive microbiology are not set up to deal with these low numbers.[72] Instead of modelling actual numbers, the probability of presence should be taken into account. In addition, the composition of the natural growth medium in MA packages (being the fruit and vegetables) is not well defined and highly variable. To predict growth of microbes in MA packages, both the biological interaction between produce and microbes and the direct effect of changed atmospheric conditions on the growth rates of microbes should be taken into account. Research in this field is still developing[7] and given its complex nature, mechanistic models integrating the outlined microbial aspects will not be readily available.

Although (biological) variation is hard to model, the effect of variation in a system can be easily demonstrated once a model of that system is available. Simply by running the model multiple times, taking randomly distributed values for one or more of the model parameters, the effect of variation becomes clear. Such a so-called Monte-Carlo approach can be applied to a MAP model by drawing

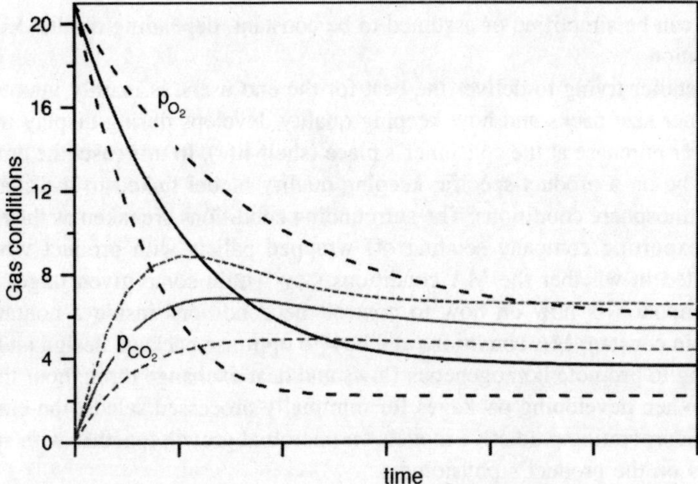

Fig. 16.6 For a given MA package the effect of ±25% biological variation on gas exchange and ±10% variation in packaged biomass were calculated. The average MA conditions developing during time (solid lines) and their 95% confidence intervals (dotted lines) were plotted based on 200 simulations.

randomly distributed values for, for instance, film thickness or the product's respiration rate. Based on the results a 95% confidence interval for the gas conditions inside an MA pack can be formulated (Fig. 16.6). Variation can induce certain risks, especially when package atmospheres are targeted close to what is feasible. When aimed for O_2 levels close to the fermentation threshold,[74] the risk is that some of the packages, depending on the variation in gas exchange rate, result in O_2 levels dropping below the fermentation threshold.[16] This results in packages with unacceptable fermented produce. Biological variation is generally much larger (±25% is not exceptional) than the physical sources of variation (generally less than ±10%) as the physical factors are generally easier to control. Biological variation also comes back in the initial quality of the packaged product resulting in different length of keeping quality or shelf-life. Some quality change models try to account for these sources of variation.[36,32,59,60]

16.4 Dedicated MAP models

Though it is possible to develop a model covering all facets of MAP at all levels, such a model would be impossible to operate. Before one is able to use such a model it needs to be fully parameterised. Generally, this information is not available in all situations. Moreover, it is not always relevant to go to such a level of completeness. Depending on the specific issues involved in a particular application, dedicated MAP models can be extracted from the overall conceptual model. Some elements need to be worked out in more detail while others can be simplified or assumed to be constant, depending on the dedicated application.

A retailer trying to deliver the best for the end users, is mainly interested in consumer size packs and how keeping quality develops during display in retail and after purchase at the consumer's place (shelf-life). In this case, the emphasis would be on a product-specific keeping quality model linked to the change in pack atmosphere conditions. The surrounding conditions are taken as they are. A large exporting company sending off wrapped pallets with product would be interested in whether the MA conditions stay within some given target limits. The emphasis is now on how to control the conditions inside a container to maintain constant MA conditions and how to optimise package design and pallet stacking to promote homogeneous flows and heat exchange throughout the bulk load. When developing packages for minimally processed salads, the emphasis is on incorporating predictive models on microbial growth together with specific models on the product's physiology.

16.5 Applying models to improve MAP

The previous section mentioned some potential applications for MAP models. In this section, we explore some of this potential to enhance the practical implementation of MAP and to lift it beyond the phase of 'pack-and-pray'.

16.5.1 Dimensioning MAP

There is more than one 'right' solution to the search for a suitable MA package for a specific product. Assuming the product is known, including its gas exchange characteristics and some optimum target MA conditions, and the external storage conditions are set but beyond control, there are still a number of degrees of freedom through which the MA package can be manipulated for better or for worse.

To realise the target MA conditions the total permeance of the package has to be dimensioned in relation to the amount of product packaged. Besides choosing a different film material with a higher or lower permeability, film thickness and film area can be changed as well. A film that is too permeable to be used as a wrapping can give good results when used to seal the top of an impermeable tray because of the reduced diffusion area. A film that is suitable for a small consumer pack can be too impermeable to be used as a liner in a carton because of the increased amount of biomass per unit of available diffusion area. Trying to influence this ratio, by packing less produce in a package, results in an increased void volume. This increases the time needed for the product to bring the package gas levels to the target MA conditions. This is not favourable, as it takes longer before the product gets the maximum benefit of the optimum MA conditions. However, the buffering capacity of a relative large void volume can have its positive effects when the MA package has to survive short periods of sub-optimal conditions. During a short warm period, a package with a small void volume could rapidly generate anaerobic conditions while a package with a large void volume could have been transferred to cooler conditions before becoming anaerobic.

Dimensioning an MA package appears to be extremely complex due to the many different interactions involved. A MAP model can considerably enhance this search for a package with a fast enough dynamic phase, resulting in steady state values close to the target gas conditions and enough buffering capacity to be applicable in practice.

16.5.2 Developing new films

In the case of a company wanting to bring a new MA pack on the market for a specific product with the package dimensions already set by other market requirements, one needs to search for the right film to complete the MA pack. Normally, film permeance is used as input in the MAP model. However, the model formulation can be turned around to calculate the required film permeance based on the product's gas exchange characteristics, assuming some known optimum target MA conditions and given the external storage conditions. As the storage and transport conditions throughout a chain will not be constant, this exercise should be repeated over a range of temperatures or a number of different temperature scenarios. MA conditions that are optimal at one temperature do not need to be optimal at another temperature. For instance, the tolerance to low oxygen levels decreases with increasing temperature.[5,74]

Once models are available to describe the effect of humidity and condensation on film permeance, they can be used to predict humidity levels inside the package and to predict how film permeability is affected by this. This will help to set detailed specifications for films with regard to this aspect. This will be especially usefully in the MAP of minimally processed produce, soft fruits and leafy vegetables, because of the high humidity levels occurring in these packages.

16.5.3 Optimising logistic chains

Given the ultimate MA package for a certain product, its eventual success mainly depends on temperature control between the moment of packing and the moment of opening the package by the consumer. In a logistic chain where temperature is not controlled throughout, application of MAP is a waste of time, money and produce. Using a MAP model to simulate a package going through a logistic chain will give insight into the strong and weak parts of that chain.[35] It will make clear which parts of the chain are responsible for the largest quality losses of the packaged product and therefore need improvement. It enables the optimisation of a whole chain considering the related costs and benefits.

To get the most out of such an exercise, a MAP model should be used that includes a keeping quality or quality change model. Assessing the benefits and losses in terms of product quality gives much more insight than just the observation that the MA conditions dropped below or above their target levels. The question that should always be asked is how these deviations affect the quality and keeping quality. The product quality gives static information on the status of the product at a certain moment, for instance at the point of sale. Keeping quality provides dynamic information on how long a product can be stored, kept for sale, transported to distant markets or remains acceptable after sale to the consumer.

16.5.4 Sensitivity studies

Generally, an MA package is developed based on some average product characteristics, assuming an average amount of product packaged, using the specifications of an average sample of film and assuming the MA pack will be handled and stored at certain average conditions. As the average MA pack does not exist, the question arises how the non-average package will behave at non-average conditions and how far the MA conditions will diverge from the ideal target levels. Sensitivity studies are ideal to test how sensitive a system of MAP is to changes in one or more parameters or conditions. Using a MAP model, sensitivity studies can be easily conducted by running the model multiple times, using a range of values for the different conditions under study. This will help identifying which aspects of MAP should be more strictly controlled because of their potential impact on the system as a whole.

The results strongly depend on the MA pack under study. For instance, depending on the gas exchange rate the same change in packaged biomass will have different effects on the steady state MA conditions. So, it cannot be stated in general that MAP is insensitive to a change in biomass. Also, the gas conditions in an MA pack where film and produce have comparable temperature dependencies are insensitive to temperature. Using this same pack for packaging a produce with a different temperature dependency can result in gas conditions extremely sensitive to temperature. Even if the MA conditions in an MA pack are insensitive to temperature due to the balanced combination of film and product, this does not mean that the quality of the packaged product is insensitive to temperature. These are just two different ways of assessing MAP. If a good quality change model is lacking, the 'optimum MA conditions' are the only criteria to apply when judging MA packs. When a good quality change model is available, sensitivity studies can be performed on what is really important: product quality.

16.6 The risk and benefits of applying models

Applying models to improve MAP has, like every technique, its pros and cons. Some of the advantages have already been mentioned implicitly in the previous sections describing the areas of application. By applying models, the development phase of MAP can be shortened. Numerous experiments can be done behind the laptop checking all possible situations that would take weeks to test in practice. With a good conceptual model in mind and the mathematical equivalent available at the fingertips, developing MAP can be lifted beyond the phase of 'pack-and-pray'.

When developing a model, continued balancing should be going on between the completeness and relevance of the described phenomena and the level of detail and complexity of the model needed to realise this. For scientific purposes, the ultimate model would be a mechanistic one describing all relevant underlying processes. For practical purposes, one should start from such a detailed mechanistic approach and simplify as far as possible without affecting the explanatory power of the model for that specific dedicated application. In the practice of post-harvest physiology, that detailed mechanistic model is not available and the best one can do is to develop a hypothetical mechanism in agreement with the observed phenomena and in agreement with current general physiological and biochemical concepts. Such a mechanistic model can still be extremely valuable to develop concepts by verifying or falsifying hypotheses. Developing, for instance, a quality change model forces the expert to formulate a conceptual model and to realise where the gaps in his knowledge are. This is probably the most valuable and general advantage of developing mechanistic models as it enhances the understanding of a complex system and directs future research to fill the gaps.

In spite of the advantages, one should stay aware of some potential traps. One of them being the risk of forgetting about real life, simply because not everything goes according to the model. The books can prescribe transport at 2ºC but cannot prevent the driver turning off the cool unit when delivering early in the morning in an urban region not wanting to wake up its inhabitants. Kiwifruit could last another week according to the developed firmness model but in practice are already lost due to spoilage.

This brings us to the fact that you cannot get anything out of a model you did not include to start with. When condensation is not included it is impossible to assess sudden temperature drops on their potential to induce condensation with all the consequences for the omnipresent microbes. If a quality change model leans heavily on one single limiting quality attribute, the user of the model should stay aware of specific situations turning another quality attribute into the limiting factor. At the same time it should be recognised that it is impossible to include everything in the model as it would be impossible to validate it completely.

In conclusion, one should always be alert when applying models outside the range for which they were validated. In the case of empirical models especially, this can result in unrealistic predictions. A model can easily process unrealistic data without getting into a moral conflict. The user should always stay alert to recognise such anomalies.

16.7 Future trends

Some of the trends needed to safeguard the future of MAP are very basic while others are on the level of refining existing knowledge. One of the most embarrassing gaps in current knowledge is a good database on permeance data of packaging films that includes their temperature dependency. The packaging film industry should develop a standard certificate for this that comes with each film they produce for MAP.

To enable a fundamental approach to MAP the gas exchange of the different products should also be systematically characterised as a function of at least O_2, CO_2 and temperature. This knowledge, essential for the success of MAP, is still very fragmented. To improve the models on MAP of minimally processed produce involving high humidity levels, a better understanding is needed of the effect of humidity and condensation on film permeances. A great deal of work has still to be done to integrate the expertise from microbiology within the field of MAP.

Although models on gas exchange are becoming well established, models on how the physiology underlying quality is linked to the metabolism are not readily available. Their development is hampered by gaps in the knowledge of post-harvest physiology. However, to assess MA packages on the quality of their actual turnout, quality change models are needed. The ultimate goal would be to develop generic models that can be validated for a wide range of commodities.

The last issue that needs to be covered in the near future is the characterisation of biological variation and its impact on product behaviour in general and on MAP in particular. Although this issue is important for the post-harvest industry as a whole, MAP would greatly benefit from a more fundamental approach.

16.8 Sources of further information and advice

This chapter was first published as part of a book on food process modelling.[66] People interested in the applications of modelling and its potential in the food sector will find in this book a wide range of examples, providing a blend of conceptual and mathematical approaches, incorporating both developments of general principles and specific applications. Over the last 15 year several books on practical applications and technologies of MAP have been published.[9,49,50,21,11,25] These contain important sources of information for the industry covering a wide range of food products (fruit, vegetables, meat, fish, pre-cooked foods and bakery products), including regulations and guidelines.

World-wide, research groups are trying to identify optimum storage conditions for particular products. The University of California in Davis especially, has invested huge amounts of energy disseminating this kind of information on horticultural products through publications, courses and conferences. Most of this up-to-date information is accessible through their website (http://postharvest.ucdavis.edu) covering product-specific post-harvest information, optimum gas and temperature conditions, bibliographies, etc. The main occasion at which scientists exchange information in the area of MA techniques is at the four-yearly international 'Controlled Atmosphere Research Conference' with the results being disseminated through its detailed conference proceedings.[53,54,55]

16.9 References

1. ANDRICH G, ZINNAI A, BALZINI S, SILVESTRI S and FIORENTINI R, 'Modelling of aerobic and anaerobic respiration rates of golden delicious apples as a function of storage conditions', *Acta Horticulturae*, 1998 **476** 105–12.
2. BANKS N H, CLELAND D J, YEARSLEY C W and KINGSLEY A M. 'Internal atmosphere composition – a key concept in responses of fruits and vegetables to modified atmospheres', *Proceeding of the Australasian Postharvest Conference*, 1993.
3. BANKS N H, DADZIE B K and CLELAND D J. 'Reducing gas exchange of fruits with surface coatings', *Posth. Biology Technol.*, 1993 **3** 269–84.
4. BEAUDRY R and LAKAKUL R. 'Basic principles of modified atmosphere packaging', *Washington State University Tree Fruit Postharvest Journal*, 1995 **6**(1) 7–13.
5. BEAUDRY R M, CAMERON A C, SHIRAZI A and DOSTAL-LANGE D L. 'Modified-atmosphere packaging of blueberry fruit: effect of temperature on package O_2 and CO_2', *J. Amer. Soc. Hort. Sci.*, 1992 **117** 436–41.
6. BEEVER D J and HOPKIRK G. 'Fruit development and fruit physiology', in: Warrington I J, Weston G C (eds), *Kiwifruit, science and management*, Ray Richards publisher in association with NZSHS, Auckland NZ, 1990.
7. BENNIK M H J, *Biopreservation in modified atmosphere packaged vegetables*, Thesis Wageningen Agricultural University. ISBN 90-5485-808-7, 1997.
8. BLACKMAN F F, *Analytic studies in plant respiration*, Cambridge University Press., 1954.
9. BLAKISTONE B A, *Principles and applications of modified atmosphere packaging of foods* (2nd edn), Kluwer Academic Publishers, London UK, 1999.
10. BRASH D W, CHARLES C M, WRIGHT S and BYCROFT B L. 'Shelf-life of stored asparagus is strongly related to postharvest respiratory activity', *Posth. Biology Technol.*, 1995 **5** 77–81.
11. BRODY A L, *Controlled-Modified Atmosphere/Vacuum Packaging of Foods*, Food & Nutrition Press, Trumbull CT, 1989.
12. BURG S A and BURG E A. 'Gas exchange in fruits', *Physiol. Plant.*, 1965 **18**: 870–84.
13. BURTON W G. 'Biochemical and physiological effects of modified atmospheres and their role in quality maintenance', in: Hultin H O, Milner M (eds), *Postharvest Biology and biotechnology*, FNP, Westport CT, 1978.

14. CAMERON A C and YANG F S. 'A simple method for the determination of resistance to gas diffusion in plant organs', *Plant Physiol.*, 1982 **70** 21–23.

15. CAMERON A C, BEAUDRY R M, BANKS N H and YELANICH M V. 'Modified-atmosphere packaging of blueberry fruit: Modeling respiration and package oxygen partial pressures as a function of temperature', *J. Amer. Soc. Hort. Sci.*, 1994 **119** 534–9.

16. CAMERON A C, PATTERSON B D, TALASILA P C and JOLES D W. 'Modeling the risk in modified-atmosphere packaging: a case for sense-and-respond packaging', in: Blanpied G D, Bartsch J A, Hicks J R (eds), *Proc. 6th Intl. Controlled Atmosphere Research Conference*, Ithaca NY, 1993.

17. CHAMBROY Y, GUINEBRETIERE M H, JACQUEMIN G, REICH M, BREUILS L and SOUTY M. 'Effects of carbon dioxide on shelf-life and post harvest decay of strawberries fruit', *Sciences des Aliments*, 1993 **13** 409–23.

18. CHAU K V, ROMERO R A, BAIRD C D and GAFFNEY J J. 'Transpiration coefficients for certain fruits and vegetables', *Trans. ASHRAE*, 1988 **94** 1553–62.

19. CHEVILLOTTE P. 'Relation between the reaction cytochrome oxidase-oxygen and oxygen uptake in cells in vivo. The role of diffusion', *J. Theor. Biol.*, 1973 **39** 277–95.

20. CLELAND A C. '*Food refrigeration processes – Analysis, design and simulation*', Elsevier Science Publishers, London UK, 1990.

21. DAY B, *Principles and Applications of Modified Atmosphere Packaging of Foods* (2nd ed.), Aspen Publishers, Inc., 1997.

22. DE VRIES H S M, WASONO M A J, HARREN F J M, WOLTERING E J, VAN DER VALK H C P M and REUSS J. 'Ethylene and CO_2 emission rates and pathways in harvested fruits investigated, in situ, by laser photothermal deflection and photoacoustic techniques', *Postharvest Biol. Technol.*, 1996 **8** 1–10.

DOYON G, GAGNON J, TOUPIN C and CASTAIGNE F. 'Gas transmission properties of Polyvenyl Chloride (PVC) films studied under subambient and ambient conditions for modified atmosphere packaging applications', *Packaging Technology and Science*, 1991 **4** 157–65.

24. EXAMA A, ARUL J, LENCKI R W, LEE L Z and TOUPIN C. 'Suitability of plastic films for modified atmosphere packaging of fruits and vegetables', *J. Food Sci.*, 1993 **58**(6) 1365–70.

25. FARBER J M and DODDS K L, *Principles of Modified-Atmosphere and Sous Vide Product Packaging*, Technomic Pub Co, Lancaster PA, 1995.

26. FISHMANN S, RODOV V and BEN-YEHOSHUA S. 'Mathematical model for perforation effect on Oxygen and water vapor dynamics in modified-atmosphere packages', *J. Food Sci.*, 1996 **61**(5) 956–61.

27. FONSECA S C, OLIVEIRA F A R and BRECHT J K, 'Modelling respiration rate of fresh fruits and vegetables for modified atmosphere packages: a review', *J. Food Engin.*, 2002 **52** 99–119.

28. GAFFNEY J J, BAIRD C D and CHAU K V. 'Methods for calculating heat and mass transfer in fruits and vegetables individually and in bulk', *Trans. ASHRAE*, 1985 91(2B) 333–52.

29. GIESE J. 'How Food Technology covered modified atmosphere packaging over the years', *Food Technology*, 1997 **51**(6) 76–7.
30. GORRIS L G M and PEPPELENBOS H W. 'Modified atmosphere and vacuum packaging to extend the shelf-life of respiring food products', *Horttechnology*, 1992 **2** 303–9.
31. HERTOG M L A T M and TIJSKENS L M M. 'Modelling modified atmosphere packaging of perishable produce: keeping quality at dynamic conditions', *Acta Alimentaria*, 1998 **27** 53–62.
32. HERTOG M L A T M, BOERRIGTER H A M, VAN DEN BOOGAARD G J P M, TIJSKENS L M M and VAN SCHAIK A C R. 'Predicting keeping quality of strawberries (cv. 'Elsanta') packaged under modified atmospheres: an integrated model approach', *Posth. Biology Technol.*, 1999 **15** 1–12.
33. HERTOG M L A T M, NICHOLSON S E and BANKS N H. 'The effect of modified atmospheres on the rate of firmness change in 'Braeburn' apples', *Posth. Biol. Technol.*, 2001, **23** 175–84.
34. HERTOG M L A T M, PEPPELENBOS H W, EVELO R G and TIJSKENS L M M. 'A dynamic and generic model on the gas exchange of respiring produce: the effects of oxygen, carbon dioxide and temperature', *Posth. Biology Technol*, 1998 **14** 335–49.
35. HERTOG M L A T M, PEPPELENBOS H W, TIJSKENS L M M and EVELO R G. 'Modified atmosphere packaging: optimisation through simulation', in: Gorny J R (ed.), *Proc. 7th Intl. Controlled Atmosphere Research Conference*, Davis CA, 1997.
36. HERTOG M L A T M. 'The impact of biological variation on postharvest population dynamics', *Posth. Biol. Technol.*, 2002 **26** 253–63.
37. KADER A A. 'A summary of CA requirements and recommendations for other fruits than pomme fruits', *6th Intl. Controlled Atmosphere Research Conference*, Ithaca NY, 1993.
38. KADER A A. 'Modified atmospheres during transport and storage', Chapter 11 in: A A Kader (ed.), *Postharvest Technology of Agricultural Crops*, 2nd edition, Publication 3311. Universtity of California, California, USA, 1992.
39. KADER A A, ZAGORY D and KERBEL E L. 'Modified atmosphere packaging of fruits and vegetables', *Crit. Rev. Food Sci. Nutr*, 1989 **28** 1–30.
40. KANELLIS A K, *Biology and biotechnology of the plant hormone ethylene*, Published Dordrecht; Boston: Kluwer Academic Publishers, 1997.
41. LEE D S, HAGGAR P E, LEE J and YAM K L. 'Model for fresh produce respiration in modified atmospheres based on principles of enzyme kinetics', *J. Food Sci.*, 1991 **56** 1580–5.
42. MAGUIRE K M and BANKS N H. 'Harvest date, cultivar, orchard, and tree effect on water vapour permeance in apples', *J. Amer. Soc. Hort. Sci.*, 2000 **125** 100–4.
43. MAGUIRE K M, BANKS N H and LANG A. 'Sources of variation in water vapour permeance of apple fruit', *Posth. Biology Technol.*, 1999 **17** 11–17.
44. MAGUIRE K M, BANKS N H and OPARA L U. 'Factors affecting weight loss of

apples', *Horticultural Reviews*, 2001 **25** 197–234.

45. MAGUIRE K M, LANG A, BANKS N H, HALL A, HOPCROFT D and BENNET R. 'Relationship between water vapour permeance of apples and micro-cracking of the cuticle', *Posth. Biology Technol.*, 1999 **17** 89–96.
46. MAKINO Y, IWASAKI K and HIRATA T. 'A theoretical model for Oxygen consumption in Fresh Produce under an atmosphere with carbon dioxide', *J. Agric. Engng Res.*, 1996 **65** 193–203.
47. MAKINO Y, IWASAKI K and HIRATA T. 'Oxygen consumption model for Fresh Produce on the basis of Adsorbtion theory', *Transactions of the ASAE*, 1996 **39**(3) 1067–73.
48. MANNAPPERUMA J D and SINGH R P. 'Modelling of gas exchange in polymeric packages of fresh fruit and vegetables', in: Singh R P, Oliveira, F A R (eds), *Process optimisation and minimal processing of foods*, CRC Press, Boca Raton, 1994.
49. OORAIKUL B and STILES M E, *Modified Atmosphere Packaging of Food*, Ellis Horwood Limited, Chichester, West Sussex, UK, 1991.
50. PARRY R T, *Principles and Applications of Modified Atmosphere Packaging of Foods*. Blackie Academic & Professional, An Imprint of Chapman & Hall, Glasgow, UK, 1993.
51. PEPPELENBOS H W, *The use of gas exchange characteristics to optimize CA storage and MA packaging of fruits and vegetables*, Thesis Wageningen Agricultural University, ISBN 90-5485-606-8, 1996.
52. POLDERDIJK J J, BOERRIGTER H A M and TIJSKENS L M M. 'Possibilities of the model on keeping quality of vegetable produce in controlled atmosphere and modified atmosphere applications', *Proceedings of the 19th International congress of refrigeration*, volume II, 1995.
53. *PROC. 6TH INTL. CONTROLLED ATMOSPHERE RESEARCH CONFERENCE*, Ithaca NY, 1993.
54. *PROC. 7TH INTL. CONTROLLED ATMOSPHERE RESEARCH CONFERENCE*, Davis CA, 1997.
55. *PROC. 8TH INTL. CONTROLLED ATMOSPHERE RESEARCH CONFERENCE*, Rotterdam NL, 2001.
56. RATTI C, RAGHAVAN G S V and GARIÉPY Y. 'Respiration model and modified atmosphere packaging of fresh cauliflower', *J. Food Engin.*, 1996 **28** 297–306.
57. RATTI C, RAGHAVAN G S V, GARIÉPY Y and CRAPISTE G H. 'Long term storage of vegetables under modified atmospheres', Paper 96F-040, *AgEng*, Madrid, 1996.
58. ROBERTSON D and AGUSTI J, *Software Blueprints: Lightweight Uses of Logic in Conceptual Modelling*, Addison Wesley Longman Inc., 1999.
59. SCHOUTEN R E, KESSLER D, ORCARAY L and VAN KOOTEN O. 'Predictability of keeping quality of strawberry batches', *Posth. Biol. Technol.*, 2002 **26** 35–47.
60. SCHOUTEN R E, TIJSKENS L M M and VAN KOOTEN O. 'Predicting keeping quality of batches of cucumber fruit based on a physiological mechanism',

Posth. Biol. Technol., 2002 **26** 209–20.

61. SLOOF M, TIJSKENS L M M and WILKINSON E C. 'Concepts for modelling the quality of perishable products', *Trends in Food Science & Technology*, 1996 **7**(5) 165–71.
62. SONG Y, KIM H K and YAM K L. 'Respiration rate of blueberry in modified atmosphere at various temperatures', *J. Amer. Soc. Hort. Sci.*, 1992 **117** 925–9.
63. TIJSKENS L M M and EVELO R G. 'Modelling colour of tomatoes during postharvest storage', *Posth. Biology Technol.*, 1994 **4** 85–98.
64. TIJSKENS L M M and POLDERDIJK J J. 'A generic model for keeping quality of vegetable produce during storage and distribution', *Agricultural Systems*, 1996 **51** 431–52.
65. TIJSKENS L M M. 'A model on the respiration of vegetable produce during postharvest treatments', in: Fenwick G R, Hedley C, Richards R L and Khokhar S. (eds), *Agri-food quality, an interdisciplinary approach*, The Royal Society of Chemistry, Cambridge, UK, 1996.
66. TIJSKENS L M M, HERTOG M L A T M and NICOLAÏ B, *Food Process Modelling*, Woodhead Publishing Ltd., Cambridge UK, 2001.
67. TIJSKENS L M M, KONOPACKI P, SIMČIČ M and HRIBAR J. 'Biological variance in agricultural products. Colour of apples as an example', in: Artés F, Gil M I and Conesa M A (eds), *Improving Postharvest Technologies Of Fruits, Vegetables And Ornamentals,* IIR conference, Murcia 2000, Vol. 1 25–42.
68. TIJSKENS L M M, VAN SCHAIK A C R, HERTOG M L A T M and DE JAGER A. 'Modelling the firmness of Elstar apples during storage and transport', *Acta Horticulturae*, 1997 **485** 363–71.
69. VAN DER SMAN R G M, *Latice Boltzmann schemes for convection-diffusion phenomena; application to packages of agricultural products*, Thesis Wageningen, ISBN 90–5808–048-X, 1999.
70. VAROQUAUX P, ALBAGNAC G, NGUYEN C and VAROQUAUX F. 'Modified atmosphere packaging of fresh beansprouts', *J. Sci. Food Agric.*, 1996 **70** 224–30.
71. VERMEIREN L, DEVLIEGHERE F, VAN BEEST M, DE KRUIJF N and DEBEVERE J. 'Developments in the active packaging of foods', *Trends in Food Science & Technology*, 1999 **10** 77–86.
72. WIJTZES T, *Modelling the microbial quality and safety of foods*, Thesis Wageningen Agricultural University, ISBN 90-5485-578-9, 1996.
73. YANG C C and CHINNAN M S. 'Modeling the effect of O_2 and CO_2 on respiration and quality of stored tomatoes', *Transactions of the ASAE*, 1988 **31** 920–5.
74. YEARSLEY C W, BANKS N H and GANESH S. 'Temperature effects on the internal lower oxygen limits of apple fruit', *Posth. Biol. Technol.*, 1997 **11** 73–83.
75. ZAGORY D and KADER A A. 'Modified atmosphere packaging of fresh produce', *Food Technology*, 1988 **42** 70–7.

Part III

Novel packaging and particular products

17

Active packaging in practice: meat

C.O. Gill, Agriculture and Agri-Food Canada

17.1 Introduction

Preservative packagings for fresh meats should maintain acceptable appearance odour and flavour for product, while allowing the development of desirable characteristics associated with ageing, and retarding the onset of microbial spoilage (Taylor, 1985). Such effects can be achieved by packaging meats under various atmospheres of oxygen, carbon dioxide, carbon monoxide and/or nitrogen. The atmosphere within a pack may alter during storage, because of reactions between components of the atmosphere and the product, and/or because of transmission of gases into or out of the pack through the packaging film (Stiles, 1991). Packagings of that type are termed Modified Atmosphere Packs (MAP), which are distinguished from Controlled Atmosphere Packs (CAP) within which invariant atmospheres are maintained throughout the time of storage (Brody, 1996).

Both MAP and CAP can take various forms, depending on the type of meat that is packaged, the form of the meat, and the commercial uses for the product. Obviously, a commercial user of preservative packagings would usually seek the simplest, and presumably least expensive packaging that would give a storage life and organoleptic quality suitable to the trading envisaged for a particular product. Thus, the optimum packaging for a product can be decided only with knowledge of how the qualities of the particular meat are affected by the various atmospheres to which it might be exposed, and the conditions the packaged product will have to tolerate during commercial storage, distribution and display.

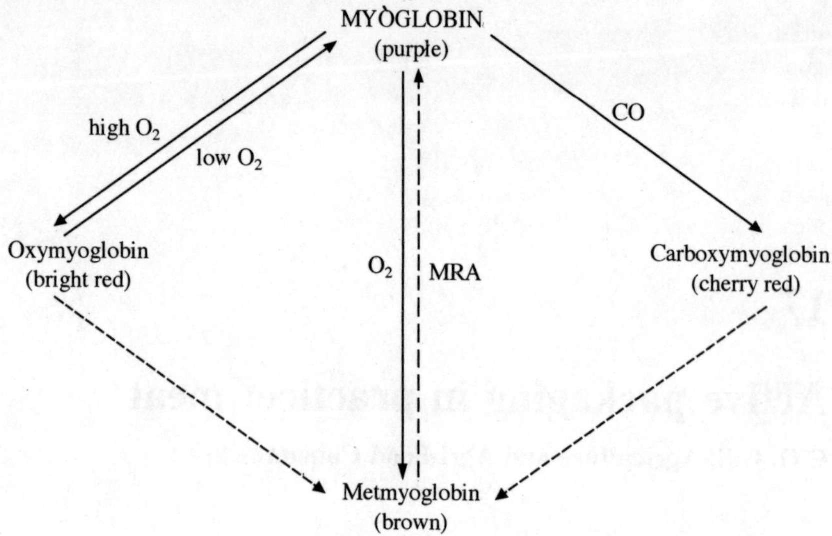

Fig. 17.1 Reactions of myoglobin with oxygen and carbon monoxide.

17.2 Control of product appearance

The appearance of raw meat has major effects on the purchasing decisions of consumers (Cornforth, 1994). For red meats, consumers much prefer bright, red muscle tissue and white rather than yellow fat. When bone is present in a retail cut, consumers prefer that any exposed spongy bone appears bright red also. For poultry, bright, white flesh and skin are preferred.

The colour of muscle tissue in red meat is determined by the quantity and chemical state of the muscle pigment myoglobin (Fig. 17.1). The deoxy form is a dull, purple colour that consumers consider unattractive. The function of myoglobin is to transfer oxygen from blood to the muscle tissue cells. Myoglobin therefore reacts rapidly and reversibly with oxygen to give the bright red form oxymyoglobin. The fraction of pigment in the oxymyoglobin form is dependent on the partial pressure of oxygen to which the pigment is exposed (Livingston and Brown, 1981). Myoglobin can also react with oxygen to give the stable, oxidised form metmyoglobin (Faustman and Cassens, 1990). Meat with the dull, brown colour of metmyoglobin is considered undesirable by most consumers (Renerre, 1990).

Although metmyoglobin is stable, it is slowly reduced to deoxymyoglobin by enzymic reactions involving reduced co-enzymes (Echevarne *et al.*, 1990). Those reactions are termed metmyoglobin reduction activity. Muscle tissue with high metmyoglobin reduction activity can generally maintain a bright red colour when exposed to oxygen for longer than tissue with little or none of the activity, although high respiratory activity tends to accelerate discolouration (O'Keefe and Hood, 1982). Different muscles vary considerably in their metmyoglobin reduction and respiratory activates, and so vary in their colour stabilities during the first days after slaughter. For example, the *longissimus dorsi* usually has good colour stability while the colour stability of the *psoas major* is poor (Hood, 1980). However, enzymic activities in muscle tissue decay with time, so after storage for several days all muscle tissue has similar, low colour stability (Ledward, 1985). The colour stability of ground meat is similarly low because both respiratory and metmyoglobin reduction activates are rapidly lost when meat is ground (Madhavi and Carpenter, 1993).

Both deoxymyoglobin and oxymyoglobin can oxidise to metmyoglobin. However, the rate of the oxidation reaction is considerably faster with deoxy- than with oxymyoglobin (Ledward, 1970). Consequently, when oxygen tensions are low, and most of the myoglobin is in the deoxy form, oxidation of the pigment occurs rapidly; while oxidation is retarded when the oxygen tension is high and most of the pigment is in the oxy form. Haemoglobin visible in cut, spongy bone reacts similarly with oxygen. Thus, increasing the oxygen in a pack atmosphere above atmospheric concentrations will stabilise the desirable red colours of muscle tissue and cut spongy bone surfaces. In addition, high concentrations of oxygen will increase the depth of the oxymyoglobin layer at the tissue surface, and so enhance the red colour of the muscle tissue (Young *et al.*, 1988).

Although high oxygen concentrations will retard pigment oxidation they do not prevent it. Pigment oxidation is prevented only if oxygen is stripped from the pack atmosphere and subsequently prevented from entering the pack (Gill, 1989). When a pack is first filled with a gas or gases other than oxygen, at least some traces of oxygen will be present in the atmosphere (Penney and Bell, 1993). The residual oxygen will react with the muscle pigment to form metmyoglobin. However, provided that the metmyoglobin reduction capacity of the muscle tissue is not exceeded, the metmyoglobin will be reconverted to deoxymyoglobin during the first few days of storage (Gill and Jones, 1994a). After that, the pigment will remain in the deoxy form until it is exposed to air or a high oxygen atmosphere (Table 17.1). Then, the tissue will bloom to the bright red colour of freshly cut meat as oxymyoglobin is rapidly formed at tissue surfaces. Such a desirable colour will, however, be maintained for a relatively short time if the tissues have little if any metmyoglobin reduction activity to counteract the unavoidable oxidation of the pigment.

In addition to the discolouration of the muscle tissue, exposed spongy bone in cuts that have been stored under anoxic atmospheres tend to darken and finally blacken relatively rapidly when the cuts are exposed to air. That intense discolouration appears to be due to the accumulation of haemoglobin at cut bone surfaces during storage (Gill, 1990). In air, the pigment oxidises as it would in freshly cut tissue, but because the amount of pigment is so much greater, the final colour is dark brown or black, rather than the lighter brown colours that spongy bone will develop after meat is cut when fresh.

As an alternative to using high oxygen concentrations to stabilise meat colour, or oxygen depleted atmospheres to prevent discolouration, red colours for muscle and bone tissues can be maintained by exposing the tissues to carbon

Table 17.1 Fractions of metmyoglobin in the muscle pigment of beef steak surfaces after display in air for 1 h, following storage at -1.5°C under N_2, CO_2 or, 67% O_2+33% CO_2 (Gill and Jones, 1994a)

Storage time (days)	Metmyoglobin (%) Storage atmosphere N_2	CO_2	$O_2 + CO_2$
1	7	60	4
2	25	25	7
4	23	14	0
6	0	2	3
8	8	6	9
12	0	0	17
16	0	6	10
20	7	6	23
24	8	0	42
60	0	0	–

monoxide. Carbon monoxide reacts with myoglobin to form the cherry red pigment carboxymyoglobin, which is stable and oxidises only slowly (Lanier *et al.*, 1978). Therefore, exposure of meat to low concentrations of carbon monoxide in a pack atmosphere will result in the tissues developing persistent red colours.

The above comments about the colour of red meats are not wholly applicable to poultry muscle. Poultry muscle generally has low concentrations of myoglobin and high rates of oxygen consumption. Consequently, little oxymyoglobin is formed when poultry muscle is exposed to air and consumers are accustomed to the tones imparted to poultry meat by muscle pigment in the deoxy- and metmyoglobin forms (Millar *et al.*, 1994). Therefore, the colour of poultry meat is not enhanced by storage under high oxygen atmospheres, while the appearance of the meat is not grossly degraded by its exposure to low concentrations of oxygen that would rapidly discolour red meats.

17.3 Control of flavour, texture and other characteristics

Other undesirable, non-microbiological changes that can occur during the storage of raw meats are oxidation of lipids that impart stale and rancid odours and flavours to the product; loss of exudate from the muscle tissue; and loss of texture and development of liver-like flavours as results of the breakdown of proteins. A desirable change is the increase of tenderness with ageing of the muscle tissue.

In the absence of oxygen, lipids will not oxidise. Thus, rancidity does not develop when meat is packaged under an oxygen depleted atmosphere. Oxidation will occur with meat in air or oxygen enriched atmospheres. Although it would be expected that the rates of lipid oxidation would increase with increasing oxygen concentration, it has been reported that rates of oxidation in air and oxygen enriched atmospheres are similar (Ordonez and Ledward, 1977). Antioxidants naturally present in or added to raw meats will retard the development of rancidity and oxidation of myoglobin, but grinding of meat can greatly accelerate lipid oxidation (Sanchez-Escalante *et al.*, 2001). Lipid oxidation is also accelerated by iron and iron containing compounds. Consequently, when mechanically separated meats, which contain relatively large amounts of iron, are included in comminuted products the oxidative stability of the products is greatly reduced (Cross *et al.*, 1987).

Loss of exudate from meat is undesirable, because of ill-effects upon the appearance and handling qualities of cuts, and because of loss of saleable weight when cuts must be divided and repackaged. Exudate losses are unavoidable, but tend to be less with muscle tissue of higher than normal pH. Exudate losses are exacerbated by cutting of meat to smaller portions and pressure on the product (Offer and Knight, 1988). Therefore, in practice, the only options for containing the adverse effects of exudate loss are the avoidance of pressure on product and the inclusion in packs of absorbent pads or wraps of sufficient capacity to hold all the exudate that may be released.

Unlike most changes that occur in meat with age, increased tenderness is generally desirable (Jeremiah *et al.*, 1993). The rate of tenderisation declines approximately exponentially with time of storage. For beef stored at 2°C, 80% and 100% of maximum tenderisation have been reported to be achieved after about 9 and 17 days, respectively (Dransfield *et al.*, 1992). Rates of tenderisation are seemingly affected little if at all by the compositions of pack atmospheres, and with most meats, as with beef, tenderising apparently does not continue indefinitely. Even so, the breakdown of proteins can continue, with the accumulation of peptides and free amino acids that impart liver-like flavours to the meat (Rhodes and Lea, 1961). Consumers may find such flavours objectionable (Gill, 1988a). With lamb it has been observed that tenderising can continue until the fibrous texture of muscle tissues is lost. That undesirable loss of texture and development of liver-like flavours do not occur when lamb is stored under an atmosphere of carbon dioxide (Gill, 1989). No other effects of carbon dioxide on tenderising processes have been reported.

17.4 Delaying microbial spoilage

Spoilage bacteria will grow on meat that is not frozen under both aerobic and anaerobic conditions (Lowry and Gill, 1984). When the initial numbers of bacteria are relatively low, the spoilage flora will be dominated by those species of bacteria that grow most rapidly in the environment provided by the meat and the surrounding atmosphere (Gill, 1986). When initial numbers are high, slower growing species may persist as substantial fractions of a flora, as the maximum numbers may be approached before they are overgrown by the usually dominant species. The meat will be spoiled when the metabolic activities of the spoilage bacteria cause changes in the appearance, odour or flavour of the product that are unacceptable to the consumers (Gill, 1981). The stage of development of the spoilage flora at which such changes occur depends on both the composition of the spoilage flora and the intrinsic qualities of the tissues on which the bacteria are growing.

When fresh meat is stored in air, the spoilage flora is dominated by species of *Pseudomonas*, which are strictly aerobic (Gill and Newton, 1977). Those organisms preferentially utilise glucose, which is present in small quantities in muscle tissue of normal pH (5.5) and usual higher values. When glucose is exhausted the bacteria metabolise amino acids and produce offensive by-products such as ammonia, amines and organic sulphides (Nychas *et al.*, 1988). Thus, on normal pH muscle tissue the onset of aerobic spoilage occurs abruptly when bacterial numbers are about $10^8/cm^2$. However, on muscle tissue of high pH (> 6.0) and moist fat tissue, little or no glucose may be available (Gill and Newton, 1980). Then, aerobic spoilage will occur when bacterial numbers are about $10^6/cm^2$.

The pseudomonads grow at their maximum rates when oxygen concentration in the atmosphere is as low as 1% (Clark and Burki, 1972). Therefore, increasing the oxygen concentration in a pack atmosphere to preserve meat colour does not accelerate microbial spoilage. However, if the storage life of meat is to be extended the rapid growth of pseudomonads must be suppressed.

Growth of pseudomonads is inhibited by carbon dioxide. The growth rate decreases with increasing concentrations of carbon dioxide in the atmosphere up to about 20% (Gill and Tan, 1980). Further increases in carbon dioxide concentration do little more to slow the rate of growth. Thus, with an aerobic atmosphere, a doubling of the time before the onset of microbial spoilage is the most that can be achieved by the inclusion of carbon dioxide in a pack atmosphere.

When growth of pseudomonads is inhibited by carbon dioxide, the flora of meat in an aerobic atmosphere is usually dominated by lactic acid bacteria, with more or less large fractions of strict aerobes, such as pseudomonads and acinetobacteria, and facultative anaerobes, such as *Brochothrix thermosphacta* and enterobacteria (Gill and Jones, 1996). If meat is held in air after storage under a modified atmosphere the lactic acid bacteria, which are of low spoilage potential, may continue to predominate in the flora. However, the fractions of the strict aerobes and facultative anaerobes will usually increase as the flora proliferates; and spoilage will develop as a result of the activities of those latter organisms (Gill and Jones, 1994b).

Under anaerobic conditions, the strictly aerobic pseudomonads cannot grow and again the spoilage flora is usually dominated by lactic acid bacteria (Egan, 1983). Those bacteria can grow to maximum numbers about $10^8/cm^2$ without spoilage of the meat. Thereafter, spoilage will develop only slowly as the by-products of the lactic acid bacteria's metabolism impart acid, dairy flavours to the meat (Dainty *et al.*, 1979). The spoilage process can differ if the tissue pH is >5.8 or the atmosphere contains traces of oxygen. Then, facultative anaerobes such as *B. thermosphacta*, enterobacteria and *Shewanella putrefaciens* may grow to spoil the meat as the flora approaches maximum numbers (Blickstad, 1983;

Grau, 1983). However, in a controlled atmosphere of carbon dioxide alone, the growth of some of those organisms is inhibited or prevented when temperatures are at the lower end of the chill temperature range (Gill and Harrison, 1989). Inclusion of small amounts of carbon monoxide in anaerobic atmospheres does not affect development of the spoilage flora (Sørheim *et al.*, 1999). If meat is held in air after storage under an anaerobic atmosphere, spoilage by facultative anaerobes or strictly aerobic organisms is likely to occur although lactic acid bacteria continue to predominate in the flora (Gill and Jones, 1996).

17.5 The effects of temperature on storage life

All changes that occur in chilled meat during storage are likely to be accelerated by increasing temperature. As most changes are deleterious, it follows that the optimum temperature for storing chilled meats is the minimum that can be maintained indefinitely without freezing the muscle tissue. In practice, that temperature is found to be -1.5 ± 0.5°C (Gill *et al.*, 1988).

When red meats are displayed in aerobic atmospheres, discolouration rather than microbial spoilage is likely to limit the useful life of the product. The rate at which discolouration develops in muscle tissue exposed to air appears to increase linearly with temperature for all muscles, but the rate of increase differs between muscles (Hood, 1980). The rate of increase seems to be less for colour stable than for colour unstable muscles, as discolouration of the colour stable *longissimus dorsi* and the colour unstable *psoas major* muscles are reported to be, respectively, twice and five times as rapid at 10°C than at 0°C. The effect of temperature on the rate of discolouration of meat stored in modified atmospheres rich in oxygen does not appear to be well identified in the literature, but it seems likely that discolouration with increasing temperature accelerates much as for meat stored in air.

When meat is stored anaerobically, the colour stability of muscle tissue increases at first, and then declines (O'Keefe and Hood, 1980–81). The initial increase of stability is probably related to the relatively rapid loss of respiratory activity, while the subsequent decrease in stability reflects the decay of metmyoglobin reduction activities. The rate at which colour stability degrades is reported to be twice as fast at 5°C and four times as fast at 10°C as at 0°C (O'Keefe and Hood, 1982).

Rates of lipid oxidation in air and oxygen enriched atmospheres are apparently similar, but the effect of storage temperature on the rate of development of rancidity does not seem to have been established. Exudate losses are reported to be about 30% and 100% more, respectively, at 5°C and 10°C than at 0°C (O'Keefe and Hood, 1980–81). The rate at which muscle tenderises is over twice as fast at 10C as at 0C (Dransfield, 1994).

Spoilage bacteria will grow on meat that is not frozen at temperatures down to -3°C under both aerobic and anaerobic conditions. Thus, storage at chiller temperatures can delay but not prevent the ultimate onset of microbial spoilage.

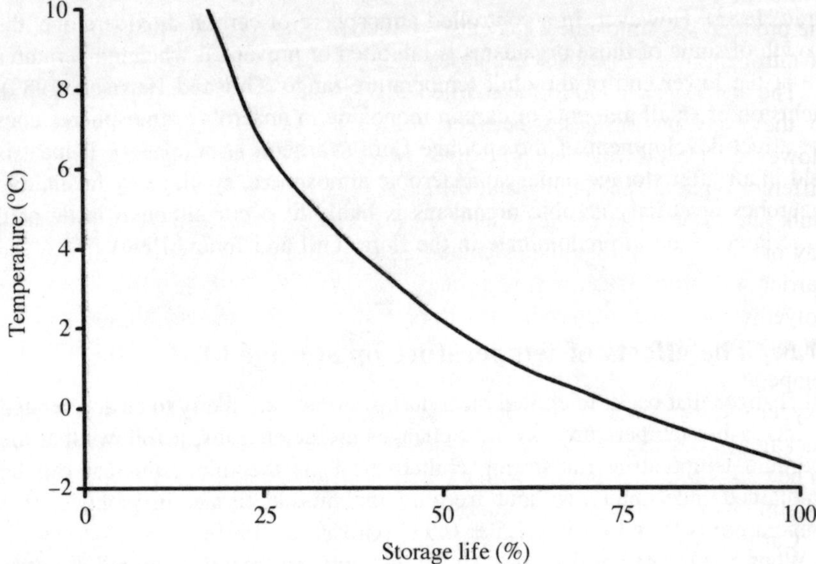

Fig. 17.2 Effects of storage temperature on the storage life of chilled meat limited by microbial spoilage.

Although the rates of growth of different species of spoilage bacteria differ considerably the rates of all increase rapidly with small increases in temperature above the optimum for storage of chilled meat (Gill and Jones, 1992; Gill *et al.*, 1995). The proportional loss of storage life for the same increase in storage temperature is then broadly similar for all types of spoilage flora. Thus, it is found that the storage life of meat in any or no packaging at 0, 2 and 5ºC is about 70, 50 and 30%, respectively, of the storage life that would be obtained for the product stored at −1.5ºC (Fig. 17.2).

In view of the substantial effects of small increases in temperature on rates of discolouration and bacterial growth, it is apparent that any storage life ascribed to a fresh meat product must be accompanied by a statement of the storage temperature if the storage stability of the product is to be properly understood.

17.6 MAP technology for meat products

Modified atmosphere packagings may be used for bulk or retail ready product. Several trays of retail ready product may be placed in a master pack which is filled with the modified atmosphere, or individual, sealed trays may contain the modified atmosphere. Modified atmospheres invariably contain substantial fractions of carbon dioxide to retard the growth of aerobic spoilage organisms. In addition, atmospheres used with red meats will usually contain a high concentration of oxygen to preserve the meat colour or the initial atmosphere may contain a small amount of carbon monoxide to impart a stable red colour to the product. An atmosphere may also contain a more or less substantial fraction of nitrogen, to prevent pack collapse.

The materials used to form modified atmosphere packs must provide a barrier to the exchange of gases between the pack and the ambient atmosphere. However, the gas barrier properties of the packaging materials differ for different types of packaging and differing commercial functions of the packs. Bulk and master packagings which are expected to contain product for only a day or two are often laminates composed of a strong material with limited gas barrier properties, such as nylon, and a sealable layer of a material such as polyethylene. Such materials may have nominal oxygen transmission rates of more than 100 $cc/m^2/24h/atm$ under stated conditions of humidity and temperature. However, films used for modified atmosphere packs usually have oxygen transmission rates between 10 and 100cc $O_2/m^2/24h/atm$, while packagings designed to contain product for the longest possible times are likely to be composed of materials with oxygen transmission rates less than 10 $cc/m^2/$ 24/atm (Jenkins and Harrington, 1991).

Carbon dioxide, the essential component of any effective modified atmosphere for meat is highly soluble in both muscle and fat tissues (Gill, 1988b). The solubility in muscle tissue decreases with decreasing pH and increasing temperature but, within the chill temperature range, solubility in fat increases with increasing temperatures (Fig. 17.3). Because of the dissolution of carbon dioxide in the product, the initial atmosphere in a pack should contain a higher concentration of carbon dioxide than the 20% that it is desirable to maintain after equilibration for maximum inhabitation of the aerobic spoilage bacteria. The smaller the volume of the atmosphere in relation to the product mass, the higher the carbon dioxide concentration needed in the input gas, and the greater the decrease in the volume of the atmosphere as carbon dioxide dissolves in the tissues after the pack is sealed.

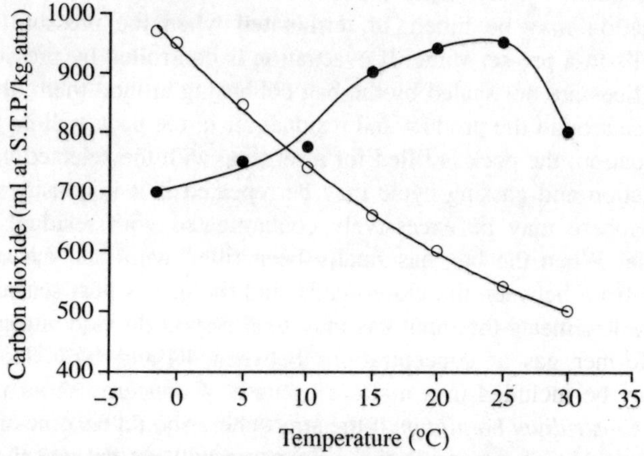

Fig. 17.3 Effects of temperature on the solubility of carbon dioxide in normal pH muscle tissue (○) and fat tissue (•) of beef.

Unlike carbon dioxide, the solubility of oxygen in muscle and fat tissues is low. However, oxygen is converted to carbon dioxide by the respiratory activities of both muscle tissue and bacteria. Although both gases are lost through packaging films when both are at concentrations above those of air, the carbon dioxide dissolved in tissue buffers decreases in carbon dioxide concentrations. Thus, with modified atmospheres rich in oxygen it is usually found that oxygen concentrations decline with time of storage, but that carbon dioxide concentrations alter little after the initial dissolution of the gas in the tissues (Nortje and Shaw, 1989). If packs with oxygen-rich atmospheres are to be stored for relatively long times, the volume of the pack atmosphere should be about three times the volume of the product, to avoid excessive decreases of oxygen concentrations (Holland, 1980).

The solubility of nitrogen in tissues is low, and the gas is metabolically inert. Thus, the only function of nitrogen in a pack atmosphere is to buffer against changes in the volume of the atmosphere that could lead to pack collapse, with crushing of the contained product. If carbon monoxide is included in a pack atmosphere it is at concentrations less than 1%. The gas will be removed from the pack atmospheres as it reacts rapidly and essentially irreversibly with myoglobin. The changes in pack atmosphere volumes as a result of the binding of carbon monoxide are trivial in comparison with the volume decreases arising from dissolution of carbon dioxide. Modified atmosphere packagings for bulk meats are usually intended only to enhance stability for short times during the distribution of product from slaughtering or carcass breaking facilities to retail packing facilities. Protection of the product from crushing by its being in a pillow-pack is often considered to be as important as any effects of the atmosphere on the colour or microbiological condition of the product.

Bulk packs are usually formed using equipment with two flattened tubes (snorkels) that are inserted into the mouth of each bag. Sprung guides at each side of the mouth prevent bunching. The mouth is held closed around the snorkels by padded jaws. Air is evacuated from the bag through the snorkels. The evacuation may be timed, or terminated when the pressure within the snorkels falls to a pre-set value. If evacuation is controlled by pressure and the snorkel orifices are not sealed by the bag collapsing around them, then the bag will collapse around the product and residual air in the pack will be minimised. After evacuation, the pack is filled for a set time with the selected gas mixture. The evacuation and gassing cycle may be repeated if it is considered that the pack atmosphere may be excessively contaminated with residual air after a single cycle. When the bag has finally been filled with gas, the snorkels are withdrawn from between the closed pads, and the bag is heat sealed.

With poultry meats the input gas may be a carbon dioxide/nitrogen mixture with the former gas at concentrations between 40 and 60%. However, 5% oxygen may be included in a mixture because of concerns about the possible growth of *Clostridium botulinum* if the atmosphere should become anaerobic. In fact, the inclusion of oxygen in the atmosphere will not prevent the growth of botulinum organisms, as anaerobic niches that could permit the growth of such organisms exist in any package of raw meat, irrespective of the surrounding atmospheres (Lambert *et al.*, 1991). For red meats the input gas would preferably be 70% oxygen and 30% carbon dioxide. However, nitrogen is often included in a mixture although that gas will serve no useful function when, as in these circumstances, the pack is flexible and the volume variable, and any undesirable pack collapse may be countered by simply increasing the volume of input gas.

Snorkel type equipment is also used for master packaging of retail ready product, with master packs being filled with the same gas mixtures that are used with bulk product. Retail ready product that is master packaged is usually in conventional, expanded polystyrene trays, which are overwrapped with a clinging film of oxygen permeability between 5,000 and 10,000 cc/m^2/24h/atm. The trays usually contain plastic covered paper pads, to absorb exudate from the meat.

Because collapse of the bag around the trays when the master pack bag is evacuated could easily lead to crushing of the trays, evacuation is usually timed. Evacuation of the bag is then highly uncertain, as the amount of air in the bag when the mouth is closed around the snorkels can vary greatly. Moreover, the overwrapped trays will contain more or less large amounts of air that cannot be removed during evacuation. Consequently, the master pack atmospheres are diluted with air to varying extents. Carbon dioxide and oxygen concentrations in master pack atmospheres are then often much below the concentrations optimal for preservation of the product. However, irrespective of the gas atmosphere, master packs provide mechanical protection for filled trays during their distribution from central cutting facilities to retail outlets.

Various types of equipment have been developed for preparing different forms of lidded trays that each contain a modified atmosphere. The atmosphere used for such trays is typically 60% oxygen, 30% carbon dioxide and 10% nitrogen. Storage/display lives of up to two weeks are often claimed for product in such trays. However, to attain such useful life, temperature during display as well as during storage must be well controlled, and the volume of the pack atmosphere must be large in relation to the amount of product in the pack. Control of product temperatures during display is often uncertain (Bøgh-Sørensen and Olsson, 1990), and many retailers consider that small quantities of product in large packs are unattractive to consumers. Therefore, retailers often select modified atmosphere packs to provide an attractive packing in which a high concentration of oxygen, and thus an enhanced meat colour, are maintained for a limited time. Adequate display stability for the product is obtained by control of product temperatures near −1.5°C and by frequent, often daily delivery of freshly packaged product to retail outlets (Gill *et al.*, 2002a). The success of many current distribution systems for master packed product is achieved similarly.

The use of carbon monoxide in modified atmosphere is not permitted in most countries, because of the highly poisonous nature of that gas. Despite that, the risks to consumers from the presence of small amounts of carboxymyoglobin in raw meat appear to be small, and carbon monoxide is a common component of the modified atmospheres used with raw meats in Norway (Sørheim *et al.*, 1997). As carboxymyoglobin confers a red colour on meat irrespective of the presence of oxygen, a modified atmosphere with carbon monoxide need contain no oxygen. The input gas then typically contains 60% carbon dioxide and 40% nitrogen, with carbon monoxide at 0.3 to 0.5%. The major components of the input gas are at concentrations that will give the maximum carbon dioxide concentration after equilibration without the risk of pack collapse. Thus, the carbon dioxide concentration can be maintained at levels above that required for maximum inhibition of aerobic spoilage organisms for relatively long times, without resort to volumes of atmosphere much greater than the volumes of product. Therefore carbon monoxide/high carbon dioxide atmospheres stabilise meat colour and delay microbial spoilage, and so preserve the product in an acceptable condition even when delivery is relatively infrequent and display is prolonged.

17.7 Controlled atmosphere packaging for meat products

The only types of controlled atmosphere packagings currently used with raw meats are those in which an anaerobic atmosphere is maintained indefinitely. Controlled atmosphere packagings may be used for bulk product or items of irregular shape, such as whole lamb carcasses, or as master packs for retail-ready product. Controlled atmosphere packaging is not suitable for individual trays of retail-ready product because of the undesirable colour of anoxic meat, and because packaging materials that are impermeable to gases are mostly opaque. Readily available films that are essentially gas impermeable are laminates that incorporate a layer of aluminum foil, laminates with two layers of a metallised film, or laminates with unusually thick layers of plastics with high barrier properties (Kelly, 1989).

Controlled atmospheres may be of carbon dioxide or nitrogen, or mixtures of the two gases. Nitrogen can provide an anaerobic atmosphere, but does not otherwise affect the muscle tissue or the microflora. Thus, the storage life of meats in a controlled atmosphere of nitrogen is similar to that of meats in vacuum pack; although in a gas impermeable, controlled atmosphere pack there is no oxidation of myoglobin in exudate or muscle tissue, which eventually become evident with meat in vacuum packs as the result of small quantities of oxygen permeating the packaging films (Jeremiah *et al.*, 1992).

Atmospheres of carbon dioxide have inhibitory effects on some organisms of the anaerobic spoilage flora, and can apparently retard the excessive tenderising of at least lamb. The inhibiting effects of carbon dioxide on the microflora appear to reduce rapidly with reducing concentrations of carbon dioxide in the atmosphere, so an atmosphere of or near 100% carbon dioxide is required if the storage stability of the product is to be substantially increased over that attainable with a nitrogen atmosphere (Gill and Penney, 1988). When an atmosphere rich in carbon dioxide is used, the high solubility of the gas in meat tissues must be taken into account. In an atmosphere of 100% carbon dioxide, meat will absorb approximately its own volume of the gas. Thus, the initial gas volume must exceed the required final volume by the volume of the enclosed meat.

Specialised equipment for forming controlled atmosphere packs is available. With such equipment, snorkels are inserted in the mouth of a filled bag, and the mouth is closed around them; then a hood is placed over the bag, with enclosure of the snorkels and bag sealing elements of the equipment. Air is withdrawn from the bag through the snorkels while the hood is simultaneously evacuated. The bag inflates in the evacuated hood, which ensures that no part of the bag collapses to entrap air. Some air is then admitted into the hood to give a low pressure which will collapse that bag around the product without crushing it. Thus, the volume of the bag is minimised before it is filled with gas. A pack may be flushed with the input gas one or more times before it is sealed. That relatively elaborate filling procedure is adopted to minimise the amount of residual oxygen in the pack. Even so, residual oxygen concentrations after pack sealing are usually about 100ppm (Penney and Bell, 1993).

Snorkel equipment without a hood and even tray gassing equipment have been used, at least experimentally, for the production of controlled atmosphere packs. The residual oxygen in such packs is apparently often about 1%, which can have grossly adverse affects upon the colour of red meats. Even 100 ppm of oxygen can result in discolouration of product. However, in those latter circumstances discolouration is usually transient, as the metmyoglobin is reduced to myoglobin, usually within four days, as anoxic conditions are established and maintained (Gill and Jones, 1994a).

Various studies have been conducted to determine if oxygen scavengers might be used to prevent permanent discolouration of red meats in atmospheres with initial concentration about 1%, or transient discolouration of meats in atmospheres with very low concentrations of residual oxygen. Although some success with the atmospheres of the former type have been reported (Doherty and Allen, 1998), the general utility of such an approach must be doubted because the muscle tissue itself acts as a very efficient oxygen scavenger (Table 17.2). Certainly, findings with the use of oxygen scavengers in atmospheres of very low initial oxygen concentration have been that numerous, fast reacting oxygen scaverages must be employed if transient browning is to be prevented (Tewari *et al.*, 2002).

17.8 Future trends in active packagings for raw meats

In most developed countries, sales of raw meat at supermarkets have tended to increase at the expense of sales at specialised butchers' stores (Mannion, 1995). The maintenance of butchering facilities at supermarkets is increasingly seen as undesirable, both because of the use of costly floor space that might otherwise

Table 17.2 Half life of oxygen in packs containing 4L of atmospheres with < 1% oxygen when packs contained either four trays of ground beef or 32 oxygen scavengers each with a capacity of 200 mlg oxygen (Gill and McGinnis, 1995)

Temperature (C)	O_2 half life (h)	
	With meat	With O_2 scavengers
−1.5	4.7	0.6
0	3.8	0.6
2	2.9	0.5
5	1.4	0.5
10	1.6	0.5

be used for selling foods, and because of difficulties with obtaining staff skilled in butchery. Therefore, supermarket operators have been for some years generally inclined to move toward the preparation of display ready product at central butchering facilities (Lazar, 2001). Although modified atmosphere packagings of various types have been used with mixed results in central cutting operations, most successful operations now rely on the frequent preparation of retail packs, with frequent and speedy delivery of product held at temperatures near the optimum for chilled meat, rather than the preservative capabilities of modified atmospheres.

Simultaneously there has been a trend towards consolidation of slaughtering facilities so that in some regions, such as North America, most animals are now slaughtered at relatively few large plants. Given the move towards central preparation of retail ready meat, it would seem economically advantageous to prepare retail packs at slaughtering plants. That would avoid the double handling, and double packaging of product that now occurs with the consignment of vacuum packaged primal cuts or bulk packed product from slaughtering plants to central cutting facilities. The retail packaging of product of compromised colour stability, and loss of product weight as exudate after prolonged storage in vacuum pack could also be avoided.

Although retail preparation of product at slaughtering plants is increasing, particularly with poultry, the need for frequent and speedy delivery limits the area of distribution. Thus, for the largest plants from which product is widely distributed preparation of retail product is a minor activity at most.

A general conflation of slaughtering with preparation of retail-ready product at a few large plants would seem to be practicable only if the useful life of retail-ready product reliably exceeds the storage and display times usual in current commercial practice (Fig. 17.4). When meat is stored and displayed at normal commercial temperatures, such storage stability can be attained by master packaging meat under controlled, anoxic atmospheres, as has been demonstrated with commercial systems for the distribution of lamb in the USA or by use of carbon monoxide in modified atmospheres, as has been demonstrated with commercial systems in Norway.

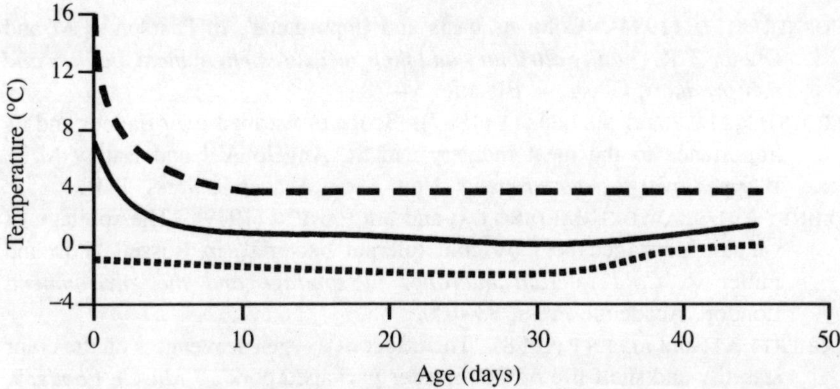

Fig. 17.4 Median temperatures of all (—), the coldest 2% (▪▪▪) and the warmest 2% (– –) of vacuum packaged beef distributed in Canada (Gill *et al.*, 2002b).

Despite the commercial advantages of, and the trivial risks associated with, the use of carbon monoxide, it is unlikely that many countries will sanction meat being treated with a recognised poison. A trend towards increasing use of controlled atmosphere packaging for retail ready product might then be anticipated. However, the complexities of meat trading are likely to ensure that such a trend develops only slowly.

Although controlled atmosphere packaging could be used for continental distribution of retail-ready meat, it is unlikely to be used for global distribution of such product. Storage life would not necessarily constrain global distribution but the low packing density of retail packed product as compared with bulk product could render shipment of meat by sea uneconomical. Thus, controlled atmosphere packing is unlikely to replace vacuum packing in trading of chilled meats to distant markets, unless there is a move to retail portioning but not retail packaging at exporting plants. Otherwise, use of controlled atmosphere packing for trading meat over long distances is likely to remain restricted to products that cannot be successfully vacuum packaged, such as whole lamb carcasses.

17.9 References

BLICKSTAD E (1983), 'Growth and end product formation of two psychrotrophic *Lactobacillus* spp. and *Brochothrix thermosphacta* ATCC 115009 at different pH values and temperatures'. *Appl Environment Microbiol*, 46, 1345–50.

BØGH-SØRENSEN L and OLSSON P (1990), 'The chill chain', in Gormley T R *Chilled foods: the state of the art*, London, Elsevier, 245–67.

BRODY A L (1996), 'Integrating aseptic and modified atmosphere packaging to fulfil a vision of tomorrow', *Food Technol*, 50 (4), 56–66.

CLARK D S and BURKI T (1972), 'Oxygen requirements of strains of *Pseudomonas* and *Achromobacter*', *Canad J Microbiol*, 18, 321–26.

CORNFORTH D. (1994), 'Color-its basis and importance', in Pearson A M and Dutson T R, *Quality attributes and their measurement in meat, poultry and fish products*, Glasgow, Blackie, 34–78.

CROSS H R, LEU R and MILLER M F (1987), 'Scope of warmed-over flavour and its importance to the meat industry', in St. Angelo A J and Bailey M E, *Warmed-over flavour of meats*, New York, Academic Press, 1–18.

DAINTY R M, SHAW B G, HARDING C O and MICHANIE S (1979), 'The spoilage of vacuum packaged beef by cold tolerant bacteria', in Russell A D and Fuller R, *Cold tolerant microbes in spoilage and the environment*, London, Academic Press, 83–100.

DOHERTY A M and ALLEN P (1998). 'The effect of oxygen scavengers on the color stability and shelf-life of CO_2 master packaged pork', *J Muscle Foods*, 9, 351–63.

DRANSFIELD E (1994), 'Tenderness of meat, poultry and fish', in Pearson A M and Dutson T R, *Quality attributes and their measurement in meat, poultry and fish products*, Glasgow, Blackie, 289–315.

DRANSFIELD E, WAKEFIELD D K and PARKMAN I D (1992). 'Modeling post-mortem tenderization. 1: Texture of electrically stimulated and non-stimulated beef', *Meat Sci*, 31, 57–73.

ECHEVARNE C, RENERRE M and LABAS R (1990), 'Metmyoglobin reductive activity in bovine muscle', *Meat Sci*, 27, 161–72.

EGAN A F (1983), 'Lactic acid bacteria of meat and meat products', *Antonie van Leeuwenhoek*, 49, 327–36.

FAUSTMAN C and CASSENS R G (1990), 'The biochemical basis for discoloration in fresh meat: a review', *J Muscle Foods*, 1, 217–43.

GILL C O (1981), 'Meat spoilage and evaluation of the potential storage life of fresh meat', *J Food Prot*, 46, 444–52.

GILL C O (1986), 'The control of microbial spoilage in fresh meats', in Pearson A M and Dutson T R, *Advances in Meat Research, vol. 2*, Westport, AVI Publishing pp. 49–88.

GILL C O (1988a) 'Microbiology of edible meat by-products', in Pearson A M and Dutson T R, *Edible meat by-products*, Barking, Elsevier, 47–82.

GILL C O (1988b), 'The solubility of carbon dioxide in meat', *Meat Sci*, 22, 65–71.

GILL C O (1989), 'Packaging meat for prolonged chilled storage: the Captech process', *Brit Food J*, 91, 11–15.

GILL C O (1990), 'Meat and modified atmosphere packaging', in Hui Y H, *The encyclopedia of food science and technology*, New York, Wiley, 1678–83.

GILL C O and HARRISON, J C L (1989), 'The storage life of chilled pork packaged under vacuum or carbon dioxide', *Food Microbiol*, 26, 313–24.

GILL C O and JONES T (1992), 'Assessment of the hygienic efficiencies of two commercial processes for cooling pig carcasses', *Food Microbiol*, 9, 335–43.

GILL C O and JONES T (1994a), 'The display life of retail-packaged beef steaks after their storage in master packs under various atmospheres', *Meat Sci*,

38, 385–96.

GILL C O and JONES T (1994b), 'The display of retail-packs of ground beef after their storage in master packs under various atmospheres', *Meat Sci*, 37, 281–95.

GILL C O and JONES T (1996), 'The display life of retail-packaged pork chops after their storage in master packs under atmospheres of N_2, CO_2 or O_2+CO_2', *Meat Sci*, 42, 203–13.

GILL C O and NEWTON K G (1977), 'The development of spoilage flora on meat stored at chill temperatures', *J Appl Bacteriol*, 43, 189–95.

GILL C O and MCGINNIS J C (1995), 'The use of oxygen scavengers to prevent the transient discolouration of ground beef packed under controlled, oxygen-depleted atmospheres', *Meat Sci*, 41,19–27.

GILL C O and NEWTON K G (1980), 'Development of bacterial spoilage at adipose tissue surfaces of fresh meat'. *Appl Environ Microbiol*, 39, 1076–7.

GILL C O and PENNEY N (1988), 'The effect of the initial gas volume to meat weights ratio on the storage life of chilled beef packaged under carbon dioxide', *Meat Sci*, 22, 53–63.

GILL C O and TAN K H (1980), 'Effect of carbon dioxide on growth of meat spoilage bacteria', *Appl Environ Microbiol*, 39, 317–19.

GILL C O, PHILLIPS D M and HARRISON J C L (1988), 'Product temperature criteria for shipment of chilled meats to distant markets', in *Refrigeration for food and people*, Paris, International Institute of Refrigeration, 40–7.

GILL C O, FRISKE M, TONG A K W and MCGINNIS J C (1995), 'Assessment of the hygienic characteristics of a process for the distribution of processed meats, and of storage conditions at retail outlets', *Food Res Int*. 28, 131–8.

GILL C O, JONES T, RAHN K, HOUDE A, MCGINNIS J C, CAMPBELL S, HOLLEY R A and LEBLANC DI (2002a), 'Control of product temperatures during the distribution of retail ready beef to stores and vacuum packaged beef to restaurants', *Dairy Food Environ Sanit*, 22, 422–8.

GILL C O, JONES T., RAHN K, CAMPBELL S, LEBLANC D I, HOLLEY R A and STARK R. (2002b). 'Temperatures and ages of boxed beef packaged and distributed in Canada', *Meat Sci,* 60, 401–10.

GRAU F H (1983), 'Microbial growth on fat and lean surfaces of vacuum packaged beef', *J Food Sci*, 48, 326–9.

HOLLAND G C. (1980), 'Modified atmospheres for fresh meat distribution', *Proc 33rd Meat Ind Res Conf*, Chicago, Am Meat Sci Asn, 1980.

HOOD D E (1980), 'Factors affecting the rate of metmyoglobin accumulation in pre-packaged beef', *Meat Sci*, 4, 247–65.

JENKINS W A and HARRINGTON J P (1991), 'Fresh meat and poultry', in *Packaging foods with plastics*, Lancaster, Technomic Publishing, 109–22.

JEREMIAH L E, PENNEY N and GILL C O (1992), 'The effects of prolonged storage under vacuum or CO_2 on the flavour and texture profiles of chilled pork', *Food Res Int*, 25, 9–19.

JEREMIAH L E, TONG A K W, JONES S D M and MCDONELL C (1993), 'A survey of Canadian consumer perceptions of beef in relation to general perceptions regarding foods', *J Consum Stud Home Econom*, 17, 13–37.

KELLY R S A (1989), 'High barrier metalized laminates for food packagings', in Findlayson K M, *Plastic film technology, vol. 1*, Lancaster, Technomic Publishing, 146–52.

LAMBERT A D, SMITH J P and DODDS K L (1991), 'Effects of initial O_2 and CO_2 and low-dose irradiation on toxin production by *Clostridium botulinum* in MAP fresh pork', *J Food Prot*, 54, 939–44.

LANIER T C, CARPENTER J A, TOLEDO R T and REAGAN J O (1978), 'Metmyoglobin reduction in beef systems as affected by aerobic, anaerobic and carbon monoxide-containing environments', *J Food Sci,* 43, 1788–92.

LAZAR V (2001), 'You've got it!' *Meat Proces*, 40 (10), 22–31.

LEDWARD D A (1970), 'Metmyoglobin formation in beef stored in carbon dioxide enriched and oxygen-depleted atmospheres', *J Food Sci* 35, 33–7.

LEDWARD D A (1985), 'Post-slaughter influences on the formation of metmyoglobin in beef muscle', *Meat Sci,* 15, 149–71.

LIVINGSTON D J and BROWN W D (1981), 'The chemistry of myoglobin and its reactions', *Food Technol*, 35(5), 244–52.

LOWRY P D and GILL CO (1984), 'Mould growth on meat at freezing temperatures', *Int J Refrig,* 7, 133–6.

MADHAVI D L and CARPENTER C E (1993), 'Aging and processing affect color, metmyoglobin reductase and oxygen consumption of beef muscles', *J Food Sci* 58, 939–47.

MANNION P (1995), 'Meat retail: major change', *Meat Int,* 3 (4), 10–13.

MILLAR S, WILSON R, MOSS B W and LEDWARD D A (1994), 'Oxymyoglobin formation in meat and poultry', *Meat Sci,* 36, 397–406.

NORTJE G L and SHAW B G (1989), 'The effect of ageing treatment on the microbiology and storage characteristics of beef in modified atmospheres packs containing 25% CO_2 plus 75% O_2', *Meat Sci,* 25, 43–58.

NYCHAS G J, DILLON V M and BOARD R G (1988), 'Glucose, the key substrate in the microbiological changes occurring in meat and certain meat products', *Biotechnol Appl Biochem*, 10, 203–31.

OFFER G and KNIGHT P (1988), 'The structural basis of water holding in meat; part 2, drip loss', in Lawrie R A, *Developments in meat science, vol. 4*, London, Elsevier, 173–243.

O'KEEFE M and HOOD D E (1980–81), 'Anoxic storage of fresh beef. 2: Colour stability and weight loss', *Meat Sci*, 5, 267–81.

O'KEEFE M and HOOD D E (1982), 'Biochemical factors influencing metmyoglobin formation on beef from muscles of differing colour stability', *Meat Sci*, 7, 204–28.

ORDONEZ J A and LEDWARD D A (1977), 'Lipid and myoglobin oxidation in pork stored in oxygen and carbon dioxide-enriched atmospheres', *Meat Sci,* 1, 41–8.

PENNEY N and BELL R G (1993), 'Effect of residual oxygen on the colour, odour and taste of carbon-dioxide-packaged beef, lamb and pork, during short term storage at chill temperatures', *Meat Sci*, 33, 245–52.

RENERRE M. (1990), 'Factors involved in the discolouration of beef meat', *Int J Food Sci, Technol*, 25, 613–30.

RHODES D N and LEA C M (1961) 'Enzymic changes in lamb's liver during storage', *J Sci Food Agric.*, 12 211–27.

SÁNCHEZ-ESCALANTE A, DJENANE D, TORRESCANO G, BELTRAN J A and RONCALÉS P (2001), 'The effects of ascorbic acid, taurine, carrosine and rosemary powder on colour and lipid stability of beef patties packaged in modified atmosphere', *Meat Sci*, 58, 421–9.

STILES M E (1991), 'Modified atmosphere packing of meat, poultry and their products', in Oriakul B and Stiles M E, *Modified atmosphere packagings of foods*, West Sussex, Ellis Herwood, 118–47.

SØRHEIM O, AUNE T and NESBAKKEN T (1997), 'Technological, hygienic and toxicological aspects of carbon monoxide use in modified-atmosphere packaging of meat', *Trend Food Sci Technol*, 8, 307–12.

SØRHEIM O, NISSEN H and NESBAKKEN T (1999), 'The storage life of beef and pork packaged in an atmosphere with low carbon monoxide and high carbon dioxide', *Meat Sci*, 52, 157–64.

TAYLOR A A (1985), 'Packaging fresh meat', in Lawrie R, *Developments in meat science, vol. 3*, Barking, Elsevier, 89–113.

TEWARI G, JEREMIAH L E, JAYAS D S and HOLLEY R A (2002), 'Improved use of oxygen scavengers to stabilize the colour of retail-ready meat cuts stored in modified atmospheres', *Int J Food Sci Technol*, 37, 199–207.

YOUNG L L, REVIERE R D and COLE A B (1988), 'Fresh red meats: a place to apply modified atmospheres', *Food Technol*, 42, 65–9.

18

Active packaging in practice: fish

M. Sivertsvik, NORCONSERV, Norway

18.1 Introduction

Fresh fish products are usually more perishable than most other foodstuffs due to their high a_W, neutral pH, and presence of autolytic enzymes. The spoilage of fish and shellfish results from changes caused by oxidation of lipids, reactions due to activities of the fishes' own enzymes, and the metabolic activities of microorganisms (Ashie *et al.*, 1996). The rate of deterioration is highly temperature dependent and can be inhibited by the use of low storage temperature (e.g. fish stored on ice). The spoilage of fresh fish is usually dominated by microbial activities, however, in some cases chemical changes, such as autoxidation or enzymatic hydrolysis of the lipid fraction may result in off-odours and flavours and, in other cases, tissue enzyme activity can lead to unacceptable softening of the fish (Huss *et al.*, 1997). The degree of processing and preservation together with product composition and storage temperature will decide whether fish undergoes microbial spoilage, biochemical spoilage, or a combination of both. These factors contributed to difficulties when using different technologies to extend the storage of fresh fish above that obtained by traditional ice storage.

The term fish product covers a wide range, and includes fish that differ widely in composition, origin, shelf-life and applicability to novel packaging technologies. The range covers fish from temperate waters with a microbial flora adapted to psychrotrophic conditions to fish from tropical waters with different microflora, just as freshwater fish differ from seafish. There is also a wide distribution in the chemical composition of fish, for example, lean fish almost without fat, such as cod, and fatty fish, such as salmon; which often contains 20% fat. Some fish have a long natural shelf-life on ice (e.g. halibut) but others, like the pelagic species (e.g. mackerel and herring) have a very short shelf-life. Fresh fish is very different from the various processed fish products that need packaging: heat-treated fish products (ready meals, fish pudding/balls), smoked, dried, or salted fish. We still have not taken into account the numerous different species of crustaceans and molluscs.

This chapter will cover active packaging of fish products including the use of atmosphere modifiers such as oxygen scavengers and carbon dioxide emitters, packaging that controls water or with anti-microbial and anti-oxidative properties, and indicator mechanisms. Modified atmosphere packaging (MAP) is regarded by some as an active packaging technology. This is by now a well established method to extend the shelf-life of foods, including fish products, and will not be covered in this chapter except for the MA methods different from traditional MAP using gas flushing.

Obviously, having such a broad spectrum of products, it is unlikely that one specific novel or active packaging technology will be a success for all, just as not all fishery products benefit from MAP when compared to vacuum packaging (Sivertsvik *et al.*, 2002a). So, the potential for an active packaging technology to be successful for a product would depend on the technology's ability to control and inhibit the shelf-life deteriorating spoilage reactions (e.g. bacterial growth of specific bacteria, oxidative rancidity, colour changes) in the specific product.

18.2 The microbiology of fish products

As mentioned, the deterioration of fresh fish is usually microbial so controlling the microbial growth is usually the most important parameter for an active packaging technology to be successful. Fish normally have a particularly heavy microbial load owing to the method of capture and transport to shore, slaughtering method, evisceration and retention of skin in retail portions. The microorganisms associated with most seafood reflect the microbial population in their aquatic environment (Colby *et al.*, 1993; Liston 1980; Gram and Huss 1996). Microorganisms are found on all the outer surfaces (skin and gills) and in the intestines of live and newly caught fish. The total numbers of organisms vary enormously and Liston (1980) states a normal range of 10^2–10^7 cfu (colony forming units)/cm^2 on the skin surface. The gills and the intestines both contain between 10^3 and 10^9 cfu/g (Huss 1995). The fish muscle is sterile at the time of slaughtering/catch, but quickly becomes contaminated by surface and intestinal bacteria, equipment, and humans during handling and processing. The microflora of temperate-water fish is dominated by psychrotrophic, aerobic or facultative anaerobic Gram-negative, rod-shaped bacteria, but Gram-positive organisms can also be found in varying proportions. The flora on tropical fish often carries a slightly higher load of Gram-positive and enteric bacteria.

The composition of fresh fish flesh makes it favourable to microbial growth. The muscle is composed of low collagen, low lipid, and high levels of soluble non-protein-nitrogen (NPN) compounds. Trimethylamine-oxide (TMAO), a part of the NPN compounds, can be broken down to trimethylamine (TMA) by endogenous enzymes. However, at chilled temperatures TMA is produced by the bacterial enzyme TMA oxidase. TMA is recognised as the characteristic 'fishy' odour of spoiled fish. When the oxygen level is depleted, many of the spoilage bacteria can utilise TMAO as a terminal hydrogen acceptor, thus allowing them to grow under anoxic conditions. Towards the end of shelf-life, various malodorous low molecular-weight sulphur-compounds such as H_2S and CH_3SH, together with volatile fatty acids and ammonia are produced because of bacterial growth.

During chilled storage, there is a shift in bacterial types. The part of the microflora, which will ultimately grow on the products, is determined by the intrinsic (e.g. *post mortem* pH in the flesh, the poikilothermic nature of fish, and presence of TMAO and other NPN components) and extrinsic parameters (e.g. temperature, processing, and packaging atmosphere). When a product is microbial spoiled, the spoilage microflora will usually consist of a mixture of species, many of which can be completely harmless both in terms of health hazards and in terms of ability to produce off-odours and off-flavours. The bacterial group causing the important chemical changes during fish spoilage often consists of a single species; the specific spoilage organisms (SSO). Little is known of the SSOs of different fish from various aquatic environments under different packaging conditions. However, for many fish stored under aerobic conditions in ice, *Shewanella putrefaciens* has been identified as the main spoilage bacteria (Gram *et al.*, 1987). *S. putrefaciens* produce very intense and unpleasant off-odours, reduce TMAO to TMA and produce H_2S.

Under anaerobic conditions (MAP, vacuum packaging, active packaging technologies) the spoilage bacteria differ from aerobic spoilage. The Gram-negative organism *Photobacterium phosphoreum* has been identified as the organism responsible for spoilage in VP and in MA packs (Dalgaard 1995). The growth rate of this organism is increased under anaerobic conditions and in contrast to *S. putrefaciens*, *P. phosphoreum* is shown to be highly resistant to CO_2. It was also shown that the growth of this bacteria corresponds very well with the shelf-life of packed fresh cod. *P. phosphoreum* reduces TMAO to TMA at 10–100 times the amount per cell than *S. putrefaciens* probably due to the large size of the former (diameter $5\mu m$) while very little H_2S is produced during growth in fish substrates (Dalgaard *et al.*, 1996). Spoiled MAP cod is characterised by high levels of TMA, but little or no development of the putrid or H_2S odours typical for some aerobically stored spoiled fish. *P. phosphoreum* is widespread in the marine environment and it seems likely that this organism or other highly CO_2 resistant microorganisms are responsible for spoilage of packed seafood products. Lactic acid bacteria and *Brochothrix thermosphacta* have been identified as the typical SSOs of freshwater fish and fish from warmer waters.

To obtain a longer shelf-life for fresh fish than obtained by ice or chilled MAP/vacuum, the approach is to inhibit the SSO limiting shelf-life for the technology chosen. For example *P. phosphoreum* is sensitive to freezing and is totally inactivated in thawed chilled MAP cod fillets after frozen storage at −20°C and −30°C for 6–8 weeks. This approach has been used to further extend the shelf-life of MAP cod (Guldager *et al.*, 1998) and salmon (Emborg *et al.*, 2002) at 2°C, and should also be the approach for successful use of active packaging technologies to control microbial spoilage.

18.3 Active packaging: atmosphere modifiers

Many of the most used active packaging technologies are closely related to modified atmosphere packaging. Together with anaerobic conditions, carbon dioxide is the active gas of MAP because it inhibits growth of many of the normal spoilage bacteria (Sivertsvik *et al.*, 2002b). The effect of CO_2 on bacterial growth is complex and four activity mechanisms of CO_2 on microorganisms has been identified (Farber 1991; Daniels *et al.*, 1985; Dixon and Kell 1989; Parkin and Brown 1982): Alteration of cell membrane function includes effects on nutrient uptake and absorption; direct inhibition of enzymes or decreases in the rate of enzyme reactions; penetration of bacterial membranes, leading to intracellular pH changes; and direct changes in the physico-chemical properties of proteins. Probably a combination of all these activities accounts for the bacteriostatic effect (Sivertsvik *et al.*, 2002a). The CO_2 is usually introduced into the MA-package by evacuating the air and flushing the appropriate gas mixture into the package prior to sealing, typically using automatic form-fill-seal or flow-packaging machines. Two other approaches to create a modified atmosphere for a product are either to generate the CO_2 and/or remove O_2 inside the package after packaging or to dissolve the CO_2 into the product prior to packaging. Both methods can give appropriate packages with smaller gas/product ratio, and thus decrease the package size that has been a disadvantage of MAP from the start.

The first approach involves the most commercialised active packaging technology, namely oxygen scavengers, that by now are available from several manufacturers (Mitsubishi Gas Chemical Co., ATCO, Bioka, Sealed Air/Cryovac, Multisorb a.o.), in different forms (sachets, packaging film, closures) with different active ingredients (iron, enzymes, dye). Some of the same companies have also developed CO_2 emitters, using the O_2 in the package headspace to produce CO_2 and to develop a CO_2/N_2 atmosphere inside a package without the use of gas flushing. Other methods for generating the CO_2 gas inside the packages after closure include the use of dry ice (solid CO_2) (Sivertsvik *et al.*, 1999) or carbonate possibly mixed with weak acids (Bjerkeng *et al.*, 1995).

The second approach is to dissolve the CO_2 into the product prior to packaging. Since the solubility increases at lower temperatures and at higher CO_2 pressures, a sufficient amount of CO_2 can be dissolved into the product during 1–2 hours prior to packaging using elevated pressures. This method is called soluble gas stabilisation (SGS) (Sivertsvik 2000). This is not an active packaging technology by definition, but it is a novel alternative to MA and it has been used successfully alone and in combination with O_2 scavengers (see below).

The commercial use of atmosphere modifiers, and O_2 scavengers in particular, with fish products has been mostly limited to the Japanese market and to dried (seaweed, salmon jerky, sardines, shark's fin, rose mackerel, cod, squid) or smoked (salmon) products (Ashie *et al.*, 1996). These ambient stored products have low a_W (<0.85) so the microbial deterioration is not shelf-life limiting, therefore the effect of the O_2 scavengers is to prevent oxidative reactions, discolouration, and mould growth. Other commercial products are fresh yellow-tail, salmon roe, and sea urchin all stored at superchilling conditions packaged with O_2 scavenger primarily to prevent oxidation and discolouration, but also to inhibit bacterial growth to a lesser degree (Ashie *et al.*, 1996). Different O_2 scavengers are chosen dependent on the amount of O_2 to scavenge (pack size and material) and product a_W. O_2 scavengers for high a_W foods react faster compared to scavengers for dry foods but in general the absorption is slow and exothermic.

Removal of oxygen from package interiors improves shelf-life by sub-optimising the environment for aerobic microbiological growth and for adverse oxidative reactions such as rancidity. Ferrous ironbased oxygen scavengers rely on the presence of moisture for activation, with a water activity of at least 0.7 required, and 0.85–0.9 being preferred (Brody 2001). Oxygen absorbers are designed to reduce oxygen levels to less than 100 ppm in package head-space. In iron-based oxygen scavengers the oxygen is removed by oxidation (rusting) of powdered iron forming non-toxic iron oxide (Ashie *et al.*, 1996). Oxygen absorbers could be used to create oxygen-free conditions in head-space of packages of medium barrier properties. The sachet will absorb residual oxygen and oxygen permeated through the packaging material during storage. More inexpensive or environmentally 'friendly' packaging materials with lower oxygen barriers could be used in combination with an oxygen absorber instead of high-cost barrier materials (Sivertsvik, 1997). However, not all oxygen absorbers can be combined with MAP. Some of them, like the iron-based Ageless SS-type from Mitsubishi meant for use in high a_W foods, will unintentionally absorb some of the carbon dioxide present, and decrease some of the inhibitory effects of CO_2 on bacterial growth. This is caused by a reaction of iron with CO_2 to form ferrous carbonate, and secondarily this ferrous carbonate reacts with O_2. This reaction will also slow down the O_2 absorption (Sivertsvik, 1997 and 1999).

The use of O_2 absorbers (Ageless SS-100) had only a marginal effect on microbial growth in packages of fishcakes, fish pudding and mackerel fillets, and far less than the significant effect obtained by MAP (Sivertsvik 1997). However, a signficant effect of the O_2 absorber was observed in packages with salmon fillets. The use of O_2 absorbers inhibited development of rancidity (TBARS) in both mackerel and salmon fillets (Sivertsvik, 1997), but in no higher degree than O_2-free MAP.

The effect of SGS treatment, different O_2-absorbers/CO_2-emitters and combinations of these on growth of psychrotrophic bacteria in salmon fillets is shown in Fig. 18.1 (Sivertsvik, 1999). The best microbial quality was

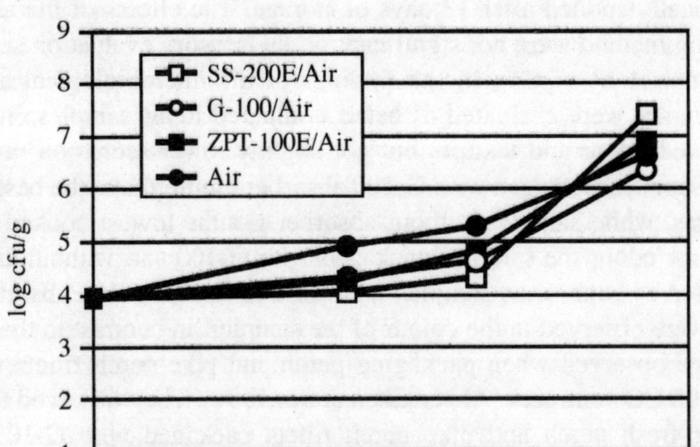

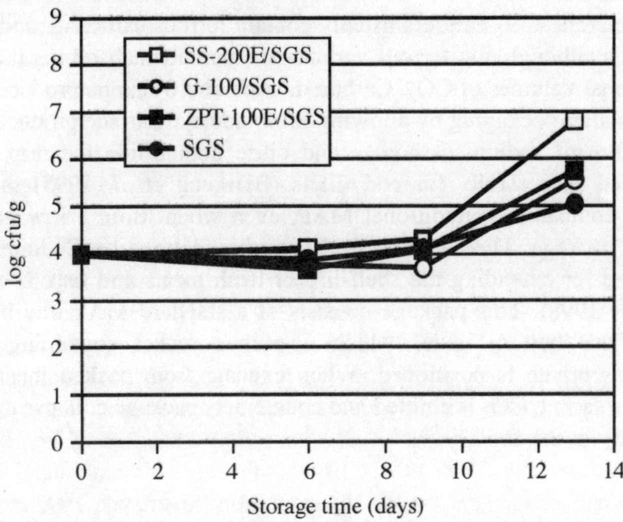

Fig. 18.1 Effects of different oxygen absorbers on the growth of psychotrophic bacteria in salmon fillets stored at 1ºC a) packaged in air or b) SGS treatment with CO_2 prior to packaging ❑ = Ageless SS oxygen absorber for high a_w foods; ■ = Ageless ZPT oxygen absorber for dry foods; ○ = Ageless G combined oxygen absorber and carbon dioxide emiiter; and ● = package without absorber (Sivertsvik *et al.* 1999)

observed in packages combining SGS with the combined O_2 absorber and CO_2 emitter (Agelss G-100) i.e., the packages with most CO_2 inside the package. The fastest microbial growth was observed in salmon stored in air without absorbers and in air with Ageless SS-200 and ZPT-100 O_2-absorbers. These samples were microbiologically spoiled after 13 days of storage. The effects of the absorbers and packaging method were not significant on the sensory evaluation scores but multiple comparisons confirmed the findings of the microbiological analyses. The SGS samples were evaluated as better compared to air samples on cooked flavour, cooked odour and texture, but got slightly lower scores on raw odour evaluation. Samples packaged with G-100 absorber/emitter gave the best cooked sensory scores, while samples without absorber got the lowest cooked sensory scores. On raw odour the samples packaged with G-100 and without absorbers were evaluated as better than samples with SS-200 and ZPT-100 absorbers. No differences were observed in the colour of the samples, in contrast to the reddish colour change observed when packaging perch and pike perch fillets with the Ageless G-100 CO_2-emitters (Ahvenainen *et al.*, 1997). They observed the same shelf-life for fresh perch and pike perch fillets packaged with G-100 as for traditional MAP using an anoxic high CO_2 atmosphere, and 2–4 days longer shelf-life when compared with over-wrap or vacuum packaging. However, colour change and a smell of raw liver in the raw fillets in the active packages was observed. This was not observed in the traditional MA-packages. No differences between the two packaging technologies were observed after cooking of the fish.

The commercial CO_2 emitters usually contain ferrous carbonate and a metal halide catalyst although non-ferrous variants are available, absorbing the O_2 and producing equal volumes of CO_2. Carbon dioxide could also be produced inside the packages after packaging by allowing the exudates from the product to react with a mixture of sodium carbonate and citric acid inside the drip pad, an approach used successfully for cod fillets (Bjerkeng *et al.* 1995) increasing shelf-life as compared to traditional MAP, even when using a low gas head-space in the package. The Verifrais package manufactured by Codimer, which has been used for extending the shelf-life of fresh meats and fish, is a similar concept (Day 1998). This package consists of a standard MAP tray but has a perforated false bottom under which a porous sachet containing sodium bicarbonate/ascorbate is positioned. When exudate from packed meat or fish drips onto the sachet, CO_2 is emitted and counteracts package collapse due to the CO_2 solubility in the food.

18.4 Active packaging: water control

Excess moisture is a major cause of food spoilage and different humidity absorbers are used to protect dried products from humidity damage. However, these absorbers have a limited effect on fish products. Several companies manufacture moisture drip absorbent pads, sheets and blankets for liquid water control in watery foods such as meat, fish, poultry, fruit and vegetables. Moisture drip absorber pads or false-bottomed trays are commonly placed under packaged fresh meat, fish, poultry and prepared fruit to absorb unsightly tissue drip discharge. Larger sheets and blankets are used for absorption of melted ice from chilled seafood during airfreight transportation (Day 1998). Commercial moisture absorber sheets, blankets and trays include Toppan Sheet (Toppan Printing, Japan), Thermarite (Thermarite, Australia) and Fresh-R-Pax (Maxwell Chase, US).

An approach to extending shelf-life of chilled fresh fish is to decrease the water activity at the surface. The Showa Denko Co. (Tokyo, Japan) has developed a film (Pichit film), which is in the form of a pillow with entrapped propylene glycol between layers of polyvinyl alcohol (PVA). The PVA-film is very permeable to water but is a barrier to the glycol. When placed in contact for several hours with the surface of meat or fish by wrapping the film around, it absorbs water and causes injury to spoilage bacteria. This technique can increase the shelf-life of ocean fish by 2–4 days (Labuza 1993). The action is due to an a_w difference between the fish (0.99) and the glycol (0.0); thus the water is rapidly drawn out of the fish surface. This surface dehydration not only inhibits some microbes but also may injure others without causing change in fish quality (Labuza and Breene 1989). It is most likely that some glycol also transfers to the food surface and slows microbial growth. The pichit film has been shown to maintain the colour of tuna, veal, pork and beef (Arakawa *et al.* 1990), since the colour in these products is related to the myoglobin content in the meat and a dehydration of the surface will lead to increased myoglobin concentration. The effect of the pichit film on the shelf-life of fish has been little exploited but for salmon fillets the effect of 2 and 4 hours of pichit pre-treatment on microbial growth and sensory spoilage was non-existent (Sivertsvik 2000).

18.5 Active packaging: anti-microbial and anti-oxidant applications

Some commercial anti-microbial films and materials have been introduced, again primarily in Japan. For example, one widely reported product is a synthetic silver zeolite which has been directly incorporated into food contact packaging film. The major potential food applications for anti-microbial films include meat, fish, bread, cheese, fruit and vegetables. Several antimicrobial compounds might have potential to be incorporated into package structures to convert them into active packaging: chlorine dioxide, silver salts, bacteriocins, ozone, and natural spices such as rosemary and its derivatives (Brody 2001) but few have been investigated to be used in or on packaging material of fish products. One exception is benzoic acid anhydride on PE-film used on fish fillets (Han 2000).

One anti-microbial packaging application used commercially for semi-moist and dried fish products in Japan uses ethanol emitters (e.g. Ethicap, Antimold 102 and Negamold (Freund Industrial), Oitech (Nippon Kayaku), ET Pack (Ueno Seiyaku) and Ageless type SE (Mitsubishi Gas Chemical)) (Day 1998). These films and sachets contain absorbed ethanol in a carrier material that allows the controlled release of ethanol vapour.

Essential oils have anti-microbial effects and oils of oregano and cinnamon have the strongest antimicrobial activity, followed by lemongrass, thyme, clove, bay, marjoram, sage and basil oils (Mejlholm and Dalgaard 2002). Oregano oil (0.05% v/w) reduced growth of *P. phosphoreum*, the SSO in naturally contaminated MAP cod fillets and extended shelf-life from 11–12 days to 21–26 days at 2ºC (Mejlholm and Dalgaard 2002). Obviously, essential oils can extend the shelf-life of MAP seafood but because of the volatile nature of these components incorporating them into an active packaging could be a challenge.

Another component with potential as an active packaging ingredient for fresh fish is acetate buffer that can extend the shelf-life of MAP packaged cod fillets by spraying it onto the fillets prior to packaging (Boskou and Debevere 2000). Production of total volatile bases and TMA was inhibited in treated fillets for ten days' storage under modified atmospheres. Inhibition of TMA production could be attributed to growth inhibition of H_2S-producing bacteria, inhibition of the TMAO-dependent metabolism of TMAO-reducing bacteria and the stable pH during storage. The shelf-life, at 7ºC, of treated cod fillets, based on cooked flavour score, was almost 12 days, approximately 8 days more than the shelf-life of the control fillets (Boskou and Debevere 2000). Potassium sorbate is another preservative shown to increase shelf-life of fish products (Fey and Regenstein 1982; Drosinos *et al.*, 1997) and could be an active ingredient in packaging materials for fishery products.

Incorporating anti-oxidants, such as vitamin C and E, in packaging film may potentially reduce oxidative reactions such as the development of rancid flavour and odour in fatty fish products. The degradation of texture, flavour, and odour of stored seafood is attributed to the oxidation of unsaturated lipids. Processing operations such as salting, cooking, and mincing promote oxidation while smoking, dehydration, and freezing retard oxidation. The rate and degree of lipid degradation in frozen fish depends upon the fish species and muscle type, dark or white. Lipid oxidation proceeds in the following decreasing order: skin, dark muscle, and white muscle. Lipid oxidation within a given species will vary with season and location within the tissue. Metal ions affect oxidation in the following decreasing order: Fe2+, haemin, Cu2+, and Fe3+. Oxidation can be reduced through the use of single or combined antioxidants, however, vacuum packaging has a greater reduction on oxidation than the presence of additives (Flick *et al.*, 1992).

Lipid oxidation in fish fillets wrapped with butylated hydroxytoluene (BHT) anti-oxidant incorporated PE-film was inhibited as compared to non-wrapped fish fillets (Huang and Weng 1998). The BHT-incorporated PE film was able to inhibit lipid oxidation in both fish muscle and oil.

18.6 Active packaging: edible coatings and films

Edible films can be looked upon as an active packaging technology, and many of the anti-microbials or anti-oxidants mentioned above incorporated in the packaging material could as well be incorporated in an edible film meant to be eaten together with the product. An edible film might be a better approach to ensure good contact between the active component and the food.

Different edible coatings have been developed to be used on fish products. Methyl cellulose and hydroxypropyl cellulose reduce uptake of fat during frying, important to the preparation of many seafood products. Alginates reduce moisture loss from fresh fish, while palmitates reduce moisture loss from frozen fish. Other edible films demonstrated to reduce moisture transfer, especially out of frozen fish, include whey protein isolates, coconut oil, and acetylated mono and diglycerides (Brody 2001).

The chitosan coating of fresh fillets of cod (*Gadus morhua*) and herring (*Clupea harengus*) reduced moisture losses, lipid oxidation, headspace volatiles (total volatile basic nitrogen, TMA, and hypoxanthine), and growth of microorganisms as compared to uncoated samples. The preservative efficacy and the viscosity of chitosan were interrelated, the efficacy of chitosans with higher viscosities superior to that of lower viscosity. Thus, chitosan as an edible coating could enhance the quality of seafoods during storage (Jeon *et al.*, 2002). Skinless tilapia (*Dreochromis niloticus x D. aureus*) fillets were covered with a gelatin coating containing benzoic acid as an antimicrobial agent. After seven days of storage under refrigeration, tilapia fillets coated with gelatin containing benzoic acid had acceptable VBN contents, increased moderately in microbial loads, and showed no significant sensory difference ($P < 0.05$) from fresh fillets. The results indicate that an antimicrobial gelatin coating is suitable for preservation of tilapia fillets (Ou *et al.*, 2002). Glucose oxidase in a alginate coating extended shelf-life of winter flounder as compared to coating without enzyme (Field *et al.*, 1986).

18.7 Active packaging: taint removal

During storage of packaged fish microbial metabolites and protein breakdown products, such as amines and aldehydes, accumulate in the head-space of the package, leading to, for example, putrid (H_2S) and fishy (TMA) odours. Removal of these components would therefore often enhance the initial perception of the products upon package opening, and also to some degree increase sensory shelf-life.

The effectiveness of an innovative foam plastic tray provided with absorbers for volatile amines and liquids on the shelf- life of different fish products packed under a modified atmosphere (40%CO_2:60%N_2), was evaluated in comparison with a standard tray (Franzetti *et al.*, 2001). Fillets of sole (*Solea solea*), steaks of hake (*Merluccius merluccius*) and whole cuttlefish (*Sepia fillouxi*), placed in the two different kinds of tray, were kept at 3°C. The novel packaging associated with a rigorous control of storage temperature, increased the shelf-life up to ten days. In fact, the innovative tray sequestrated the greater part of TMA from the headspace, and led to delayed microbial growth, especially of Gram-negative and H_2S- producing bacteria. In addition it favoured the growth of bacterial strains such as *Moraxella phenylpiruvica* which are not involved in off-flavouring production because of their lypolitic activity (Franzetti *et al.*, 2001).

A Japanese patent based on the interactions between acidic compounds (e.g. citric acid) incorporated in polymers and off-odours claims amine-removing capabilities (Vermeiren *et al.*, 1999). Another approach to remove amine odours has been provided by the Anico Co. (Japan). The Anico bags made from a film containing ferrous salt and an organic acid such as citric or ascorbic acid are claimed to oxidise the amine or other oxidisable odour-causing compounds as they are absorbed by the polymer (Vermeiren *et al.*, 1999).

Some commercialised odour-absorbing sachets, e.g. MINIPAX1 and STRIPPAX1 (Multisorb technologies, USA) absorb the odours developing in certain packaged foods during distribution due to the formation of mercaptans and H_2S (Vermeiren *et al.*, 1999). 2-in-1, from United Desiccants (USA), is a combination of silica gel and activated carbon packaged together for use in controlling moisture, gas and odour within packaged products. Profresh1 is claimed to be a freshness keeping and malodour control masterbatch used in packaging materials (PE and PS). The active component ADI50 (composition not revealed) is claimed to absorb ethylene, ethyl alcohol, ethyl acetate and H_2S. Whether these commercial odour absorbers are used and feasible for fish products is not known.

18.8 Intelligent packaging applications

Smart packaging, such as Time Temperature Indicators (TTI) is a technology that appears to have a potential, especially with chill-stored products under anaerobic conditions where microbial safety is not otherwise ensured. Strict temperature control is necessary to ensure microbial safety, and temperature abuse should be avoided both because of safety issues and shortening of shelf-life. TTIs could be applied to monitor the temperature and to detect temperature abused packages (Labuza *et al.*, 1992; Labuza 1993; Labuza 1996). Otwell (1997) evaluated the enzyme based Vitsab (Cox Technologies, Belmont, NC) TTI for use in MAP of seafood. Results demonstrated that the colour change from enzyme-based Vitsab labels correlated to spoilage of packaged salmon and other fish. The labels changed colour before formation of botulism toxin. Furthermore, the Vitsab indicator reflected the condition of the contents within the package. (Otwell 1997). Good correlation between colour change in different commercial TTIs on microbial growth/spoilage in smoked rainbow trout has also been found by Hurme and Smolander 2002. However, not such good correlation was found. , between colour change and sensory spoilage.

A systematic approach for fish shelf-life modelling and TTI selection in order to plan and apply an effective quality monitoring scheme for the fish chill chain was developed by Taoukis *et al.*, 1999, who modelled the growth of the SSOs of the Mediterranean fish boque (*Boops boops*).

Cox Technologies has also developed a spoilage indicator, FreshTag, meant to be affixed to the package surfaces. This indicator rapidly detects, measures and signals the presence of decomposition volatile bases such as ammonia, TMA, and dimethylamine in the headspace of the package (Kruijf *et al.*, 2002). Results have been promising for shrimp, scallops, and coldwater fresh fish but not for fatty fish such as salmon or tuna (Brody 2001). The reason for the latter is probably because the shelf-life of fatty fish is not only limited by microbial growth alone, but also by oxidative reactions (Sivertsvik *et al.*, 2002a).

Related to the spoilage indicator is the Toxin Alert indicator (Toxin Alert, Mississauga, Canada). This indicator can possibly detect the growth of pathogenic microorganisms in real time without waiting (Brody 2001).

18.9 Future trends

It is useful to distinguish between two categories of packaged fishery products. Those eaten without any heat treatment immediately prior to consumption, such as ready to eat products, sashimi/sushi, smoked salmon, cooked shellfish, and those products that will be subjected to heat treatment sufficient to kill all vegetative pathogens before serving (e.g. most fresh fish). Safety concerns regarding pathogenic microorganisms are of primary importance and deserve first priority during manufacture of fishery products, and the first category in particular.

Fish and shellfish are vehicles for transmission of foodborne diseases (Huss *et al.*, 1997). Pathogens found on fish are either naturally present in the fish from the aquatic environment (*Clostridium botulinum* type E and non-proteolytic types B and F; pathogenic *Vibrio* spp.; *Aeromonas hydrophila*; *Plesiomonas shigelloides*), or frequently present (*Listeria monocytogenes*, *C. botulinum* proteolytic types A, B; *C. perfringens; Bacillus* spp.), or originating from the animal/human reservoir (*Salmonella, Shigella, Escherichia coli, Staphylococcus aureus*). The pathogens of major concerns when packaging fish under anaerobic conditions have been and still are *C. botulinum* type E and non-proteolytic type B and *Listeria monocytogenes* i.e. those able to grow and multiply at chilled temperatures and that are more or less unaffected by CO_2 atmospheres (Sivertsvik *et al.*, 2002a).

Many of the above-mentioned active packaging technologies do not improve microbial safety of the products above that obtained by traditional MAP. O_2 scavengers and CO_2 emitters give little or no additional shelf-life to fresh fishery products compared to MAP and vacuum packaging, but the technologies could in some cases replace the traditional packaging technologies. Different anti-microbial components can extend shelf-life through inhibiting spoilage organisms. However, the inhibition of the spoilage bacteria reduces also bacterial competition that may permit growth and toxin production by non-proteolytic *C. botulinum* or growth of other pathogenic bacteria. The anti-microbial packaging should therefore also be able to inhibit growth of pathogens as well as the specific spoilage organisms to ensure microbial safe products. The risks from botulism in MAP fish have been widely reviewed (Sivertsvik *et al.*, 2002a) and even if the results are not conclusive, there is a potential threat for a packaged fish product to become toxic prior to spoilage especially at storage temperature of 8ºC or above.

A mathematical model has been developed for prediction of lag time prior to *C. botulinum* toxigenesis (Baker and Genigeorgis 1990). The model revealed that about 75% of experimental variation was explained by storage temperature while the size of the *C. botulinum* spore inoculum explained 7.5%. Collectively factors such as MA compositions, type of fish and *C. botulinum* spore type attributed to only 2.3% of the variation. The data confirms the importance of temperature control. Under temperature abuse conditions most fish, independent of packaging method, can become toxic. TTIs and possibly toxin indicators are therefore maybe among the few methods that can be used to ensure the safety of fresh fishery products. For processed or prepared fish products changes in the a_W, pH, salt, or heat-treatment is used to control the threat, and hopefully there will be active packaging technologies that could do the same without adversely changing the flavour, odour, colour and texture of a fresh fish product.

Many of the active packaging technologies increase shelf-life of fish only marginally and usually not the initial prime quality. A better effect can be obtained by increasing the raw material quality, i.e. reducing the initial spoilage counts or lowering the storage temperature by one or two degrees. Combining active packaging with superchilling (sub-zero (−1 to −2ºC) storage) might reveal synergistic effects as observed for MAP combined with superchilling on the quality of salmon fillets (Sivertsvik *et al.*, 2003). Superchilling is one of the few preservation technologies able to extend the prime quality phase of fresh fish (Haard 1992), a phase limited by autolytic enzyme activity and not microbial growth. A technology able to control and prolong this phase should be preferred. However, none of the existing active packaging technologies is able to do this today for fresh fish. Active packaging has therefore a greater potential to be a success for fish products with added value, for example, fish based ready meals. This is also possibly the segment with highest growth potential. Adding value to the raw material will drive increased seafood consumption. Value is not merely in gutting, trimming, and cleaning but in striving for ready-to-eat or ready-to-heat-and-eat with all the accompaniments of sauces and toppings (Brody 2001). These innovations will have to be supported by packaging that incorporates convenience; just as is offered by many minimally processed prepared foods.

18.10 References

AHVENAINEN, R., LYIJYNEN, T., RANDELL, K., SKYTTA, E., HURME, E. and HATTULA, T. 1997. The shelf-life of perch and pike-perch fillets in different ratail packages. Poster presented at the 3rd Nordfood conference 'Sustainable food production and competitive industries', 23–25 November 1997, Copenhagen, Denmark.

ARAKAWA, N., CHU, Y.-J., OTSUKA, M., KOTOKU, I. and TAKUNO, M. 1990. Comparative study on the color of animal and tuna muscle after using a contact-dehydrating sheet. *Home Econ. Jpn.* **41:** 947–50.

ASHIE, I.N., SMITH, J.P. and SIMPSON, B.K. 1996. Spoilage and shelf-life extension of fresh fish and shellfish. *Crit. Rev. Food Sci. Nutr.* **36:** 87–121.

BAKER, D.A. AND GENIGEORGIS, C. 1990. Predicting the safe storage of fresh fish under modified atmospheres with respect to *Clostridium botulinum* toxigenesis by modelling length of the lag phase of growth. *J. Food Prot.* **53:** 131–40, 153.

BJERKENG, B., SIVERTSVIK, M., ROSNES, J.T. and BERGSLIEN, H. 1995. Reducing package deformation and increasing filling degree in packages of cod fillets in CO_2-enriched atmospheres by adding sodium carbonate and citric acid to an exudate absorber. In *Foods and Packaging materials – Chemical Interactions*. Edited by P. Ackermann, M. Jägerstad, and T.Ohlsson. The Royal Society of Chemistry, Cambridge, UK. pp. 222–7.

BOSKOU, G. AND DEBEVERE, J. 2000. Shelf-life extension of cod fillets with an acetate buffer spray prior to packaging under modified atmospheres. *Food Addit. and Contam.* **17:** 17–25.

BRODY, A.L. 2001. Is something fishy about packaging? *Food Technol.* **55:** 97–98.

COLBY, J.-W., ENRIQUEZ-IBARRA, L. and FLICK, G.J. JR. 1993. Shelf life of fish and shellfish. In *Shelf life studies of foods and beverages – Chemical, biological, physical and nutritional aspects*. Edited by G.Charalambous. Elesevier Science Publishers B.V., Amsterdam, The Netherlands. pp. 85–143.

DALGAARD, P. 1995. Qualitative and quantitative characterization of spoilage bacteria from packed fish. *Int. J. Food Microbiol.* **26:** 319–33.

DALGAARD, P., MEJLHOLM, O. and HUSS, H.H. 1996. Conductance method for quantitative determination of *Photobacterium phosphoreum* in fish products. *J. Appl. Bact.* **81:** 57–64.

DANIELS, J.A., KRISHNAMURTHI, R. and RIZVI, S.S.H. 1985. A review of effects of carbon dioxide on microbial growth and food quality. *J. Food Prot.* **48:** 532–7.

DAY, B. P. F. *Active packaging of foods.* 1998. Chipping Campden, Glos., UK, Campden and Chorleywood Food Research Association. New technologies bulletin, No. 17, pp. 1–12.

DIXON, N.M. and KELL, D.B. 1989. The inhibition of CO_2 of the growth and metabolism of micro-organisms. *J. Appl. Bact.* **67:** 109–36.

DROSINOS, E.H., LAMBROPOULOU, K., MITRE, E. and NYCHAS, G.J.E. 1997. Attributes of fresh gilt-head seabream (*Sparus aurata*) fillets treated with potassium sorbate, sodium gluconate and stored under a modified atmosphere at 0 +/- 1 degrees C. *J. Appl. Microbiol.* **83:** 569–75.

EMBORG, J., LAURSEN, B.G., RATHJEN, T. and DALGAARD, P. 2002. Microbial spoilage and formation of biogenic amines in fresh and thawed modified atmosphere-packed salmon (*Salmo salar*) at 2 degrees C. *J. Appl. Microbiol.* **92:** 790–9.

FARBER, J.M. 1991. Microbiological aspects of modified-atmosphere packaging technology – a review. *J. Food Prot.* **54:** 58–70.

FEY, M.S. AND REGENSTEIN, J.M. 1982. Extending shelf-life of fresh wet red hake and salmon using CO_2-O_2 modified atmosphere and potassium sorbate ice at 1ºC. *J. Food Sci.* **47:** 1048–54.

FIELD, C.E., PIVARNIK, L.F., BARNETT, S.M. and RAND, A.G. 1986. Utilization of Glucose Oxidase for extending the shelf-life of fish. *J. Food Sci.* **51:** 66.

FLICK, G.J., HONG, G.P. and KNOBL, G.M. 1992. *Lipid Oxidation of Seafood During Storage*. Acs Symposium Series **500:** 183–207.

FRANZETTI, L., MARTINOLI, S., PIERGIOVANNI, L. and GALLI, A. 2001. Influence of active packaging on the shelf-life of minimally processed fish products in a modified atmosphere. *Packaging Technol. Sci.* **14:** 267–74.

GRAM, L. and HUSS, H.H. 1996. Microbiological spoilage of fish and fish products. *Int. J. Food Microbiol.* **33:** 121–37.

GRAM, L., TROLLE, G. and HUSS, H.H. 1987. Detection of specific spoilage bacteria from fish stored at low (0ºC) and high (20ºC) temperatures. *Int. J. Food Microbiol.* **4:** 65–72.

GULDAGER, H.S., BØKNÆS, N., ØSTERBERG, C., NIELSEN, J. and DALGAARD, P. 1998. Thawed Cod Fillets Spoil Less Rapidly Than Unfrozen Fillets When Stored under Modified Atmosphere at 2ºC. *J. Food Prot.* **61:** 1129–36.

HAARD, N.F. 1992. Technological Aspects of Extending Prime Quality of Seafood: A Review. *J. Aqua. Food Prod. Tech.* **1:** 9–27.

HAN, J.H. 2000. Antimicrobial food packaging. *Food Technology* **54:** 56–65.

HUANG, C.H. AND WENG, Y.M. 1998. Inhibition of lipid oxidation in fish muscle by antioxidant incorporated polyethylene film. *J. Food Proc. Preserv.* **22:** 199–209.

HURME, E. and SMOLANDER, M. 2002. Intelligent systems in food packaging: effectiveness and advances in the supply chain. Paper presented at 3rd Nordic Foodpack Seminar, September 4–6, KCL, Espoo, Finland.

HUSS, H.H. 1995. *Quality and quality changes in fresh fish*. FAO, Rome.

HUSS, H.H., DALGAARD, P. and GRAM, L. 1997. Microbiology of fish and fish products. In *Seafood from producer to consumer, integrated approach to quality*. Edited by J.B. Luten, T. Børresen, and J. Oehlenschläger. Elsevier Science B.V., Amsterdam, The Netherlands. pp. 413–30.

JEON, Y.J., KAMIL, J.Y.V.A. and SHAHIDI, F. 2002. Chitosan as an edible invisible film for quality preservation of herring and Atlantic cod. *J. Agric. Food Chem.* **50:** 5167–78.

KRUIJF, N.D., BEEST, M.V., RIJK, R., SIPILÄINEN-MALM, T., LOSADA, P.P. and MEULENAER, B.D. 2002. Active and intelligent packaging: applications and regulatory aspects. *Food Addit. Contam.* 19 (Supplement): 144–62.

LABUZA, T.P. 1993. Active packaging technologies for improved shelf life and quality. In *Science for the Food Industry of the 21st Century, Biotechnology, supercritical fluids, membranes and other advanced technologies for low calorie, healthy food alternatives*. Edited by M.Yalpani. ATL Press, Inc., Mount Prospect, Il. pp. 265–84.

LABUZA, T.P. 1996. An Introduction to active packaging for foods. *Food Technology* **50:** 68–71.

LABUZA, T.P. and BREENE, W.M. 1989. Applications of 'active packaging' for improvement of shelf-life and nutritional quality of fresh and extended shelf-life foods. *J. Food Proc. Preserv.* **13:** 1–69.

LABUZA, T.P., FU, B. and TAOUKIS, P.S. 1992. Prediction for shelf life and safety of minimally processed CAP/MAP chilled foods: a review. *J. Food Prot.* **55:** 741–50.

LISTON, J. 1980. Microbiology in fisheries science. In *Advances in fish science and technology*. Edited by J.J. Connell. Fishing News Books ltd, Farnham, Surrey, UK. pp. 138–57.

MEJLHOLM, O. and DALGAARD, P. 2002. Antimicrobial effect of essential oils on the seafood spoilage micro-organism Photobacterium phosphoreum in liquid media and fish products. *Lett. Appl. Microbiol.* **34:** 27–31.

OTWELL, S. 1997. Time, Temperature, Travel – a quality balancing act. *Seafood Int.* (Sept) 57–61.

OU, C.Y., TSAY, S.F., LAI, C.H. and WENG, Y.M. 2002. Using gelatin-based antimicrobial edible coating to prolong shelf-life of tilapia fillets. *J. Food Qual.* **25:** 213–22.

PARKIN, K.L. and BROWN, W.D. 1982. Preservation of seafood with modified atmospheres. In *Chemistry & Biochemistry of Marine Food Products.* Edited by R.E. Martin, G.J. Flick, C.E. Hebard, and D.R. Ward. AVI Publishing Company, Westport, Connecticut. pp. 453–65.

SIVERTSVIK, M. 1997. Active packaging of seafood – an evaluation of the use of oxygen absorbers in packages of different seafood products. Paper presented at 27th WEFTA Conference, October 19–22, 1997, Instituto del frio (CSIC), Madrid, Spain.

SIVERTSVIK, M. 1999. Use of O_2-absorbers and soluble gas stabilisation to extend shelf-life of fish products, Nordic Foodpack Symposium, 31 May–2 June 1999, Dansk Teknologisk Insitutt, HøjeTåstrup, Danmark.

SIVERTSVIK, M. 2000. Use of soluble gas stabilisation to extend shelf-life of fish. In *Proceedings of 29th WEFTA-meeting*, October 10–14, 1999, Leptocarya, Pieria, Greece. *Edited by* S.A.Georgakis. Greek Society of Food Hygienists and Technologists, Thessaloniki, Greece. pp. 79–91.

SIVERTSVIK, M., ROSNES, J.T., VORRE, AA., RANDELL, K., AHVENAINEN, R. and BERGSLIEN, H. 1999. Quality of whole gutted salmon in various bulk packages. *J. Food Quality* **22:** 387–401.

SIVERTSVIK, M., JEKSRUD, W.K. and ROSNES, J.T. 2002a. A review of modified atmosphere packaging of fish and fishery products – significance of microbial growth, activities and safety. *Int. J. Food Sci. & Technol.* **37:** 107–27.

SIVERTSVIK, M., ROSNES, J.T. and BERGSLIEN, H. 2002b. Modified atmosphere packaging. In *Minimal processing technologies in the food industry.* Edited by T. Ohlsson and N. Bengtsson. Woodhead Publishing Ltd, Cambridge, UK. pp. 61–86.

SIVERTSVIK, M., ROSNES, J.T. and KLEIBERG, G.H. 2003. Effect of modified atmosphere packaging and superchilled storage on the microbial and sensory quality of Atlantic salmon (*Salmo salar*) fillets. *J. Food Sci.* (in press).

TAOUKIS, P.S., KOUTSOUMANIS, K. and NYCHAS, G.J.E. 1999. Use of time-temperature integrators and predictive modelling for shelf life control of chilled fish under dynamic storage conditions. *Int. J. Food Microbiol.* **53:** 21–31.

VERMEIREN, L., DEVLIEGHERE, F., BEEST, M.V., KRUIJF, N.D. and DEBEVERE, J. 1999. Developments in the active packaging of foods. *Trends Food Sci. Technol.* **10:** 77–86.

19

Active packaging and colour control: the case of meat

M. Jakobsen and G. Bertelsen, The Royal Veterinary and Agricultural University, Denmark

19.1 Introduction

The colour stability of meat products is influenced by a large number of factors: some of biochemical nature, some due to handling during slaughter and processing, and others due to packaging and storage conditions. This chapter focuses on modelling how colour shelf-life is affected by the external factors applied during packaging and storage. However, meat from different sources shows different tendencies to undergo colour deterioration and this variation in internal factors influences the developed models. Therefore some consideration will also be given to discussing how internal factors like, e.g., muscle type and addition of nitrite in cured meat affect the models. Modelling can be used to identify the most important factors/interaction of factors affecting quality loss during storage and to define critical levels of these factors, thereby forming the basis for proposing the optimal packaging and storage conditions or the best compromise if several deteriorative reactions need to be considered. Caution in choosing the optimal packaging and storage conditions can largely improve the colour shelf-life of meat products.

Modelling of MAP systems shows great potential for optimising/tailoring storage and packaging parameters to maintain product quality, in this case a good meat colour stability (Jakobsen and Bertelsen, 2000; Lyijynen *et al.*, 1998; Pfeiffer *et al.*, 1999). It will be demonstrated how modelling can be used to identify the most important factors affecting colour shelf-life. Multivariable experimental design is necessary to be able to investigate the large number of influencing factors on several levels as well as the interactions between factors.

19.2 Packaging and storage factors affecting colour stability

Modified atmosphere packed meat is a complex and dynamic system where several factors interact (Zhao et al., 1994). Models can be used to describe how the initial package atmosphere changes over time and how these changes affect product quality and shelf-life. The dynamic changes in headspace gas composition during storage can be modelled as a function of:

- gas transmission rates of the packaging material
- initial gas composition
- product and package geometry
- gas absorption in the meat.

The knowledge of changes in gas composition can be combined with models on quality changes in the meat as a function of packaging and storage conditions such as:

- storage time
- storage temperature
- gas composition
- light exposure

and predictions of product shelf-life can be made. Pfeiffer *et al.* (1999) developed simulations of how product shelf-life changes with different packaging and storage conditions for a wide range of food products (primarily dry products). However, at present sufficient models for many quality deteriorative reactions are lacking and only few attempts have been made to model chemical quality changes in meat products, in contrast to modelling of microbial shelf-life, where extensive work has been performed (McDonald and Sun, 1999).

19.2.1 Modelling dynamic changes in headspace gas composition

Permeability of the packaging film

Headspace gas composition changes dynamically due to several factors. Gas exchange with the environment occurs over the packaging film if the partial pressure of a gas differs on the two sides of the film. The amount of gas that permeates over the film can be calculated from equation 19.1 (Robertson, 1993):

$$Q = P \cdot \Delta p \cdot t \cdot A \qquad 19.1$$

Q = the amount of gas that permeates over the film (cm^3)
P = the permeability of the packaging film ($cm^3/m^2/24h/atm$)
Δp = the difference in gas partial pressure on the two sides of the film (atm)
t = the storage time (24h)
A = the area of the package (m^2)

Different gases have different permeability through the same film. For conventional films, the permeability of CO_2 is generally 4–6 times larger than that of O_2 and 12–18 times larger than that of N_2. The permeability of a plastic film is roughly proportional to the thickness of the film. Doubling film thickness approximately halves film permeability.

Permeability is also influenced by storage temperature and relative humidity. Pfeiffer *et al.* (1999) found that the empirical equation 19.2 fitted well with literature data for oxygen permeability.

$$P(T, RH) = \exp(c_0 + c_1/T + c_2 \cdot RH + c_3 \cdot RH^2) \qquad 19.2$$

$P =$ the permeability of the packaging film
$T =$ storage temperature
$RH =$ storage relative humidity
C_{0-3} are experimental derived coefficients.

Gas exchange over the packaging film is of particular importance when the film needs to maintain a narrowly defined gas concentration as shown in the example in section 19.3.2, where the permeability of even small amounts of O_2 into a package containing a cured meat product is considered a critical packaging parameter.

Gas absorption in the meat

Headspace gas composition can also change due to gas absorption in the meat. Packaging in elevated levels of CO_2 can result in large amounts of CO_2 absorbed in the meat (Jakobsen and Bertelsen, 2002; Zhao *et al.*, 1994) and thereby large changes compared to the initially applied gas composition. Absorption of O_2 and N_2 is negligible compared to the absorption of CO_2 (Jakobsen and Bertelsen, 2002). Models for CO_2 solubility as a function of packaging and storage parameters such as product to headspace volume ratio, temperature and initial CO_2 level were developed by Zhao *et al.* (1995) and Devlieghere *et al.* (1998). Fava and Piergiovanni (1992) developed models of CO_2 solubility as a function of different compositional parameters, a_w, pH, protein, fat and moisture content. The amount of absorbed CO_2 ranges from 0–1.8 L CO_2/Kg meat, depending on the applied CO_2 partial pressure, temperature, pH of the meat, etc. (Jakobsen and Bertelsen 2002). As regards gas absorption, equilibrium is obtained during the first one or two days. Microbial or meat metabolism can also cause slight changes in gas composition by using O_2 and producing CO_2.

19.3 Modelling the impact of MAP

When it is understood how the gas atmosphere can change from the initially applied atmosphere under different packaging and storage conditions, this knowledge can be used to evaluate the effect on quality deteriorating reactions. Besides microbial growth, the primary concern when packaging both fresh meat and cured meat products is colour stability. The mechanisms of colour changes in fresh meat and cured meat products are completely different as can be seen from the examples on modelling given in the following two sections.

When packaging fresh meat products an elevated oxygen partial pressure needs to be maintained to keep the meat pigment myoglobin in its oxygenated bright red state. By modelling a MAP system for fresh beef, the most critical external factors are identified to be storage temperature and gas composition (Jakobsen and Bertelsen, 2000). By modelling a MAP system for cured meat products the most critical external factors are identified to be low availability of oxygen combined with exclusion of light to prevent degradation of nitrosylmyoglobin by photo oxidation (storage temperature was kept constant at 5ºC) (Møller *et al.*, 2003). However, low availability of oxygen is not ensured solely by reducing the residual oxygen level in the headspace during the packaging process. Other equally critical factors are a high product to headspace ratio and a packaging film of low oxygen transmission rate (OTR) of the packaging film (Møller *et al.*, 2003).

19.3.1 Optimising colour stability of fresh beef

Jakobsen and Bertelsen (2000) and Bro and Jakobsen (2002) modelled colour stability of fresh beef under different packaging and storage conditions. In all cases colour measurements were performed with a Minolta Colorimeter using the L, a, b coordinates. Red colour was expressed as the a-value, the higher the a-value the redder the sample. When packaging fresh red meats elevated O_2 partial pressures are used to stabilise myoglobin in its bright red oxygenated form (oxymyoglobin). However, elevated O_2 levels may increase some deteriorative reactions e.g. lipid oxidation. Consequently it is interesting to investigate if a level of O_2 exists that is acceptable when considering both colour stability and lipid oxidation. Jakobsen and Bertelsen (2000) investigated different packaging and storage conditions (Table 19.1) and developed a regression model/response surface model predicting the colour a-value as a function of storage time, storage temperature and O_2 level based on steaks of *Longissimus dorsi* muscles from four different animals.

The resulting model (equation 19.3) contains the main effects of the three factors plus two-way interactions and two squared effects. Interpretation of the model is best done by exploring the response surface plot (Fig. 19.1).

$$\text{a-value} = \beta_0 + \beta_1 \cdot \text{Day} + \beta_2 \cdot \text{Temp} + \beta_3 \cdot \text{O2} + \beta 4 \cdot \text{Day} \cdot \text{Temp} + \beta 5 \cdot \text{Day} \cdot \text{O2} + \beta_6 \cdot \mathit{Temp} \cdot \text{O2} + \beta_7 \text{Day} \cdot \text{Day} + \beta_8 \cdot \text{Temp} \cdot \text{Temp} \qquad 19.3$$

where the betas are regression coefficients.

Figure 19.1 shows a response surface plot varying the two factors, temperature and O_2 level, while keeping the third factor, storage time, constant at day no. 6. Figure 19.1 also reveals an interval of approximately 50–80% O_2, where the O_2 level does not affect the colour a-value significantly (the nearly

Table 19.1 Packaging and storage conditions used in the models developed in Jakobsen and Bertelsen (2000)

Modelling factor	Abbreviation	No. of levels	Setting of levels
Storage time (days)	Day	5	2, 4, 6, 8, 10
Temperature (°C)	Temp	3	2, 5, 8
O_2 level (%)	O2	5	20, 35, 50, 65, 80

horizontal lines in this interval means that the a-value is depends only on the temperature). The borders of this interval change a little depending on the setting of the day. But it is evident that the O_2 level can be reduced from the normally used 70–80% without adverse effect on the colour shelf-life.

The complexity of the interactions/squared terms in equation 19.3 called for further search for adequate models. A novel approach called GEMANOVA (Generalized Multiplicative ANOVA) was therefore used in Bro and Jakobsen (2002). In this study the effect of different packaging and storage conditions (Table 19.2) on the colour stability of steaks of *Longissimus dorsi* muscles from three different animals was investigated. The effect of light was evaluated as the time of exposure to a fluorescent tube commonly used for retail display (1000 lux at the package surface for 0, 50 or 100% of the storage time).

Even when considering only two factor interactions a traditional ANOVA

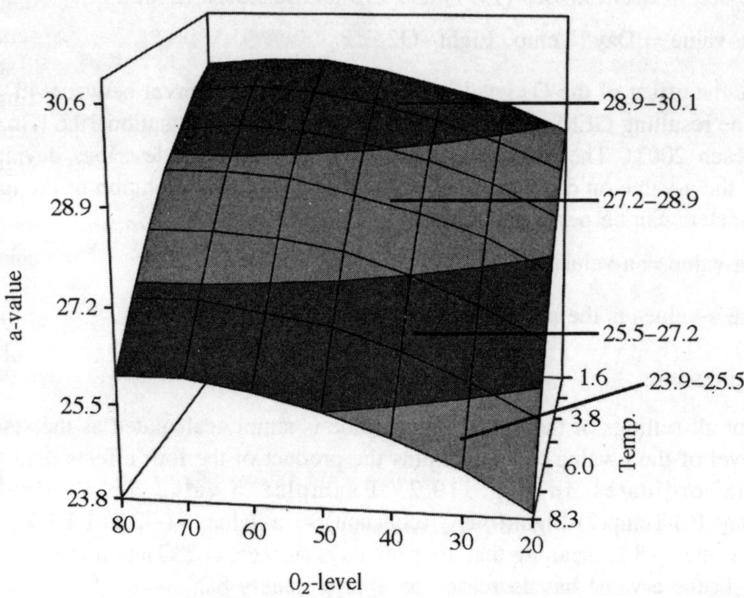

Fig. 19.1 Response surface plot of predicted a-values (average of four animals) after six days storage at different temperatures and different oxygen levels. (Adapted from Jakobsen and Bertelsen, 2000).

Table 19.2 Packaging and storage conditions used in the models developed in Bro and Jakobsen (2002)

Modelling factor	Abbreviation	No. of levels	Setting of levels
Storage time (days)	Day	4	3, 7, 8, 10
Temperature (ºC)	Temp	3	2, 5, 8
Light exposure (%)	Light	3	0, 50, 100
O_2 level (%)	O2	3	40, 60, 80

model for the experiment in Table 19.2 would look like equation 19.4 (before removal of any insignificant effects).

$$\text{a-value} = \beta_0 + \beta_1 \cdot \text{Day} + \beta_2 \cdot \text{Temp} + \beta_3 \cdot \text{Light} + \beta_4 \cdot \text{O2} + \beta_5 \cdot \text{Day} \cdot \text{Temp} + \beta_6 \cdot \text{Day} \cdot \text{Light} + \beta_7 \cdot \text{Day} \cdot \text{O2} + \beta_8 \cdot \text{Temp} \cdot \text{Light} + \beta_9 \cdot \text{Temp} \cdot \text{O2} + \beta_{10} \cdot \text{Light} \cdot \text{O2} \qquad 19.4$$

where betas are regression coefficients.

On the other hand, when applying the GEMANOVA model the interactions are modelled as one higher-order multiplicative effect, resulting in equation 19.5 (before removal of any insignificant effects). The interpretation of the GEMANOVA model is much more simple than the ANOVA model as is discussed in detail in Bro (1997) and Bro and Jakobsen (2002).

$$\text{a-value} = \text{Day} \cdot \text{Temp} \cdot \text{Light} \cdot \text{O2} \qquad 19.5$$

Since the effect of the O_2 level is insignificant in the interval between 40–80% O_2, the resulting GEMANOVA model can be written as equation 19.6 (Bro and Jakobsen 2002). The interaction term DayTempLightc_{O2} describes deviations from the a-value on day 0 in a very simple way, and interpretation of the model parameters can be performed from Fig. 19.2.

$$\text{a-value} = \text{a-value}_0 + \text{Day} \cdot \text{Temp} \cdot \text{Light} \cdot c_{O2} \qquad 19.6$$

where a-value$_0$ is the a-value at day 0 and c_{O2} is a constant.

Interpretation:

- For all settings of the factors the a-value is simply calculated as the starting level of the a-value (a-value$_0$) plus the product of the four effects read from the ordinates in Fig. 19.2. Example: a-value = a-value$_0$ + Day(10)·Temp(2)·Light(0)·c_{O2} (constant) ≈ a-value$_0$ +(−2.3)·1.1·1.7·1.9 ≈ a-value$_0$ −8.2, meaning that after ten days storage, at 2ºC and no exposure to light the a-value has decreased by approximately 8.2.
- The interaction term is 0 on day 0 (the factor Day is 0).
- All changes in colour a-value during storage are negative (colour becomes less red) compared to the starting colour. The change is calculated as the

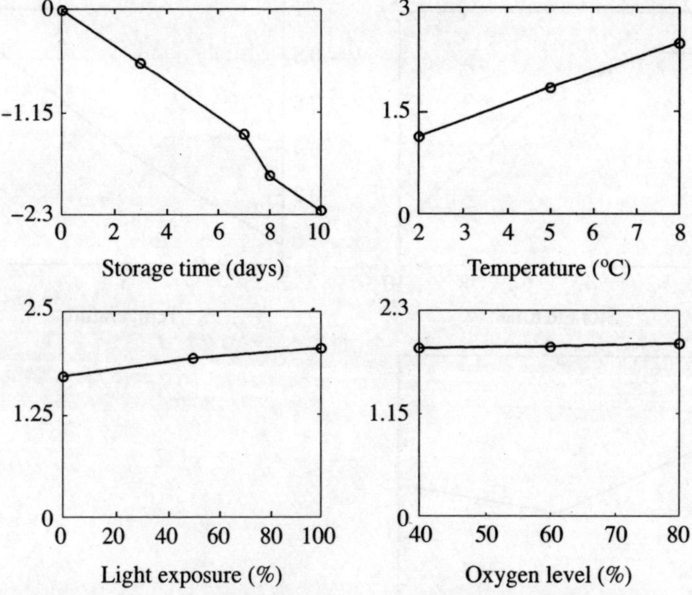

Fig. 19.2 Parameter levels for the interaction term (Day·Temp·Light·c_{O2}) in equation 19.6. (Adapted from Bro and Jakobsen, 2002).

product of the four parameters Day, Temp, Light and O_2, which consist of one negative value (Day) and three positive values.

- The changes are relative and the effect of the individual factors can be interpreted individually. For example when going from 2°C to 8°C the Temp loading increases from 1.2 to 2.4, meaning that regardless of all other factors the decrease in a-value at 8°C is twice the decrease at 2°C.
- The effects of storage time and temperature are most important.
- The effect of light is minor, although an increase in time of exposure to light seems to result in a decreased colour a-value.
- The effect of the O_2 level is insignificant in the interval from 40–80% and is therefore contained in the model as a constant (the value of the constant c_{O2} can be read from Fig. 19.2).

The GEMANOVA model confirms the results from Jakobsen and Bertelsen (2000) by emphasising the importance of keeping a low storage temperature and showing no effect of O_2 level in the interval between approximately 40–80%. However, the interpretation of the model is much more simple, since the effect of each factor can be interpreted individually. Likewise the GEMANOVA model can be applied to the data set in Table 19.1 resulting in equation 19.7 which is much more simple to interpret than equation 19.3.

$$\text{a-value} = \text{a-value}_0 + \text{Day} \cdot \text{Temp} \cdot \text{O2} \qquad 19.7$$

where a-value$_0$ is the a-value at day no. 0.

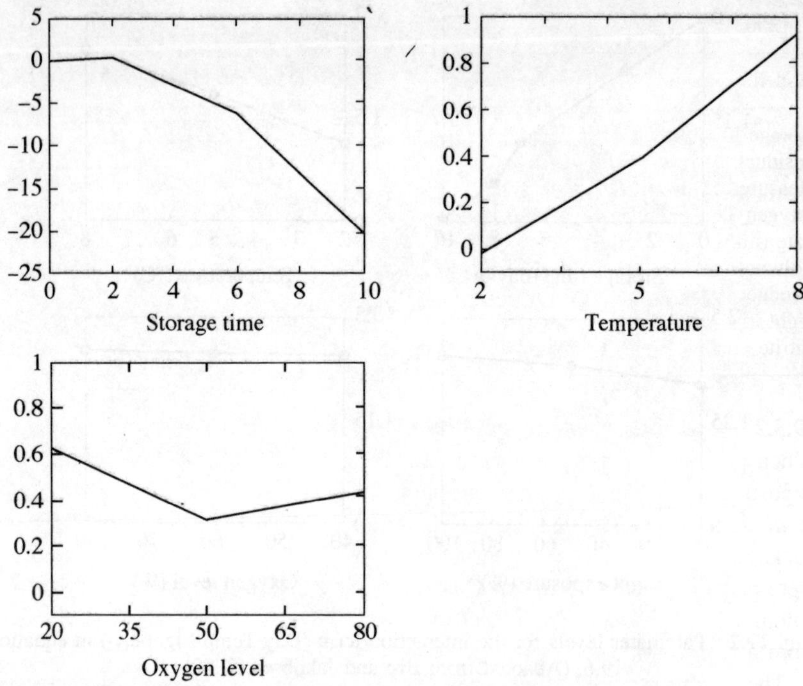

Fig. 19.3 Parameter levels for the interaction term (Day·Temp·O_2) in equation 19.7 (unpublished data).

From Fig. 19.3 the effect of the factors can be interpreted individually, and the stable interval between approximately 40–80% O_2 is evident as an elbow in the plot.

It is rather surprising that 40% O_2 is sufficient to ensure the stability of the bright red meat colour, as an O_2 level of 70–80% is commonly used in the industry. The applied product to headspace volume ratio for the experiments in Tables 19.1 and 19.2 was approximately 1:9. The large headspace volume might be the cause for only minor changes in headspace gas composition (oxygen partial pressure) taking place during storage. However, when packaging fresh meat products for retail sale, a large headspace volume is common. Furthermore, large amounts of oxygen have to permeate over the film or be used for meat/ microbial metabolism before a noteworthy change in oxygen partial pressure takes place and the meat colour will be affected.

The GEMANOVA model is an excellent tool for choosing the optimal setting for the individual factors. The knowledge of the stable interval can be used when optimising headspace gas composition or the permeability of the packaging material as a reduction in the applied oxygen level leaves the possibility of using more carbon dioxide or nitrogen in the package headspace.

Table 19.3 Packaging and storage conditions used in the models developed in Møller *et al.* (2003)

Modelling factor	Abbreviation	No. of levels	Setting of levels
Storage time (days)	Time	5	1, 3, 6, 9, 14
Residual O_2 level (%)	ResO2	3	0.1, 0.25, 0.5
Measured O_2 level (%)	MeasO2	–	Continuously
Oxygen Transmission Rate (ml/m^2/24h/atm)	OTR	3	0.5, 10, 32
Volume ratio (product to headspace)	Vol	3	1:1, 1:3, 1:5
Light intensity (Lux)	Light	2	500, 1000
Nitrite content (ppm)	Nit	2	60, 150

19.3.2 Optimising colour stability of cured ham

When packaging cured meat products it is important to keep the O_2 and light exposure at a minimum to prevent photo oxidation of nitrosylmoglobin. Møller *et al.* (2003) investigated the colour stability of cured ham under different packaging and storage conditions according to Table 19.3. Colour measurements were performed with a Minolta Colorimeter using the a-value to express the red colour of the product. The effect of light was evaluated as the light intensity from a fluorescent tube measured on the package surface.

The resulting model (after removal of insignificant effects) considering only two-factor interactions is shown in equation 19.8.

$$\text{a-value} = \beta_0 + \beta_1 \cdot \text{ResO2} + \beta_2 \cdot \text{Vol} + \beta_3 \cdot \text{Light} + \beta_4 \cdot \text{Nit} + \beta_5 \cdot \text{Time} + \beta_6 \cdot \text{MeasO2} + \beta_7 \cdot \text{Res)2} \cdot \text{Light} + \beta_8 \cdot \text{ResO2} \cdot \text{Time} + \beta_9 \cdot \text{ResO2} \cdot \text{MeasO2} + \beta_{10} \cdot \text{Vol} \cdot \text{MeasO2} + \beta_{11} \cdot \text{Light} \cdot \text{MeasO2} + \beta_{12} \cdot \text{Time} \cdot \text{MeasO2} \qquad 19.8$$

where betas are regression coefficients.

As expected, the a-value decreases with increased time, increased residual O_2 level, increased OTR, increased light intensity and decreased nitrite content. However the study also shows the importance of interactions between factors. The interaction between O_2 level and product to headspace volume ratio is especially interesting. Normally, the focus is on the residual O_2 level (%) in the package and it is commonly overlooked that also the total amount of available oxygen molecules is important. The total amount of oxygen molecules available for colour deteriorative reactions is determined by the residual oxygen level after packaging, the meat to headspace volume ratio, and the amount of oxygen that permeates into the package headspace in combination. It is not sufficient to keep a low O_2 level in the package headspace. If the headspace volume is large there will still be plenty of oxygen molecules for colour deterioration.

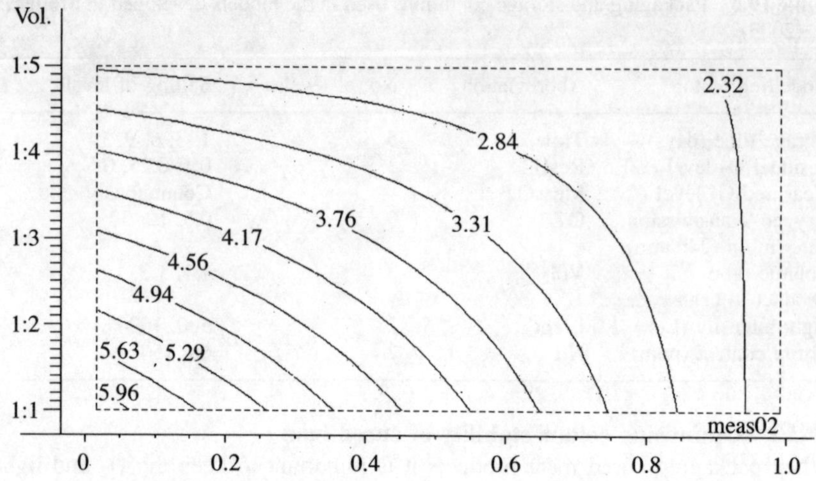

Fig. 19.4 Contour plot of the interaction effect between volume ratio (product:headspace) and measured O_2 level (%) after nine days storage. (Adapted from Møller *et al.*, 2003).

Figure 19.4 shows a contour plot of the interaction between 'measured O_2 level' and 'volume ratio' (the remaining factors are fixed to the following settings: residual O_2 level = 0.25%, light intensity = 1000 lux, nitrite = 60 ppm, storage time = 9 days).The 'measured O_2 level' is the actual O_2 level measured during storage and therefore takes into account both the 'residual O_2 level' right after packaging and the oxygen permeated into the package over the packaging material (OTR). The a-value of the product for a given combination of 'measured O_2 level' and 'volume ratio' can be found from the plot by reading the a-value from the corresponding contour line, e.g., applying 0.10% 'measured O_2 level' and a 'volume ratio' of 1:1.3 results in an a-value of 5.6 after nine days of storage. It appears that to maintain a high a-value, it is necessary to keep both the oxygen level and the headspace volume low (lower left corner of the plot), solely keeping the O_2 level low is not sufficient. The interaction between O_2 level and light intensity is also important (Møller *et al.*, 2003) but more complex. It is evident that increased light intensity accelerates colour deterioration but the degree of colour deterioration is dependent on the amount of available oxygen in the package (which again is dependent on the three factors: 'residual O_2 level', 'volume ratio' and 'OTR'). Work is in progress to clarify matters on this issue. Modelling clearly demonstrates that the product to headspace volume ratio should be given far more attention when optimising MAP of cured ham.

19.4 Pre- and post-slaughter factors

The examples in section 19.3 clearly demonstrate the usefulness of modelling for identification of the most important factors/interaction of factors affecting colour deterioration. For fresh beef it is recognised that keeping a low storage temperature is the key parameter to obtain a long colour shelf-life. In addition a wide interval of oxygen partial pressure exists that results in optimal colour stability, leaving the possibility of optimising the gas composition with respect to other quality deteriorating reactions, e.g., lipid oxidation without compromising colour stability. With respect to cured meat products it is important to realise that several factors influence the total amount of O_2 molecules available for oxidation. However, due to large biological differences between meat from different sources and to differences in handling and processing of the meat it is also important to recognise that internal factors have an effect on the developed models. The described models can be used to predict the general response of a meat product to changes in external factors, but not to predict the exact a-value for a certain piece of meat. That would require much more specific models (for each product type) and incorporation of knowledge on the internal factors into the models (Jakobsen and Bertelsen, 2000).

19.4.1 Fresh meat

Large variations in colour stability between meat of different origins can strongly influence the developed models. Different meat types show large variability due to different myoglobin content and different metabolic type (Renerre, 1990). The content of myoglobin is, e.g., largest in beef followed by lamb and pork, and the colour of pork is more stable than the two other species. Steaks of *Longissimus dorsi* muscles have high colour stability and steaks of *semi-membranosus* muscles have medium colour stability. Animals of different age, breed, feeding, etc., will also show differences in colour stability (Renerre, 1990; Jensen *et al.*, 1998).

It appears from Figs 19.5 and 19.6 that there is a huge variation in colour stability between meat from different sources. A range of intrinsic factors influences the oxidative balance in raw meat and thereby the colour stability of the meat (Bertelsen *et al.*, 2000). Thus the oxidative stability of muscles is dependent on the composition, concentrations, and reactivity of (i) oxidation substrates (lipids, protein and pigments), (ii) oxidation catalysts (prooxidants such as transition metals and various enzymes) and (iii) antioxidants, e.g., vitamin E and various enzymes. For a review see Bertelsen *et al.* (2000).

Meat from different origins shows a different tendency to undergo colour deterioration. It is therefore necessary to investigate meat from a large number of sources to be able to make general conclusions. Despite the large variations in the colour stability of meat from different animals and muscle types investigated in section 19.31 the pronounced effect of temperature and the constant interval of O_2 are common. Only the rate of colour deterioration differs.

19.4.2 Cured meat

A range of intrinsic factors affects the colour stability of nitrite cured meat products. The most important are the level of nitrite and the content of vitamin E

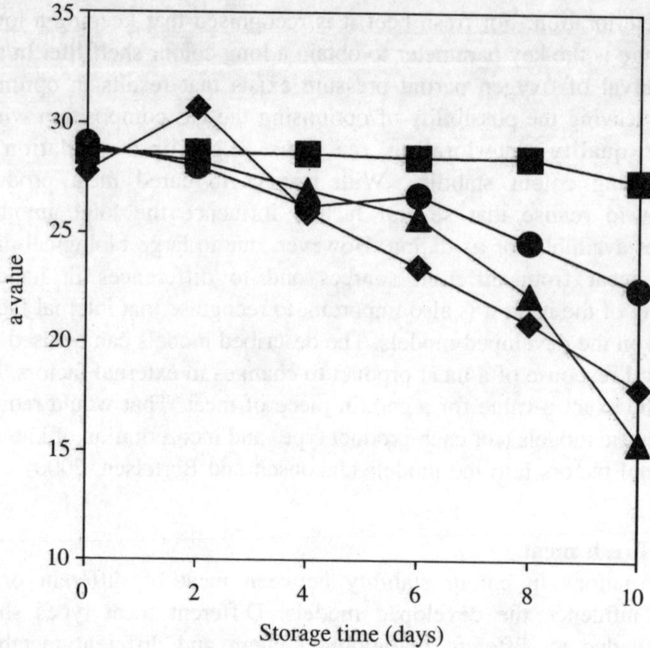

Fig. 19.5 Measured a-values for four different animals stored in 80% O_2 at 8°C. (Adapted from Jakobsen and Bertelsen, 2000).

(Weber *et al.*, 1999). Thus, optimum colour stability can be achieved only by using a multifactorial approach, where both intrinsic and extrinsic factors are considered (Bertelsen *et al.*, 2000). From Fig. 19.7 the effect of the nitrite content on the rate of colour deterioration is evident. Increasing the nitrite content improves the colour. This emphasises the necessity of investigating the specific product of interest in order to define critical levels of packaging and storage factors.

19.5 Future trends

The obvious tool for the optimisation of product shelf-life by controlling packaging and storage conditions is computer simulation. Models of changes in headspace gas composition should be combined with models describing changes in the most important quality parameters. A computer program should be given inputs on:

- permeability of the different packaging films to be compared
- storage temperature
- relative humidity during storage
- gas composition measured after packaging
- the headspace and product volumes
- light conditions during storage.

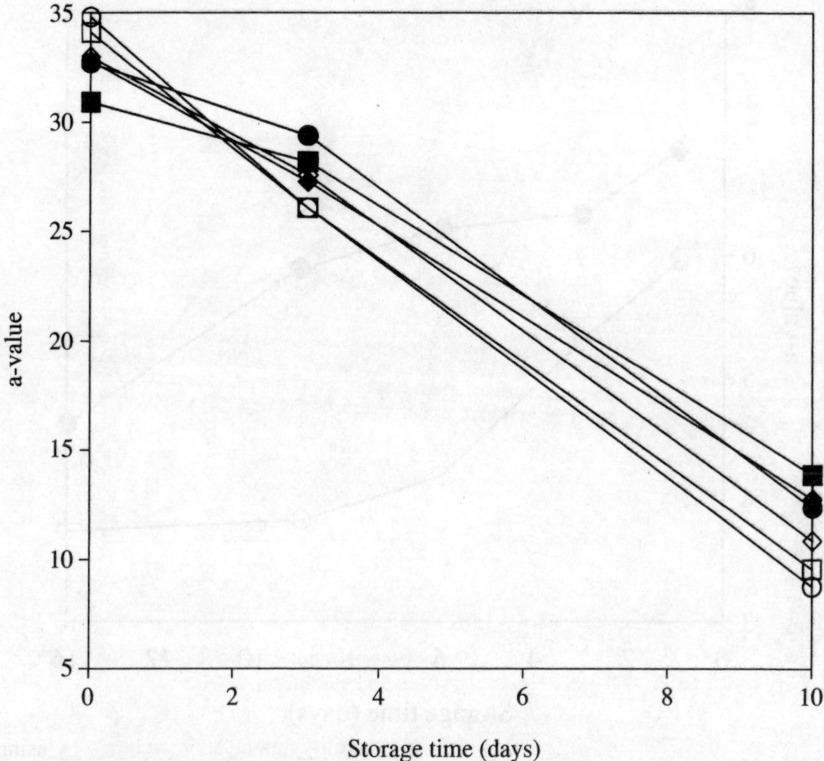

Fig. 19.6 Measured a-values for two muscle types from three different animals, stored in 80% O_2 at 8°C. *Longissimus dorsi* muscles (filled symbols) and *Semi-membranosus* muscles (open symbols) (unpublished data).

By using computer simulations the time for reaching, e.g., an oxygen content critical for the colour stability of a given product can be predicted. Furthermore, the permeability of the packaging material can be specified, or the shelf-life using a specific packaging film can be predicted. Such computer simulations were developed by Pfeiffer *et al.*, (1999) primarily for predicting quality changes, moisture gain and lipid oxidation in several dry products. The models described in the earlier sections are well suited for defining critical factors and levels for maintaining good meat colour stability for fresh and cured meat products.

Computer simulations are an attractive supplement to storage experiments since it will not be necessary to test all combinations of the factors before the optimal packaging and storage conditions can be chosen, considering both the product shelf-life and minimisation of the packaging material.

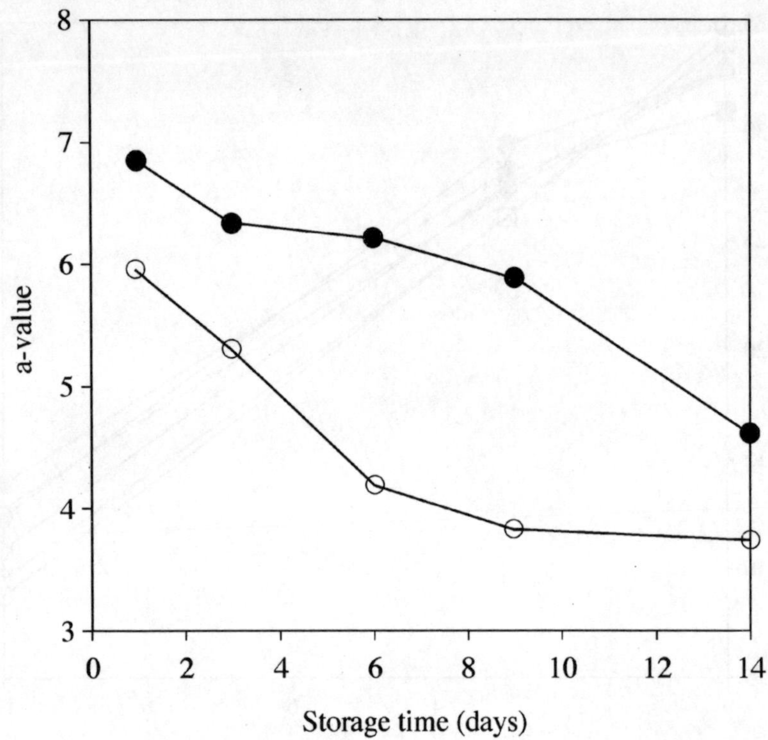

19.7 Measured a-values for cured ham containing 150ppm (filled circles) and 60ppm (open circles) nitrite (each point is an average of 42 samples) (unpublished data).

19.6 References

BERTELSEN G, JAKOBSEN M, JUNCHER D, MØLLER J, KRØGER-OHLSEN M, WEBER C and SKIBSTED L H (2000), Oxidation, shelf-life and stability of meat and meat products, *Proceedings 46th ICoMST, Argentina.*

BRO R (1997), PARAFAC. Tutorial and applications, *Chemometrics and Intelligent Laboratory Systems,* 38, 149–71.

BRO R and JAKOBSEN M (2002), Exploring complex interactions in designed data using GEMANOVA. Color changes in fresh beef during storage, *Journal of Chemometrics*, 16, 294–304.

DEVLIEGHERE F, DEBEVERE J and VAN IMPE J (1998), Concentration of carbon dioxide in the water-phase as a parameter to model the effect of a modified atmosphere on microorganisms. *International Journal of Food Microbiology,* 43, 105–13.

FAVA P and PIERGIOVANNI L (1992), Carbon dioxide solubility in foods packaged with modified atmosphere. II: Correlation with some chemical-physical characteristics and composition. *Industrie Alimentari,* 31, 424–30.

JAKOBSEN M and BERTELSEN G (2000), Colour stability and lipid oxidation of fresh beef. Development of a response surface model for predicting the effects of temperature, storage time, and modified atmosphere composition, *Meat Science*, 54, 49–57.

JAKOBSEN M and BERTELSEN G (2002), The use of CO_2 in packaging of fresh red meats and its effect on chemical quality changes in the meat: A review. *Journal of Muscle Foods*, 13, 143–68.

JENSEN C, FLENSTED-JENSEN M, SKIBSTED L H and BERTELSEN G (1998), Effects of rape seed oil, copper(B) sulphate and vitamin E on drip loss, colour and lipid oxidation of chilled pork chops packed in atmospheric air or in a high oxygen atmosphere, *Meat Science*, 50(2), 211–21.

LYIJYNEN T, HURME E, HEISKA K and AHVENAINEN R (1998), Towards precision food packaging by optimisation, *VTT research notes 1915, Finland.*

MCDONALD K and SUN D. (1999), Predictive food microbiology for the meat industry: a review, *International Journal of Food Microbiology*, 52, 1–27.

MØLLER J K S, JAKOBSEN M, WEBER C J, MARTINUSSEN T, SKIBSTED L H and BERTELSEN G (2003), Optimization of colour stability of cured ham during packaging and retail display by a multifactorial design, *Meat Science*, 63, 169–75.

PFEIFFER C, D'AUJOURD'HUI M, WALTER J, NUESSLI J and ESCHER F (1999), Optimizing food packaging and shelflife, *Food Technology*, 53, 6, 52–59.

RENERRE M (1990), Review: Factors involved in the discoloration of beef meat, *International Journal of Food Science and Technology*, 25, 613–30.

ROBERTSON G L (1993), *Food packaging: Principles and practice*, Chapter 4, Marcel Dekker, Inc. New York.

WEBER J C; ALDANA L and BERTELSEN G (1999). Inhibition of oxidative processes by low level of residual oxygen in modified atmosphere packaged pre-cooked cured and non-cured meat products during chill storage. *Proceedings of the 11th IAPRI World Conference on Packaging*. Singapore. P. 114–20.

ZHAO Y, WELLS J H and MCMILLIN K W (1994), Applications of dynamic modified atmosphere packaging systems for fresh red meats: Review, *Journal of Muscle Foods*, 5, 299–328.

ZHAO Y, WELLS J H and MCMILLIN K W (1995), Dynamic changes of headspace gases in CO_2 and N_2 packaged fresh beef. *Journal of Food Science*, 60 (3), 571–57.

20

Active packaging and colour control: the case of fruit and vegetables

F. Artés Calero, Technical University of Cartagena, Spain and P. A. Gómez, National Institute for Agricultural Technology, Argentina

20.1 Introduction

Consumer satisfaction is related to fresh product quality. This quality is generally associated with visual appearance, colour being one of the most important aspects in the consumer's purchase decision. The association of certain colours with the acceptance of fruits and vegetables begins early and is maintained through life. For instance, when the red colour of fruit is enhanced, the perceived sweetness level increases. Colour is normally used to determine acceptable limits for a given grade of product and to define colour tolerances for both harvest and trade. Combined with other characteristics it can be used to establish indices of maturity, enabling us to know whether a commodity can be harvested and to predict postharvest life of the product. For this reason colour requirements are more and more prevalent in retailer's specifications.

A knowledge of fruit and vegetable pigment composition allows us to evaluate the input of postharvest treatments on colour and quality. In fresh as well as in minimally processed products it is crucial to know the main factors affecting pigment stability as well as the main changes associated with processing. Analysing pigment composition of fruit and vegetables and their derivatives is important for optimising postharvest treatments during harvest, handling, storage and distribution. In fact, lowering O_2 and increasing CO_2 around fruit and vegetables by using controlled atmosphere (CA) or active or passive modified atmosphere packaging (MAP) techniques is commonly a good method for keeping colour stability. On the other hand, one of the main problems that reduces shelf-life of minimal processed fruit and vegetables is the enzymatic browning that occurs on the cut surface area. In this review, an update on the main tools for controlling colour changes is given. To prevent adverse changes, physical treatments, specially MAP to minimise enzymatic activity as well as combinations with antibrowning agents are considered.

20.2 Colour changes and stability in fruit and vegetables

The colour of fruit and vegetables is a direct consequence of their natural pigment composition resulting mainly from three families of pigments, chlorophylls and carotenoids, located in the chloroplasts and chromoplasts respectively, and the water-soluble phenolic compounds anthocyanins, flavonols and proanthocyanins, located in the vacuole. Betalains (e.g. betacyanins and betaxanthins) are the fourth family of plant pigments and are responsible for the red and yellow colours that occur only rarely. Chlorophylls and their derivatives are responsible for green, blue green and olive brown colours, while carotenoids are responsible for red-yellow colours. Anthocyanins are responsible for orange, red, blue, and purple and black, and intermediate colours. It is very useful to know the composition of fruit and vegetable pigments in order to evaluate the possible incidence of postharvest treatments for keeping colour and quality and extending their shelf-life, as well as that of their derived products (Artés *et al.*, 2002c; Kidmose *et al.*, 2002; Lancaster *et al.*, 1997).

During ripening chloroplasts are gradually replaced by chromoplasts containing only carotenoids, although exceptionally in some fruits, such as avocado, chlorophyll is retained in the pulp of the ripe fruit. However, in most fruits carotenoids become unmasked when chlorophyll disappears upon ripening, and usually this is accompanied by a marked biosynthesis of carotenoids. In many fruits (apple, apricot, artichoke, asparagus, blackberry, blueberry, red-carrot, cherry, cranberry, eggplant, fig, grape, red-lettuce, nectarine, olive, red-onion, 'Sanguine' orange, peach, pear, plum, pomegranate, red-skinned potato, radish, raspberry, red and black-currant, purple sweet potato, strawberry, etc.) ripening is associated with an intense anthocyanins biosynthesis. Although all these colour and pigment composition changing processes occur at the same time, very different biochemical pathways are involved for each class of pigment (Artés *et al.*, 2002c).

External colour is also influenced by physical factors, such as the presence of waxes and geometry of the fruit surface. That is the reason why colorimeters work better for liquids than they do for whole fruits and vegetables. The key problem that prevents accurate colour measurement of them is that they have non-uniform surfaces. This has a pronounced effect on how light and colour are reflected and perceived. For example, the colour measurement of a bean depends on where on the curvature of the bean's surface the measurement is made. If the angle of measuring is different from reading to reading, the quantitative colour reading will be different. If significant texture or granulation is present on the sample's surface, some light coming from the equipment may be scattered at different angles and escape detection. To compensate for this problem, specific colorimeters have been constructed using a spherical geometry that diffusely illuminates samples, eliminating the directionality of the light (Marsili, 1996). Placing the head-reader of the apparatus on the skin of the fruit or vegetable is preferable to measuring at a distance. From 200 measurements on ten 'Golden Delicious' apples, measuring at a distance of 4mm from the fruit, produced higher standard deviations in colour parameters than placing the device on the apple (Madieta, 2002).

20.3 Colour measurement

Internal and external colour can be both subjectively and objectively determined, in the latter case employing accurate devices. For determining pigment composition and defining colour quality indices in fruit and vegetables some methods are currently available, including the use of colour charts and chromatographic (HPLC, TLC) and spectrophotometric (UV-vis, colourimetry, etc.) analytical techniques. In the past few years, there has been a trend to use colorimetric rather than chemical analysis of pigment for describing colour changes and characterisation. Tristimulus colorimetric measurements are quicker and cheaper than conventional methods (Francis, 1969) and, overall, they are of a non-destructive nature. Colour is monitored in a three-dimensional colour space in terms of the chromatic colour coordinates L* (lightness), a* and b*, based on the CIELAB colour measurement system (Commission Internationale de l'Eclairage – International Commission on Illumination, CIE, 1986). In fact the CIE specified two colour spaces; one of these was intended for use with self-luminous colours and the other for use with surface colours. These notes are principally concerned with the latter known as CIE 1976 (L* a* b*) colour space or CIELAB (McGuire, 1992).

The quantification of tristimulus data is based upon trigonometric functions. These coordinates, after a correct manipulation, provide an indication of several aspects of colour. Values of a*/b* ratio have been considered a good indicator of changes in ripening in tomatoes and *citrus* (Arias *et al.*, 2000; Artés and Escriche, 1994; Artés *et al.*, 2000b). A more accurate measurement of colour can be obtained indicating that angle, named hue angle ($h^\circ = \text{arctg } b^*/a^*$), which represents the basic tint of a colour, and chroma $[(a^{*2}+b^{*2})^{1/2}]$, an index analogous to colour saturation or intensity (McGuire, 1992). On average, the human eye perceives hue differences first, chroma or saturation differences second, and lightness/darkness last (Marsili, 1996).

Hue index is adequate to predict colour when pigment degradation has taken place. It could be used to follow dilution, heat effects, browning, etc. The analysis of colour is used in those cases to determine the efficacy of postharvest treatments, including packaging, storage and distribution. However, there are conflicting reports in the literature on the correlation between colour measurements and pigment composition. For example, tristimulus colour measurements did not correlate well with changes in pigment composition of several apple cultivars (Lister, 1994). The β-carotene pigment, an important nutritional component as a precursor of vitamin A and the main carotenoid in green leafy vegetables and responsible for the orange colour in fruit and vegetables, was one of the first studied when trying to find a relationship between pigment content and colour (Francis, 1969).

The applicability of using skin colour measurements to predict changes in pigment composition was investigated by analysing a wide range of fruit and vegetables. There were linear relationships between hue and anthocyanin concentration and between L* and log of chlorophyll concentration. However, there was not a unique linear combination of pigments that gave a unique point in the colour space and, at the same time, a given set of colour coordinates could be achieved by many combinations of pigments (Lancaster *et al.*, 1997).

Colour measurement in cranberry products is a good example of the interrelation between colour and pigment. The colour of cranberry juice is due to four anthocyanin pigments. There are other minor red pigments as well as six yellow flavonoid pigments but their contributions are less important. In fresh juice fruits, where pigments are homogeneously distributed, the relationship is stronger than for the whole fruit, where pigments are unevenly located in the cell layers below the epidermis (Francis, 1969).

A recent study revealed that colour parameters were not good estimators of anthocyanin levels in raspberry, a highly perishable fruit with a storage life limited by decay and darkening of the typical red colour (Haffner *et al.*, 2002). However, it was found that values of a*/b* ratio were well related with changes in lycopene (the predominant carotenoid) content in tomatoes (Arias *et al.*, 2000). This agrees with cautions given in previous reports for interpreting changes in colour coordinates as simple changes in pigment composition (Lancaster *et al.*, 1997).

It could be concluded that there is a wide range in the degree of correlation between colour measurements and pigment composition. In order to find a high correlation each pigment would have to be carefully weighted for its contribution to the colour. For precise predictions, colour values should be checked against a chemical method to make sure that changes in colour are actually due to these pigments.

20.4 Process of colour change

The structure of chlorophyll present in fruits and vegetables is affected during development, ripening, and senescence, and throughout postharvest treatments, with a consequent effect on external and internal colour. During fruit ripening and leaf senescence chlorophyll catabolism takes place. In fact chlorophyll degradation is a normal process of the ageing phenomenon in leafy vegetables and occurs to provide energy for the senescing leaves (O'Hare and Wong, 2000). In this way, because leaves are still alive after harvest and continue to respire using energy in the process, chlorophyll is metabolised to maintain life.

Colour change is primarily related to a reduction in the amount of chlorophyll, which highlights other pigments such as carotenoids and anthocyanins. During fruit ripening the chlorophyll usually disappears due to chloroplast degeneration to gerontoplast. In leaves the chloroplasts commonly disintegrate but some of them remain, masking the yellow carotenoid colour. However, in ripe fruits chloroplasts degenerate into chromoplasts, concomitantly with a massive biosynthesis of carotenoids (Matile *et al.*, 1997; Hörtensteiner, 1999). This change from chloroplast to chromoplast is particularly important in the case of fruits called *carotenogenic* (e.g. pepper, tomato, orange and persimmon), characterised by this extensive new synthesis of carotenoids, usually accompanied by a change in the carotenoid profile of the fruit (Artés *et al.*, 2002c).

In climacteric fruits, the maximum degradation of chlorophyll takes place during the climacteric rise, although generally slight quantities of chlorophyll are always present in the internal tissues. It has been found in apples and pears that degradation of chlorophyll could be mainly due to hydrolytic activity of chlorophyllase enzyme (EC 3.1.1.14) that transforms chlorophyll into phytol and porphyrin and the resultant chlorophyllide has no effect on colour changes. However, this effect was not found in tomatoes and disintegration of chloroplast membranes occurs before the loss of green colour (Pantastico, 1979). Pigment changes during tomato ripening imply a loss of chlorophyll and an accumulation of lycopene. If ripening proceeds under sub-optimal conditions for lycopene synthesis, β-carotene accumulates resulting in yellow fruit (Shewfelt *et al.*, 1988). As tomatoes turn from green to red, changes in the L*a*b* parameters during ripening are characterised by a decrease in hue and a concomitant increase in chroma.

Non-climacteric fruits do not ripen off the tree and should be picked when fully ripe to ensure their best flavour. During ripening of non-climacteric fruits like *citrus* or sweet peppers, the process of natural colour break from green to the typically ripe orange/yellow/red is called degreening and takes place very gradually. During degreening of *citrus* the loss of chlorophyll accumulated into the chromoplasts of the epidermis (flavedo) and vesicles and the concomitant manifestation and new biosynthesis of carotenoids generally occur very slowly (Eaks, 1977). Shippers usually accelerate the degreening process of harvested *citrus* or sweet peppers (Fig. 20.1) both to advance the marketing period, when prices are higher, and to make fruits more attractive to consumers. The industrial technique commonly consists of applying low concentrations (5–50 ppm) of exogenous ethylene at 18–24°C and 90–95 % RH for two to four days (Artés *et al.*, 2000b; Gómez *et al.*, 2002).

Yellowing of green minimally processed products is not appealing to consumers and has a negative effect on sales of the product. It has been demonstrated that colour change from bright green to brown in fresh as well as in minimally fresh processed green vegetables is related to the presence of pheophytin, formed when chlorophyll loses its bound magnesium atom, which is substituted by hydrogen (Schwartz and von Elbe, 1983). More than 50%

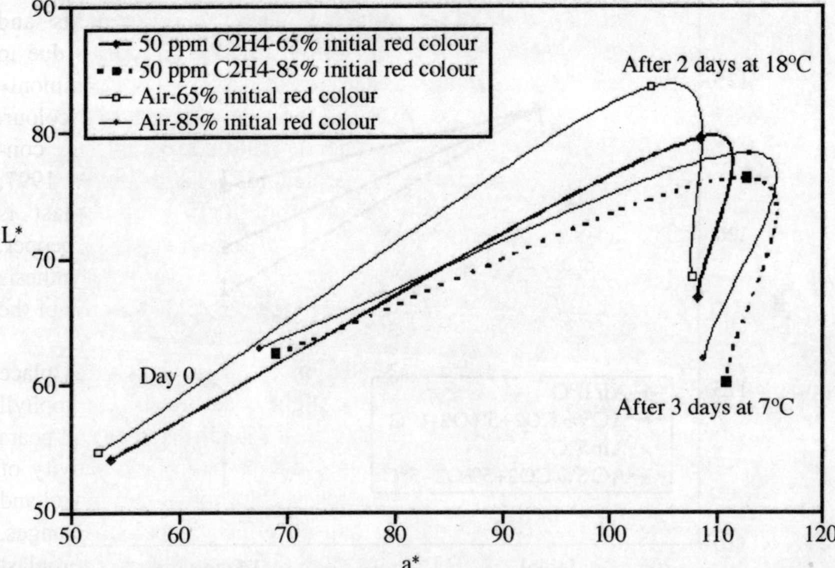

Fig. 20.1 Changes in L* and a* parameters of bell peppers with 65% and 85% of initial red colour degreened with 50ppm c_2H_4 during 2 days at 18ºC followed by 3 days at 7ºC in air (Gómez *et al.*, 2002).

conversion of the chlorophyll to pheophytin can occur before a change of colour from bright green to olive brown could be observed (Lau *et al.*, 2000).

The rate of chlorophyll degradation can be lowered by several means. However, a combination of them is more effective than one single method. The most important tool is chilling although there is a limit to temperature because some fruit and vegetables are susceptible to damage caused by low temperatures below their freezing point, suffering chilling injuries commonly accompanied by undesirable colour changes.

Ethylene greatly accelerates chlorophyll degradation. Usually, leafy vegetables do not produce much ethylene, but can be affected from ethylene coming from other sources. The inclusion of ethylene scavengers within packages containing these vegetables could provide protection against ethylene action (O'Hare and Wong, 2000). Operations involved in fresh processing, like cutting, grating or peeling, stimulate ethylene biosynthesis that could cause physiological disorders, lowering the quality of the products. Some changes affecting colour include accumulation of phenolic compounds in carrots, red discoloration in chicory and endive, or russet spotting in lettuce (Artés, 2000b; Van de Velde and Hendrickx, 2001; Verlinden *et al.*, 2001). It has been reported that ethylene produced during cutting of fresh processed spinach notably accelerates the loss of chlorophyll and damage is proportional to the ethylene level reached (Abe and Watada, 1991). Also celery sticks stored in atmospheres where ethylene is present showed a decrease in hue (Fig. 20.2) as colour changed from dark green to yellowish-green (Artés *et al.*, 2002b). On the other

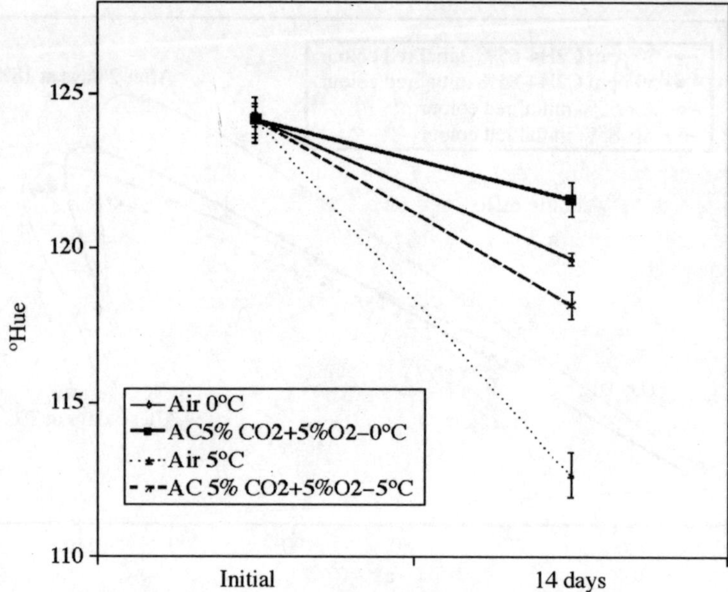

Fig. 20.2 Colour changes (°Hue) of celery sticks stored for 14 days in air and in controlled atmosphere (5% CO_2 + 5% O_2) free of ethylene at 0 and 5°C (Artés *et al*,. 2002b).

hand, antioxidants are related to chlorophyll retention in leafy products. Two antioxidants commonly present in fruits and vegetables are ascorbic acid and β-carotene, which protect chlorophyll by inhibiting the reactions that degrade it, retarding yellowing (Schwartz and von Elbe, 1983).

20.4.1 Anthocyanin degradation

Anthocyanins are very unstable pigments, particularly once removed from their natural environment and the protection provided by co-pigmentation, leading to unattractive yellowish and brownish pigments. This is particularly evident when minimal fresh processed products are prepared. When conditioning fruit and vegetables by techniques like peeling, cutting, slicing, etc., cell membranes are disrupted, allowing the mixing of phenolic substrates located in the vacuole and specific polyphenol oxidases enzymes (PPO; EC 1.14.18.1) associated to cell membranes, (mainly in the plastids). Washing the product immediately after cutting removes sugars and other substrates at the cut surfaces minimising reactions responsible for changes in colour and nutritional quality.

It is well known that colour due to anthocyanins is particularly degraded by the enzymic hydrolysis in harvested products as recently reviewed (Artés *et al.*, 2002c). Anthocyanins are oxidised in the vacuole of the plant cells in the presence of molecular O_2 and under appropriate conditions of pH, temperature and water activity, by the action of the enzyme tyrosinase (EC 1.10.3.1) or polyphenol oxidase

(PPO). But anthocyanins are not direct substrates for PPO, which catalyses the hydroxylation of monophenols to o-diphenols (cresolase activity EC 1.14.18.1) and the oxidation of o-diphenols to o-quinones (catecholase activity EC 1.10.3.1). Catecholases were considered as the main PPO enzymes responsible for browning in fruit and vegetables. These o-quinones are very reactive molecules that rapidly condense by combining with amino or sulfhidril groups of proteins and with reducing sugars, producing different brown, black or red polymers of high molecular weight and unknown structure known as melanines (Artés *et al.*, 1998). In contrast to the ethylene effect on chlorophyll, anthocyanin synthesis and ethylene production seem to be correlated. In fact, red cherries stored in air reached high ethylene levels and the highest anthocyanin content by the end of cold storage (Remón *et al.*, 2000).

The increase in pH and decrease in titratable acidity induced by high CO_2 during CA storage of fruit and vegetables have a strong effect in anthocyanin expression and stability. The red flavylium cation (AH+) remains stable only in acidic conditions. Changes in anthocyanin stability can result from nucleophilic attacks by water molecules on the anthocyanin molecule to form a colourless pseudobase, hemiacetal, or carbinol. The flavylium form can be restored by acidification. The colourless carbinol can form chalcone (a yellow pigment) by the opening of the ring structure. As pH increases above 4, a blue quinonoidal base is formed. Increase in pH above 7 can result in the loss of a proton from the hydroxyl group to form a second quinonoidal base (Holcroft and Kader, 1999b). In addition, these authors reported that phenylalanine ammonia lyase (PAL, EC 4.3.1.5) and flavonoid glucosyltransferase (GT, EC 2.4.1.28), two key enzymes in the synthetic pathway of anthocyanins in strawberry, were adversely affected by high CO_2 levels during cold storage.

On the other hand, it has been demonstrated that the degrading effect of vitamin C on anthocyanin stability leads to undesirable colour changes in model solution and in natural pomegranate juice systems (Martí *et al.*, 2001). Exposure to light and heat also induced these degrading reactions. However, glucosylation provides protection against photodegradation and the formation of intermolecular copigmentation complexes and ion-pairs lowered the degradation of anthocyanins (Brouillard *et al.*, 1997).

20.4.2 Browning

Browning is the result of a chain of reactions that very often occurs in fruit and vegetables. The first step of that process takes place in the vacuole and it is the deamination of the amino acid phenylalanine by PAL. The product of that reaction is the cinnamic acid which is hydroxylated into various phenolic compounds. When O_2 is present, the PPO located in the cytoplasm (plastids) oxidises the compounds to *o*-quinones, which polymerise into brown compounds (Siriphanich and Kader, 1985). The relationship between PPO and browning was reported when it was found that CO_2 competitively inhibited PPO activity in mushrooms retaining their colour, although at high concentrations increasing browning (Murr and Morris, 1974).

Peroxidases (POD; EC 1.11.1.7.) could also be involved in browning although to a lesser extent, due to low availability of H_2O_2 within the plant cell (Artés *et al.*, 1998; Sánchez-Ferrer *et al.*, 1995). Lipoxygenase (EC 1.13.11) and lipase (EC 3.1.1.3) have been considered as the main causes for the breakdown of some vegetables like cucumbers. The reaction between lipoxygenase and lipids substrates generates hydroperoxides that are related to senescence and scald induction. Fatty acid radicals induced by peroxidase can react with cell components leading to further breakdown. Particularly, bleaching of β-carotene and chlorophyll *a* occurs as a consequence of lipoxygenase catalysed reactions.

Browning could easily be evaluated by colorimetric methods. For example, visual scores for browning of cut lettuce were well correlated with hue (values decreased as browning occurred) and a*, although correlation with b* was lower while it was not significant with L*, Hue angle values decreased as browning occurred (Peiser *et al.*, 1998).

20.5 Colour stability and MAP

To remain competitive in the fruit and vegetable market, suppliers must offer products with an optimal overall quality. Thus, the entire chain from producers and processors to retailers must be increasingly sensitive to consumer requirements, particularly as they relate to colour. In fact, perception of sweetness, sourness and flavour intensity was highly correlated to skin colour as has been reported for sweet cherries; full dark red cherries, measured by both visual and colourimetry, had higher consumer acceptance than full bright red (Crisosto *et al.*, 2002).

Atmospheres with reduced O_2 and/or elevated CO_2 concentrations are known to extend the storage life of fruit and vegetables. MAP can bring the lowering of respiratory activity and ethylene production, delay in ripening and softening, limiting weight losses and reduced incidence of physiological disorders and decay-causing pathogens (Ahvenainen, 1996; Artés, 2000b). As MAP slows the rate at which energy reserves are used it can be applied in combination with chilling storage for improving shelf-life of fruit and vegetables. At the same time, MAP affects biochemical reactions related to pigment synthesis and degradation (Artés, 1993 and 2000a), although responses to MAP depend on the kind of fruit or vegetable. In addition to this, the effect of respiratory gases on the metabolic behaviour of plant materials depends on temperature of application due to its influence on solubility of these gases. The effects of low O_2 and/or high CO_2 on colour changes in packaged fruit and vegetables will be examined using several examples recently reported.

20.5.1 Low oxygen effects

It has been observed that the activity of tyrosinase, responsible for mushroom browning, is dependent on O_2 concentration. MAP induced higher L* values and lowered the difference between ideal mushroom target and sample than those observed for mushrooms stored in conventional packages (non-MAP). The improved colour might also be due to lower microbial growth resulting from low O_2 (Roy *et al.*, 1996).

Strawberries stored under low O_2 2kPa showed a better colour, high anthocyanin concentration and organic acids content than those stored in air. Fruits became darker red and accumulated anthocyanin, although O_2 was not as effective at high CO_2 levels in reducing decay (Holcroft and Kader, 1999b).

Freshly harvested white asparagus spears stored in air or in CA having increased O_2 concentrations (1 to 15kPa) showed a concomitant increment in anthocyanin content, resulting in an intense purple colour of the tips. Hue values were lower than those at harvest, with a decline in L* values. The lowest anthocyanin accumulation was observed at the tips of the spears stored under the lowest O_2 level (Siomos *et al.*, 2000).

Red discolouration of chicory after seven days of storage at 12ºC was strongly reduced from 90% in air to 35% under 2kPa O_2 and 0kPa CO_2 (Verlinden *et al.*, 2001). Fresh processed potato slices stored in MAP with low O_2 showed a better colour retention when the O_2 level was lowered from 3.5 to 1.4kPa, probably due to reduction of oxidase activity such as PPO, ascorbic acid oxidase (AAO, EC 1.10.3.3) and glycolic acid oxidase (GAO, EC 1.1.3.15). Slices in air showed a decrease in L* compared to MAP. It was advantageous to have almost no initial O_2 within packages by flushing N_2. This active MAP was very important for the keeping quality of slices, taking into consideration that residual O_2 in the packages was enough to prevent anaerobiose (Gunes and Lee, 1997).

To prevent browning of minimally processed potatoes, dipping in some chemical agents was essential because MAP alone did not avoid this disorder. Browning is closely related with O_2 and CO_2 levels in the package and O_2 must be decreased to an acceptable minimum level as soon as possible, it being advantageous to have almost no O_2 initially within packages (Gunes and Lee, 1997).

The intensity of browning of ready-to-eat apples depends on the atmosphere composition. Apple cubes in MAP were efficiently preserved from browning and showed the lowest colour losses when initially displacing O_2 by injecting 100kPa N_2 and a film with low O_2 permeability was used. This atmosphere was the main factor affecting lightness, and L* changes occurred four times more slowly than when O_2 was about 2kPa or when medium O_2 permeability films were used (Soliva-Fortuny *et al.*, 2001).

It has been suggested that O_2 levels greater than 21kPa may influence the postharvest life of intact and fresh processed fruit and vegetables and PPO may be substrate-inhibited by high O_2 levels (Day, 1994). Superatmospheric O_2 could have an effect on respiratory activity and ethylene synthesis and action, although response depends on the commodity, ripening stage, O_2 level, length of storage and temperature. Levels of CO_2 and C_2H_4 should also be considered. When focusing on pigment changes, Kader and Ben-Yehoshua (2000) reported that 40–50kPa O_2 accelerated ripening of tomatoes, with 60–100kPa stimulating synthesis of lycopene in *rin* varieties. 40–80kPa O_2 improved the colour of endocarp and juice of orange cultivars, increasing the deepest of orange colour. However, an undesirable change from yellow to orange has been observed in grapefruit. An increase in the red colour of flesh and juice of blood-orange 'Sanguine' *cv* has been related to anthocyanin synthesis.

Superatmospheric O_2 was particularly effective in inhibiting browning of different fresh processed vegetables including mixed salads and chicory endive (Jacxsens *et al.*, 2001, Allende *et al.*, 2002). However, 'Bartlett' pear slices kept in 40, 60 or 80kPa O_2 exhibited similar severity of cut surface browning during storage at 10ºC (Gorny *et al.*, 2002).

20.5.2 High carbon dioxide effects

Tolerance to high CO_2 is commonly reduced in fruit and vegetables. As an example, levels of 3–5kPa induced superficial scald and abnormal flavour in *citrus* during cold storage (Artés, 1995; Kader, 1990). For this reason, the use of CO_2 for keeping colour in plant materials must be particularly adapted for each species.

Exposure of mango fruits to 50kPa CO_2 at 40–44ºC for 160 minutes used as a quarantine treatment affected colour during cold storage. After 20 days at 10ºC control fruits showed a higher decrease in hue than those heated under CO_2 enriched CA. Decrease in hue was a good estimator of the green to yellow colour turn. Yellowing increased in the absence of CO_2 and chroma (colour intensity) decreased (Ortega-Zaleta and Yahia, 2000).

MAP stored peaches for 21 days at 2ºC, with equilibrium CO_2 level about 20kPa, showed a residual effect of high CO_2 during the subsequent three days at 20ºC. Colour development was slow and L* and chroma ground values were maintained as at harvest (Fernández-Trujillo *et al.*, 1998). However, very high CO_2 levels (73kPa in MAP) destabilised cyanidin derivatives in the skin of 'Starkimson' apples (Remón *et al.*, 2000).

Despite the benefits of CO_2 enriched atmospheres in controlling postharvest decay, anthocyanin concentration is affected and more particularly in the internal tissues. During storage of strawberries in air plus 10 or 20kPa CO_2, skin chroma increased with time and was not as affected by gas composition as flesh colour which turned pale. The hue of berries held in air was lower than that in CA and chroma under 10kPa CO_2 was slightly higher than under 20kPa CO_2 (Holcroft and Kader, 1999a).

In contrast to the effect observed in strawberries, CO_2 did not affect anthocyanin in pomegranate fruit. Juice red colour increased in intensity during postharvest storage (8 weeks, 5ºC) when 5kPa CO_2 was combined with 5kPa O_2 while air-stored fruits showed a slight pale red (low a* value) colour (Artés *et al.*, 1996). Chroma of the skin was better maintained when fruits were stored in air plus 10kPa CO_2 while L* of the internal integuments decreased with time due to browning. Integuments were darker after six weeks at 10ºC and 20kPa

CO_2, as a consequence of CO_2 injury, while chroma and hue did not change (Holcroft *et al.*, 1998).

Inconsistent results have been found in sweet cherries. For red and purple cherries 11kPa of CO_2 inhibited anthocyanin synthesis and when initially packaged with a high CO_2 level maintained their anthocyanin content during storage (Remón *et al.*, 2000). In order to avoid risk of high CO_2 on undesirable colour changes 'Ambrunés' sweet cherries were treated for 26 days at 1–2ºC with CO_2 shocks (air plus 20% CO_2) once a week for 24 hours every week followed by active aeration, or in continuous 20% CO_2 enriched air atmosphere. Air control and both CO_2 treatments were followed by three days in air at 13ºC. At the end of shelf-life an increase of cyanidin derivatives and in total anthocyanins content was detected in fruit under both CO_2 treatments without differences between them. In all treatments L* increased while a*, hue and chroma decreased, indicating that no undesirable changes in colour were induced by high CO_2. This CO_2 shocks technique could be useful as a commercial alternative to continuous CO_2 for controlling decay in sweet cherries, being easier and cheaper to apply than continuous CA storage, without adverse collateral effects (Artés *et al.*, 2002d).

The presence of anthocyanins in white asparagus spears, evident from purple colour development in their tips, is an undesirable change that affects quality. Anthocyanin accumulation was slow in MAP (1kPa O_2 and 5–7kPa CO_2) during six days at 2.5, 10 and 20ºC, and tips were still white (a* did not change) at the end of storage. On the contrary, for spears stored in air, the anthocyanin content increased at all temperatures, resulting in an intense purple colour of the tips, the presence of light having little or no effect on any of these treatments (Siomos *et al.*, 2000). Later studies revealed that the increase in the anthocyanin concentration was prevented when spears were stored under higher than 5kPa CO_2 in the dark or than 10kPa in the light. Hue showed no changes from harvest values. Brief exposure to 100kPa CO_2 before air storage was as effective as continuous storage under the above conditions in preventing development of red colour (Siomos *et al.*, 2001).

Since pH has a marked effect on anthocyanin stability and colour expression, changes in pH induced by CA could cause significant losses in colour. The effect of CO_2 enriched atmospheres on colour and anthocyanin concentration depends on fruit species. Those that have a high buffering capacity prevent pH changes and maintain pigmentation better. However, it is possible to correlate reduced pericarp pH with low PPO activity. A pH of 4.2 or less may almost abolish PPO activity (Jiang *et al.*, 1997).

It has been reported that high CO_2 can improve colour and chlorophyll retention during storage of broccoli. MAP with 11kPa O_2 plus 7.5kPa CO_2 throughout 96h at 5ºC resulted in no differences in green colour retention of florets evaluated by hue, compared to the level at harvest (Barth and Zhuang, 1996). Experiments combining low temperature and CA revealed that increasing the CO_2 level to 2.6 or 4.7kPa (O_2 21kPa) maintained the appearance, as well as chlorophyll content of stored snow peas (Ontai *et al.*, 1992).

High CO_2 levels produced injury in lettuce by increasing PAL activity, but were effective for avoiding brown stain. Under air plus 15kPa CO_2 production of phenolic compounds was blocked and did not change during storage. If CO_2 was replaced by air, phenolics content increased, quickly at 20ºC and slowly at 0ºC. Previous reports indicated that in lettuce tissue, CO_2 prevented browning by both blocking the production of new phenolic compounds and inhibiting PPO (Siriphanich and Kader, 1985). For determining browning in 'Baby' and 'Romaine' lettuce midribs L* was a good parameter (a decrease indicates darkening), and an increase in a* was associated with reddish colours due to browning (Castañer *et al.*, 1999).

Prepared diced yellow onion did not always develop enzymatic browning during storage. However, discoloration resulting from yellowing has been reported (Mencarelli *et al.*, 1990, Blanchard *et al.*, 1996). Keeping O_2 at 2kPa, while CO_2 increased from 0 to 15kPa, the b* values were lowered. CO_2 effect was also observed after cooking. While control diced onion showed an intensification of browning and blackening of the cut surfaces, browning of cooked onion purée (expressed as b*) was slowed down slightly by lowering O_2, and more strongly by increasing CO_2 (Blanchard *et al.*, 1996).

Ready-to-eat apple is susceptible to browning and several methods have been tried to inhibit PPO. By using active MAP of 2.5kPa O_2 and 7kPa CO_2 a depletion of 62% of PPO activity was found. A first-order model fitted with enough accuracy was proposed for predicting colour changes. Browning was better described through decrease in L* and increase in total colour difference with respect to the initial values (Soliva-Fortuny *et al.*, 2001).

On the other hand, CO_2 injury in whole and fresh processed apple has been reported. Concentrations over 10kPa for 'Golden Delicious', 'Renatta' and 'Imperatore' *cv* produced flesh browning (Gorini, 1979). CA of air plus 10 or 20kPa CO_2 accelerated tissue browning and necrosis of fresh cut pear slices compared to those stored in air. Necrosis and severe cut surface browning first occurred in the flesh tissue closest to the core and spread radially (Gorny *et al.*, 2002).

A level of 39kPa CO_2 with O_2 close to 1kPa considerably affected anthocyanin concentration in apple tissue. The decline was attributed to non-enzymatic browning of anthocyanins with free amino acid released by damaged tissue under high CO_2, indicating that CO_2 was a key factor in destabilising those anthocyanins (Lin *et al.*, 1989).

Sliced persimmon stored in CA containing 12kPa CO_2 resulted in colour retention. It maintained good visual quality when stored under air plus 12kPa CO_2 or 2kPa O_2 plus 12kPa CO_2. However, air and 2kPa O_2-stored fruit developed areas of black pigmentation on the cut surfaces at the limit of marketability (Wright and Kader, 1997).

20.6 Combining low oxygen, high carbon dioxide and other gases

The colour development of tomatoes is influenced by gas composition. Several authors have developed empirical models to calculate colour parameters for tomatoes stored in constant or varying gaseous environments. When fruits were exposed to 20 different gas atmospheres involving O_2 and CO_2 levels of 5–20kPa and 0–20kPa respectively, it was found that changes in O_2 had a greater effect on colour development than changes in CO_2. Models require modifications depending on initial maturity stage, temperature and cultivar. Delayed tomato ripening at 3–5kPa CO_2 could in part be attributed to their effect on inhibiting ethylene synthesis (Yang and Chinnan, 1987).

For cherries, it was observed that skin colour exhibited a similar trend in four MAP (all of them with an equilibrium atmosphere of 1–3kPa O_2 and 9–12kPaCO_2). The hue shifted from red to blue/red and then back to red. Thus, at the end of three weeks storage, the skin colour remained effectively unchanged. Since any change was temporary, it is difficult to explain. Variation in pH due to MAP can affect the colouration of anthocyanins producing qualitative differences in this pigment (Remón *et al.*, 2000, Gil *et al.*, 1996b).

The anthocyanin level of raspberry fruit increased during storage in normal air while no changes were found after storage in CA of 10kPa O_2 plus 15 or 30kPa CO_2. Air stored fruits were darker (lower L*), less intense in red (lower a*) and less yellow (lower b*) and more red-bluish (lower hue). CA with 10kPa O_2 plus 30kPa CO_2 resulted in darker berries but CA with 10kPa O_2 plus 15kPa CO_2 had no influence on any colour parameters (Haffner *et al.*, 2002).

Sweet pomegranate stored at 2 or 5ºC for 12 weeks in MAP (15kPa O_2 and 3kPa CO_2 at the steady state) was compared with fruits packaged with perforated film. Air atmosphere maintained red skin-colour, with anthocyanin levels decreasing in all treatments. Chroma and hue values of the arils in air stored pomegranates suffered slight or no changes. However, MAP stored fruits showed a decrease in chroma (Artés *et al.*, 2000a).

Changes in skin and flesh colour of 'Paraguayo' peaches stored for three weeks at 0.5ºC were analysed, and hue was the best index. Intermittent warming (IW) for one day at 20ºC every six days of cold storage and MAP was applied. The high CO_2 (19.5kPa) and low O_2 (5.1kPa) generated by MAP alone or combined with IW, delayed colour development even during post storage ripening. Ground colour measured by hue angle was the best index for monitoring colour evaluation (Fernández-Trujillo and Artés, 1997).

Shrinkable packaging has been used to create MAP with lower O_2 and higher CO_2 inside the products than in the surroundings. This inhibits ethylene production and action. In this way, the hue of apples stored unwrapped decreases, while the hue of shrink-wrapped apples increased indicating that the unpacked apples were more reddish and less yellowish (Lin *et al.*, 1989).

Low O_2 (0.25–0.5kPa), high CO_2 (air enriched with 5, 10 or 20kPa CO_2) or superatmospheric O_2 (40–80kPa) applied alone did not prevent cut surface browning of fresh cut 'Bartlett' pears slices. However, a post cut dipping in a solution of ascorbic acid, calcium lactate and cysteine (adjusted to pH 7) reduced browning and extended shelf-life of pear slices for up to eight days at 0ºC (Gorny *et al.*, 2002).

Each vegetable responds differently to MAP. For example mizuna (an Asian vegetable of the *Brassica* family) kept its green colour with less than 2kPa O_2, but not with CO_2 higher than 5kPa. Pak choy (another Asian leafy vegetable) responds well to either enhanced CO_2 (up to 15kPa) or reduced O_2, while Chinese mustard responds well to a combination of reduced O_2 and enhanced O_2 (O'Hare and Wong, 2000).

High CO_2 (29kPa) and low O_2 (1kPa) led to anaerobic respiration, causing a great degradation of chlorophyll in snow pea pods (Ontai *et al.*, 1992). However, bagging pea pods with a polymeric film (5kPa O_2 and 5kPa CO_2) and stored 28 days at 5ºC, led to maintenance of appearance and colour, as well as internal quality (chlorophyll, ascorbic acid and sugar contents). MAP reduced the breakdown of chlorophyll to pheophytin, and control pods had increased yellowing and were less intensely green (Pariasca *et al.*, 2001).

For keeping quality and avoiding browning of fresh-cut butter head lettuce, Varoquaux *et al.* (1996) found that levels of O_2 lower than 3kPa combined with CO_2 lower than 10kPa were optimal. For generating these atmospheres in MAP, plastic polymers with low permeability to O_2 and high to CO_2 must be applied. To ensure that MAP salad products have no brown edges (the major quality defect) in commercial production, less than 0.5kPa O_2 and high CO_2 (more than 7kPa) are used. However, these conditions may lead to fermentative metabolism with production of ethanol and acetaldehyde, and off-odours development (Cantwell and Suslow, 2002)

A corn zein based polymer was evaluated to investigate its ability to perform as biodegradable MAP for avoiding yellowing of fresh broccoli. Atmospheres of 2–8kPa O_2 and 5–15kPa CO_2 maintained original colour and texture for six days at 5ºC. This behaviour was attributed to the low respiration rate induced by high CO_2 that also inhibited the metabolic reactions leading to colour and texture loss (Rakotonirainy *et al.*, 2001).

Endives stored for five days at 7ºC were highly susceptible to red discoloration and browning, both factors reducing their shelf-life, and minimal processing enhances quality degradation by increasing the respiration rate. A gas mixture of 10kPa CO_2 and 10kPa O_2 avoids colour changes and showed the best quality from consumers' point of view (Van de Velde and Hendrickx, 2001). When several CA were applied to chicory in O_2 levels ranging from 2–21kPa and CO_2 from 0–19kPa, the best composition to discourage red discoloration was provided by low O_2 combined with high CO_2. The effect of O_2 was found to be more important than that of CO_2. In addition to atmosphere composition effects, the higher the storage temperature (5, 12 and 20ºC) or the storage duration (3, 7, 10 and 14 days) the more red discoloration developed (Verlinden *et al.*, 2001).

Minimally processed pomegranate seeds have a greatly reduced postharvest life compared to whole fruit, and MAP is an excellent method for extending their shelf-life. The quality of ready-to-eat pomegranate seeds has been evaluated by using active (5kPa O_2 and 0kPa CO_2) and passive MAP at 0–5ºC. In general, a slight decrease in the juice anthocyanin content was found. L* increased under MAP, showing an opposite effect to unpacked seeds, and compared to air atmosphere, visual appearance was improved (Gil *et al.*, 1996a; Villaescusa *et al.*, 2001). However, MAP (11–13kPa O_2 and 9–12kPa CO_2) did not show any effect on delaying browning of fennel dice stored for 14 days at 0ºC followed by four days in air at 15ºC, very probably due to the relatively high O_2 levels reached. A decrease of hue and chroma for MAP and air treatments, without differences among them, was found. During storage in air or MAP yellowing occurred, probably due to the fact that enzymatic browning on fennel dice was only a surface reaction. After storage, colour of fennel juice changed without differences among control and MAP treatments although in contrast with results in dice, chroma increased (Artés *et al.*, 2002a).

Active (flushed with 4kPa O_2 plus 10kPa CO_2) MAP maintained the quality of fresh cut cantaloupe cubes for nine days at 5ºC better than passive MAP. Among the benefits of MAP better colour retention was reported. A high lightness and bright orange colour on cube surfaces were observed. The initial L* hue and chroma values remained unchanged in MAP but decreased in the control (Bai *et al.*, 2001). In treatments having higher than 15kPa O_2 and in treatments without CO_2, translucency (a typical cause of low quality in cantaloupes) occurred (O'Connor-Shaw *et al.*, 1995).

20.6.1 Other gases

It has been suggested that argon and other noble gases could be used for MAP applications. Apparently, although they are chemically inert, they are biochemically active. Their solubility in water is higher compared with N_2 and O_2. Probably, they interfere with cell membrane fluidity and O_2 receptor sites of enzymes. It was observed that Ar suppress enzyme activity and controls adverse chemical reactions, reducing respiration rates and, consequently, having a direct effect on extending shelf-life (Day, 1994). However, there is a lack of conclusive evidence supporting the view that partial or total substitution of N_2 with Ar has beneficial effects in terms of quality for MAP systems. Comparative experiments using both Ar and N_2 on two enzymes, one involved in the respiration process (malic dehydrogenase EC 1.1.1.37) and the other involved in the oxidative browning of fruit (tyrosinase) confirmed that there was only a slight reduction in activity using Ar compared with N_2 (Zhang *et al.*, 2000).

The combination of CO at concentrations between 5–10kPa with O_2 levels lower than 5kPa has been shown to be able to retard browning and extending shelf-life in fresh-cut lettuce and other products (Cantwell, 1992). The use of exogenous NO by initial short-term fumigation could extend the postharvest life of fresh horticultural products through inhibiting ethylene production and action. Thus disadvantageous changes in colour of harvested fruits and vegetables could be delayed, although NO application was more effective in non-climacteric than in climacteric species (Leshem, 2000).

20.7 Future trends

Increasingly, efforts to understand and control colour changes in fruit and vegetables must be focused to ensure that they satisfy consumer demand. Only then can emergent packaging techniques for avoiding undesirable colour changes be developed on a commercial scale. The use of physical treatments like active MAP (including the use of CO and superatmospheric O_2), or edible coatings that restrict O_2 entrance, and moderate thermal treatments must be developed. The combination of physical treatments with natural anti-browning agents like organic acids or derivatives of 4-hexilresorcinol should also be explored (Artés *et al.*, 1998). Some examples follow.

For cut apple and potato, the use of a cellulose-based edible coating, as carrier of antioxidants, acidulants and preservatives, prolonged storage life by one week when stored in overwrapped trays at 4°C. Ascorbic acid and some of its derivatives formulated as a coating delayed browning more effectively than when applied in an aqueous solution (Baldwin *et al.*, 1996).

Combination of several browning inhibitors (like potassium sorbate and isoascorbic acid) was more effective to reduce browning of fresh cut mangoes than those applied individually. Browning in air stored product was attributed to degradation of the tissue induced by dryness of the surface, and use of MAP to maintain high humidity was helpful (González Aguilar *et al.*, 2000).

The quality of litchi greatly depends on postharvest treatments to suppress peel browning. In order to replace current standard treatment with SO_2, a potentially hazardous chemical, spraying with hot water (55°C, 20s) while being brushed was evaluated. The PPO activity was reduced and fruits maintained the anthocyanins in their red-pigmented form, showing a uniform red colour for at least 35 days (Lichter *et al.*, 2000; Underhill and Critchley, 1992).

The development of genetic technology and engineering for optimising colour during postharvest life of fruit and vegetables is presently a key objective of several research groups in the world.

20.8 References

ABE K A and WATADA A E, (1991), 'Ethylene absorbent to maintain quality of lightly processed fruits and vegetables', *J Food Sci*, 56, 1493–96.

AHVENAINEN R, (1996), 'New approaches in improving the shelf life of fresh-cut fruits and vegetables', *Trends Food Sci Technol.*, 7, 179–86.

ALLENDE A, JACXSENS L, DEVLIEGHERE F, DEBEVERE J and ARTÉS F, (2002), 'Effect of superatmospheric oxygen packaging on sensorial quality, spoilage, and *Listeria monocytogenes* and *Aeromonas caviae* growth in fresh processed mixed salads', *J Food Protect*, 65(10), 1565–73.

ARIAS R, LEE TC, LOGENDRA L and JANES H, (2000), 'Correlation of lycopene measured by HPLC with the L*, a*, b* color readings of the hydroponic tomato and the relationship of maturity with color and lycopene content'. *J Agric Food Chem*, 48, 1697–1702.

ARTÉS F, (1993), 'Diseño y cálculo de polímeros sintéticos de interés para la conservación hortofrutícola en atmósfera modificada', in: *Nuevo curso de ingeniería del frío,* Edit A. Madrid. Ch. 16, 427–54.

ARTÉS F, (1995), 'Innovaciones en los tratamientos físicos modulados para preservar la calidad hortofrutícola en la postrecolección. III. Tratamientos gaseosos', *Rev Esp Ciencia Tecnol Alim*, (*Food Sci Technol Inter*), 35 (3) 247–69.

ARTÉS F, (2000a), 'Conservación de los productos vegetales en atmósfera modificada', in: Lamúa M. (ed.), *Aplicación del frío a los alimentos,* A. Madrid-Mundi-Prensa, Madrid, Spain, Ch. 4. 105–25.

ARTÉS F, (2000b), 'Productos vegetales procesados en fresco', in: Lamúa M. (ed.), *Aplicación del frío a los alimentos*. A. Madrid-Mundi-Prensa, Madrid, Spain, Ch. 5. pp. 127–41.

ARTÉS F and ESCRICHE A, (1994), 'Intermittent warming reduces chilling injury and decay of tomato fruit', *Journal of Food Science*, 59, 1053–56.

ARTÉS F, MARÍN G and MARTÍNEZ J A, (1996), 'Controlled atmosphere storage of pomegranate', Z. *Lebensm. Unters Forsch*, 203, 33–37.

ARTÉS F, CASTAÑER M and GIL M I, (1998), 'Enzymatic browning in minimally processed fruit and vegetables', *Food Sci Technol Inter*, 4 (6), 377–89.

ARTÉS F, VILLAESCUSA R and TUDELA JA, (2000a), 'Modified atmosphere packaging of pomegranate', *J Food Sci*, 65 (7), 1112–16.

ARTÉS F, MARÍN J G, PORRAS I and MARTíNEZ J A, (2000b), 'Evolución de la calidad del limón, pomelo y naranja durante la desverdización', *Rev Iberoamericana Tecnología Postcosecha*, 1 (2), 71–9.

ARTÉS F, ESCALONA V H and ARTÉS-HDEZ F, (2002a), 'Modified atmosphere packaging of fennel', *Journal of Food Science*, 64, 1550–4.

ARTÉS F, GÓMEZ P and ALLENDE A, (2002b), 'Calidad y actividad fisiológica de apio procesado en fresco y conservado en atmósfera controlada', *I Congreso Español Ciencias y Técnicas del Frío. CYTEF 2002*, Cartagena, Spain, p. 52.

ARTÉS F, MÍNGUEZ M I and HORNERO D, (2002c), 'Analysing changes in fruit pigments', in: *Colour in food. Improving quality*. Ed. D. B. MacDougall. CRC Press and Woodhead Publishing Ltd. Ch. 10. 248–82.

ARTÉS F, ARTÉS-HDEZ F, TUDELA JA and THILINGUIRIAN D, (2002d). Action du CO_2 sur la qualité des cerises réfrigérés. *Rev Générale Froid*. 1021: 17–21.

BAI J, SAFTNER R, WATADA A and LEE Y, (2001), 'Modified atmosphere maintains quality of fresh cut cantaloupe (*Cucumis melo* L.)', *J Food Sci* 66 (8), 1207–12.

BALDWIN E A, NISPEROS M O, CHEN X and HAGENMAIER R D, (1996), 'Improving storage life of cut apple and potato with edible coating', *Postharvest Biol Technol.*, 9, 151–63.

BARTH M and ZHUANG H, (1996), 'Packaging design affects antioxidant vitamin retention and quality of broccoli florets during postharvest storage', *Postharvest Biol Technol*, 9, 141–50.

BLANCHARD M, CASTAIGNE F, WILLEMOT C and MAKHLOUF J, (1996), 'Modified atmosphere preservation of freshly prepared diced yellow onion', *Postharvest Biol Technol* 9, 173–85.

BROUILLARD R, FIGUEIREDO P, ELHABIRI M and DANGLES O, (1997), 'Molecular interactions of phenolic compounds in relation to the colour of fruit and vegetables', in: *Phytochemistry of Fruit and Vegetables*. Edited by F A Tomás-B. and R J Robins. Oxford Science Publications 3, 29–48.

CANTWELL M, (1992), 'Postharvest handling systems: Minimally processed fruits and vegetables', in: *Postharvest Technology of Horticultural Crops*. (Kader, A A ed.) 277–281, 2nd edn. publ. 3311. University of California. USA.

CANTWELL M and SUSLOW T, (2002), 'Postharvest handling systems: fresh-cut fruits and vegetables', in: *Postharvest Technology of Horticultural Crops*. (Kader, A A ed.) 277–281, 3rd edn. publ. 3311. University of California. USA. Ch 36, 445–63.

CASTAÑER M, GIL M I, RUIZ M and ARTÉS F, (1999), 'Browning susceptibility of minimally processed Baby and Romaine lettuces', *Eur Food Res. Technol*, 209, 52–6.

CIE, (1986), *Colorimetry*, Central Bureau of the Commission Internationale de l'Éclairage. Vienna, 2nd edn. Publication CIE 15, 2.

CRISOSTO C, CRISOSTO G and RITENOUR M, (2002), 'Testing the reliability of skin color as an indicator of quality for early season 'Brooks (*Prunus avium* L.) cherry', *Postharvest Biol Technol*, 24, 147–54.

DAY B, (1994), 'High oxygen modified atmosphere packaging for fresh prepared produce', *Postharvest News Information*, 7 (3), 31N-4N.

EAKS IL, (1977), 'Physiology of degreening – summary and discussion of related topics', *Proc. Int. Soc. Citriculture*, 1, 223–6.

FERNÁNDEZ-TRUJILLO J P and ARTÉS F, (1997), 'Quality improvement of peaches by intermittent warming and modified-atmosphere packaging', *Z Lebesm Unters Forsch A*, 205, 59–63.

FERNÁNDEZ-TRUJILLO J P, MARTÍNEZ J A and ARTÉS F, (1998), 'Modified atmosphere packaging affects the incidence of cold storage disorders and keeps "flat" peach quality', *Food Res Intern*, 31 (8), 571–9.

FRANCIS F J, (1969), 'Pigment content and color in fruits and vegetables', *Food Technol*, 23, (32), 32–36.

GIL M I, MARTÍNEZ J A and ARTÉS F, (1996a), 'Minimally processed pomegranate seeds', *Lebensm.-Wiss. u. Technol.*, 29, 708–13.

GIL M I, ARTÉS F, and TOMÁS BARBERÁN F, (1996b), 'Minimal processing and modified atmosphere packaging effcts on pigmentation of pomegranate seeds', *J Sci Food Agri*, 80, 1545–52.

GÓMEZ P, CONESA A, ARTÉS-HDEZ F, MÉNDEZ M, AGUAYO E and ARTÉS F, (2002), 'Influencia de la desverdización en la calidad de pimiento recolectado en dos estados de madurez'. *Maduración y Postrecolección 2002. VI Simposio Nacional y III Ibérico.* Madrid, 2–5 October 2002.

GONZÁLEZ-AGUILAR GA, WANG CY and BUTA JG, (2000), 'Maintaining quality of fresh-cut mangoes using antibrowning agents and modified atmosphere packaging', *J Agric Food Chem.* 48, 4204–8.

GORINI F, (1979), 'La frigoconservazione dei prodotti ortofrutticoli', *Manuale Ireda*, 285 pp.

GORNY J, HESS-PIERCE B, CIFUENTES R and KADER A A, (2002), 'Quality changes in fresh cut pear slices as affected by controlled atmospheres and chemical preservatives', *Postharvest Biol Technol* 24, 271–8.

GUNES G and LEE C, (1997), 'Color of minimally processed potatoes as affected by modified atmosphere packaging and antibrowning agents', *J Food Sci* 62 (3), 572–5.

HAFFNER K, ROSENFELD H, SKREDE G and WANG L, (2002), 'Quality of red raspberry *Rubus idaeus* L. cultivars after storage in controlled and normal atmospheres', *Postharvest Biol Technol* 24, 279–89.

HOLCROFT D M and KADER A A, (1999a), 'Carbon dioxide-induced changes in color and anthocyanin synthesis of stored strawberry fruit', *HortScience* 34 (7), 1244–8.

HOLCROFT D M and KADER A A, (1999b), 'Controlled atmosphere-induced changes in pH and organic acid metabolism may affect color of stored strawberry fruit', *Postharvest Biol T*echnol 17, 419–32.

HOLCROFT D M, GIL M I and KADER A A , (1998), 'Effect of carbon dioxide on anthocyanins, phenylalanine ammonia lyase and glucosyltransferase in the arils of stored pomegranates', *J Amer Soc Hort Sci* 123 (1), 136–40.

HÖRTENSTEINER S (1999), 'Chlorophyll breakdown in higher plants and algae', *Cell Mol Life Sci.* 56, 330–47.

JACXSENS L, DEVLIEGHERE F and DEBEVERE J, (2001), 'Effect of high oxygen modified atmosphere packaging on packaging on microbial growth and sensorial qualities of fresh-cut produce', *Int. J. Food Microbiol.*, 71(2–3), 197–210.

JIANG Y, ZAUBERMAN G and FUCHS Y, (1997), 'Partial purification and some properties of polyphenol oxidase extracted from litchi fruit pericarp', *Postharvest Biol Technol* 10, 221–8.

KADER A A, (1990). 'Modified atmospheres during transport and storage of fresh fruits and vegetables', *I Intern. Cong Food Technology Development*. Murcia. Spain. 1: 149–63.

KADER A and BEN-YEHOSHUA S, (2000), 'Effects of superatmospheric oxygen levels on postharvest physiology and quality of fresh fruits and vegetables', *Postharvest Biol Technol*, 20, 1–13.

KIDMOSE U, EDELENBOS M, NORBAEK R and CHRISTENSEN L P. (2002) 'Colour stability in vegetables', in: *Colour in food. Improving quality*. Ed. D B Mac Dougall. CRC Press and Woodhead Publishing Ltd. Ch.8. 179–232.

LANCASTER J E, LISTER C E, REAY P F and TRIGGS C M, (1997), 'Influence of pigment composition on skin in a wide range of fruit and vegetables', *J Amer Soc Hort Sci.*, 122, 4, 594–8.

LAU M, TANG J and SWANSON B, (2000), 'Kinetics of textural and color changes in green asparagus during thermal treatments', *J Food Eng*, 45, 231–6.

LESHEM Y Y, (2000), 'Nitric oxide in plants. Occurrence, functions and use'. Ed. Kluver Academic Pub. 154 pp.

LICHTER A, DVIR O, ROT I, AKERMAN M, REGEV R, WIESBLUM A, FALLIK E, ZAUBERMAN G and FUCHS Y, (2000), 'Hot water brushing: an alternative method to SO_2 fumigation for color retention of litchi fruits', *Postharvest Biol Technol* 18, 235–44.

LIN T Y, KOEHLER P E and SHEWFELT R L, (1989), 'Stability of anthocyanins in the skin of Starkrimson apples stored unpackaged, under heat shrinkable wrap and in-package modified atmosphere', *J Food Sci* 54 (2), 405–7.

LISTER C E, (1994), 'Biochemistry of fruit color in apples (*Malus pumila* mill.)', in: Lancaster, Lister, Reay and Triggs (eds), 'Influence of pigment composition on skin in a wide range of fruit and vegetables'. *J Amer Soc Hor. Sci* 122; (4), 594–98.

MCGUIRE R G, (1992), 'Reporting of objective color measurements', *HortScience* 27, 12, 1254–5.

MADIETA E, (2002), 'Apple colour measurements. Some metrological approaches', *Postharvest Unlimited*. Leuven, Belgium, June.

MARSILI R, (1996), *Food color: more than meets the eye. A guide to understanding color tolerancing*, X-Rite, Inc., 3100 44th St. SW, Grandville, MI 49418.

MARTÍ N, PÉREZ-VICENTE A and GARCÍA-VIGUERA C, (2001), 'Influence of storage temperature and ascorbic acid addition on pomegranate juice', *J Sci Food Agric* 82, 217–21.

MATILE P, SCHELLENBERG M and VICENTINI F, (1997), 'Localization of chlorophyllase in the chloroplast envelope', *Planta* 201, 96–9

MENCARELLI F, DE SANTIS D, MASSANTINI R and CONTINI M, (1990), 'Impiego delle ridotte concentrazione di ossigeno nella breve conservazione della cipolla in fette', *Ind. Alim,* 29, 982–4.

MURR D P and MORRIS L L, (1974), 'Influence of O_2 and CO_2 on O-diphenol oxidase activity in mushrooms', *J Amer Soc Hort Sci*, 99, (2), 155–8.

O'CONNOR-SHAW R E, ROBERTS R, FORD, A L and NOTTINGHAM S E, (1995), 'Shelf life of minimally processed honeydew melon, kiwifruit, papaya, pineapple and cantaloupe', *J Food Sci,* 59 (5), 1–6.

O'HARE T and WONG R, (2000), 'Slowing leaf yellowing in Asian vegetables', *Access to Asian Foods*, 7, 2–4.

ONTAI S, PAULL R and SALTVEIT M, (1992), 'Controlled atmosphere storage of sugar peas', *HortScience*, 27, 39–41.

ORTEGA-ZALETA D and YAHIA E, (2000), 'Tolerance and quality of mango fruit exposed to controlled atmospheres at high temperatures', *Postharvest Biol Technol*, 20, 195–201.

PANTASTICO E R B, (1979), 'Fisiología de la postrecolección, manejo y utilización de frutas y hortalizas tropicales y subtropicales', Edit. Cia Editorial Continental. 663 pp.

PARIASCA J A, MIYAZAKI T, HISAKA H, NAKAGAWA H and SATO T, (2001), 'Effect of modified atmosphere packaging and controlled atmosphere storage on the quality of snow pea pods (*Pisum sativum* L. var *saccharatum*)', *Postharvest Biol Technol*, 21, 213–23.

PEISER G, LÓPEZ-GALVEZ G, CANTWELL M and SALTVEIT M, (1998), 'Phenylalanine ammonia lyase inhibitors control browning of cut lettuce', *Postharvest Biol Technol*., 14, 171–7.

RAKOTONIRAINY A, WANG Q and PADUA G, (2001), 'Evaluation of zein films as modified atmosphere packaging for fresh broccoli', *J Food Sci*, 66, 8, 1108–11.

REMÓN S, FERRER A, MARQUINA P, BURGOS J and ORIA R, (2000), 'Use of modified atmosphere to prolong the postharvest life of Burlat cherries at two different degrees of ripeness', *J Sci Food Agric*, 80, 1545–52.

ROY S, ANANTHESWARAN C AND BEELMAN R (1996), 'Modified atmosphere and modified humidity packaging of fresh mushrooms', *J Food Sci*, 61, (2), 391–7.

SÁNCHEZ-FERRER A, RODRÍGUEZ-LÓPEZ J N, GARCÍA-CÁNOVAS F and GARCÍA-CARMONA F, (1995), 'Tyrosinase: a comprehensive review of its mechanism', *Biochim Biophys Acta*, 1247, 1–11.

SCHWARTZ S J and VON ELBE J, (1983), 'Kinetics of Chlorophyll degradation to pyropheophytin in vegetables', *J Food Sci*, 48, 1303–6.

SHEWFELT R L, THAI C N and DAVIS J W, (1988), 'Prediction of changes in color of tomatoes during ripening at different constant temperatures', *J Food Sci*., 53, (5), 1433–7.

SIOMOS A S, SFAKIOTAKIS E M and DOGRAS C, (2000), 'Modified atmosphere packaging of white asparagus spears: composition, color and textural quality responses to temperature and light', *Scientia Horticulturae*, 84, 1–13.

SIOMOS A, DOGRAS C and SFAKIOTAKIS E, (2001), 'Color development in harvested white asparagus spears in relation to carbon dioxide and oxygen concentration', *Postharvest Biol Technol*, 23, 209–14.

SIRIPHANICH J and KADER A, (1985), 'Effects of CO_2 on total phenolics, phenylalanine ammonia lyase, and polyphenol oxidase in lettuce tissue', *J Amer Soc. Hor. Sci,* 110, (2), 249–53.

SOLIVA-FORTUNY R, GRIGELMO-MIGUEL N, ODRIOZOLA-SERRANO I, GORINSTEIN S and MARTÍN-BELLOSO O, (2001), 'Browning evaluation of ready to eat apples as affected by modified atmosphere packaging', *J Agric. Food Chem*, 49, 3685–90.

UNDERHILL S J and CRITCHLEY C, (1992), 'The physiology and anatomy of lychee pericarp during fruit development', *J Hort. Sci*., 67, 437–44.

VAN DE VELDE M and HENDRICKX M, (2001), 'Influence of storage atmosphere and temperature on quality evolution of cut belgian endives', *J Food Sci*.

66, (8), 1212–18.

VAROQUAUX P, MAZOLLIER J and ALBAGNAC G, (1996), 'The influence of raw material characteristics on the storage life of fresh-cut butter head lettuce'. *Postharvest Biol Technol.*, 9:127–39

VERLINDEN B E, GILLIS N, LAMMERTYN J and NICOLAÏ B, (2001). 'Red discoloration of chicory under modified atmosphere storage conditions', in: *Improving Postharvest Technologies of Fruits, Vegetables and Ornamentals*. Artés F, Gil M I and Conesa M A (eds). International Institute of Refrigeration. 2: 650–7.

VILLAESCUSA R, TUDELA J A and ARTÉS F, (2001), 'Influence of temperature and modified atmosphere packaging on quality of minimally processed pomegranate seeds', in: *Improving Postharvest Technologies of Fruits, Vegetables and Ornamentals*. Artés F, Gil M I and Conesa M A (eds). International Institute of Refrigeration. 1: 445–50.

WRIGHT K P and KADER A, (1997), 'Effect of controlled-atmosphere storage on the quality and carotenoid content of sliced persimmons and peaches', *Postharvest Biol Technol*, 10, 89–97.

YANG CC and CHINNAN M S, (1987), 'Modelling of color development of tomatoes in modified atmosphere storage', *Transactions of the ASAE*, 30,2, 548–53.

ZHANG D, QUANTICK P, WIKTOROWICZ R and IRVEN J, (2000), 'Noble gases for modified atmosphere packaging of fresh fruits and vegetables', *Critical Rev in Food Sci*.

Part IV

General issues

21

Optimizing packaging

T. Lyijynen, E. Hurme and R. Ahvenainen, VTT Biotechnology, Finland

21.1 Introduction

Package design has great significance for the success of foodstuffs nowadays. Packages are clearly an integral part of the manufacturing and distribution processes. As clothes speak for their wearers, so too packages speak for the packed food product. Packages are developed not only to make weekdays easier for the consumer, but also to make times of celebration more festive. Many food products would not be in shops and on dining tables, if they had not been packed. Nowadays packages face difficult challenges and roles. They have to create the ambience that hitherto was forged by personal service. Packages replace the salesman.

In addition, packaging has many other functions and requirements which it has to fulfil more and more effectively and economically. These functions and requirements are changing all the time and their importance in ensuring the success of the product is growing. The aim is to make the optimal package that satisfies all functional requirements in addition to meeting environmental and cost demands as well as possible. The answer to these complex demands is precision packaging.

VTT Precision Packaging Concept has been developed in the Technical Research Centre of Finland and it is a new and unique tool to optimize packaging for foodstuff. Elsewhere optimization methods for transport packages have been developed, but not for primary packages, which are in direct contact with foodstuff. The VTT Precision Packaging Concept is based on the predetermined minimum shelf-life needed to allow the packed foodstuff to reach the consumer's table from the factory. The shelf-life is naturally chosen to suit the market and the business strategy of the company. A longer shelf-life is needed for an export market than for a home market. Depending on the business strategy, the company can favour either short or long best-before times. It should be noted that the selling time of the product does not necessarily have any other connection to the real shelf-life but, naturally, the selling time is always shorter than the shelf-life. This chapter gives a description of the VTT Precision Packaging Concept for the optimization of food packages. At the end of the chapter, examples of optimization are given using different gas-packed foodstuffs.

21.2 Issues in optimizing packaging

The basic aim in the optimization of food packages is to create a tool for decision-making policy for launching new products. The packaging optimization concept should not only advantageously help food manufacturers, but also packaging material and packaging manufacturers to evaluate and compare the feasibility of their new and present products. In today's competitive market, packaging innovation can be a big advantage in efforts to persuade the consumer to buy a certain brand, and the packaging optimization concept may help to reduce the economic support and time needed for package development. Several factors should be considered in the optimization process for packages, covering the performance in logistics, marketing properties, consumer convenience, costs, and environmental stresses (see Fig. 21.1). The general goals in package optimization are the cost-effectiveness of the packaging process and environmental issues, e.g. source reduction of packaging materials. The importance of these factors is discussed below.

21.2.1 Performance in logistics

One of the most important tasks of a food package is to afford protection from environmental conditions, like oxygen, light and moisture. This is crucial for maintaining the quality and safety of most packaged foods. Therefore, it is essential that a package has sufficient mechanical strength to protect the packaged product from environmental stresses during distribution and storage. Environmental issues, however, demand that packaging material consumption is kept to a minimum and that packaging materials be recoverable. This requires the food industry to use thinner materials that still have both sufficient mechanical strength and barrier properties. An optimized, downsized package may also reduce wholesaler and retailer costs. On the other hand, in some cases it may also be favourable to use relatively thick, recyclable monomaterial layers (e.g. polyethylene) suitable for energy recovery. In other words, an optimized food package minimizes the waste in the overall packaged product.

New packaging solutions should also be technically feasible. That is, they may not set any limitations on either packaging speeds or the quality of seals. Package dimensions need to be logistically congruent with the secondary

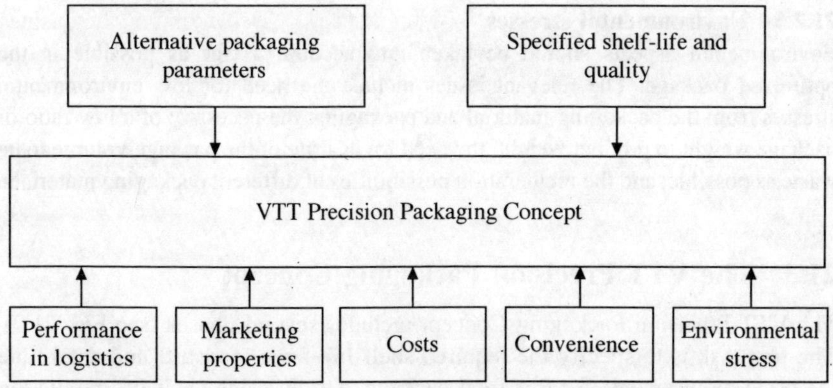

Fig. 21.1 The VTT Precision Packaging Concept is a customized tool for optimization of food packaging.

package and pallet. The ratio of packaged product to package volume should be as high as possible.

21.2.2 Marketing properties

The marketing properties of a package should be fulfilled as well as necessary, not as well as possible. This means that, optimally and cost-effectively, only adequate investment is needed to fulfil the need for, e.g., package design in terms of packaging material consumption, decoration and information. Keeping up the brand image of a product should be taken into account when making packaging decisions. It is important that package designers, manufacturers and users can co-operate closely to achieve an optimum package.

21.2.3 Costs

From a food manufacturer's point of view, especially, the costs of primary packaging materials as well as indirect packaging costs (storage, transportation, energy consumption, and labour costs) should certainly be as low as possible. A cost-competitive package is, of course, a benefit for both the packaging and the food manufacturers. A packaging innovation requiring a minimum of investment and giving consumers products they like, at affordable prices, can be seen to be a very attractive goal. Such innovations could be, e.g., easy-open seals or thinner materials which consumers find more environmentally friendly.

21.2.4 Consumer convenience

Such properties of a package that make it convenient for the consumer include it being easy to handle, carry, store and dispose of/re-use, as well as its openability, resealability, and microwaveability.

21.2.5 Environmental stresses

Environmental aspects should be taken into account as far as possible in the optimized package. The relevant issues include the need for low environmental stresses from the packaging material and packaging, the necessity of a low ratio of package weight to product weight, the need for as little of the package volume to be waste as possible, and the incineration possibilities of different packaging materials.

21.3 The VTT Precision Packaging Concept

The VTT Precision Packaging Concept includes several phases (see Fig. 21.2). The first task is to specify the required shelf-life for a foodstuff and determine the basic requirements for this food package, e.g. by using shelf-life prediction models. After that it is possible to choose different combinations of package types and to optimize the package. Optimization is performed in four steps: (i) scoring the tested package types, (ii) evaluating the importance of package characteristics, (iii) calculating the coefficients for each of the characteristics, and (iv) calculating the optimization result of each of the tested package types.

21.3.1 Determination of basic requirements for packaging

Precision packaging is based on a pre-determined minimum shelf-life needed to allow the packed foodstuff to reach the consumer's table from the factory. The shelf-life is naturally chosen to suit the market and the business strategy of the company. A longer shelf-life is needed for an export market than for a domestic market.

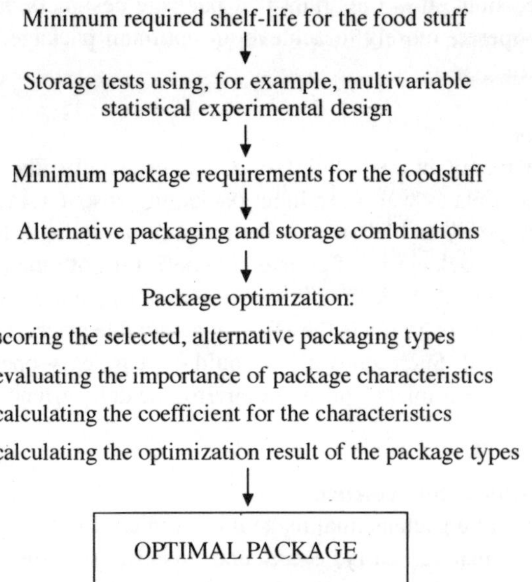

Fig. 21.2 A scheme for determining the optimal package.

The VTT Concept uses mathematical modelling to determine the minimum package requirements for the foodstuff. The modelling of the shelf-life of foodstuffs is performed in the following steps: (i) selecting factors and responses, (ii) selecting the experimental design method (screening or more extensive method), (iii) carrying out tests, (iv) analysing results, (v) making shelf-life predictions and (vi) determining the minimum package requirements. The factors that might be relevant and might affect the quality deterioration of the foodstuff are chosen, for example, oxygen transmission rate of a packaging material, carbon dioxide concentration of a package, package volume and storage conditions (temperature, illumination, etc.). The quality parameters in shelf-life testing of food can include sensory and chemical analyses and microbiological determinations. The minimum package requirements can also be based both on the information of the shelf-life tests and on the literature data.

21.3.2 Determining different package and storage combinations

After the required packaging parameters for a foodstuff at certain storage conditions have been determined, it is possible to select the different alternatives of packages and packaging materials, packaging methods and package conditions all of which give equally acceptable food quality at the end of the required shelf-life of the product (e.g. vacuum or gas-flushed package with specified packaging material and gas concentrations).

21.3.3 Scoring the selected alternative package types

The selected, alternative package types giving the same minimum required shelf-life, are first scored by rating different characteristics of each package type. As an example, the following package attributes can be used in a Precision Packaging Concept:

- mechanical strength of a package
- suitability with respect to packaging standards
- ratio of package weight to product weight
- volume of package waste
- possibility of incineration
- life cycle assessment of packages
- marketing properties of a package
- consumer convenience
- consumer attitudes
- cost of packaging material
- indirect packaging costs.

All these characteristics are scored on a scale from 1 to 5, with 1 corresponding to poor and 5 to excellent quality. Fractional numbers are also allowed. The scoring is performed in co-operation with experts representing wholesalers, food manufacturers, packaging material manufacturers and independent specialists of packaging technology and plastics industry.

Mechanical strength of a package

This is determined using a simulated transportation test. For example in the VTT transportation test, a whole pallet positioned on a vibration table is subjected to a simulated road transportation of about 1000 km. As an example, during this 45-minute test run, the table is vibrated at a constant acceleration of 0.5 g, with the frequency of the table sweeping between 5 and 55 Hz. Since each package type resonates at its own specific frequency, every package tested is exposed to more or less high accelerations during the test procedure. In general, the packages in the top layer of a pallet are exposed to the highest accelerations, which can be even more than 6 g. After the test, all the possible flaws, such as flexes, fractures, open seals and product disorientation originating from the transportation tests are recorded, and the test packages are scored as follows:

- 1 point: if even one tested package in the pallet is leaking, the whole sample group is rejected.
- 3 points: there are some minor dents or flexes.
- 5 points: no package is affected by the vibration test.

Suitability with respect to packaging standards

This is evaluated by examining the volume occupied by the primary packages tested in a secondary package, which had the dimensions of the pallet area. Scores between 1 and 5 are given on the following basis:

- 1 point: relatively large empty space left in the secondary package.
- 5 points: primary packages tight together, no empty space in the secondary package.

Ratio of package weight to product weight

The score for this ratio (= *Sm*) is calculated using eqn 1

$$Sm = 0.444 \cdot (1 - m_p/m_f \cdot 100) + 5 \tag{1}$$

where m_p = weight of the empty package and m_f = weight of the packaged food. Equation 1 was formulated on the following basis:

- 1 point corresponds to an empty package weighing 10% of the weight of the packaged product, and
- 5 points correspond to an empty package weighing 1% of the weight of the packaged product.

Volume of package waste

This is scored from 1 to 5 on the following basis:

- 1 point: a rigid package, e.g. tray, which cannot be scrunched and which would require a large space in a household waste basket.
- 5 points: a package that is made of a thin flexible material, which can easily be put in a household waste basket without the need for stuffing or scrunching.

Suitability of the empty package for incineration
This is evaluated as follows:

- 1 point: cannot be incinerated under any conditions, e.g., metal can.
- 2 points: can be incinerated in a plant specialized for problem waste.
- 3 points: can be incinerated in a plant for municipal waste.
- 4 points: can be incinerated in a power plant for fossil fuel resources.
- 5 points: can be incinerated in the home.

Marketing properties of a package
These are evaluated by awarding scores between 1 and 5 on the following basis:

- 1 point: package design or packaging material does not meet the requirements of brand/company image and logistics, it has poor printability, insufficient space for labels and poor visibility on the shelf.
- 5 points: excellent package design, proper packaging material, excellent printability, enough space for labels, excellent visibility of the packaged product on the shelf.

Consumer convenience
This is evaluated by awarding scores between 1 and 5 points as follows:

- 1 point: package dimensions and sharp corners make it difficult to handle, carry, store and dispose of, it is difficult to open without a tool and/or spillage of the product, it is not resealable, and is not suitable for microwave ovens.
- 5 points: package is easy to handle, carry, store and dispose of/re-use, as well as being easy to open, and reseal, and is suitable for microwave ovens, if necessary.

Scored costs of packaging material
These (= *Sc)* are calculated using eqn 2 based on the annual production reports given by the food manufacturers and the corresponding cost information for the necessary primary packaging materials given by the different packaging material manufacturers. The cost of the packaging material is first scored by the food manufacturers, as follows:

- 1 point represents a packaging material cost that is far too high compared to the total manufacturing cost of the product, whereas
- 5 points equals a very economical ratio of packaging material costs to the total manufacturing cost.

$$Sc = [4/(c_1 - c_5)]^{-1} \cdot (c_5 - c_x) + 5 \tag{2}$$

where c_1 and c_5 are the costs corresponding to 1 and 5 points, respectively, awarded by the food manufacturer, and c_x is the cost of the packaging material of the sample. For example, if the food manufacturers rate packaging materials costs of 0.03 €/package for a specified product as very economical whereas 0.13

€/package is far too expensive, 5 points would be awarded to 0.03 €/package, 3 points to 0.08 €/package and 1 point 0.13 €/package.

Indirect packaging costs
These are estimated by including storage, transportation, energy consumption and labour costs. The estimations are made qualitatively. That is, different package types (flowpacks, form-fill-sealed-packs, preformed trays) are ranked against each other on a scale of 1 to 5.

21.3.4 Evaluating the importance of package characteristics

After the characteristics of the selected, alternative packages have been scored, the importance of the characteristics is evaluated by the food and packaging material manufacturers, the wholesaler and independent packaging technology and plastics industry specialists. As an example, the importance of the characteristics evaluated by the food and packaging material manufacturers is presented in Fig. 21.3.

21.3.5 Calculating the coefficients

The coefficients I for each of the characteristics are then calculated as follows (eqn 3):

$$I = 0.4 \cdot i_f + 0.25 \cdot i_p + 0.25 \cdot i_w + 0.1 \cdot i_s \tag{3}$$

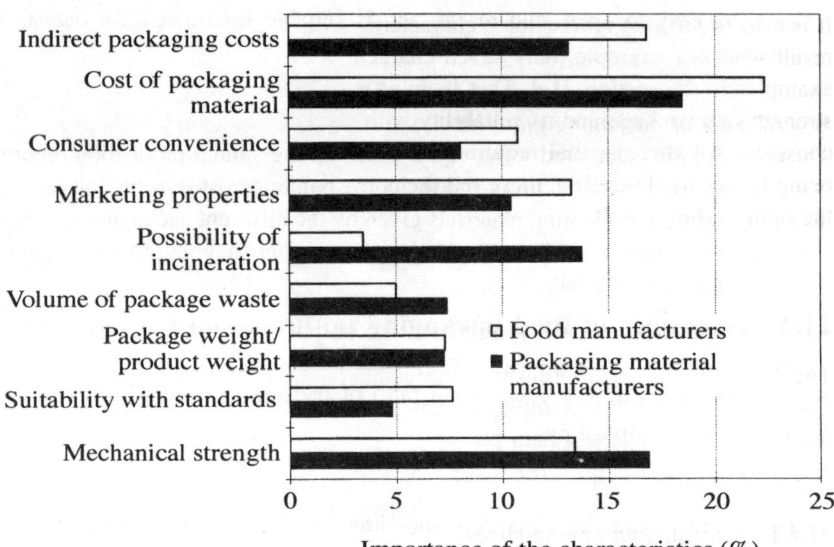

Fig. 21.3 The importance of the various characteristics of packages evaluated by Finnish food manufacturers and packaging material manufacturers.

where i_f is the importance of a given package characteristic as evaluated by the food manufacturer of the product, i_p is the mean rating for the importance of the package characteristic as evaluated by the packaging material manufacturers of the product, i_w is the importance of the package characteristic as evaluated by the wholesaler, and i_s is the mean importance of the package characteristic as evaluated by the independent specialist for the packaging technology and plastics industry. Since food manufacturers have a very significant role in the decision-making process in terms of package performance, applicability and functionality, the weighting of i_f with a coefficient of 0.4 ensures that its influence on the coefficient I is the greatest. The significance of the influence of both the wholesaler and the packaging material manufacturers is rated as being quite important and so each of their importance scores is weighted with a coefficient value of 0.25. The influence of the independent specialists is rated the lowest and the corresponding importance score is, therefore, weighted with a coefficient value of 0.1. The coefficients used in eqn 3 were agreed between all the participants referred to above and VTT.

21.3.6 Calculating the optimization result

The score, s, given to each of the characteristics of the tested packages is multiplied by the corresponding coefficient, I, given to these nine characteristics, where OR is the optimization result of a certain tested package type (eqn 4):

$$OR = s_1 \cdot I_1 + s_2 \cdot I_2 + s_3 \cdot I_3 + \ldots + s_9 \cdot I_9 \tag{4}$$

It is also possible to ignore some characteristics, and calculate the optimization result with, for example, only seven characters. In fact, this is the case in the examples in the section 21.4. That is, in an optimized package, the mechanical strength of a package and its suitability with respect to packaging standards are considered as the essential requirements that all packages should fulfil before being launched. Therefore, these requirements can be ignored when calculating the optimization result.

21.4 Examples of food packaging optimization

The following examples illustrate the using of the VTT Precision Packaging Concept. The foodstuff examples are gas-packed cheese slices, raw chicken legs, roasted chicken balls and ham pizza.

21.4.1 Gas-packed cheese slices

Storage and packaging are particularly important factors for the shelf-life of cheese. There are some basic packaging requirements that are the same for all types of cheese: firstly, oxygen must be excluded to prevent mould growth and rancidity, and secondly moisture must be retained to preserve the texture and avoid weight loss. The effect of light must also be considered in the packaging of cheese. Modified atmosphere packaging can be used to package soft and crumbly textured cheeses without damaging them. Modified atmosphere packages are easier to open than vacuum packages and they also extend the shelf-life of cheeses. The example of packaging optimization was for sliced matured cream cheese. The amount of cheese slices in packages was 150 g. Modified atmosphere packaged cheese slices have a last day of use 13 weeks after packaging when the storage temperature is below +8ºC.

According to the results of the modelling test, an ideal gas-package for sliced matured cream cheese, considering a shelf-life of eight weeks (temperature 6ºC) and keeping its quality to the maximum, may have following properties:

- protection from light during distribution
- oxygen transmission rate of packaging material: 5–95 $cm^3/m^2 \cdot 24\,h$ (at 23ºC, 50% RH and latm)
- carbon dioxide concentration: 20%
- residual oxygen level: $< 1\%$
- headspace volume: 55–275 ml/100 g cheese.

The next step was to choose the different packaging alternatives for sliced cheese (see Table 21.1). Then the characteristics of selected, alternative packages scored and the importance of the characteristics evaluated by food and packaging material manufacturers, the wholesalers and independent packaging specialists (see Table 21.2). The most important factors affecting the optimization result were indirect packaging costs, marketing properties, and cost of packaging materials. The optimization results suggest that the selected, alternative packages differed most in the costs of their packaging materials.

To compare the best (Package 4), see Table 21.2 and worst rated (Package 2), the following estimations can be made. If the annual production of the product is 10 million packages, Package 4 saves about 30 000 kg of packaging material annually compared to Package 2. Similarly, Package 4 saves about €109 000 annually in packaging material costs compared to Package 2.

21.4.2 Gas-packed raw chicken legs

The shelf-life of cold stored fresh poultry is mainly restricted by microbial growth, especially by *Pseudomonas* and *Acinetobacter*. Psychrotrophs contaminate easily during slaughter, because psychrotrophs appear often in feathers and legs. Growth of these bacteria can effectively be slowed down using gas packaging. Spoilage can be noticed in sensory quality when the amount of bacteria is 10^6–10^8 cfu/cm^2, and slime formation exists when the amount of bacteria is 10^6–10^8 cfu/cm^2.

The example of packaging optimization was for fresh industrially slaughtered chicken legs. The average weight of one chicken leg was 277 g and volume 262 ml^3. The last date of use is seven days after packaging when the storage

Table 21.1 The sliced matured cream cheese packages used in the package optimization. Package information: 150 g cheese slices in the package and the package volume about 350 ml^3. Secondary package was a cartonboard box

Feature	Package 1	Package 2	Package 3	Package 4	Package 5	Package 6
Package type	tffs	tffs	tffs	tffs	tffs	tffs
Monomaterials used in package	PA, PE, EVOH	PA, PE, EVOH, mPET	PA, PE, PP, mPET	PA, PE PP, EVOH	PA, PE, PP, EVOH	PA, PE, EVOH
Weight of an empty package (g)	7.6	7.7	4.9	4.8	5.4	5.0
Transparency	clear lid and tray	metallized lid, clear tray	metallized lid, clear tray	white lid, clear tray	white lid, clear tray	clear lid and tray

Abbreviation: tffs = thermoformed tray/fill/seal

Table 21.2 The scores of various characteristics of tested cheese slices packaging types, the calculated coefficients for each characteristic and the optimization results for tested packages

Characteristic	*I*	Scores					
		Package 1	Package 2	Package 3	Package 4	Package 5	Package 6
Mechanical strength	*0.10*	*1*	*5*	*5*	*5*	*5*	*5*
Suitability with standards	*0.08*	*5*	*5*	*5*	*5*	*5*	*5*
Package weight/product weight	0.05	3.2	3.2	4	4	3.8	4
Volume of package waste	0.06	5	4.5	4.5	5	5	5
Possibility of incineration	0.08	3	3	3	3	3	3
Marketing properties	0.17	3.09	3.67	3.25	3	2.92	3.59
Consumer convenience	0.11	3	3	3	3	3	3
Cost of packaging material	0.15	0.55	0.3	3.1	3.45	3	2
Indirect packaging costs	0.20	3	3	3	3	3	3
Optimization result		2.2 *	2.3	2.7	2.7	2.6	2.6

* Package 1 failed the simulated transportation test, and therefore, it was not acceptable.

Note: The scores given for the mechanical strength of a package and its suitability for packaging standards have been ignored.

Table 21.3 The raw chicken leg packages used in the package optimization. Package information: 1 kg raw chicken legs in the package and the package volume about 1800 ml^3. Secondary package was an open plastic stackable and reusable box

Feature	Package 1	Package 2	Package 3	Package 4	Package 5
Package type	preformed tray	preformed tray	tffs	tffs	preformed tray
Monomaterials used in package	PA, PE, PP	PA, PE, PP	PA, PE, PP	PA, PE, EVOH	PA, PE
Weight of an empty package (g)	40	31	16	10	65
Transparency	clear lid, white tray	clear lid, white tray	clear lid, and tray	clear lid, and tray	clear lid, black tray

Abbreviation: tffs = thermoformed tray/fill/seal

temperature is below +6°C. According to the results of modelling shelf-life test, to achieve a maximum quality for raw chicken legs for seven days under the existing distribution conditions (temperature +6°C) the package parameters could be the following:

- oxygen transmission rate of package: 10–14 ml/package · 24 h (at 23°C, 50% RH and latm
- carbon dioxide concentration: 50–80 %
- residual oxygen level: < 1 %
- protection from light: not necessary
- headspace volume: 50–110 ml/100 g product.

The next step was to choose the different packaging alternatives for raw chicken legs. Five different packaging types were chosen as described in Table 21.3. The characteristics of selected, alternative packages were scored and the importance of the characteristics were evaluated by food and packaging material manufacturers, the wholesalers and independent packaging specialists (see Table 21.4).

The most important factors affecting the optimization result were indirect packaging costs, marketing properties and cost of packaging materials. The optimization results suggest that the selected packages differed most in their marketing properties and consumer convenience: rigid preformed trays were evaluated significantly better than semi-rigid/flexible trays which were thermoformed just before packaging. In addition, the mechanical strength of the thermoformed trays was not satisfactory. To compare the best rated preformed tray (Package 2) and Package 5, the following estimations can be made. If the annual production of the product is 1 million packages, Package 2 saves about 35 000 kg packaging material annually compared to Package 5. In addition, Package 2 saves about €32 000 annually in packaging material costs compared to Package 5.

Table 21.4 The scores of various characteristics of tested raw chicken legs packaging types, the calculated coefficients for each characteristic and the optimization results for tested packages

Characteristic	*I*	Scores				
		Package 1	Package 2	Package 3	Package 4	Package 5
Mechanical strength	*0.11*	*5*	*3*	*1*	*1*	*3*
Suitability with standards	*0.09*	*1*	*1*	*5*	*5*	*3*
Package weight/product weight	0.07	3.7	4.1	4.7	5	2.6
Volume of package waste	0.05	1	1	5	5	3
Possibility of incineration	0.08	3	3	3	3	3
Marketing properties	0.13	4	3.5	1.5	1	4
Consumer convenience	0.10	4.5	4.5	2	1.5	4.5
Cost of packaging material	0.13	3.75	4.2	5	5	3.15
Indirect packaging costs	0.25	3	3	3	3	3
Optimization result		2.7**	2.8**	2.6*	2.5*	2.7

* Packages 3 and 4 failed the simulated transportation test, and therefore, they were not acceptable.
** The dimensions of Packages 1 and 2 were not suitable for the secondary package used for the product.
Note: The scores given for the mechanical strength of a package and its suitability for packaging standards have been ignored.

21.4.3 Gas-packed roasted chicken balls

Minced chicken balls belong to the group of ready-to-eat foods that are supposed to be heated before eating. The cooked prepared foods typically have low initial microflora levels and they are very sensitive to post-preparation contamination. Unpacked and air-packed cooked minced meat or poultry products usually spoil due to yeast and mould growth. The growth of these micro-organisms can be strongly retarded by using gas-packaging. The ingredients of the chicken balls were minced chicken meat, onion, breadcrumbs, potato, soy protein products, egg, potato flour, vegetable oil, salt, powdered chicken soup, seasonings, glucose and flavour intensifier. The last day for use is ten days after packaging when the storage temperature is kept below +6°C. Each package contained 400 g of chicken balls.

The next step was to choose the different packaging alternatives for roasted chicken balls. Eight different alternatives were selected as described in Tables 21.5a and 21.5b. The characteristics of selected, alternative packages were scored and the importance of the characteristics were evaluated by food and packaging material manufacturers, the wholesalers and independent packaging specialists (see Tables 21.6a and 21.6b).

The most important factors affecting the optimization result were indirect packaging costs, cost of packaging materials, and marketing properties. The optimization results indicate that flowpacks obtained the best scores, especially

Table 21.5a The roasted chicken balls packages used in the package optimization. Package information: 400 g roasted chicken balls in the package. Secondary package was an open plastic stackable and reusable box

Feature	Package 1	Package 2	Package 3	Package 4
Package type	flowpack	flowpack	tffs	tffs
Monomaterials used in package	PET, PE, EVOH	PE, PET	PA, PE	PA, PET, PE
Package volume (ml^3)	≈1000	≈1000	≈800	≈1000
Weight of an empty package (g)	6	7	10	22
Transparency	clear	clear	clear lid and tray	clear lid and tray

Abbreviation: tffs = thermoformed tray/fill/seal

Table 21.5b The roasted chicken balls packages used in the package optimization. Package information: 400 g roasted chicken balls in the package. Secondary package was an open plastic stackable and reusable box

Feature	Package 5	Package 6	Package 7	Package 8
Package type	tffs	preformed tray	preformed tray	preformed tray
Monomaterials used in package	PA, PET, PE, EVOH	PA, PE, PP	PA, PE, PP	PE, PS, EVOH
Package volume (ml^3)	≈1100	≈1300	≈1200	≈1500
Weigth of an empty package (g)	18	26	27	42
Transparency	clear lid, yellow tray	clear lid, opaque tray	clear lid, opaque tray	clear lid, and tray

Abbreviation: tffs = thermoformed tray/fill/seal

in terms of packaging material costs and indirect packaging costs. The marketing properties of the flowpacks, however, were not evaluated as being as good as the properties of the other package types in general. Leaking seals were found in one type of flowpack, and from one type of preformed tray. Therefore, a modification in package/packaging material structure should be made before the mechanical strength of these package types will be at an acceptable level.

To compare the best rated flowpack (Package 1), thermoformed tray (e.g. Package 5) and preformed tray (Package 6), the following estimations can be made. If the annual production of the product is 1 million packages, the annual savings in package material consumption for Packages 1 and 5 are 20 000 kg and 8 000 kg, respectively, when compared to Package 6. Similarly, the annual savings in packaging material costs for Packages 1 and 5 are about €48 800 and €32 000, respectively, when compared to Package 6.

Table 21.6a The scores of various characteristics of tested roasted chicken balls packaging types, the calculated coefficients for each characteristic and the optimization results for tested packages

Characteristic	I	Scores: Package 1	Package 2	Package 3	Package 4
Mechanical strength	*0.10*	*5*	*1*	*5*	*3*
Suitability with standards	*0.09*	*5*	*5*	*5*	*3*
Package weight/product weight	0.07	4.7	4.6	4.3	3
Volume of package waste	0.07	5	5	4	3
Possibility of incineration	0.07	3	3	3	3
Marketing properties	0.14	2	2	1.5	3
Consumer convenience	0.10	3.5	3.5	2	2.5
Cost of packaging material	0.14	5.14	5.57	5.53	4.33
Indirect packaging costs	0.23	5	5	3	3
Optimization result		3.4	3.4*	2.6	2.6

* Package 2 was not acceptable since it failed the simulated transportation test.
Note: The scores given for the mechanical strength of a package and its suitability for packaging standards have been ignored.

Table 21.6b The optimization results for four different roasted chicken balls package types

Characteristic	I	Scores: Package 5	Package 6	Package 7	Package 8
Mechanical strength	*0.10*	*5*	*3*	*5*	*1*
Suitability with standards	*0.09*	*3*	*3*	*1*	*3*
Package weight/product weight	0.07	3.4	2.6	2.4	0.8
Volume of package waste	0.07	3	1	1	1
Possibility of incineration	0.07	3	3	3	3
Marketing properties	0.14	2.5	3	2.5	3
Consumer convenience	0.10	2.5	3.5	2.5	3
Cost of packaging material	0.14	4.5	3.7	3.15	0
Indirect packaging costs	0.23	3	2	2	2
Optimization result		2.6	2.2	1.9**	1.5*

* Package 8 was not acceptable since it failed the simulated transportation test.
** The dimensions of Package 7 were not suitable for the secondary package used for the product.
Note: The scores given for the mechanical strength of a package and its suitability for packaging standards have been ignored.

21.4.4 Gas-packed pizza

Pizza is a typical ready-to-eat food that must be heated before eating. Pizza is a complex food with a variety of components (carbohydrates, proteins and fats), so there are many substrates for microbial and chemical attack during spoilage. Quality impairment of pizzas is mainly due to mould growth and staling of bottom (pizza crust). Furthermore, exposure to light causes fading of colours, cheese becoming yellowish, and rapid oxidation of fat.

The example foodstuff for package optimization was ham pizza. The ingredients of the ham pizza were wheat flour, crushed tomato, tomato ketchup, cheeselike vegetable oil product, cooked ham, onion, vegetable oil, water, salt, thickening agent and seasonings. The last date of use is 16 days after packaging when storage temperature is below +6°C. According to the results of the shelf-life modelling test and the other literature data, an ideal package for ham pizza, considering the present shelf-life (16 days) and keeping its quality to the maximum level, may have the following properties:

- protection from light
- oxygen transmission rate of the packaging material: 5–95 $cm^3/m^2 \cdot 24\,h$ (at 23°C, 50% RH and latm
- carbon dioxide concentration: 10–40 %
- residual oxygen level: < 1 %.

Next step was to choose the different packaging alternatives for ham pizza. Five different alternatives were selected as described in Table 21.7. The characteristics of selected, alternative packages were scored and the importance of the characteristics was evaluated by food and packaging material manufacturers, the wholesalers and independent packaging specialists (see Table 21.8).

The most important factors affecting the optimization result were indirect packaging costs, marketing properties, and cost of packaging materials. The

Table 21.7 The ham pizza packages used in the package optimization. Package information: package volume about 600 ml^3. Secondary package was an open plastic stackable and reusable box

Feature	Package 1	Package 2	Package 3	Package 4	Package 5
Package type	flowpack	flowpack	flowpack	tffs	tffs
Monomaterials used in package	PA, PE, carton-PP-plate	mPET, PE, carton-PP-plate	PET, PE, carton-PP-plate	PA, PET, PE, EVOH	PA, PET, PE, EVOH
Weight of an empty package (and plate) (g)	18	18	17	16	20
Transparency	white film, white plate	metallized film, white plate	clear film, white plate	clear lid and tray	clear lid, white tray

Table 21.8 The scores of various characteristics of tested ham pizza packaging types, the calculated coefficients for each characteristic and the optimization results for tested packages

Characteristic	*I*	Scores Package 1	Package 2	Package 3	Package 4	Package 5
Mechanical strength	*0.11*	*5*	*5*	*5*	*3*	*3*
Suitability with standards	*0.08*	*5*	*5*	*5*	*1*	*1*
Package weight/product weight	0.07	1.4	1.4	1.7	1.8	1.1
Volume of package waste	0.07	3	3	3	1	1
Possibility of incineration	0.07	3.5	3.5	3.5	3	3
Marketing properties	0.15	3	3.5	4	1.5	1.5
Consumer convenience	0.11	3.5	3.5	4	2	2
Cost of packaging material	0.14	2	1.2	2.6	3	5
Indirect packaging costs	0.23	5	5	5	3	3
Optimization result		2.7	2.7	3.0	1.9*	2.1*

* The dimensions of Packages 4 and 5 were not suitable for the secondary package used for the product.
Note: The scores given for the mechanical strength of a package and its suitability for packaging standards have been ignored.

optimization results indicate that flowpacks were better than thermoformed trays, the only benefit of the trays being the lower cost of packaging material. To compare the best rated flowpack (Package 3) and thermoformed tray (Package 5), the following estimations can be made. If the annual production of the product is 10 million packages, Package 3 saves 20 000 kg of packaging material annually compared to Package 5. On the other hand, the packaging material costs for Package 5 are about €168 000 less than for Package 3.

21.5 Conclusion: improving decision-making

The VTT Precision Packaging Concept can benefit the packaging industry for example by providing the means for comparing different package types (e.g. flowpacks, form-fill-sealed-packages and preformed trays). The concept can help the food industries in choosing plastic packaging materials and considering the demands of logistics, cost savings and shelf-life and improving competitiveness. In addition, it gives packaging material exporters a method of optimizing materials in the different conditions of exporting countries and helping in the after-treatment of plastic-based packaging materials.

22

Legislative issues relating to active and intelligent packaging

N. de Kruijf and R. Rijk, TNO Nutrition and Food Research, The Netherlands

22.1 Introduction

Major technological developments in food packaging can introduce many benefits to consumers and food and food-packaging industries, but at the same time they are liable to the introduction of new problems. Although active and intelligent packaging continues to broaden in scope and these new packaging systems are already being successfully applied in the USA, Japan and Australia, its penetration in the European marketplace has been quite limited thus far. This is partly due to the strict European regulations for food contact materials, which fail to keep up with technological innovations and currently prohibit the application of many of these systems. In addition, a lack of knowledge of consumer acceptance, of economic aspects and of the environmental impact of these novel concepts and, in particular, the lack of hard evidence of their effectiveness demonstrated by independent investigators has inhibited their commercial usage.

Within the Actipak project active and intelligent packaging systems were defined as follows:[1]

- Active packaging actively changes the condition of the packaged food to extend shelf-life or improve food safety or sensory properties while maintaining the quality of the packaged food.
- Intelligent packaging systems monitor the condition of packaged foods to give information about the quality of the packaged food during transport and storage.

In Europe, no specific regulation governing active and intelligent food packaging exists to date. Most active and intelligent agents are not considered as food additives but rather as food contact material constituents, and therefore these food packaging systems should comply with the existing regulations for food contact materials. When these regulations were drafted, no allowance was made for active and intelligent packaging as these systems were not applied as food contact materials in Europe at that time. The current packaging regulations require that all components used for the manufacture of food contact materials are covered by so-called positive lists. These lists of approved compounds usually include components required to manufacture the packaging material. Constituents used for other purposes such as extending or monitoring the shelf-life of packaged foods are not included. Therefore, most active and intelligent agents are not listed. In addition, active and intelligent systems should comply with relevant overall and specific migration limits. The overall migration limit of 60 mg per kg food is a major hurdle to the application of active packaging in Europe, especially when the system is designed to release active ingredients into foods to extend their shelf-life or improve their quality. Moreover, current migration tests are not always suitable for these new packaging systems because the conventional ratio of 6 dm^2 to 1 kg food is generally much smaller and, in addition, they often differ in contact mode from conventional packaging. Therefore, a new approach to food packaging regulations is required, and new migration test methods should be developed and validated for some of these new food packaging systems.

No single European regulation currently covers specifically the use of active and intelligent packaging systems. The food-contact application of active and intelligent packaging systems is covered by a range of EU regulations, each having its specific requirements, such as regulations for food-contact materials, food additives, biocides, modified-atmosphere packaging, hygiene of foodstuffs, labelling and packaging waste. Some of these regulations may be, unintentionally, an obstacle to the introduction of active and intelligent systems in Europe. Therefore, a few years ago, two initiatives were taken to implement active and intelligent packaging within the European regulations.

In 1999, a pan-European project was started within the framework of the EU FAIR R&D programme. The study aims at initiating amendments to European legislation for food contact materials to establish and implement active and intelligent systems within the current relevant regulations for packaged food in Europe.[1, 2] In 2000, a comprehensive report on legislative aspects of active and intelligent food packaging was published by a project group under the Nordic Council of Ministers. The report describes some types of active and intelligent food contact materials, the legislation the project group found to be relevant to consider, as well as some conclusions and proposals for administrators for future work with recommendations and interpretations of existing legislation. Also, the possibility of establishing new specific legislation for active and intelligent packaging is considered.[3] Both initiatives will now be discussed in more detail below.

22.2 Initiatives to amend EU legislation: European project

In 1999, a European study was started to enable the safe application of active and intelligent packaging systems throughout Europe by initiating amendments to European legislation for food contact materials in order to establish and implement these systems in current relevant regulations for packaged food in Europe. The study was entitled 'Evaluating safety, effectiveness, economic-environmental impact and consumer acceptance of active and intelligent packagings' ('Actipak'). The Actipak project was co-ordinated by TNO Nutrition and Food Research and was jointly carried out by nine research organizations and three industrial companies.[1] The research project consisted of five key tasks. The study was completed by the end of 2001. For each task the main results and conclusions are summarized below.

Task 1: Inventory

At the start of the project an overview of all existing commercial and promising but not (yet) commercially available active and intelligent packaging systems was prepared. The review contains information on technology, market trends, consumer needs and current legislation in Europe and relevant countries outside Europe. Part of the review has been described in detail in a separate publication.[2] The main conclusion to be drawn from the review is that no European regulation currently covers the use of active and intelligent packaging. The traditional European regulations for food contact materials, the overall migration limit and lists of approved compounds may be inconsistent with some of the objectives of active and intelligent packaging. In addition, some 25 packaging systems were selected for compositional analysis and overall migration study (Task 2).

Task 2: Classification of active and intelligent systems

In this task the composition and migration behaviour of selected active and intelligent packaging systems were investigated to identify conflicts with current legislation. A total of 20 active systems and 6 intelligent systems were investigated. The composition was investigated in view of the EU positive list and positive lists of national regulations. Determination of the composition focused on active ingredients and relevant reaction products. The compositional analysis of some active packaging systems has been described in detail.[4, 5] Some typical results are shown in Table 22.1.[1]

The compositional analysis revealed that many active and intelligent packaging systems are very complex in composition. Apart from plastics, other materials such as paper, metals, adhesives, printing and minerals are being used. Existing EU legislation for food contact materials such as the EU Directive for polymeric food contact materials (Directive 90/128/EEC and its amendments) applies to only a minority of the materials tested. In addition, the overall migration behaviour of the active and intelligent packaging systems was investigated. Some relevant results of the overall migration study obtained for oxygen scavengers and moisture absorbers

Table 22.1 Composition of some active and intelligent packaging systems[1]

Packaging system	Ingredients identified
Oxygen scavengers	Iron powder Silicates Sulfite Chloride Polymeric scavenger Elements: Fe, Si, Ca, Al, Na, Cl, K, Mg, S, Mn, Ti, Co, V, Cr, P
Antimicrobial releasers	Acids Silicates Ethanol Zinc Elements: Si, Na, Al, S, Cl, Ca, Mg, Fe, Pd, Ti
Indicators	Methylene blue and other colour indicators Acids Antioxidants Mineral oil Sugars Elements: Na, Ca, K, Si, Al, Mg

are presented in Table 22.2.[1] A complete overview of all migration values obtained in this study has been reported by De Meulenaer *et al.*[5] Quite a few migration values obtained exceed the overall migration limit. Some of the high levels observed were supposed to be attributable to the use of inappropriate liquid migration simulants. Solid migration simulants such as agar gels could be an alternative.[6] The three time-temperature indicators were not included in the overall migration study. As the current systems are generally applied on the outside of the packaging and for relatively short periods of time, the packaging material can be considered to be a functional barrier, and therefore migration testing of time-temperature indicators is not relevant.

Based on the results of the evaluation of the composition and the migration behaviour, the active and intelligent systems were classified in view of restrictions of current regulations into five categories (A–E) according to the scheme shown in Fig. 22.1. These categories are:

Category A: Systems that comply with the current legislation (i.e. composition and migration).

Category B: A system belongs to category B if it contains components not listed in the positive lists of the EC (90/128/EEC and amendments) but which are food additives and/or natural components and/or other components of which toxicological data are available. The migration behaviour of the category-B systems is in compliance with the migration limits as set by the EC.

Table 22.2 Overall migration from oxygen scavengers and moisture absorbers[1]

Sample	Type	Test condition	Overall migration (mg/sample) into: Water	3% Acetic acid	10% Ethanol	15% Ethanol	95% Ethanol	Iso-octane	Olive oil
Oxygen scavenger	Sachet	10 days at 40°C	620[b]	1700[c]	–	800[a]	210[c]		–
		2 days at 20°C						1.9[c]	
Oxygen scavenger	Cap	10 days at 40°C	74[c]	98[c]	80[c]	–	43[c]		–
		2 days at 20C						0.9[c]	
Oxygen scavenger	Crown	30 min. at 70°C	1.0[c]	1.7[c]	1.5[a]	–	–	–	27.8[a]
		+ 10 days at 40°C							
Moisture absorber	Sachet	10 days at 40°C	<0.1[a]	970[c]	–	0.6[c]	2.3[c]		
		2 days at 20°C						<0.1[a]	–
Moisture absorber	Pad	10 days at 40°C	9.3[b]	46[c]	–	7.2[b]	21[c]		–
		2 days at 20°C						18[c]	
Moisture absorber[d]	Film	10 days at 40°C	260[a]	300[a]	–	300[b]	8.2[b]		–
		2 days at 20°C						0.1[c]	

[a] Standard deviation <5% (n = 3 or 4)
[b] Standard deviation >5% and <10% (n = 3 or 4)
[c] Standard deviation > 10% (n = 3 or 4)
[d] Overall migration in mg/dm^2 instead of mg/sample
– Not measured

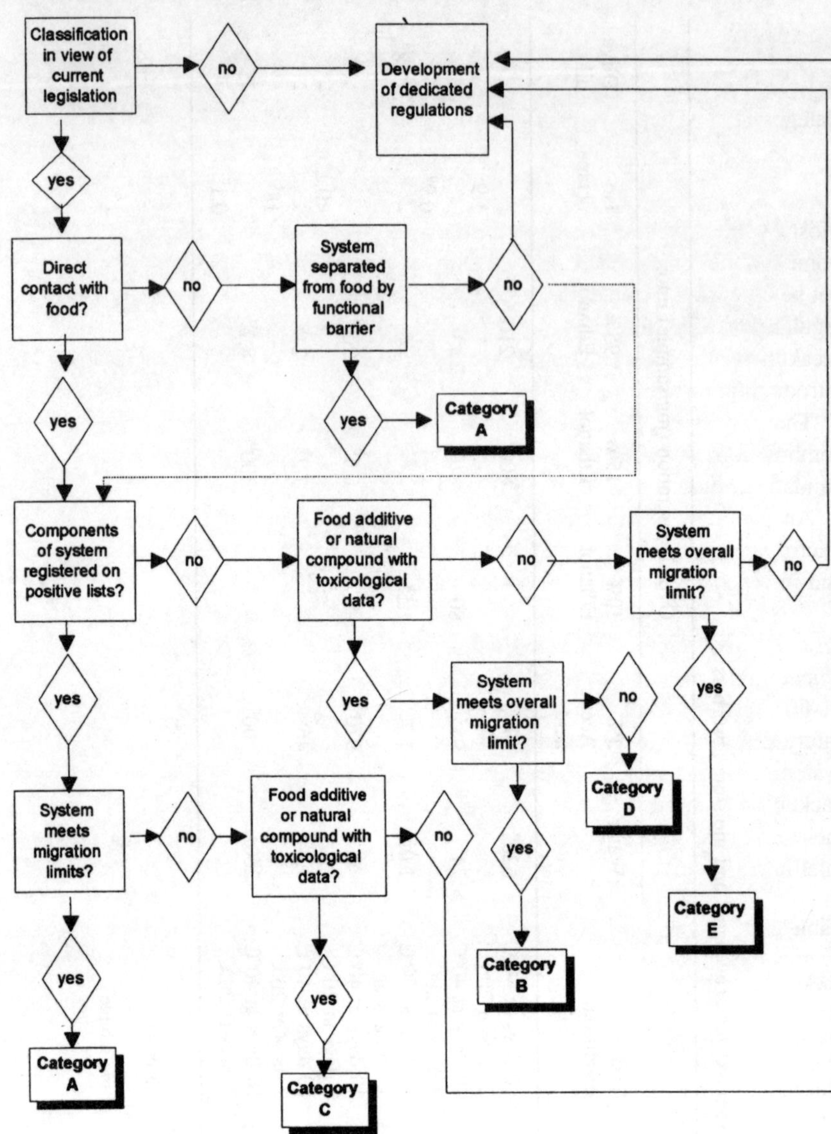

Fig. 22.1 Classification of active and intelligent food-packaging systems in view of current legislation. For a description of categories A–E, see the text (reproduced with permission from Food Additives and Contaminants, July 2002. http:/www.tandf.co.uk).

Category C: These systems contain components that are included in the positive lists of the EC, but the migration exceeds the migration limit(s) set in the current legislation.

Category D: These systems contain components that are not included in the positive lists of the EC but are food additives or natural components or other components for which toxicological data are available. In addition, the migration from the systems exceeds the migration limit(s) set by the EC.

Category E: These systems contain components that neither are listed nor are food additives or natural components or other components for which no toxicological data are available.

Most of the systems investigated could be classified into categories A and B. Some fall into categories C and D. Only a carbon dioxide-releasing system could not be classified.[5] Generally, it could be concluded that an extension of existing regulations with dedicated requirements seems to be necessary to permit the breakthrough of these materials on the EU market and to guarantee their safe introduction and use in Europe.

The results of the classification have been used to select representative combinations of foods and active and intelligent packaging systems for further validation studies.

An overview of the food-packaging combinations selected for evaluation of microbiological safety, shelf-life-extending capacity and efficacy of the active and intelligent systems is presented in Table 22.3.

Task 3: Evaluation of microbiological safety, shelf-life-extending capacity and efficacy of active and intelligent systems

In this task an overall evaluation of the capability (including effectiveness, safety and shelf-life-extending capacity) of the active and intelligent packaging systems was conducted. To this end, the microbiological safety of the test food, packed and stored in active packaging systems, was determined by analyzing their microbiological condition. In addition, the risk of false indication of intelligent systems was examined. Furthermore, the effectiveness of active

Table 22.3 Food-packaging combinations selected for validation studies

Packaging system	Food
Oxygen-scavenging film	Fresh pasta
Moisture-absorbing film	Fish
Moisture-absorbing pad	Fresh meat
Ethylene-absorbing film	Bananas
Antimicrobial film	Cheese
Antimicrobial film	Meat
Antimicrobial film	Fruit
Aldehyde-absorbing film	Cereal
Oxygen-scavenging sachet	Milk powder
Oxygen-scavenging sachet	Biscuits
Moisture-absorbing sachet	Milk powder
Antimicrobial sachet	Sandwich bread
Oxygen-scavenging crown	Beer
Time-temperature indicators	Fish
Oxygen indicators	Sliced meat
Carbon dioxide indicator	Sliced meat

Table 22.4 Effectiveness and shelf-life extending capacity of some food/active packaging test combinations

Active packaging	Food product	Effective	Shelf-life extension *
Oxygen-scavenging film	Fresh pasta	Yes	Yes, longer microbiological shelf-life not due to O_2 absorption but to barrier characteristics of the active film
Moisture-absorbing pad	Pork	Yes	No, same microbiological and sensory shelf-life
Antimicrobial film	Cheese/ bread	Possibly	No, same microbiological shelf-life
Aldehyde-absorbing film	Cereals	Yes	Yes, longer sensory and chemical shelf-life
O_2-absorbing sachet	Milk powder	Yes	No, but a good alternative (same sensory and chemical shelf-life) to MAP can packaging
O_2-absorbing sachet	Cooked ham	Yes	Yes, longer sensory shelf-life/same microbiological shelf-life
O_2-absorbing crown cork	Beer	Yes	No, same sensorial shelf-life

* Compared with a food/packaging combination without an active packaging system.

packaging systems to improve the microbiological stability of food, as compared to traditional packaging systems, was tested. Also the extension of sensory and chemical shelf-life was investigated for different active packaging/food combinations.

In total, 12 studies were performed to investigate the effectiveness and shelf-life-extending capacity of selected food/active packaging combinations. Some typical results are presented in Table 22.4. Most of the active systems investigated appeared to be effective as claimed by their manufacturers. From the shelf-life studies it can be concluded that a number of active systems indeed prolong shelf-life. The indication capacity of three time-temperature indicators, two oxygen indicators and a carbon dioxide indicator was investigated. The indicators investigated indicated relatively well the conditions they were meant for (time-temperature history, package headspace oxygen or carbon dioxide).

Task 4: Toxicological, economic and environmental evaluation of active and intelligent systems

Intelligent devices and some active systems may contain substances that are not food additives and have not been evaluated by the EU Scientific Committee on Food (SCF) for use in food contact materials. Within the Actipak project it was therefore agreed to study the consequences when a substance is not on the positive list of the directives on food contact materials and to collect and interpret available toxicological data. Examination of existing toxicity data of one substance with oxygen absorption capacity indicated the substance to be potentially mutagenic. This demonstrates that substances used in active and intelligent packaging systems should be evaluated by SCF before allowing them to come in contact with foodstuffs. In other words, they should be evaluated like all other substances used in food contact materials.

To establish acceptance among European consumers of active and intelligent systems that have been proved to be suitable and safe, these systems were subjected to an international study on consumers' attitudes towards application of these systems. This study also provides insights into national differences and general attitudes. Consumer focus groups consisting of 8–12 people of mixed age and sex were formed in six European countries, namely the UK, Italy, Germany, the Netherlands, Finland and Spain. The results demonstrated that for active and intelligent devices to be readily accepted in Europe in the immediate future, their introduction to the marketplace should be supported by a substantial information campaign clarifying their benefits and how they function. They will not gain acceptance purely by virtue of extension of shelf-life. Also, to avoid confusion, some standardization, at least of indicators, would be preferable. Attitudes are fairly consistent in Europe with the exception of Spain and possibly Italy. Consumers in Spain were much more ready to accept both active devices (absorbers, including sachets) and indicators, and responded very positively to them. Italy also seemed slightly keener than the rest of Europe.

The economic consequences and environmental implications of active and intelligent systems were evaluated as part of the project. The shelf-life-extending capacity of active packaging is expected to reduce food waste due to spoilage. Consequently, energy and packaging materials may be saved. Multi-layer barrier packaging materials might be replaced by less complicated packaging materials, thus reducing packaging waste. In addition, from the study the conclusion can be drawn that the use of intelligent packaging such as time-temperature indicators will decrease the waste generated in the long term.

Task 5: Recommendations for legislative amendments

Finally, all results of the project and the requirements of all relevant EU regulations were evaluated. Based on this evaluation recommendations were drafted for the implementation of suitable active and intelligent systems in relevant European Directives. These recommendations were discussed informally with several national and European authorities. In addition to food packaging regulations, other relevant European regulations were studied such as regulations for food additives, biocides, pesticides, modified-atmosphere packaging, flavouring, food hygiene, labelling, product safety and packaging waste. These regulations generally do not form a serious hurdle to the safe introduction of active and intelligent food packaging systems in Europe. The directive on food hygiene even appeared to be an incentive to the use of active and intelligent packaging.

The first proposal for changing the framework Directive 89/109/EEC has resulted in a draft amendment of the this directive in which active packaging is included in the scope as described in Article 1. It is expected that this amendment will be approved by the end of 2003. This will remove the first barrier to the introduction of active packaging systems in Europe. A more detailed description of the results of this task will be given in section 22.4.

22.3 Initiatives to amend EU legislation: Nordic report

The Nordic countries (Denmark, Finland, Iceland, Norway and Sweden) have a long tradition of co-operation in the food packaging area, and these countries have similar legislation for food contact materials. A project group under the Nordic Council of Ministers has discussed the legal aspects of active and intelligent systems. The project group was chaired by Dr Fabech of the Danish Veterinary and Food Administration. In 2000, the project group published a report on legislative aspects of active and intelligent food packaging.[3] This so-called 'Nordic Report' aimed at contributing to a solution of legislative problems related to active and intelligent food contact materials. In the first chapter of that report an overview is given of different types of active and intelligent food packaging. The effectiveness of these systems and the test requirements are discussed. The most important part of the report is a comprehensive overview of European legislation relevant to active and intelligent packaging. In section 22.4, a description of these EU directives is given and their relevance to active and intelligent packaging is discussed.

In the Nordic report recommendations are also given as to which parts of the EU legislation should be reviewed and which questions could be solved through interpretation of existing legislation. Preferably, harmonized legislation should be interpreted on a European basis to avoid divergence in interpretation, which could lead to barriers to trade. Proposals are given for solutions to problems by interpretation. According to the Nordic group, it is not necessary to introduce new EU legislation. Instead, amendments should be made to existing legislation and guidelines on how to interpret existing legislation should be given. Finally, initiatives are proposed to be taken by legislators, both on a national and on an EU level, when drafting new or revising existing legislation on active and intelligent packaging.

22.4 Current EU legislation and recommendations for change

For this study of relevant European regulations, a schedule was made of the scope of active and intelligent packaging systems. Definitions of active and intelligent systems are proposed. Based on that principle an overview of the physical appearance of the systems is required as well as a division by functionality of the various systems.

22.4.1 Scope of active and intelligent systems

Active systems

Active packaging systems may differ in appearance. Active packaging systems may be packaging materials to wrap foodstuffs, but may also be added to the packed food in the form of a sachet, label, box, etc. A correct description, which will be used in regulatory amendments, would be 'active food contact material systems'. For practical reasons, the term 'active packaging systems' will be used here.

Conventional packaging materials are considered passive, and their main function is protection against the environment. Active packaging systems intentionally absorb or release substances from or to the food or its environment. Ingredients required to achieve the effect may be incorporated in the packaging material itself or packed in a sachet or label inserted into the package. The total contact area of active packaging systems may be the same as for conventional packaging material, such as a film. But, in case of sachets or labels, the ratio may be significantly smaller than 6 dm^2/kg food. This may influence migration requirements and testing protocols. Both absorption and release of substances should not endanger human health. For this purpose many regulations at the EU and the national level are in force, which should be taken into account to judge the acceptability of an active packaging system.

Intelligent systems

Intelligent systems are only occasionally packaging materials. They usually are packed together, inside or outside the primary packaging, with the food in the form of a label, a pill, etc. As there is potential contact with food they should be called 'intelligent food contact material systems' but, for practical reasons they will be called 'intelligent packaging systems' here.

Intelligent packaging systems provide the user with information on the conditions of the food. Intelligent systems do not influence the food but provide information to consumers, retailers, manufacturers, etc. Intelligent packaging systems should not release their constituents to the food. In many cases a so-called functional barrier, which prevents migration, is present. However, attention must be paid to the fact that intelligent systems may contain all kinds of chemicals required for detection of the intended information. Attention should also be paid to the acceptance of the use of these substances, particularly for packed foods presented directly to the consumer. Starting from the requirement that safety of the food and subsequently safety of the consumer shall never be endangered, the legal restrictions as well as the possibilities for the use of active and intelligent systems were studied in depth. Solutions for existing barriers are proposed.

22.4.2 Identification of relevant regulations

Active and intelligent packaging systems in contact with foods should comply with regulations on food contact materials. In addition, the composition of the food can be influenced by the use of active packaging systems. The following regulations are considered and further discussed:

- food contact materials
- food additives
- flavouring
- hygiene
- biocides
- pesticides
- labelling
- product safety
- weight and volume
- waste.

22.5 Food contact materials

The requirements for food contact materials (FCM) are formulated in general terms in Framework Directive 89/109/EEC;[7] some materials are regulated in detail in specific directives. Directive 89/109/EEC is under revision and will be published in 2003.

22.5.1 Framework Directive 89/109/EEC

Directive 89/109/EEC specifies the definition of FCM and general requirements. Article 2 requires production of FCM according to good manufacturing practice, while application of FCM shall not endanger human health or change the composition or sensory properties in an unacceptable way. Article 6 describes the requirements for labelling and a demonstration of compliance with specific directives.

Relevance to active and intelligent packaging systems

Undoubtedly, active and intelligent packaging systems are intended to come into contact with food, although some may be separated by a 'functional barrier' from the food. Therefore, active and intelligent packaging systems fall within the scope of framework Directive 89/109/EEC. According to article 2, they shall not endanger human health, nor change the food's sensory characteristics. The latter requirement may be influenced by personal preferences and could be an issue of discussion. In addition, in further specific directives like 2002/72/EC[8] an overall migration limit of 60 mg/kg food is established as a purity requirement. Active systems developed to release certain components most likely will not comply with this requirement. To provide clarity, the scope of Directive 89/109/EEC should be extended to allow intentional migration from food contact materials at levels exceeding 60 mg/kg.

Intentional migration of substances has an effect on the composition of the food. It should be emphasized that the released substances are subject to various relevant regulations pertaining to food ingredients, food additives, labelling, etc. Intelligent packaging systems shall comply with Article 2, so no additional provisions in the framework directive are considered necessary. Specific measures may be required to regulate the chemicals used in the intelligent packaging systems, but this is a subject of specific directives.

Recommendations for extending Directive 89/109/EEC

Based on the results of the Actipak project, amendment of Directive 89/109 has been proposed, and the proposals have been adopted for implementation. A revised Directive will include an extended scope that mentions the allowed use of active and intelligent food contact materials. Special attention will be given to releasing packaging systems. The food in contact with such systems shall comply with any relevant food or food additive regulation. The releasing active packaging systems will be limited to materials that release substances added for that purpose. This means that natural materials, for example wooden barrels for wine or whisky storage, are excluded from the definition of active food contact materials.

Proper labelling will also be required. This includes the conditions (time and temperature) in which the system can be brought into contact with the food and the food that may be in contact with a releasing system. As food additive regulations have to be obeyed the food packer should be informed about the amount of substance released from one object. Annex I of the Directive will be extended with active and intelligent systems. Annex I contains a list of materials, covered by specific measures. This means that in the future a specific directive will be drafted on active and intelligent packaging systems.

22.5.2 Directive 80/590/EEC[9]

Symbol for food contact materials

In Directive 80/590/EEC the symbol to be used for food contact materials not already in contact with foodstuffs is introduced. The symbol shall be used according to the requirements of Directive 89/109/EEC. Alternatively, subjects may be accompanied with the words ‘suitable for food contact’.

Relevance to active and intelligent packaging systems

Both active and intelligent packaging systems will not be available to consumers, as they usually require special care before bringing them into contact with foodstuffs. The final user of the A&I systems has to be informed that the subject is suitable for food contact, and thus the systems have to be labelled accordingly. Options are to print the symbol on the system or, at the wholesale stage, to add documentation with this symbol or proper wording. In those cases where a system as such is available to consumers the system should also be labelled in accordance with the requirements of this directive.

Recommendations
Directive 80/590/EEC should be followed. There is no need for amendment of this directive.

22.5.3 Plastics directives

Directive 2002/72/EC sets requirements for food contact materials manufactured solely from plastics. The composition of plastics permitted as food contact materials is based on the principle of a positive list. Maximum allowed migration limits of plastic components are based on the toxicological properties of substances. An overall migration limit of 60 mg/kg food or 10 mg/dm^2 is set to prevent contamination of the food to an unacceptable level.

The directive is intended to harmonize certain classes of substances such as monomers, starting materials and additives. Polymerization regulators are not covered by the directive, but they shall not endanger human health according to framework Directive 89/109/EEC. In some countries, including the Netherlands and Germany, these substances are regulated at a national level. Article 8 of Directive 2002/72/EC requires verification of compliance with the requirements of the directive in accordance with the rules laid down in Directives 82/711/EEC and 85/572/EEC. In addition, the materials and articles shall be accompanied with a declaration of compliance at the marketing stage rather than the retail stage.

Relevance to active and intelligent packaging systems
Active packaging systems manufactured solely from plastics must comply with the requirements of Directive 2002/72/EC, meaning that composition and migration behaviour must be in compliance with the positive list and the migration restrictions. Active packaging systems, such as some types of oxygen absorbers, based on active ingredients that are incorporated in the backbone of the polymer shall comply with the directive. It is argued that these materials may be used to wrap the food in the same way as conventional packaging materials. This means that all substances used should have been evaluated by the SCF and be added to the positive list with or without a specific migration limit.

Plastic materials and articles containing a substance intentionally released to the food should be treated differently. The base polymer should comply with Directive 2002/72/EC, whereas the released substance should be an approved food ingredient or food additive. Listing of the released substances on the positive list for plastics seems unnecessary as these substances should be allowed as food ingredients or food additives. However, allowance of the presence of such substances should be provided for in the plastics directive. Overall migration from releasing materials or articles may conflict with the overall migration limit. A requirement of enforcement authorities may be the possibility to check the overall migration of the polymer itself. In principle, this could be determined by the classic determination of the overall migration and subsequent subtraction of the specific migration of the released substance. However, in many cases the amount of released substance may be much higher than the overall migration limit and the analytical error may be even higher than the overall migration limit itself. As a matter of fact, the plastic material can only be verified for compliance if the plastic is available without the releasing substance. This would require either enforcement of the plastic material at an early stage or demonstration of compliance by a reliable and acceptable certification procedure.

Homogeneous intelligent systems manufactured from plastics only (mono- and multi-layers) and in which the intelligent ingredients are immobilized in the polymer backbone or blended as an additive in the plastic should comply with the requirements of Directive 2002/72/EC provided they are intended to come directly into contact with the food.

Two types of composed materials and articles can be identified. First, there are systems that are manufactured by packing the active or intelligent ingredients in a plastic bag or box. Such a system is usually inserted into the primary package with the foodstuff. The plastic part of the system should be in compliance with Directive 2002/72/EC. However, the ingredients packed inside cannot be considered as plastic. These systems should be considered an entity of a food contact material and hence the whole system is excluded from the plastics regulation. Special provisions will be required to include this type of food contact materials.

Systems of a second type are composed of various types of packaging materials, such as plastic, paper, metal, printing, adhesives, varnish and active or intelligent ingredients. Usually the individual components of the final system are hard to recognize. Such composite materials and articles are not covered by Directive 2002/72/EC. It is most likely that no EU regulation exists on the individual parts of the system. Regulations for paper, metal, printing inks and varnishes exist at the national level of some member states, waiting for harmonization at the EU level. This means that these systems are subject to national regulations and to the framework Directive 89/109/EEC. Both for enforcement authorities and for manufacturers this is an uncomfortable situation as it is difficult to establish the safety of these types of food contact materials. It seems realistic to assume that fully harmonized legislation on all types of food contact materials will not be available in the short term.

Concerning active and intelligent ingredients, it was found that some are already included in a positive list, such as iron oxide used in oxygen absorbers, but many others are not. However, all these substances need to be regulated to avoid their use is forbidden without firm grounds and the possibility that unsafe situations may occur. Therefore, if there is direct contact with the food, the system should be submitted to migration testing protocols and the relevant substances should be toxicologically evaluated and subsequently added to a positive list.

Frequently applied intelligent systems, such as time/temperature indicators, are positioned on the outside of the primary food packaging. In addition, these systems are usually made of plastic and connected to the packaging by an adhesive layer. Use of time/temperature indicators almost automatically implies that storage times are relatively short and temperatures are low. Taking this into account the probability of migration through the primary packaging into the food is negligible. These types of intelligent systems should not be considered as food contact materials with respect to migration testing. Nevertheless, it may be necessary to include 'intelligent substances' in a positive list for which toxicological evaluation can be kept to a minimum.

Recommendations

- Active and intelligent packaging systems manufactured from plastics only shall comply with the compositional and migration requirements (except for intentionally released substances).
- Substances intentionally released from an active releasing system shall comply with relevant requirements for food and food additives. Provisions should be made for allowance of migration values higher than the overall migration of 60 mg/kg food.
- It is proposed to draft a specific directive in which active and intelligent packaging systems are regulated. For regulation of composite materials reference to existing national regulations with regard to the base packaging materials and separate listing of the active and intelligent ingredients seems the best solution for the time being.
- Active and intelligent packaging systems should be accompanied with a declaration of compliance provided the provisions proposed have been realized.
- It will remain very difficult and laborious for enforcement laboratories to prove violation of Article 2 of Directive 89/109/EEC for complex systems. For manufacturers it may be difficult to demonstrate compliance with the rules, as they are usually not aware of the composition of all parts of the final article. A proper certification system may provide a better guarantee of the safety of the packaging system. Proper rules and guidelines, as well as the appointment of recognized certification laboratories would be required for that purpose. The scheme given in Fig. 22.2 could be a starting point for drafting a certification procedure.

22.5.4 Basic rules for migration tests

At the EU level, rules for testing plastic food contact materials are given in 82/711/EEC[10] as amended by 93/8/EEC[11] and 97/48/EEC.[12] Directive 85/572/EEC[13] provides a list of simulants that could replace real foodstuffs in migration testing. Simulants prescribed for compliance testing are water, 3% acetic acid, 10% ethanol or olive oil. In some cases olive oil may be replaced with the substitute food simulants 95% ethanol and iso-octane. In Directive 85/572/EEC it is recognized that a fat simulant may be a stronger extractive than the food. Depending on the food and its fat content reduction factors are included in the list. This means that the migration value obtained with a fat simulant should be divided by the value indicated for that particular food. The reduction factors vary from 2 to 5.

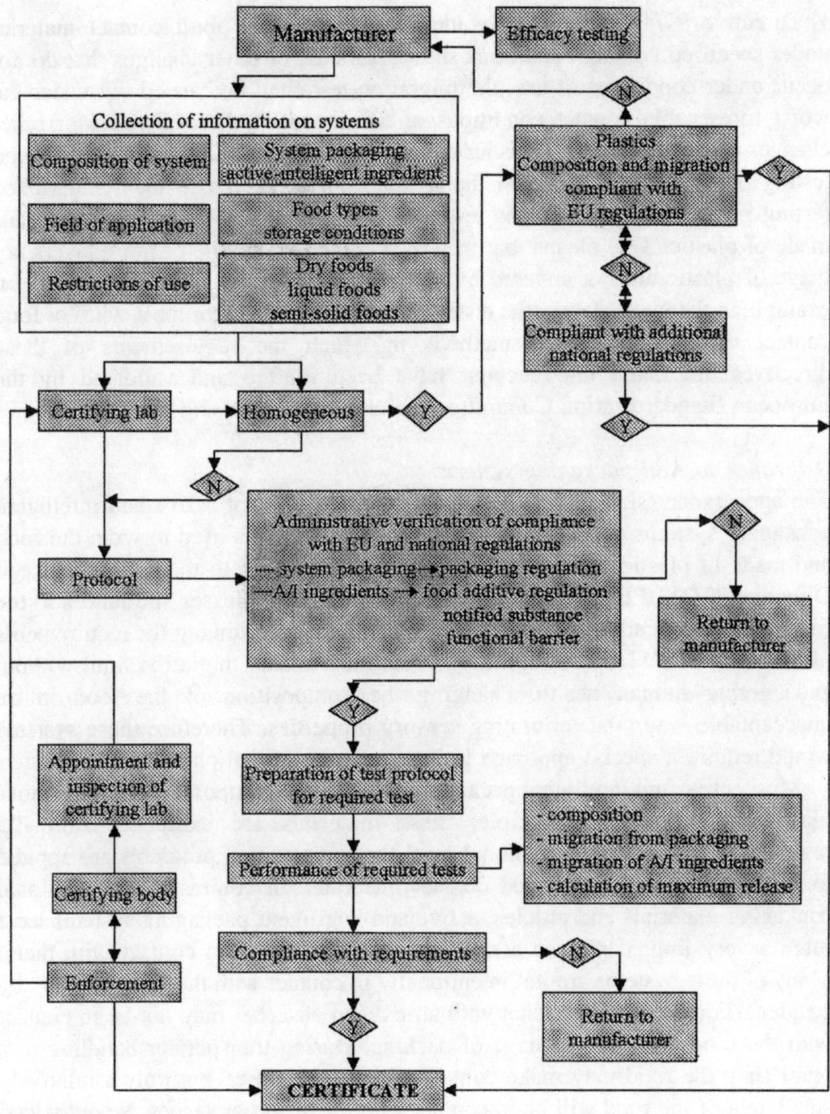

Fig. 22.2 Scheme for certification procedure.

In Directive 97/48/EC detailed conditions of time and temperature are given to demonstrate compliance with the limits set in Directive 2002/72/EC. The test conditions to be applied shall represent the worst foreseeable conditions of use in case of contact with foodstuffs. Food contact materials and articles should be accompanied with a statement indicating the restrictions of use, if any, with respect to the types of food and the maximum contact conditions of time and temperature, according to Article 6 of Directive 89/109/EEC.

Directive 97/48/EC explicitly mentions that, if the food contact material under specified contact conditions shows physical or other changes that do not occur under conditions of use, the migration test shall be carried out under the worst foreseeable contact conditions of use in which these physical or other changes do not take place. This article allows for the use of specially developed testing protocols depending on the problems encountered in the standardized testing protocols. However, the test protocols are applicable only to materials made of plastic. This means that materials composed of one or more layers not made of plastic are not covered by the EU regulation. At a national level, for example in the Netherlands, the testing protocols are used for most types of food contact materials. Detailed methods in which the requirements of these directives are taken into account have been drafted and validated by the European Standardisation Committee (CEN) in EN 1186 and EN 13130.

Relevance to A&I packaging systems

The appearance (size and shape) and the composition of active and intelligent packaging systems depend on their application. Systems used to wrap the food and made of plastics solely can be examined according to the requirements of Directive 82/711/EEC. If such a system intentionally releases substances to the food, then technically the system can be examined according to the requirements of Directive 82/711/EEC, but it may exceed the overall migration limit without endangering human health, changing the composition of the food in an unacceptable way or deteriorating sensory properties. Therefore these systems would require a special approach in interpretating migration values.

Most active and intelligent packaging systems are composed of various (non-plastic) materials. In principle, these materials are excluded from EU regulations. However, at a national level the same testing protocols are applied to most other non-plastic food contact materials. In contrast to conventional packaging materials and articles, active and intelligent packaging systems have often a very limited surface area compared to the food in contact with them. Many of these systems are not intentionally in contact with the food but only by accident. For example, a sachet with an oxygen absorber may not be in contact with the food at all at the stage of packing. During transport or handling in a retail shop the food may make contact with the absorber, but only a relatively small area of the food will be in contact with the absorber sachet. Nevertheless, migration may occur and migration testing is required to guarantee food safety. The test conditions of time and temperature can be selected from Directive 82/711/EEC, and the appropriate simulants from Directive 85/572/EEC. For active and intelligent packaging systems in contact with dry foodstuffs (without free fat on the surface) no migration tests with simulants are prescribed. If necessary, the specific migration of substances should be measured in the food itself.

Systems in contact with aqueous or fatty foods require testing with simulants. In principle, the protocols prescribe that food contact materials are brought into contact with a food simulant. This can be achieved by total immersion or by one-sided contact of the material with the food simulant. One-sided contact of plastic materials is achievable by filling an article such as a bottle or by using a migration cell for one-sided contact. Due to the construction of many active and intelligent packaging systems this approach is not feasible, and only submersion of the article is an option. The conditions of contact and, as a consequence, the migration of substances during submersion in food simulant may deviate severely from the conditions of contact occurring under real conditions of contact. For example, the conditions of contact of a small oxygen absorber with roasted nuts are not comparable to submersion in a fat simulant, not even when the allowed reduction factor is applied. When submersing the oxygen absorber in oil the whole article is soaked with oil, which does not happen when it is in contact with nuts. Comparable situations were observed when using systems in contact with meat, for which Directive 85/572/EEC requires testing with water and oil. The tests with water and, in case of processed meat products, with 3% acetic acid by total immersion results in excessive migration of iron ions into the food simulant. After the migration period the food simulant is usually brown-coloured by iron oxide. This phenomenon does not occur with foodstuffs; otherwise, the food contaminated with brown spots would not be acceptable from a sensory point of view.

In the case of moisture absorbers, submersion of the absorber leads to contact conditions significantly different from those occurring in contact with food under real conditions as well. Active and intelligent packaging systems are in contact with the foodstuff under different conditions from conventional packaging materials. In addition, the composition (multi-layer) of the system, as well as the presence of an active ingredient, are reasons for high migration when testing under conventional conditions. Therefore, there is a need for extending the existing test protocols with so-called dedicated test methods. Within the Actipak project some experiments with dedicated tests have been performed. Oxygen-absorbing labels were tested by sandwiching the label between layers of filter paper immersed in iso-octane as the fatty food simulant. After the migration period the paper was extracted and the overall migration was determined. Migration from a paper fibre-based moisture absorber was determined with a block of agar. The agar immobilizes the water in a comparable way as water bound in meat, for example. To demonstrate potential migration the absorber was first partly saturated with water containing a fluorescent label. After the contact period the migration of the fluorescent label was measured. This test could be useful to demonstrate whether or not migration may occur. Similar tests were performed with a moisture regulator based on the hygroscopic properties of sugar solutions. Migration of iron and sodium chloride from an oxygen absorber in real food, food simulants and alternative simulants has been determined as well. The results are very promising, but need further standardization and validation.

Intelligent systems placed on the outside of the primary packaging may form a separate group. These intelligent systems are connected to the packaging material by means of an adhesive. Many intelligent systems are composed of plastic material that contains the intelligent ingredients as one of the layers of the system or in a plastic sachet. There is no direct contact with the food. In addition, the shelf-life of foods with an intelligent system on the outside is relatively short. Even if a polyolefin is used for the primary packaging the lag time will prevent any migration. There is no need yet to require migration testing of intelligent systems connected to the outside of the primary packaging.

Recommendations

- Active and intelligent packaging systems composed of only plastic shall be tested according to Directives 82/711/EEC and 85/572/EEC.
- Substances intended to be released from an active system could be quantified by migration testing or by determination of the total amount present, while assuming that the total amount of substance present in the active system will be released to the packed food.
- The annex of Directive 97/48/EC should be extended to allow testing with foodstuffs too.
- Article 1 (4) of Directive 82/711/EEC should be amended to allow application of the provisions of the Directive to active and intelligent packaging systems not composed of plastics only.
- Intelligent systems placed on the outside of the primary packaging should be excluded from migration testing. Clause 4 of Chapter II of the Annex of Directive 97/48/EEC should be extended for that purpose.
- An additional Chapter V in the Annex of Directive 97/48/EC should be inserted to allow for dedicated test protocols for some types of active and intelligent packaging systems.
- Dedicated test protocols need further development and standardization.

22.5.5 Other directives on food contact materials

Other specific directives concerning food contact materials have been published. However, these directives do not influence the use of active and intelligent packaging systems and are hence not discussed here in detail. For the sake of completeness, these directives are listed below.

93/10/EEC[14] regenerated cellulose film
93/111/EC[15] 1st amendment to Directive 93/10/EEC
84/500/EEC[16] ceramic articles intended to come into contact with foodstuffs
2002/16/EC[17] use of certain epoxy derivatives

22.6 Food additives

The requirements for food additives are formulated in general terms in Framework Directive 89/107/EEC. Specific directives have been published on colours, sweeteners and food additives other than colours and sweeteners.

22.6.1 Framework Directive 89/107/EEC[18] as amended by Directive 94/34/EEC[19]

Directive 89/107/EEC specifies the definition for food additives and the scope of the directive. In simple terms, it states that food additives are not food ingredients or characteristic ingredients. Food additives are intentionally added to attain a technological effect during manufacturing, storage and distribution of the food. Various categories of food additives have been identified, each with its typical properties. Food additives are allowed only if there is a technological need, if there is no hazard to human health and if they do not mislead the consumer. Consumers should be informed about the presence of additives in foodstuffs by means of proper labelling of the food or the food additives. At a national level specific requirements on listing the ingredients as well as their traceability may exist.

Relevance to active and intelligent packaging systems

The directive on food additives is relevant only to systems that intentionally release substances into the food. The substance intentionally released from an active system should in the first place be an allowed food additive covered by one of the categories listed in Annex I of Directive 89/107/EEC. In addition, there should be a technological need that cannot be met by other means. Validity of this clause may be difficult to demonstrate but active systems fulfil a technological function in the food when food is already packed. In addition, the requirement to add the lowest level possible to achieve a desired effect may support the use of active systems. Active systems usually will be active at the surface of the packed food, whereas a food additive is often mixed into the food. As a result, the total amount of a substance may be significantly reduced when using an active system.

Foods may contain a substance that is also released from an active packaging system. In those cases, the final concentration in the food should be taken for a proper judgement of compliance with regulatory requirements. The food packer will carry that responsibility in first instance. The proper labelling of the active releasing system concerning the maximum amount of substance released from an active system avoids the possibility of that maximum limit being exceeded. Active releasing systems may release the food additive via the headspace of the packed food to obtain a distribution as uniform as possible. In other cases the transfer of substances may be caused by intense contact with the active system. In both cases the concentration at the surface may be higher than the maximum allowed concentration. However, measured on the basis of the bulk of the food the amount of food additive should be significantly below the allowed concentration limit. Taking into account that the whole bulk of the packed food is consumed this should not be a problem. In analysis of the foodstuff a proper homogenization of the food should be ensured.

Recommendations

- Directive 89/107/EEC does not form any hurdle to the use of active and

intelligent packaging systems. The substances released from active packaging systems shall comply with the requirements of this directive.
- Foods in contact with a releasing system should be homogenized before analyzing the food on the total amount of the relevant food additive.

22.6.2 Specific directives on colours, sweeteners and food additives other than colours and sweeteners

In addition to the Framework directive, specific directives on food additives have been published. Directive 95/2/EC[20] (last amended by 2001/5/EC[21]) provides a glossary of the various categories of food additives covered by the directive. Also substances not included in the directive are indicated, for example substances for the treatment of drinking water. The directive is based on the positive list principle. The substances, provided with a so-called E number, are listed in five separate annexes. The annexes list substances for general use or for use in specified foods or concentrations.

A relevant issue is the packaging gases that are allowed in all foodstuffs. In this respect, the Directive defines packaging gases as gases other than air, introduced into a container before, during or after placing a foodstuff in that container.

Packaging gases provided with an E number are carbon dioxide, argon, helium, nitrogen, dinitrogen oxide and oxygen. The additives are subject to purity requirements, which are laid down in specific directives. Requirements for colours used in foodstuffs are laid down in Directive 94/36/EC.[22] Colours allowed to add or restore colour in foodstuffs include colours of natural sources. In five annexes the permitted colours and the conditions of their use are laid down. The annexes include a positive list, a list of foodstuffs that may not be coloured, and colours with restricted uses.

Directive 94/35/EC[23] (as amended by Directive 96/83/EC[24]) concerns the use of sweeteners added to foodstuffs. Only the sweeteners listed may be used in the foodstuffs listed at a level fulfilling the intended purpose and shall not mislead consumers.

Relevance to active and intelligent packaging systems

The specific directives of the framework Directive 89/107/EEC are relevant only to releasing systems. The specific directives are detailed and do not allow deviations. Therefore, the releasing systems should comply with qualitative and quantitative requirements on food additives. Manufacturers and food packers should realize that a releasing system may not be generally applicable to all foodstuffs but only to specified ones. Therefore, it seems obvious that the manufacturer of releasing systems should give proper instructions and define conditions of use, although the final user or food packer has his own responsibility as well.

None of the specific directives mentions the removal of substances from the packed foodstuffs. This may be logical as the directives are dealing with additives. The use of an oxygen absorber, which removes oxygen from the headspace of the packed food, is excluded from the directive whereas flushing with nitrogen is included. The resultant packaging gas is, however, similar. The application of gas absorbers is not covered by any directive and remains the responsibility of the food packer. As the application of oxygen absorbers is very similar to the use of packaging gases, it seems logical that labelling and food safety are handled in the same way.

Labelling of packed food

Packaging gases used for packaging certain foodstuffs should not be regarded as ingredients and therefore should not be included in the list of ingredients on the label. However, consumers should be informed of the use of such gases inasmuch as this information enables them to understand why the foodstuff they have purchased has a longer shelf-life than similar products packaged differently. Therefore, the following text should be used on the label in the national language: 'packed under a protective atmosphere', as is required for modified-atmosphere packaging

Food safety

When the atmosphere inside a package is altered, the limiting factor for shelf-life may also change. For example, in an oxygen-free atmosphere the growth of aerobic micro-organisms is inhibited, but this atmosphere may promote the growth of anaerobic micro-organisms. The limiting factor for shelf-life may then become the growth of anaerobic micro-organisms. A similar reasoning may be valid for preservative-releasing systems. Shelf-life studies should reveal the spoilage mechanism and the actual shelf-life of the food should be established.

Recommendations

- Food additives released from active packaging systems shall comply with the requirements laid down in the framework directive and its subsequent specific directives. Limits and requirements on the total quantity of additives in foods and the purity of the additives shall be obeyed. Also the limitation of addition of substances to specified foods must be taken into account.
- Oxygen absorbers should be included in the section about modified-atmosphere packaging by amending Article 1(3 r) of Directive 95/2/EC as follows:

  'packaging gases' are gases other than air, introduced into a container before, during or after placing of the foodstuff in that container, or by selective removal of oxygen after placing of the foodstuff in that container'

- Substances released into food shall be labelled according to requirements on labelling.

22.7 Food flavouring

Framework Directive 88/388/EEC[25] (as amended by 91/71/EEC[26]) concerns flavouring substances for use in or on foodstuffs to impart odour and/or taste. The flavouring substances should be obtained from materials of vegetable or animal origin or by chemical synthesis. Flavourings should not imply addition of any element or substance in a toxicologically dangerous quantity. Maximum levels for arsenic, lead, cadmium and chromium have been set. Also the content of 3,4-benzopyrene is limited in all foods. There is a short list of substances that may be used at certain maximum concentrations in foodstuffs.

Labelling requirements concerning the description, quantity, suitability for food use and traceability have been laid down. In Council regulation EC 2232/96[27] a procedure is laid down which includes the listing of all flavouring substances in use in the EU member states. The substances will be evaluated to establish their conditions of use. Commission decision 1999/217,[28] as amended by Commission decision 2000/489,[29] lists more than 2800 substances The registration is a first step to a harmonized positive list of flavouring substances. The Nordic countries have some specific rules for the use of flavourings in certain food products.[3]

Relevance to active and intelligent packaging systems

Active packaging systems releasing flavourings are by definition an attractive way of flavouring food. A flavour added to the packed foodstuff will generate an attractive or characteristic smell when consumers open the packed food. Sausage casing may be flavoured to release the smoke flavour to the sausage in order to obtain a flavour taste and to preserve the sausage. A classic example is the use of wine barrels, which are used to store the wine but at the same time release their flavour to the wine, which may be characteristic of the wine. In modern wine making wood chips may be used to obtain the same effect. Although a wine barrel is clearly an active packaging material in the definition of active packaging systems, for historical reasons and because of the natural origin, wine barrels could be excluded from classification as an active packaging material. Application of flavour-releasing systems will not be hindered by the existing regulations on flavouring, provided the rules laid down for flavouring of foodstuffs are taken into account.

Release of flavour, can however, also be used to hide some negative aspects of the foodstuff. Directive 88/388/EEC clearly indicates that flavouring must not be allowed to mislead consumers. The use of flavouring to hide spoilage is not acceptable; it would mislead consumers and may cause serious food poisoning. But, when the flavour is added in order to overwhelm an off-flavour of the food and the use of flavouring does not cause any toxic harm it may be found acceptable. Active flavour-releasing systems are also strong tools to avoid so-called scalping of flavour of packed foods. By supplementing the flavours through the packaging material this effect could be avoided.

Another category of active packaging systems that may mislead consumers are absorbers. For instance, an absorber could be used to remove the amine smell of fish and, as a consequence, consumers will be deprived of a sensory indicator for spoilage. These types of active systems are not covered by the flavouring regulations but are actually comparable to hiding effects or, in the worst case, misleading consumers.

Recommendations

- There are no fundamental objections to the use of an active system that releases flavouring substances, provided the regulations on flavouring are followed.
- Allowed total quantities of flavourings in foods shall not be exceeded.
- Flavouring to hide spoilage is not allowed.
- Flavouring to mask natural or synthetic off-flavours should be further studied. Conditions of acceptability should be drafted.
- Removal of substances is not an issue of the flavouring regulation but needs legal attention. Appropriate provisions could be included in the specific directive on active and intelligent food contact materials.

22.8 Biocides and pesticides

Biocides are substances intended to destroy, deter, render harmless, prevent the action of, or otherwise exert a controlling effect on any harmful organism by chemical or biological means, as defined in Directive 98/8/EC.[30] Several areas are excluded from the directive. Among others, food additives subject to Directive 89/107/EEC and food contact materials subject to Directive 89/109/EEC are excluded from the biocide regulations.

22.8.1 Relevance to active and intelligent packaging systems

The biocide regulation excludes food additives and food contact materials. Thus, any substance with a biocidal effect should be listed in these regulations. Active systems intended to release a biocidally active substance into foods are limited to the use of substances allowed as food additives. All requirements and restrictions laid down in the regulations on food additives must be taken into account. Only in bulk transport is the use of biocides as well as pesticides allowed. A ship cargo space may be gassed with biocides or pesticides to protect the food. This may also be feasible with active systems of sufficient capacity. This is considered a special category of application that should comply with the rules presently valid.

Often confusion is brought about with regard to the use of biocidal substances in food contact materials. There are two reasons to add biocidal products to food contact materials. First, it may be necessary to stabilize a polymer emulsion before manufacturing the final article. This application is indispensable to allow transport and storage of the semi-manufactured product. A second application, of increasing interest, is the addition of antimicrobial substances to protect the surface of the final article from microbial contamination. In both situations the addition of the biocide should not be considered as an active system as there is no intentional influence on the food. In both cases migration of the substance should be negligible or as low as possible; anyway, there should be no effect on the food in contact with the materials.

Recommendation

The regulation on biocides excludes food contact materials and food additives. Therefore, all applications related to biocidal substances are subject to regulations on food contact materials and food additives.

22.8.2 Pesticides

Directive 91/414/EEC[31] regulates the use of pesticides, which, in short, are active substances to protect plants and plant products against harmful organisms. Plant protection products (pesticides) are used on agricultural produce and are not added to foodstuffs as preservatives. Maximum residue levels (MRLs) for each specific pesticide in agricultural produce have been defined in the Directive, either for a group of products or for individual products.

22.8.3 Relevance to active and intelligent packaging systems

The use of pesticides is legal only if approved for a specific use or on specific agricultural produce. At the pre-harvest stage the use of active packaging is unlikely even if possible. However, some products may also be treated with pesticides at the post-harvest stage; for example, the use of certain plant growth regulators for potatoes is authorized as well as some insecticides on cereal grains. The use of these pesticides is usually a matter of bulk treatment. Protective substances on potatoes or cereal grains may or may not be volatile. Treatment with non-volatile agents is unlikely as it will not be effective on the bulk of food. The protection of potatoes with volatile substances may be feasible but due to the batch treatment this is unlikely. No such applications are currently in use or under development. When active systems are developed then they should comply with the rules on treatment of food products. Impregnation with biphenyl of paper used for packaging citrus fruits has been known for many years. However, in this application biphenyl is regulated as a food additive.

Recommendation

When active systems are developed, they shall comply with the regulation on pesticides.

22.9 Food hygiene

The aim of Council Directive 93/43/EEC[32] on the hygiene of foodstuffs is to control all activities critical to food safety, and thus it covers all aspects affecting hygienic production, storage, packaging and distribution of foodstuffs, in order to ensure the safety and wholesomeness of foodstuffs. The Directive aims at establishing uniform minimum requirements for food production to ensure that only safe food is retailed. Regulations on veterinary products, such as Directive 92/5/EEC[33] for meat products, contain more detailed requirements (e.g. approval of establishment, stricter temperature conditions, official controls) for the production of some products of animal origin. Special provisions for the hygiene of quick-frozen foodstuffs are given in Council Directive 89/108/EEC[34] to protect them from microbial or other external contamination and from drying.

To achieve safe food the directive requires protection of the food within the food production chain against any contamination that renders the food unfit for consumption. Foods supporting the growth of pathogenic micro-organisms or the formation of toxins should be kept at temperatures that will not endanger health. Principles of HACCP (hazard analysis of critical control points) as given by the FAO/WHO Codex Alimentarius Commission[35] should be followed. Food packaging materials are not directly covered by the EC Directive, but hygienic conditions of the packaging materials will be a prerequisite in hygienic food production. Neither the microbiological criteria for foodstuffs nor the temperature requirements have been harmonized in the European Union. Various time-temperature requirements can therefore be found for certain food categories in different countries.[36] Where no legislation exists, the manufacturer may freely choose the best storage temperature for the product provided the product is safe for consumption.

Although legislative requirements and recommendations for temperature control during manufacturing, heating, cooling and chilled storage are abundant, there are no rules in food legislation on how long food quality should remain acceptable. Directive 2000/13/EC[37] on labelling requires pre-packed foods to bear a date of minimum durability or, for highly perishable foods, a 'use by' date. It is the manufacturer's responsibility to determine the shelf-life of the product, taking into account storage conditions, and to ensure that the product is safe throughout its assigned shelf-life. The shelf-life of foods depends on the specific properties of the food product and the environmental conditions in which the food is treated and stored. In particular, the shelf-life of microbiologically sensitive foodstuffs will depend on storage conditions of time and temperature.

22.9.1 Relevance to active and intelligent packaging systems

The Food Hygiene Directive requires that all measures be taken to ensure the safety and wholesomeness of foodstuffs during production, transport, storage and offering for sale or supply to the consumer. The use of active systems may be helpful to maintain the quality of the food and to extend its shelf-life. Intelligent systems could provide reliable information on the conditions of the food by showing, for instance, the time and temperature conditions during the life cycle of the food, or by detecting gases generated by micro-organisms.

The use of an oxygen absorber will suppress the growth of certain micro-organisms. The use of preservative-releasing systems will have a similar final effect. Foodstuffs will not only have a longer shelf-life but will also be safer at the time of consumption. The use of moisture absorbers, for example for packaged meat, has in the first instance a visual benefit as the meat juice is absorbed by the absorption pad. If, however, such a pad is treated with a selected mixture of spices, then microbial deterioration will be slowed down resulting in a longer shelf-life and safer product.

It is required today to print on packaged food the 'use by' date. Usually the 'use by' date is established on the basis of experience. For products with a long shelf-life this does not cause any problem as the storage time and temperature conditions are not very critical. For products with a long shelf-life chemical deterioration is usually the limiting factor, whereas for foods with a relatively short shelf-life microbiological conditions are often the limiting factor. Food packers may extend the safety margin to allow of some 'misuse' during transport by consumers from the retailer to their homes, or incorrect temperature settings during display. Use of a time/temperature indicator could indicate the safety of the food by indicating that the allowable storage conditions of time and temperature have not been exceeded. These time/temperature indicators could prevent unnecessary waste of food due to the elapse of the 'use by' date, which of itself is no guarantee that the food is fit for consumption. The indicator will inform consumers whether the product is still suitable for consumption. These indicators could replace the requirements of printing 'use by' dates when it is demonstrated that they are reliable and when the consumer is familiar with the use of the indicators. However, most time/temperature indicators are not capable of giving proper information on the period still to go before the 'use by' date is passed. This could be overcome by printing the production date or date of packing on the packed food instead of a 'use by' date in addition to an indication of the shelf-life. This approach would require, of course, a range of indicators with variable 'response times' to allow the use of a proper indicator.

Modified-atmosphere packaging with packaging gases is a frequently used method to preserve foods. However, if the gas-tightness of the package fails, the protective atmosphere will change and the food may become unfit for consumption. This is very difficult to observe both for the manufacturer and for consumers. Insertion of an indicator that detects, for example, oxygen will provide information not available without the indicator. Similar indicators can be inserted to detect the generation of microbial respiratory gases. The Food Hygiene Directive requires 'all measures necessary to ensure the safety and wholesomeness of foodstuffs'. The use of both active and intelligent packaging systems is a new means of meeting this requirement. Actually, the requirements of the hygiene directives strongly support the use of active and intelligent systems.

Recommendation
Allowance should be made in the Food Hygiene Directive to replace the 'use by' date with dedicated time temperature indicators.

22.10 Food labelling, weight and volume control

Labelling of foodstuffs is meant to give consumers information on the composition of the food and to protect them. In Directive 2000/13/EC requirements for labelling of foodstuffs to be delivered to the ultimate consumer are laid down. Labelling of the foodstuff should not mislead the ultimate consumer. Detailed but generally applicable requirements have been formulated as to the information to be provided. Major issues are: name, list and quantities of ingredients, shelf-life, name and address of the manufacturer, instructions for use, etc. All ingredients should be listed in descending order of quantity.

Food additives shall be designed by their category name followed by their specified name or EC number, for example 'Emulsifier E 322'. Also requirements on the minimum durability of the foodstuff should be printed by using the wording 'best before ...' or 'use by ...' depending on the perishable nature of the foodstuff. Directive 89/109/EEC lists requirements on labelling of packaging materials. This concerns, however, not the final product but the packaging material when it is not in contact with the food. In that case the packaging material should be accompanied with instructions for use such as suitability for various types of foodstuffs and maximum temperature range. In addition, it should be possible to trace back the packaging material to the manufacturer in case of a calamity.

22.10.1 Relevance to active and intelligent packaging systems

Directive 2000/13/EC requires listing of food additives used in the manufacture or preparation of foods and still present in the finished product. It may be questionable whether an additive released from the packaging material is added during manufacturing or preparation, but no doubt it will be present in the final product. Therefore, any substance intentionally released into the food while being packed should be listed according to the rules of Directive 2000/13/EC. Requirements on total quantities should be respected, irrespective of the stage at which the substance becomes part of the foodstuff. Intelligent systems could supplement the information presently given to the consumer. It is conceivable that labelling requirements could be changed due to the information given by intelligent systems. For example, 'use by' dates could be replaced by information obtained from a time-temperature indicator. However, the introduction of intelligent systems and consumer education regarding interpretation will be needed before making any changes to labelling requirements in this respect.

Active and intelligent packaging systems may be incorporated in the packaging material of the foodstuff. They can also be packed with the food in the form of a sachet, box or label. Consumers should be made aware that the object included in or on the packed food is not a part of that food. Sachets with a powder could easily be confused with ingredients like salt or pepper. Great care should be taken to prevent the consumers eating it. Labelling only by text seems not sufficient. Consumers who cannot read must be protected as well. Therefore, the introduction of a harmonized universal symbol, which indicates that the object is not part of the foodstuff, seems appropriate. According to labelling Directive 2000/13/EC, the ultimate consumer should be informed properly. Therefore, also information on the function of the inedible active or intelligent packaging system, should be printed.

Usually, active and intelligent packaging systems will not be available to the consumer as such. They will be purchased by food manufacturers and food packers. The manufacturers should also be informed about the range of applications and restrictions of use as well as about the quantity of additive that may be released from an active system. This could be achieved by means of documents attached to a batch of articles.

Recommendations

- Labelling of foods shall be in compliance with Directive 2000/13/EC. Substances released from a system should be considered a food additive added during manufacturing or preparation of the food.
- Requirements on labelling, at the retail stage, should be formulated with the aim to inform the consumer about:
  - the presence of a non-food component
  - the function of the system
  - inedibility of the system by means of written text and a pictogram
  - any possible risk upon digestion of a system.

  These requirements could be added to the directive on labelling, but it may be more appropriate to add them to the specific directive on active and intelligent food contact materials.
- At the wholesale stage, active and intelligent packaging systems should be accompanied by a certificate of compliance with regulations of food contact materials.
- Instructions on conditions and restrictions of use should be given at the wholesale stage.

22.10.2 Weight and volume control

Several EU directives deal with the weight and volume control of pre-packaged food. Directive 75/106/EEC [38] and Directive 76/211/EC[39] relate to pre-packages made up by volume and weight respectively. The pre-packages must bear an indication of the product weight or volume, known as 'nominal weight' or 'nominal volume' which they are required to contain.

22.10.3 Relevance to active and intelligent packaging systems

Active systems may influence the weight or volume of the foodstuffs. In the case of emitters of food additives (preservatives, flavouring compounds, etc.) the migration of these compounds will have a negligible effect on the weight or volume of the food. Lightweight foods, such as chips and dried herbs, may be exceptions. Moisture absorbers, such as an absorbing pad for meat drip, usually have a noticeable effect on the net weight of meat. The aim of the Directives on weight and volume control is to ensure that consumers are correctly informed on the net quantity of the food. If the active system influences the weight or volume, this must be taken into account in the declared weight or volume.

Recommendation

Active packaging systems with absorbing properties should take into account the loss of weight due to the absorber.

22.11 Product safety and waste

Directive 2001/95/EC[40] concerns general product safety. The general product safety directive dictates that all products placed on the market shall be safe. 'Safe products' mean that under normal or reasonably foreseeable conditions of use the product does not present any risk or only the minimal risks compatible with the product's use. In the judgement of safety aspects the characteristics of the product, presentation, labelling instructions and the category of consumers, in particular children, should be considered. Manufacturers are obliged to provide the relevant information to the final consumer.

22.11.1 Relevance to active and intelligent packaging systems

The general product safety directive applies to active and intelligent systems. The active or intelligent system may never endanger food safety or consumer health. To comply with safety a number of issues have to be considered before bringing systems on the market.

22.11.2 Labelling of active and intelligent systems

Several active and intelligent systems are present inside the primary packaging, such as sachets, cups and pads. It has to be made clear that these systems are not suitable for consumption. There should be no confusion with sachets or cups that contain, for example, herbs, salt or butter, which are intended to be consumed with the packaged food. Therefore, on the active or intelligent system a well legible, indelible warning has to be placed in at least the national language that the active or intelligent system is not to be consumed, for example 'DO NOT EAT'.

Functionally dyslexic people and those not able to read the national language should be able to understand the warning 'DO NOT EAT' by means of a symbol

Fig. 22.3 Proposed symbol to warn consumers not to eat the system.

printed on the label. This symbol has to express that the content of the active or intelligent system is not suitable for consumption. Harmonization of the wordings and the symbol would enhance the understanding of the wording and symbol in a short period of time, whereas the use of different indications would confuse the consumer. Within the Actipak project a symbol is proposed which is shown in Fig. 22.3. Possibly better designs could be developed, but the main issue is that only one symbol should be adopted for harmonization.

22.11.3 Size and shape of active or intelligent systems

A recommendation has been issued to member states to take action to prevent consumption of the non-food article. In this respect children, mentally disabled patients and elderly people are considered high-risk groups. It is therefore advisable that the non-food article is so large that adults cannot swallow it. For toys[41] the minimum size is determined on the basis of a defined cylinder, resulting in a size of 3.17 cm. For adults the minimum size should be increased to 5 cm. In addition, the non-food article should have a morphology distinguishing it from the packaged food. Another possibility is to thoroughly attach the active or intelligent system to the packaging.

Content of active and intelligent system

If, for any reason, the active or intelligent system releases its chemicals, no acute danger to the consumer may occur. Therefore, active and intelligent compounds present in sachets, cups or pads used in consumer packaging should not be seriously irritating, corrosive, harmful or toxic. Furthermore, these compounds shall not be carcinogenic. Directive 67/548/EEC[42] can be used to classify dangerous substances. Some active or intelligent systems consist of a film that incorporates the active compound. In these cases the Scientific Committee on Food has to assess their toxicity and migration behaviour and set limits accordingly.

22.11.4 Food imitation directive

Products referred to in the Food Imitation Directive 87/357/EEC[43] are those which, although not foodstuffs, possess a form, odour, colour, appearance, packaging, labelling, volume or size such that it is likely that consumers, especially children, will confuse them with foodstuffs and consequently place them in their mouths, or suck or ingest them, which might have serious effects.

Member states are obliged to take all measures necessary to prohibit the marketing, import and either manufacture or export of unsafe products.

22.11.5 Retail versus wholesale

Active or intelligent packaging systems can also be used in wholesale food applications, for example, during transport of wholesale packaged foodstuffs. These active or intelligent systems will not reach consumers. The final users are then professional employees, not the risk groups of children, elderly people and mentally disabled persons. Therefore, active and intelligent systems used in wholesale that are not intended to reach consumers do not have to comply with the previously described safety aspects regarding the size and content, or the food imitation directive.

Recommendations

- Measures should be taken to harmonize the text and symbol to be printed on active and intelligent packaging systems.
- Requirements on size and shape as laid down in toys regulations should be made applicable to movable objects packaged with foodstuffs.
- Quantities of substances that could have *serious health* or *lethal* effects should not be allowed.
- Directive 87/357/EEC on food imitation may be applicable to active and intelligent packaging systems depending on the appearance of the system. Manufacturers should consider this directive, in particular in the developmental phase of their system.

22.11.6 Waste

Waste Directive 94/62/EC[44] describes measures aiming at preventing the production of packaging waste. It additionally aims at reusing, recycling or recovering packaging waste to reduce the final disposal of packaging waste. The directive covers all packaging placed on the market regardless of the material used. The directive is applicable to all types of packaging waste. Member states in communication with stakeholders are encouraged to promote reuse of packaging materials. Identification codes should facilitate collection, reuse and recycling. Restrictions as to the levels of lead, cadmium, mercury and hexavalent chromium must be reduced in time. The directive mentions the recycling processes actually available. The directive also assumes that materials can be reused only when appropriate. Packaging use shall be reduced to a minimum.

Relevance to active and intelligent packaging systems

In environmental issues it is often necessary to perform a life cycle analysis. Only in that way is it possible to establish whether the use of a certain type of packaging material is the most appropriate. Addition of an active or intelligent packaging system may require an additional amount of packaging material resulting in more waste. However, if by virtue of a longer shelf-life of the packed food the waste of packaging material and food is reduced, the scale could easily be turned in favour of the use of active or intelligent packaging systems. Recovery of food contact materials, with the exception of paper, glass and metal, is limited. In recycling of plastics only recycled polyethylene terephthalate is commercially applied. For most other polymers recycling of food contact plastics into new articles is still on a modest scale. Collection, sorting, cleaning and processing is cumbersome, as the directives on plastics require that the material shall be safe and comply with the positive list. This means that recycling of complex mixtures like active and intelligent packaging systems is currently not an issue.

The presence of various chemicals, present only in relatively small quantities, may meet objections upon incineration but if the substances are of organic nature they will be incinerated. Oxygen absorbers containing iron may produce some additional slag. In general, no problems are manifesting themselves now but developers of active and intelligent packaging systems should take into account possible environmental consequences.

Recommendation

- Manufacturers should consider the use of active and intelligent packaging systems in view of environmental issues.

22.12 References

1. DE KRUIJF N, VAN BEEST M, RIJK R, SIPILÄINEN-MALM T, PASEIRO LOSADA P and DE MEULENAER B, 'Active and intelligent packaging: applications and regulatory aspects', *Food Addit Contam*, 2002 **19** 144–62.
2. VERMEIREN L, DEVLIEGHERE F, VAN BEEST M, DE KRUIJF N and DEBEVERE J, 'Developments in the active packaging of foods', *Trends Food Sci Technol*, 1999 **10** 77–86.
3. FABECH B, HELLSTRØM T, HENRYSDOTTER G, HJULMAND-LASSEN M, NILSSON J, RÜDINGER L, SIPILÄINEN-MALM T, SOLLI E, SVENSSON K, THORKELSSON À E and TUOMAALA V, *Active and intelligent food packaging – A Nordic report on legislative aspects,* Copenhagen, Nordic Council of Ministers, 2000.
4. PASTORELLI S, DE LA DRUZ C, PASEIRO P, DE MEULENAER B, HUYGHEBAERT A, DE KRUIJF N, VAN BEEST M and RIJK R, 'Chemical elucidation of the composition of some active packaging materials', *Food Addit Contam,* submitted for publication.
5. DE MEULENAER B, HUYGHEBAERT A, SIPILÄINEN-MALM T, HURME E, DE KRUIJF N, VAN BEEST M, RIJK R, PASEIRO P, PASTORELLI S, DE LA CRUZ C, WILDERVANCK C and BOUMA K, 'Classification of some active and intelligent food packaging systems with regard to current EU food

packaging legislation', *Food Addit Contam,* submitted for publication.

6. DE MEULENAER B, HUYGHEBAERT A, DE KRUIJF N and RIJK R, 'On the use of agar gels as alternative food simulants to evaluate migration from selected active packaging materials', *Innovative Food Science and Emerging Technologies,* submitted for publication.
7. Council Directive of 21 December 1988 on the approximation of the laws of the Member States relating to materials and articles intended to come into contact with foodstuffs. *Official Journal*, L040, 11/02/1989,0038-0044. Corrigendum *Official Journal*, L 347, 28/11/1989, 0037.
8. Commission Directive 2002/72/EC of 6 August 2002 relating to plastic materials and articles intended to come into contact with foodstuffs (Text with EEA relevance). *Official Journal*, L220, 15/8/2002, 0018-0058.
9. Commission Directive of 9 June 1980 determining the symbol that may accompany materials and articles intended to come into contact with foodstuffs (80/590/EEC). *Official Journal*, L151, 19/06/1980, 0021-0022.
10. Council Directive 82/711/EEC of 18 October 1982 laying down the basic rules necessary for testing migration of the constituents of plastic materials and articles intended to come into contact with foodstuffs. *Official Journal*, L297, 23/10/1982, 0026-0030.
11. Commission Directive 93/8/EEC of 15 March 1993 amending Council Directive 82/711/EEC laying down the basic rules necessary for testing migration of constituents of plastic materials and articles intended to come into contact with foodstuffs. *Official Journal*, L 090, 14/04/1993, 0022–0025.
12. Commission Directive 97/48/EC of 29 July 1997 amending for the second time Council Directive 82/711/EEC laying down the basic rules necessary for testing migration of the constituents of plastic materials and articles intended to come into contact with foodstuffs (Text with EEA relevance). *Official Journal*, L 222, 12/08/1997, 0010-0015,.
13. Council Directive of 19 December 1985 laying down the list of simulants to be used for testing migration of constituents of plastic materials and articles intended to come into contact with foodstuffs (85/572/EEC). *Official Journal*, L372, 31/12/1985, 0014-0021.
14. Commission Directive 93/10/EEC of 15 March 1993 relating to materials and articles made of regenerated cellulose film intended to come into contact with foodstuffs. *Official Journal*, L93, 17/04/1993, 0027-0036.
15. Commission Directive 93/111/EC of 10 December 1993, amending Directive 93/10/EEC relating to materials and articles made of regenerated cellulose film intended to come into contact with foodstuffs. *Official Journal*, L310 14/12/1993, 0041.
16. Council Directive of 15 October 1984 on the approximation of the laws of the Member States relating to ceramic articles intended to come into contact with foodstuffs (84/500/EC). *Official Journal*, L277 20/10/1984, 0012-0016.
17. Commission Directive 2002/16/EC of 20 February 2002 on the use of

certain epoxy derivatives in materials and articles intended to come into contact with foodstuffs (Text with EEA relevance). *Official Journal,* L 051 22/02/2002, 0027-0031.

18. Council Directive of 21 December 1988 on the approximation of the laws of the Member States concerning food additives authorized for use in foodstuffs intended for human consumption (89/107/EEC). *Official Journal,* L 040, 11/02/1989, 0027-0033.
19. European Parliament and Council Directive 94/34/EC of 30 June 1994 amending Directive 89/107/EEC on the approximation of the laws of Member States concerning food additives authorized for use in foodstuffs intended for human consumption. *Official Journal*, L 237, 10/09/1994, 0001-0002.
20. European Parliament and Council Directive No 95/2/EC of 20 February 1995 on food additives other than colours and sweeteners. *Official Journal*, L 061, 18/03/1995, 0001-0040.
21. Directive 2001/5/EC of the European Parliament and of the Council of 12 February 2001, amending Directive 95/2/EC on food additives other than colours and sweeteners. *Official Journal*, L 055, 24/02/2001, 0059-0061.
22. European Parliament and Council Directive 94/36/EC of 30 June 1994 on colours for use in foodstuffs. *Official Journal*, L 237, 10/09/1994, 0013-0029.
23. European Parliament and Council Directive 94/35/EC of 30 June 1994 on sweeteners for use in foodstuffs. *Official Journal*, L 237, 10/09/1994, 0003-0012.
24. Directive 96/83/EC of the European Parliament and of the Council of 19 December 1996 amending Directive 94/35/EC on sweeteners for use in foodstuffs. *Official Journal*, L 048 , 19/02/1997, 0016-0019.
25. Council Directive of 22 June 1988 on the approximation of the laws of the member states relating to flavourings for use in foodstuffs and to source materials for their production (88/388/EEC). *Official Journal*, L 184, 15/07/1988, 0061-0066.
26. Commission Directive of 16 January 1991 completing Council Directive 88/388/EEC on the approximation of the laws of the Member States relating to flavourings for use in foodstuffs and to source materials for their production (91/71/EEC). *Official Journal*, L 042 , 15/02/1991, 0025-0026.
27. Regulation (EC) No 2232/96 of the European parliament and of the Council of 28 October 1996 laying down a Community procedure for flavouring substances used or intended for use in or on foodstuffs. *Official Journal*, L 299, 23/11/1996, 0001-0004.
28. Commission Decision of 23 February 1999 adapting a register of flavouring substances used in or on foodstuffs drawn up in application of Regulation (EU) 2232/96 of the European parliament and of the Council of 28 October 1996 (notified under number C(1999) 399) (text with EEA relevance) (1999/217/EC). *Official Journal*, L 084, 27/03/1999,

0001-0007.
29. Commission Decision of 18 July 2000 amending Decision 1999/217/EC adopting a register of flavouring substances used in or on foodstuffs (notified under document number C(2000) 1722) (Text with EEA relevance) (2000/489/EC). *Official Journal*, L197, 03/08/2000, 0053-0056.
30. Directive 98/8/EC of the European Parliament and of the Council of 16 February 1998 on the placing of biocidal products on the market. *Official Journal*, L 123, 24/04/1998, 0001-0063.
31. Council Directive of 15 July 1991 concerning the placing of plant protection products on the market (91/414/EEC). *Official Journal*, L 230, 19/08/1991, 0001-0032.
32. Council Directive 93/43/EEC of 14 June 1993 on the hygiene of foodstuffs. *Official Journal*, L 175, 19/07/1993, 0001-0011.
33. Council Directive 92/5/EEC of 10 February 1992 amending and updating Directive 77/99/EEC on health problems affecting intra-Community trade in meat products and amending Directive 64/433/EEC. *Official Journal*, L 057, 02/03/1992, 0001-0026.
34. Council Directive of 21 December 1988 on the approximation of the laws of the Member States relating to quick-frozen foodstuffs for human consumption (89/108/EEC). *Official Journal*, L 040, 11/02/1989 0034-0037.
35. FAO/WHO Codex Alimentarius Commission (1997). Hazard Analysis and Critical Control Point (HACCP) system and guidelines for its application. Annex to CAC/RCP 1-1969, Rev. 3 (1997). In *Food Hygiene Basic Texts*. Rome: Food and Agriculture Organization of the United Nations, World Health Organization.
36. European Commission, Harmonization of safety criteria for minimally processed foods, Inventory report, FAIR Concerted Action FAIR CT96-1020.
37. Directive 2000/13/EC of the European Parliament and of the Council of 20 March 2000 on the approximation of the laws of the Member States relating to the labelling, presentation and advertising of foodstuffs. *Official Journal*, L 109, 06/05/2000, 0029-0042.
38. Council Directive of 19 December 1974 on the approximation of the laws of the Member States relating to the making-up by volume of certain prepackaged liquids (75/106/EEC). *Official Journal*, L 042, 15/02/1975, 0001-0013.
39. Council Directive of 20 January 1976 on the approximation of the laws of the Member States relating to the making-up by weight or by volume of certain prepackaged products (76/211/EEC). *Official Journal*, L 046, 21/02/1976, 0001-0011.
40. Council Directive 2001/95/EC of 3 December 2001 on general product safety (Text with EEA relevance). *Official Journal*, L 11, 15/01/2002, 0004-0017.

41. NEN-EN 71-1: 1998. 01/09/1998. Safety of toys – Part 1: Mechanical and physical properties.
42. Council Directive of 27 June 1967 on the approximation of laws, regulations and administrative provisions relating to the classification, packaging and labelling of dangerous substances (67/548/EEC). *Official Journal*, L 196, 16/08/1967, 0001-0005.
43. Council Directive of 25 June 1987 on the approximation of the laws of the Member States concerning products which, appearing to be other than they are, endanger the health or safety of consumers (87/357/EEC). *Official Journal*, L 192, 11/07/1987, 0049-0050.
44. European Parliament and Council Directive 94/62/EC of 20 December 1994 on packaging and packaging waste. *Official Journal*, L 365 , 31/12/1994, 0010-0023

23

Recycling packaging materials

R. Franz and F. Welle, Fraunhofer Institute for Process Engineering and Packaging, Germany

23.1 Introduction

Food packaging is a still growing market. As a consequence, the demand to re-use post-consumer packaging materials grows as well. Recycling of packaging materials plays an increasing role in packaging, and numerous applications can already be found on the market. Ten or twenty years ago most post-consumer packaging waste was going into landfill sites or to incineration. Traditionally, only glass and paper/board were recycled into new applications. In the case of packaging plastics the situation is quite different. Only uncontaminated in-house production waste was collected, ground and recycled into the feedstream of the packaging production line without further decontamination. With increasing environmental demands, however, post-consumer plastics packaging materials have also been considered more and more for recycling into new packaging.

A closed-loop recycling for packaging plastics is also supported by public pressure. The packaging and filling companies have to take responsibility for their packaging materials and environmental concerns. In many countries the consumer, government and the packaging companies want to have packaging materials with a more favourable ecobalance in the supermarkets. A more favourable ecobalance can be achieved with different approaches. One of these approaches is the re-use of recycled material in packaging. This development is driven by the recent strong increase in polyethylene terephthalate (PET) bottles used for soft drinks, water and other foodstuffs.[1] Today, many filling companies have decided to start using recycled plastics into their PET bottles in the near future.

But recycling of packing plastics is also a question of recycling technology and collection of packaging waste. Today many countries have established collection systems for post-consumer packaging waste, like the green dot systems. Such country-wide collecting systems guarantee increasing recovery rates. Together with new developments of recycling systems and with increasing recycling capacity the way is open for some plastics for a high value recycling of packaging waste. Due to health concerns most of the recycled post-consumer plastics are going into less critical non-food applications, but in recent years there have also been efforts to recycle post-consumer plastics like PET into new food packaging applications. This changes the situation for some packaging plastics from an open-loop recycling of packaging plastics into a closed-loop recycling into new packaging materials. However, the recycling of post-consumer plastics into direct food contact application needs much more knowledge about contamination and migration than for non-food applications, in order to assess the risk to consumers' health. Additionally a quality assurance system for post-consumer plastics should be established.

23.2 The recyclability of packaging plastics

It is generally known that food contact materials are not completely inert and can interact with the filled product.[2] In particular, interactions between packaging plastics and organic chemicals deserve the highest interest in this context. Such interactions start with the time point of filling and continue during the regular usage phase of a package and even longer, in case a consumer 'misuses' the empty packaging by filling it with chemical formulations such as household cleaners, pesticide solutions, mineral oil or others. The extent of these interactions depends on the sorption properties and the diffusion behaviour which is specific to certain polymer types or individual plastics. These physical properties together with the contact conditions ultimately determine the potential risk of food contamination from recycled packaging plastics. In other words, taking only the polymer itself into consideration and not possible recycling technologies with their special cleaning efficiencies, etc., under given conditions the inertness of the polymer is the basic parameter which determines the possibility for closed-loop recycling of packaging plastics. The inertness of common packaging polymers decreases in the following sequence:

Poly(ethylene naphthalate) (PEN), poly(ethylene terephthalate) (PET), rigid poly(vinyl chloride) (PVC) > polystyrene (PS) > high density polyethylene (HDPE), polypropylene (PP) > low density polyethylene (LDPE)

In relation to this aspect, PEN, PET or rigid PVC do possess much more favourable material properties in comparison to other packaging plastics, such as polyolefins or polystyrene and are, therefore, from a migration related point of view much better suited for being reused in packaging applications. Polymers like polystyrene and HDPE may also be introduced into closed loop recycling if the cleaning efficiency of the recycling process is high enough regarding the input concentrations of post-consumer substances. However, regarding consumers' safety, the composition and concentration of typical substances in post-consumer plastics and the ability of the applied recycling process to remove all post-consumer substances to concentrations similar to virgin materials is of interest. The incoming concentration of post-consumer contaminants can be controlled off-line with laboratory equipment like gas chromatography or HPLC or online with detecting or sniffing devices. With help of online devices nearly a 100% control of the input materials can be established. Therefore the post-consumer material is much more under control and packaging materials with high concentrations of migratable substances, or misused bottles, can be rejected and the requirements on the cleaning efficiency of the recycling process are lower. The source control is therefore the crucial point regarding of the 'worst-case' scenario of the so-called challenge test (see Section 23.4.1).

Recovery of packaging plastics into new packaging applications requires blending of recycled with virgin materials. In praxis today, the recyclate content of packaging materials varies from only a few per cent up to 50% recycled material in some packaging applications. Numerous studies have been carried out on the determination the material properties and the blending behaviour of recycled plastics. However, it is not the focus of this chapter to deal with blending of polymers but it needs to be stressed that the recycled material should be suitable for blending with virgin materials. Additionally, the mechanical properties of the recyclate should be not influenced in a negative way, so as to avoid potential consequences for the additive status of the recycled plastics.

The average number of cycles is a function of the blend ratio and the number of recycling steps carried out. In practice the average number of cycles ranges from one to three.[3] Therefore, the material is not recycled many times and the problem of accumulation of degradation products is in most cases of no concern. An inherent problem of recycling, however, is the inhomogeneity of the recovered materials. Normally various polymer additives, lubricants, etc. are used by the different polymer manufacturers or converters in order to establish the desired properties of the packaging materials, and all different polymer additives are found as a mixture in the recyclate containing packages. Modern sorting technologies are able to provide input materials for recycling which are nearly 100% of one polymer type. Taking, in addition, the additive status into account will be a sophisticated challenge of future developments. Together with the inertness of the polymers this is one reason why recent closed-loop recycling efforts are focused on polymers which have low amounts of additives e.g. PET. However, as mentioned above, the question of recyclability is mainly influenced by the source control of the input material going into the recycling process. If the recovery system considers the manufacturer or the origin of the packaging materials, usually the additive status of the input feedstock is known. An example for this will be HDPE milk bottles collected by a deposit system (see Section 23.5.2).

23.3 Improving the recyclability of plastic packaging

23.3.1 Source control

The source control is the first and most important step in closed-loop recycling of packaging plastics. There must be efficient recovery or sorting processes which are able to control the input fraction going into a closed-loop recycling process. The feedstream material should have a minimum polymer type purity of 99%. Other polymers, which may interfere, have to be sorted out of the recycling stream. Also the first life of the packaging material is of interest. In general only packages previously filled with foodstuffs should be used as an input fraction for a closed-loop recycling process. However there are exceptions, e.g. for PET due to its high inertness the first packaging application is not so important. Two studies were undertaken[4,5] to determine the impact of PET materials formerly used for non-food applications. Both studies came to the same result, that due to the low diffusivity of PET packages from non-food applications could also be used as input material for bottle-to-bottle recycling. This underlines the favourite position of PET bottles for a closed loop recycling.

It could be shown that deposit systems and recovery systems like curbside packaging collections with efficient sorting processes, are able to support input materials for high value recycling. However, as mentioned above, the higher the diffusivity of the polymer and, therefore, higher sorption of post-consumer substances the more important is the source control in order to reduce contamination with post-consumer substances or misused packages. The source control can be supplied by modern detecting or sniffing devices which are able to reduce the intake of undesired post-consumer substances into the recycling stream.

23.3.2 Contamination levels and frequency of misuse of recycled plastics

Regarding the typical contamination of post-consumer plastics most published data are available for PET bottles and corresponding recyclates. Most of them have quantified or identified substances in post-consumer PET by using different methods. Sadler *et al.*[6,7], published two studies containing data of contaminants in recycled PET. In the first study he pointed out that most compounds found in recycled PET come from PET starting materials, oligomers, flavour bases, label materials and compounds originating in base cups. Contaminants which do not fall into one of these categories are rare. In samples with high levels of contaminants the sum of all compounds was detected to be approximately 25 ppm. No single contaminant appears to be present in post-consumer PET above 1 ppm and all non-usual compounds in post-consumer PET were present below 0.1 ppm. In a second study the identity and origin of contaminants in food grade virgin and commercially washed post-consumer PET flakes were determined. A total of 18 samples of post-consumer recycled PET flakes was examined. In most cases, positive identification was possible, however, in few cases ambiguity resulted from the similarities in mass spectra of closely related compounds. Compounds identified were classified into categories associated with their chemical nature or presumed origins, e.g. small and ethylene glycol related compounds (methanol, formic acid, acetaldehyde, acetic acid), flavour compounds (limonene), benzoic acid or related benzene dicarboxylic acid substances (benzoic and terephthalic acid and corresponding esters, benzaldehyde, phthalates), aliphatic hydrocarbons and acids as well as unexpected and miscellaneous compounds (Tinuvin, nicotine).

Bayer[4] has analysed samples from five different recovery systems including PET containers from non-food applications. In these samples he identified 121 substances. The total concentration of all substances found in deposit material was 28.5 ppm. The corresponding concentrations of PET flakes coming from non-food applications were found to be 39 ppm. The key compounds identified were hexanal, benzaldehyde, limonene, methyl salicylate and 5-*iso*-propyl-2-methylphenol (the flavour compound carvacrol). In conventional washed flakes a maximum concentration of 18 ppm for limonene was determined. For PET flakes from non-food applications the major compound methyl salicylate was determined in a maximum concentration of 15.3 ppm. Additionally the material was analysed after a super-clean process. No peak could be detected in concentrations above the FDA threshold of regulation limit of 0.22 ppm.

All three published studies mentioned above found no hints for misuse of post-consumer PET bottles e.g. for storage of household cleaners etc. This is most probably due to the fact that these studies are based only on very small amounts of different flake samples. From a statistical point of view flakes from misused bottles should be extremely rare due to high dilution with non-misused PET bottles. Therefore, these published studies are not able to detect the frequency of misuse in typical post-consumer PET flakes.

In 2002 an EU project under the co-ordination of Fraunhofer IVV was finished.[8,9,10] Within this study 689 post-consumer PET flake samples from commercial washing plants were collected between 1997 and 2001. The samples are conventionally recycled deposit and curbside fractions collected in twelve European countries. In addition, 38 reprocessed pellet samples and 142 samples from super-clean recycling processes were collected. All samples were screened for post-consumer substances, and for hints of possible misuse of the PET bottles by the consumer, in order to get an overview of the quality of commercially recycled post-consumer PET. As a result the average concentrations in 689 PET flake samples for typical post-consumer compounds like limonene and acetaldehyde are 2.9 ppm and 18.6 ppm, respectively. A maximum concentration of approximately 20 ppm of limonene and 86 ppm for acetaldehyde could be determined, which is in close agreement with the above-mentioned studies. The impact of the recovery system and the country, where the post-consumer PET bottles were collected, on the nature and extent of adventitious contaminants was found not to be significant. However in three bottle flakes hints for a possible misuse of PET bottles e.g. for storage of household chemicals or fuels were found. From a statistical evaluation 0.03 to 0.04% of the PET bottles might be misused. Under consideration of the dilution of the PET flakes during washing and grinding with non-misused PET bottles average concentrations of 1.4 to 2.7 ppm for conspicuous substances from misused PET bottles were estimated from the experimental data. These concentrations can be considered as a basis for the design of challenge tests with respect to sufficiently high input concentrations of surrogates.

The frequency of misuse was also detected by two other studies. Allen and Blakistone[11] indicate that hydrocarbon 'sniffers' for refillable PET bottles rejected between 0.3 and 1% of PET bottles as contaminated. The majority of these rejections came from PET containers with 'exotic' beverages and not from harmful contaminants. Therefore the part of misused bottles on the rejection in the 'sniffer' device is less than 0.3 to 1%. Bayer *et al.*[12] reported the frequency of misuse of PET bottles is one misused bottle out of 10 000 uncontaminated bottles. Both studies are in agreement with the results of the EU project.

In conclusion for PET the predominating polymer unspecific contaminants are soft drink components where limonene plays a key role. PET unspecific contaminants such as phthalates are found far below 1 ppm. Misuse of PET bottles occurs only in a very low incidence and due to dilution with non-contaminated material the average concentration of substances originated from misuse are also in the lower ppm range. It should be noted here, that the given conclusions are only for PET bottles. If closed loop recycling of other packaging plastics is to be established similar studies on the input concentrations of post-consumer substances should be done.

Comprehensive studies on the contamination of other polymers than PET are vary rare in the literature. Huber and Franz[13] investigated 21 reprocessed HDPE pellet samples from the bottle fraction of household waste collections from five different sources. Aim of the study was to investigate the quality of the recycled HDPE samples focusing on substances which are not present in virgin polymers. The samples are recycled with conventional washing and extruding steps without a further deep cleansing recycling process. They found that the post-consumer related substances in these different samples were the same. They identified 74 substances which occur in concentrations in the polymers above 0.5 ppm. The predominant species are ester from saturated fatty acids and phthalates, hydrocarbons, preservatives, monoterpenes and sesquiterpenes including their derivatives. Most of the substances are identified as constituents from personal hygiene products, cosmetics and cleaning agents which are sorbed into the polymer material during storage. The highest concentrations were found for limonene, diethylhexyl phthalate and the isopropyl esters of myristic and palmitic acid, which are present in the concentration range of 50 ppm to 200 ppm. Many odour compounds and preservatives are determined in concentrations of about 0.5 ppm and 10 ppm. They came to the conclusion that due to the concentration and nature of contaminants found in the investigated HDPE samples the recycled material is suitable only for non-food packaging.

In a second study Huber and Franz[14] investigated a total amount 79 polymer samples (HDPE, PP, PS and PET) from controlled recollecting sources. As a result they found limonene in nearly all polymer samples independent of the polymer type in concentrations up to 100 ppm for polyolefines (HDPE and PP) and 12 ppm and 3 ppm for PS and PET, respectively. Limonene can be considered as a marker substance for post-consumer polymers. It is interesting to note that the differences in the limonene concentration are in line with the diffusion behaviour of the polymers. In addition to limonene they found phthalates esters, alkanes, 2,6-di-*tert*-butyl-4-hydroxytoluene and oligomers but no hints for misuse of the bottles for storage of toxic chemicals. They concluded that most of the investigated (conventionally recycled) polymers are excluded from closed loop recycling due to the fact that in the polymers substances can be detected which are not in compliance with the European positive list system. It should be noted here that this is an inherent problem of positive lists in view of food law compliance of recycled polymers as well as virgin polymers. A threshold of regulation concept should offer a solution of assuming that a certain concentration of non-regulated compounds is of no concern for consumers' health.

23.3.3 Recycling technology

Today a considerable diversity in recycling technologies can be found, although all of them have the same objective which is to clean up post-consumer plastics. Most of them first use a water-based washing step to reduce surface contamination and to wash off dirt, labels and clues from the labels. The material is also ground to flakes during one of the first steps in the recycling process. In most cases these washing steps are combined with separating steps where different materials like polyolefines of PET are separated due to their density. It is obvious, that the cleaning efficiency of these washing processes is normally very different, depending on time, on hot or cold water-based washing or depending on the detergents added to the washing solution. However, typical washing processes are able to remove only contaminants from the surface of the polymers.[15,16] They are not able to remove organic substances which have migrated in the polymer. Therefore the purity of washed flakes is usually not suitable for closed-loop recycling. A simple remelting or re-extrusion of the washed fakes has an additional cleaning effect,[17] however the purity is usually not sufficient for reuse in the sensitive area of food packaging.

So-called super-clean processes for closed-loop recycling of packaging materials therefore use further deep cleansing steps. Although there are many technologies commercially available the deep cleansing processes normally use heat and temperature, vacuum or surface treatment with chemicals for a certain time to decrease the concentration of unwanted substances in the polymers. The research on the cleaning efficiency of such super-clean recycling processes has shown that the existing recycling technologies are distinct in terms of rejection of unsuitable material, removal of contaminants and dilution with virgin material. Each of these stages in recycling uses special processes which have an effect on the quality of the finished recyclate containing packaging.

23.4 Testing the safety and quality of recycled material

23.4.1 Challenge test

The cleaning efficiency of super-clean processes is usually determined by challenge test. This challenge test is based on an artificial contamination of the input material going into the recycling process. Drawing up a worst-case scenario this challenge test simulates the possible misuse of the containers for the storage of household or garden chemicals in plastic containers. The first recommendations for such a challenge test are coming from the American Food and Drug Administration (FDA)[18,19] in 1992. It was a very pragmatic approach. The FDA originally suggested realistic contaminants like chloroform, diazinon, gasoline, lindane, and disodium monomethyl arsenate for use in challenge tests. However it has been shown in the past that the stability of these surrogates during recycling is in some cases not sufficient. Also the analytical methods in order to detect the surrogates are often difficult to establish and have high detection limits. It is easy to understand that the surrogates used in a challenge test should not degrade during all recycling steps. Otherwise the cleaning efficiency will be better than reality, with adverse consequences towards consumers' safety.

In the last ten years the selection of the surrogates has moved to chemicals with more model character. This development was supported by the fact that the range of chemicals available to the customers is extremely limited in practice, especially in the case of known genotoxic carcinogens. The surrogates used today in challenge tests cover the whole range of physical properties like polarity and volatility as well as the chemical nature of the compounds. Additionally, in some surrogates very aggressive chemicals towards the polymer are introduced. However, if too aggressive chemicals are used the physical properties of the polymer and the diffusion behaviour might be changed, which reduces the perception of the challenge test. Nowadays volatile chemicals like toluene, chlorobenzene, chloroform or 1,1,1-trichloroethane as well as non-volatile substances like phenyl cyclohexane, methyl stearate, tetracosane, benzophenone, methyl salicylate and methyl stearate are typically used. Of course, other substances with defined physical and chemical properties can be used for a challenge test. It should be kept in mind that such a test should challenge the recycling process in a worst-case scenario. If the resultant recyclate meets the food law requirements even under such a worst case scenario the process is able to produce recyclates suitable for reuse in packaging applications. In the last decade there have been controversial discussions between scientists, industry and authorities, in view of the worst-case character of such challenge tests. In most cases these discussions arise from the lack of information about the average contamination in the input materials for recycling. As mentioned above, the worst-case scenario depends on the concentrations of undesired substances in the post-consumer plastics as well as the frequency of misuse of plastic containers. With knowledge of contamination appropriate safety margins for each polymer type can be defined.

23.4.2 Cleaning efficiency of conventional recycling processes

Post-consumer PET which are going into packaging applications are usually recycled with super-clean recycling processes. However, these processes use conventional washing steps as well as several deep-cleansing steps in order to eliminate undesired post-consumer substances from the PET polymer matrix. Therefore the cleaning efficiency of conventional washing processes is of interest because it influences the input concentration of post-consumer substances in feedstock material going into the deep-cleansing processes.

In the literature there are a few studies on the cleaning efficiency of conventional recycling processes. These processes contain washing and surface drying steps followed in some cases by remelting of the post-consumer material. Komolprasert and Lawson[15] determined the influence of NaOH concentration, mixer speed and temperature on removal of the surrogate tetracosane from spiked PET. In this study percentages of residual tetracosane in the PET flakes which were washed in small-scale experiments using 13 different conditions were determined. The results show that the tetracosane concentration in the washed flakes was 1.4 to 3.3% of the initial spiked level. As a result only mixer speed and temperature showed a significant effect on removal of the surrogate tetracosane from the PET flakes, while the effect of NaOH concentration was insignificant. The percentage of non-volatile hydrocarbon residues in washed PET flakes varies with the initial concentration. The study determined a removal of 89 to 97% of each hydrocarbon by washing. In a second study Komolprasert and Lawson[16] determined the effect of washing and drying on the removal of surrogates in spiked PET flakes as well as in spiked PET bottles. They concluded that the combination of washing and drying removes 97 to 99% of the organic surrogates from the spiked PET bottles. The copper concentration was found to be 21% of the initial concentration after washing and drying (remark: the low cleaning efficiency for the copper organic compound is most probably due to the instability of this surrogate. It reacts during recycling to CuO which cannot be removed. This behaviour shows that metal organic compounds are in general unsuitable as surrogates for challenge tests). In case of spiked PET flakes washing and drying removes more than 99% of the initial concentration of the organic surrogates. The high cleaning efficiencies of conventional washing and drying processes are most probably due high temperatures applied during the drying step and due to the fact that contaminants rarely penetrate more than a few μm into the polymer surface. This is in agreement with the result that the initial concentrations of the surrogates in spiked bottles are much lower than those in flakes, because the surface area of flakes is higher than in bottles. A third study from Komolprasert *et al.*[17] evaluates the decontamination by remelting in a laboratory extruder. The results show that remelting can further reduce the contamination of spiked PET. However, from the data given in this paper, the amount of this reduction is very difficult to evaluate, because of the fact that the some of the applied surrogates (diazinon, malathion, metal organic copper compound) are not stable during extrusion. In addition volatile substances like toluene are almost completely removed during washing, so that an evaluation of the cleaning effect during remelting on basis of these surrogates is impossible.

In conclusion, conventional washing processes are able to reduce the input concentrations of post-consumer substances in flakes. The washing process itself most probably removes only contaminants on the surface of the flakes whereas thermal drying processes are also able to decrease substance which sorbed into the flakes. Remelting processes further reduce the contamination. Due to the fact that conventional recycling processes use a wide range of parameters and equipment, a general conclusion and a quantification of the cleaning effects for washing, drying and remelting processes is not possible on basis of the above mentioned results.

23.4.3 Cleaning efficiency of super-clean processes

In the last decade studies were undertaken to quantify the residual amounts of chemical substances in the material after deep-cleansing. Therefore the cleaning efficiency of super-cleaning recycling processes is well known. Additionally to the challenge test, the quality assurance of post-consumer recycle (PCR) PET is based on a feedstock control and an analytical quality assurance. Literature data about cleaning efficiencies of super-clean recycling processes are very rare. Three studies of the cleansing efficiency of super-clean recycling processes for PET are published by Franz and Welle.[20,21,22] The process investigated in the first two studies[20,21] contains the key steps: washing, re-extrusion and solid state polycondensation (SSP). The process was challenged with three different surrogate concentration levels. As a result the cleaning efficiencies for the different surrogates and contamination levels are between 94 and 99%.[20] The most challenging substance is benzophenone. The results show no significant dependency on the input concentration of the surrogates going into the process. It should be noted here, that this process was tested without a washing process. Including a conventional washing process, the cleaning efficiencies are increased to more than 99.3% even for benzophenone.[21] In the third study[22] a recycling process without solid stating was investigated. Except for benzophenone, the investigated recycling process reduces all surrogates by more than 95% for initial concentrations below 100 ppm and more than 90% for initial concentrations between 100 and 500 ppm. For that most challenging substance, benzophenone, a cleaning efficiency of approximately 77% at an initial contamination level of 294 ppm was obtained. In conclusion the determined cleaning efficiencies are lower than those of processes with solid stating. However, the specific migration of all surrogates from PET bottles made from contaminated and recycled PET was detected to be far below the migration limit of 10 ppb.

23.4.4 Considerations on migration evaluation

Migration from a given food/plastic package system is essentially influenced by kinetical (diffusion in plastic and food) as well as thermodynamic (equilibrium partitioning between plastic and food) factors. It is useful to start in migration evaluations as a worst assumption with total mass transfer scenarios based on knowledge of the starting concentration of a given migrant in the plastic. If this calculation leads to exceeding a migration limit, then it is advisable and necessary to refine the evaluation strategy and take account of partitioning and diffusion as the crucial parameters for migration. Complete scientific background and guidance on how to proceed can be found in the literature.[2]

The FDA suggests that dietary exposures to contaminants from recycled food contact articles at a concentration of 0.5 ppb or less generally are of negligible risk.[18,19,23] With help of so-called consumption factors (CF) these dietary exposures can be converted into migration limits. For recycled PET for food contact use, for instance, the FDA system applies a CF = 0.05 as the currently valid consumption factor for post-consumer plastics and therefore the migration limit of PET recyclate-containing packages is 10 ppb for each individual substance. On the other hand, the migration limit can be converted into a maximum bottle wall concentration for any substance occurring in post-consumer plastics (including substances from virgin polymers). For example, for PET the maximum concentration in the PET material which correlates with 10 ppb dietary intake level is 0.22 ppm for a typical PET container at a thickness of 0.5 mm. This calculation is based on very conservative assumptions that all PET bottles are contaminated and that the contaminants are assumed to migrate completely from the bottle into the foodstuff. The contaminant limits calculated above also assume 100% recycled resin content in the finished article.

It is generally known that the diffusion-controlled migration is usually much lower than the complete transfer of substances into the foodstuffs, and this especially for low diffusivity polymers like PET. Migration from virgin and post-consumer PET has been considered in numerous investigations where the low diffusion and migration rates have been reported and confirmed.[21,22,24] Therefore, diffusion models[25,26,27] provide an interesting scientific tool for a more realistic correlation between the allowed upper migration limit in the packed foodstuff and the corresponding maximum allowable concentration in the polymer.[18,19,28] A generally recognised migration model based on diffusion coefficient estimation of organic chemical substances in polymers[2] has been finished recently within the European project SMT-CT98-7513 'Evaluation of Migration Models in Support of Directive 90/128/EEC'.[29] Using this model the curves in Fig. 23.1 were calculated, which presents a correlation between migration into food and the corresponding maximum allowable concentrations of the surrogates in the bottle wall in dependency of the molecular weight for a PET bottle (assumption 1 l content with 600 cm^2 packaging surface). The migration or the corresponding residual concentration in the bottle wall was calculated for 95% ethanol at contact conditions of 40°C for 10 d. For most applications these scenarios are worst-case conditions overestimating the real migration situation by factors of at least 100. In

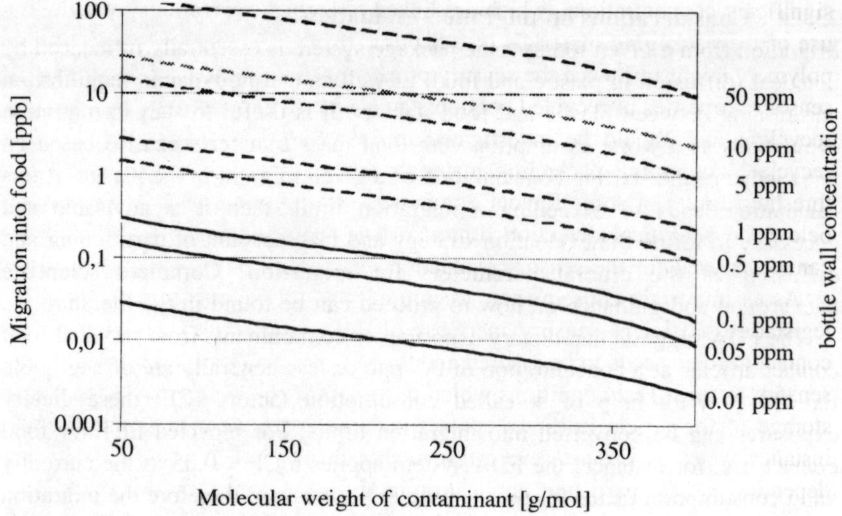

Fig. 23.1 Correlation of migration into food and calculated maximum allowable concentrations (in ppm) of the surrogates in the bottle wall in dependency of the molecular weight for a 1 l PET bottle in 95% ethanol (10 d at 40°C).[30]

view of challenge tests and particular focus on surrogates the following very conservative maximum allowable concentrations of surrogates in the bottle wall can be calculated independently of the package thickness: toluene 4.5 ppm, chlorobenzene 5.5 ppm, phenyl cyclohexane 7.5 ppm, benzophenone 8.6 ppm and methyl stearate 17 ppm. These concentrations can be considered to correlate safely with the 10 ppb migration limit for any food or simulant at test conditions of 10 days at 40°C.

Hot-fill conditions are also covered by the above mentioned modelling data. Using test conditions of 1 h at 70°C followed by 10 days at 40°C, which are the usual test conditions for hot-fill testing, instead of only 10 days at 40°C the calculated migration rises only insignificantly. Going to a more severe condition, e.g. 30 min at 100°C followed by 10 d at 40°C the factor is 1,21. This shows that the migration from PET containers into food simulants (e.g. 95% ethanol as a worst case) are very low even under hot-fill conditions.

23.4.5 Sensory Test Requirements

Huber and Franz investigated the sensory properties of conventionally recycled polymers (HDPE, PP, PS and PET).[14] They found in all samples the polymer specific odour found in virgin polymers. Nevertheless, all of the recycled polymers could be identified due to additional odour notes. For PET the lowest odour deviation was noticed and an increasing off-odour was perceptible from PS to PP and HDPE. However, the results are not surprising because in all investigated polymer samples flavour compounds like limonene can be found in significant concentrations, which are linked to sorbed compounds from the first use of the packaging materials (see Section 23.2.2). A deep cleansing of these polymers might influence the sensory properties in a positive way. However the sensory properties of recycled polymers are a crucial parameter for a closed loop recycling and should be investigated in case by case studies with the final recyclate containing packaging materials. This is due to the fact that odour threshold limits of some flavour compounds are very low and, in a few cases, below the analytical detection limits so that the results of the challenge test cannot be used for the sensory evaluation of recycled polymers.

As an important consequence, to comply with the legal requirements of any legislation sufficient sensory inertness of the recycled PET product of food contact articles needs to be assured and this can only be achieved by appropriate sensory testing. Test conditions which in many cases can serve as worst case is storage of the article in direct contact with an appropriate food simulant (for instance water as a most severe test medium) for 10 days at 40ºC. However, depending on the particular application modified tests may be more suitable.

23.5 Using recycled plastics in packaging

Technically, recycled plastics can, in principle, be applied in direct food contact applications or protected from direct food contact by a functional barrier. From a legal point of view, there may be limitations due to different regulations in the European countries and the still missing harmonised rules for EU. In any case, the use of recycled plastic materials in packaging applications has to comply with the relevant regulations and must not be at the expense of the public health, nor should it alternate the filling's quality.[31] In the following, practical examples of recycled plastics food packaging applications covered by a functional barrier as well as in direct food contact are described.

23.5.1 Indirect contact applications applying functional barriers

In the most general understanding the concept of a functional barrier can be defined as follows: A functional barrier is a layer in the package which protects the food from external influences under the applied fill and storage conditions. In most cases the functional barrier is the food contact layer or, in complex multi-layer structures, one very close to it. This layer acts as a barrier against contamination from the packaging's environment in general and, more specifically, from the recycled core layer or outer compartments of the package.

The functional barrier efficiency must not be confused with an absolute physical barrier such as glass or metal layers. It is related to a 'functional' quantity in terms of mass transfer which is dependent on the technological and application-related parameters of the respective food-package system. These parameters are:

- manufacture conditions of the package
- thickness of the functional barrier layer
- type of functional barrier plastic
- molecular weight and chemical structure of penetrants (contaminants)
- concentration and mobility of contaminants in the matrix behind the functional barrier
- time period between manufacture of package and filling
- type of foodstuff, i.e. fat content, polarity etc.
- filling conditions and storage (time, temperature) of the packed foodstuffs

More scientific background and guidance can be found in the published literature[31,32,33] as well as in the case studies described below.

Three-layer PP cups for dairy products

In 1994 a study was presented[34] where the safety in food contact use of symmetrically coextruded three-layered polypropylene (PP) cups with recycled post-consumer PP in the core layer (mass fraction 50%) and virgin food grade PP in the adjacent layers was investigated. The recycled PP, which contained about 95% PP and 5% PS, was completely under source control in the recollection system and had been used in its prior application for packaging yoghurt. The intended application for the recycled material was again packaging milk products such as yoghurt with storage for short times under refrigerated conditions.

The essential working strategy in this study was to compare the recycled plastic with new, food grade plastic material of the same type. This comparison included experimentally three investigation levels:

(1) compositional analysis of the raw materials (virgin versus recycled PP pellets),
(2) compositional analysis of the finished food contact articles (virgin versus recycled cups)
(3) migration testing on both types of cup (virgin and recycled) under regular as well as more severe test conditions.

After identification and quantification of post-consumer or recycling-related potential migrants on levels (1) and (2) these compounds were used as indicator substances to be monitored in migration measurements on level (3). The major post-consumer related compound was identified as limonene, a flavour compound which can be found in many foodstuffs and also in the non-food area. In summary it turned out that none of post-consumer or recycling-related substances could be analytically detected in the food simulants (at a detection limit of 13 ppb) under prescribed migration test conditions. However, from the results obtained under more severe test conditions, it could be concluded finally that for the compound with the highest migration, limonene, the migration into a milk product will be below 1 ppb and for other post-consumer substances far below of 1 ppb. In conclusion, based on the US

FDA threshold of regulation concept[18] the intended application was considered to be safe.

Multi-layer PET bottles for soft drink applications
In 1996 a study was published[35] in which the effectiveness of a virgin PET layer in limiting chemical migration from recycled PET was investigated. For this purpose three-layer bottles were prepared with an inner core (buried layer) of PET which was deliberately contaminated. The model contaminants used were toluene, trichloroethane, chlorobenzene, phenyl decane, benzophenone, phenyl cyclohexane and copper(II) acetylacetonate. As a result no migration was detected through a barrier of virgin PET of $186 \pm 39 \mu m$ thickness into 3% acetic acid using general migration test conditions of 10 days at 40°C and also after 6 months storage at room temperature. Also migration testing with 50% and 95% ethanol as severe contact media which are relatively aggressive to PET did not lead to measurable migration rates. Consideration of diffusion models using limonene as substance for which diffusion coefficients were available, gave estimates that for a 100μm thick PET layer a breakthrough of a substance with comparable molecular weight would take place after 7.5 years or 0.8 years at room temperature or 40°C, respectively. It was concluded that an intact PET bottle layer in contact with the food represents an efficient functional barrier against migration from any possible contaminant encapsulated in a recycled PET material under normal conditions of use for soft drinks. Today, multi-layer PET soft drink bottles have received clearance for use in Austria, Belgium, Finland, France, Norway, Sweden and the United Kingdom.

Studies on multi-layer PET and PET films for food packaging
In another project,[36] several coextruded three-layered PET films spiked in the core layer with surrogates (toluene and chlorobenzene) and having a PET barrier layer thickness between 20 to 60μm, were systematically investigated with respect to their migration behaviour under different test and contact medium conditions. It was observed that the migration measured through the different barrier layers was predictable, and a diffusion model for predicting the functional barrier properties of layered films based on Fickian diffusion was presented. Also the effects of diffusion from the core layer to a virgin barrier layer during the coextrusion process was found necessary to be considered for reliable prediction of migration. On the basis of the presented mathematical model, maximum allowable concentrations can be established for a core layer for a given barrier thickness while still fulfilling threshold or specific migration limit requirements.

In similar studies with symmetrical three-layer films spiked again in the core layer with toluene and chlorobenzene the functional barrier behaviour of high impact polystyrene (HIPS) was investigated.[27] The applied thicknesses of the HIPS barrier layer ranged higher than the above PET example and were 50μm, 100μm and 200μm. The contact medium was 50% ethanol which is a recognised medium for fatty food products for this plastic, and testing was carried out at

40°C up to 76 days. From the results it was concluded that HIPS was an appropriate functional barrier under given application parameters which need to be optimised for the particular purpose. Generally, layer thicknesses from 100μm to 200μm were found to be very efficient, and this even in the case of exaggerated test conditions as applied in this study for fatty contact. When considering aqueous food products and room or cooling temperature applications, this conclusion is still much more valid and of general character. Again, as with the PET study, the contamination effect from the core layer to the virgin barrier layers during the high temperatures of the coextrusion process was investigated. For instance, for a 50μm thick HIPS layer it turned out that the same contamination effect of the food contact surface with the surrogate toluene, which is obtained after one year's storage at room temperature, is achieved within 1 second only at the coextrusion temperature of 200°C.

23.5.2 Direct contact applications

Mono-layer PET bottle for soft drink applications

As mentioned before, PET is one of the most favoured candidates for closed-loop recycling. Due to higher costs of manufacturing multi-layer bottles, the bottle manufacturing and recycling companies started the development of recycling processes without a functional barrier of virgin PET. One decade later several super-clean recycling processes were established on an industrial scale.[1] In 2002, companies in Europe have built an overall recycling capacity of about 65.000 tons per year of super-clean post-consumer PET which can be used in direct food contact applications. The cleaning efficiencies of all the applied deep-cleansing recycling processes were investigated by challenge tests and the cleaning efficiencies are well known (see for example Lit.[20,21,22]). In Europe today, the mono-layer direct food contact approach has received clearance for use in Austria, Belgium, France, Germany, The Netherlands, Norway, Sweden and Switzerland.

Mono-layer HDPE bottles for fresh milk

In 2002 the following project was started in Northern Ireland.[37] Milk bottles were recovered by a deposit system and were subjected to a bottle-to-bottle recycling process. Due to the recovery system the recycled HDPE was completely under source control and had been used in its prior application only for packaging fresh milk. The recovered material was recycled first by a conventional washing based recycling process and then further deep-cleansed using a super-clean process. Subsequently the recycled material was used with a content of 20 to 30% without a functional barrier. The intended application for the recycled material was again bottles for fresh milk with short time storage under refrigerated conditions.

The project had several R&D phases before: after a screening of post-consumer HDPE milk bottles for compounds which are potential migrants the deep-cleansing process was evaluated and optimised. Subsequently, the cleaning efficiency of the recycling process was determined with a challenge test. As a result only HDPE related compounds such as oligomers could be detected in the recycled polymer after deep-cleansing. In addition an online sniffing device based on so-called electronic noses was integrated into the recycling process. This sniffing device is able to detect potential migrants, such as solvents or other volatile substances, which might be introduced into the recycling process. Based on the challenge test results, upper limits for the concentration of volatile substances could be defined so that any HDPE lots with higher levels can be detected, and separated for being reused in non-food packaging areas. These higher recycling efforts over-compensated the principal lower material suitability of HDPE for closed loop recycling in comparison to PET.

Another point of interest is, that the project was started in Northern Ireland at a small market scale which is under good control. The HDPE milk bottles were provided by only two bottle manufacturers, which are integrated into the project. By reading the bar code the vendors are able to separate bottles from these two manufacturing companies from other milk bottles which are rejected. Therefore, only milk bottles from bottle suppliers and filling companies which support the project are directed to the bottle-to-bottle recycling process. The recycling company, in principle, is therefore in a position to react on negative impacts of applied clues, label, colours etc. and can control its recyclate production.

23.6 Future trends

For PET, a low diffusivity polymer, closed-loop recycling is now established in several countries all over the world. The HDPE study described in Section 23.5.2 shows that the combination of source control, efficient process technology and quality control using modern sniffing devices enables manufacturing companies also to re-use packaging plastics in direct food contact which have a higher diffusivity than PET. However, it must be realised that compared to PET recycling of polyolefines will always be of much more specific character i.e. limited to a certain first application and a relatively narrow and overseeable recovery system if it is intended to reuse the plastic for sensitive applications such as direct food contact. On the other hand recycling of polyolefine plastic crates into new ones for relatively insensitive applications, such as transport containers for fresh fruits or vegetables, is a less challenging issue and can therefore be dealt with in a more general way.

Current and future technological improvements and further developments in recovery, sorting and recycling technologies will be an important basis for the expected increasing market shares of recycled polymers in the packaging area. Accompanied by increasing knowledge of possible post-consumer contaminant levels and further improving developments in analytical control systems, e.g. complete inline control of recyclate production, using appropriate sniffing devices, will enable the potential risk of exposure to the consumer of unwanted recycling-related compounds at the necessary low levels and at the expectable increasing use in the market place.

However, the food packing market is not static. Further developments leading to more complex packaging systems, e.g. introduction of new barrier coating systems, multilayers or new plastics additives or active substances, may have an impact on closed-loop recycling which needs to considered at an early stage of the packaging developments. In the PET packaging area the introduction of so-called acetaldehyde scavengers can lead for instance to a yellowing of the recycled material, a negative optical appearance of the polymer which, however, does not pose a risk for the consumer but decreases the market value of recyclate-containing packaging materials. Therefore, in the future, closed-loop recycling will be a challenge which can only be efficiently and successfully managed by collaboration between recyclers and the packaging industry chain.

23.7 Sources of further information and advice

23.7.1 European Projects 'Recycle Re-use' and 'Recyclability'

In the last decade the European Commission has supported two projects dealing with the question of recyclability and reusability of post-consumer plastics new food packaging applications. The first project AIR2-CT93-1014[3] is dealing with plastics packaging materials recovered from packaging waste. The second project FAIR CT98-4318[38] focuses on PET as the most favourable candidate plastics for direct food contact. Two other sections of the project are dealing with paper and board and recycled plastics covered with functional barriers. Both project reports provide deeper information on analytical approaches and their validation to assess the opportunity of recycling of post-consumer plastics into food packaging applications. Based on the results of the European Project FAIR CT98-4318[32] proposals for the forthcoming legislation were written and filed with the European Commission. One document is based on the results of the Europe-wide screening of post-consumer PET flakes and on migration considerations.[39] The other document gives guidance on the use of functional barriers.[33]

23.7.2 US FDA Points to Consider

In 1992 and 1995 the FDA published two guidelines for industry dealing with post-consumer plastics for direct food contact applications.[18,19] These guidelines provide recommendations about testing the cleaning efficiency of the investigated recycling process and the maximum content of post-consumer substances in recyclate-containing packaging materials as well as threshold limits for migration. Based on the agency's reviews on petitions and on own research projects the FDA now provides an update of their guidelines.[5] This update integrates new knowledge, mainly for PET, of the contamination of post-consumer material and challenge tests. Especially for PET recommendations are given for feedstock material from non-food applications, which are intended to be recycled into food packaging. The FDA also provides information about all 'non objection letters' on their internet homepage.[40]

23.7.3 European ILSI document

In 1997 an expert group under the responsibility of ILSI Europe has proposed specific guidelines on the re-use of recycled plastics in food packaging.[41] These guidelines, published in 1998, are based on the results obtained from the above-mentioned European 'Recycle Re-use' project. The intention of the document was to provide information for industry about the European view of closed-loop recycling of post-consumer plastics. The document gives recommendations for recycling or packaging companies, which want to introduce post-consumer plastics into food contact applications.

23.7.4 German BfR recommendations

The German BfR (former BgVV) published in 2000 recommendations on the mechanical recycling of post-consumer PET for direct food contact applications.[28] This document is the result of a discussion by the German 'Plastics Commission' on PET bottle-to-bottle recycling. The BfR document gives recommendations for source control, challenge test and for the quality assurance of post-consumer PET intended to come into direct food contact.

23.8 References

1. K. FRITSCH, F. WELLE, Polyethylene terephthalate (PET) for packaging, *Plast Europe*, 2002, 92(10), 40–1.
2. O.-G. PIRINGER, A. L. BANER (Editors), Plastic Packaging Materials for Food – Barrier Function, Mass Transport, Quality Assurance and Legislation, 2000, WILEY-VCH, Weinheim, New York.
3. Final report of EU funded AIR project AIR2-CT93-1014, Programme to establish criteria to ensure the quality and safety of recycled and re-used plastics for food packaging, Brussels, December 1997.
4. F. L. BAYER, Polyethylene terephthalate (PET) recycling for food contact applications: Testing, safety and technologies – A global perspective, *Food Additives and Contaminants*, 2002, *19, Supplement, 111–34.*
5. T. H. BEGLEY, T. P. MCNEAL, J. E. BILES, K. E. PAQUETTE, Evaluating the potential for recycling all PET bottles into new food packaging, *Food Additives and Contaminants*, 2002, 19, Supplement, 135–43
6. D. PIERCE, D. KING, G. SADLER, Analysis of contaminants in recycled poly(ethylene terephthalate) by thermal extraction gas chromatography – mass spectrometry, 208th American Chemical Society National Meeting. Washington DC, August 25, 1994, 458–71.

7. G. D. SADLER, Recycled PET for food contact: Current status of research required for regulatory review, Proceedings: Society of Plastic Engineering Regional Technical Conference, Schaumburg, IL, USA, November 1995, 181–91.
8. R. FRANZ, Programme on the recyclability of food-packaging materials with respect to food safety considerations – Polyethylene terephthalate (PET), paper & board and plastics covered by functional barriers, *Food Additives and Contaminants*, 2002, 19, Supplement, 93–100.
9. F. WELLE, R. FRANZ, Typical contamination levels and analytical recognition of post-consumer PET recyclates, Congress Proceedings: EU-Project Workshop 'Recyclability', Varese, February 11, 2002.
10. R. FRANZ, M. MAUER, F. WELLE, European survey on post-consumer poly(ethylene terephthalate) materials to determine contamination levels and maximum consumer exposure from food packages made from recycled PET, *Food Additives and Contaminants*, submitted for publication.
11. B. H. ALLEN, B. A. BLAKISTONE, Assessing reclamation processes for plastics recycling, 208th American Chemical Society National Meeting. Washington DC, August 25, 1994, 418–34.
12. F. L. BAYER, D. V. MYERS, M. J. GAGE, Consideration of poly(ethylene terephthalate) recycling for food use, 208th American Chemical Society National Meeting. Washington DC, August 25, 1994, 152–60.
13. M. HUBER, R. FRANZ, Identification of migratable substances in recycled high density polyethylene collected from household waste, *Journal of High Resolution Chromatography*, 1997 29, 427–30.
14. M. HUBER, R. FRANZ, Studies on contamination of post-consumer plastics from controlled resources for recycling into food packaging applications, *Deutsche Lebensmittel-Rundschau*, 1997, 93(10), 328–31.
15. V. KOMOLPRASERT, A. LAWSON, Effects of aqueous-based washing on removal of hydrocarbons from recycled polyethylene terephthalate (PETE), Congress Proceedings ANTEC'94, San Francisco, 1994, 2906-9.
16. V. KOMOLPRASERT, A. LAWSON, Residual contaminants in recycled poly(ethylene terephthalate) – Effects of washing and drying, 208th American Chemical Society National Meeting. Washington DC, 1994, 435–44.
17. V. KOMOLPRASERT, A. R. LAWSON, A. GREGOR, Removal of contaminants from RPET by extrusion remelting, *Packaging, Technology and Engineering*, 1996, September, 25–31.
18. Points to consider for the use of recycled plastics in food packaging: Chemistry considerations, US Food and Drug Administration, Center for Food Safety and Applied Nutrition (HFF-410), Washington, May 1992.
19. Guidelines for the safe use of recycled plastics for food packaging applications, Plastics Recycling Task Force document, National Food Processors Association, The Society of the Plastic Industry, Inc. March 1995.

20. R. FRANZ, M. HUBER, F. WELLE, Recycling of post-consumer poly(ethylene terephthalate) for direct food contact application – a feasibility study using a simplified challenge test, *Deutsche Lebensmittel-Rundschau*, 1998, 94(9), 303–8.
21. R. FRANZ, F. WELLE, Post-consumer poly(ethylene terephthalate) for direct food contact application – final proof of food law compliance, *Deutsche Lebensmittel-Rundschau*, 1999, 95, 424–7.
22. R. FRANZ, F. WELLE, Post-consumer poly(ethylene terephthalate) for direct food contact application – Challenge-test of an inline recycling process, *Food Additives and Contaminants*, 2002, 19(5), 502–11.
23. F. L. BAYER, The threshold of regulation and its application to indirect food additive contaminants in recycled plastics, *Food Additives and Contaminants*, 1997, 14, 661–70.
24. V. KOMOLPRASERT, A. R. LAWSON, Considerations for the reuse of poly(ethylene terephthalate) bottles in food packaging: migration study, *Journal of Agricultural and Food Chemistry*, 1997, 45, 444–8.
25. A. L. BANER, J. BRANDSCH, R. FRANZ, O. G. PIRINGER, The application of a predictive migration model for evaluation the compliance of plastic materials with European food regulations, *Food Additives and Contaminants*, 1996, 13, 587–601.
26. T. H. BEGLEY, H. C. HOLLIFIELD, Recycled polymers in food packaging: migration considerations, *Food Technology*, 1993, 109–12.
27. R. FRANZ, M. HUBER, O. G. PIRINGER, Presentation and experimental verification of a physico-mathematical model describing the migration across functional barrier layers into foodstuffs, *Food Additives and Contaminants*, 1997, 14(6–7), 627–40.
28. Use of mechanical recycled plastic made from polyethylene terephthalate (PET) for the manufacture of articles coming in contact with food, Bundesinstitut für Risikobewertung BfR, Berlin, October 2000. Also implemented into BfR Recommendation XVII.
29. O. PIRINGER, K. HINRICHS, Evaluation of Migration Models, Final Report of the EU-project contract SMT-CT98-7513, Brussels 2001.
30. R. FRANZ, unpublished results.
31. R. FRANZ, Safety assessment in modern food packaging applications, in: O.-G. Piringer, A. L. Baner, (editors), Plastic Packaging Materials for Food – Barrier Function, Mass Transport, Quality Assurance and Legislation. Wiley-VCH, Weinheim, 2000, Chapter 10.3, 336–57.
32. Final Project workshop of EU project FAIR CT98-4318, organised by the European Commission Joint Research Centre, Food Products Unit, and held on 10–11. February 2002 in Varese, Italy. A comprehensive download package of the presentations can be found at http://cpf.jrc.it/webpack/projects.htm.
33. D. DAINELLI, A. FEIGENBAUM, Guidelines for functional barrier applications, oral presentation at the workshop given under Lit.[32] and downloadable from http://cpf.jrc.it/webpack/projects.htm.

34. R. FRANZ, M. HUBER, O.-G. PIRINGER, Testing and evaluation of recycled plastics for food packaging use – possible migration through a functional barrier, *Food Additives and Contaminants*, 1994, 11(4), 479–96.
35. R. FRANZ, M. HUBER, O.-G. PIRINGER, A. P. DAMANT, S. M. JICKELLS, L. CASTLE, Study of functional barrier properties of multilayer recycled poly(ethylene terephthalate) bottles for soft drinks, *Journal of Agricultural and Food Chemistry*, 1996, 44(3), 892–7.
36. O. PIRINGER, M. HUBER, R. FRANZ, T. H. BEGLEY, T. P. MCNEAL, Migration from food packaging containing a functional barrier: mathematical and experimental evaluation, *Journal of Agricultural and Food Chemistry*, 1998, 46(4), 1532–8.
37. Personal Communication to the authors from Green Cycle, Armagh, Northern Ireland, internet http://www.greencycle.info.
38. Final report of EU project FAIR CT98-4318 'Programme on the Recyclability of Food Packaging Materials with Respect to Food Safety Considerations – Polyethylene Terephthalate (PET), Paper and Board and Plastics Covered by Functional Barriers', Brussels, 2003.
39. R. FRANZ, F. BAYER, F. WELLE, Guidance and criteria for safe recycling of post consumer polyethylene terephthalate (PET) into new food packaging applications, Guidance document prepared within EU project FAIR CT98-4318, submitted for publication.
40. Recycled Plastics in Food Packaging, US Food and Drug Administration, Center for Food Safety & Applied Nutrition, Office of Premarket Approval; internet: http://vm.cfsan.fda.gov/~dms/opa-recy.html.
41. Recycling of Plastics for Food Contact Use, Guidelines prepared under the responsibility of the International Life Sciences Institute (ILSI), European Packaging Material Task Force, 83 Avenue E. Mounier, B-1200 Brussels, Belgium, May 1998.

24

Green plastics for food packaging

J.J. de Vlieger, TNO Industrial Technology, The Netherlands

24.1 Introduction: the problem of plastic packaging waste

Polymers and plastics are typical materials of the last century and have made a tremendous growth of some hundreds of tons/year at the beginning of the 1930s to more than 150 million tons/year at the end of the 20th century with 220 million tons forecast by 2005. Western Europe will account for 19% of that amount. Today, the use of plastic in European countries is 60kg/person/year, in the US 80kg/person/year and in countries like India 2kg/person/year.[1] The basic materials used in packaging include paper, paperboard, cellophane, steel, glass, wood, textiles and plastics. Total consumption of flexible packaging grew by 2.9% per year during 1992–1997, with the strongest growth in processed food and above average growth in chilled foods, fresh foods, detergents and pet foods. Plastics allow packaging to perform many necessary tasks and provide thereby important properties such as strength and stiffness, barrier to gases, moisture, and grease, resistance to food component attack and flexibility.[2] Plastics used in food packaging must have good processability and be related to the melt flow behaviour and the thermal properties. Furthermore, these plastics should have excellent optical properties in being highly transparent (very important for the consumer) and possess good sealabilty and printing properties. In addition, legislation and consumers demand essential information about the content of the product.

Compared to the total amount of waste generated in for example the EU, packaging accounts for only a small part, about 3%. Nevertheless, the actual total amount of packaging waste in Europe is still at least 61 million tons per year and this amount has a big impact regarding the waste streams produced by households. In the Netherlands the fraction of plastics in municipal waste is nowadays 30% by volume[3] and in the US 21%. Disposal costs are high, in Europe 125 Euro per ton, in the USA 12–80 Euro per ton but in countries like Japan even 250 Euro per ton.[4]

The durability of plastics is beyond dispute. Some plastics need to be durable but many plastics have only a limited life or are used only once and therefore durability is not essential. A recent governmental action against litter in the streets in The Netherlands shows a billboard with a plastic cup lying on the highway with the message that if nobody picks it up, this cup will still be there after 90 years. The persistence of these petrochemical-based materials in the environment beyond their functional life is a problem. To bring this waste disposal under control, integrated waste management practices including recycling, source reduction of packaging materials, composting of degradable wastes and incineration have to be introduced. However, these measures will not help to decrease dependency on petroleum-based products and part of the solution can perhaps be found in the development and introduction of so-called biodegradable packaging materials that will degrade naturally into harmless degradation products at the end of their life cycle. This had led in the past to some misconceptions about how these materials could help solve the problem because policy has always been strong on supporting recycling of present plastic materials. On the other hand politicians have also reacted by introducing legislation for degradability requirements and thus providing a platform for natural polymer producers to obtain a larger market share in the non-food area.

Specific applications where biodegradability is required are sacks and bags that can be used for composting waste, foamed trays, cups and cutlery in the fast food sector, soluble foams for industrial packaging, film wrapping, laminated paper, foamed trays in food packaging, mulch films, nursery pots, plant labels in agricultural products and diapers and tissues in hygiene products.[5]

24.2 The range of biopolymers

24.2.1 Introduction

The development of biodegradable packaging alternatives has been the subject of much research and development in recent times, particularly with regard to renewable alternatives to traditional oil-derived plastics. Biopolymers, polymers synthesised by nature such as starch and polysaccharides, are an obvious alternative. However, these natural polymers on their own do not demonstrate the same material properties as traditional plastics, limiting potential applications of the technology. There are two major groups of biodegradable plastics currently entering the marketplace or positioned to enter it in the near future: polylactic acid (PLA) and starch based polymers.[6] These new polymers are truly degradable but full degradability will occur only when products made from these polymers are disposed of properly in a composting site.

24.2.2 Lactic acid

The efforts of biotechnology and agricultural industries to replace conventional plastics with plant derived alternatives have seen recently the following three approaches: converting plant sugars into plastic, producing plastic inside micro-organisms and growing plastic in corn and other crops. Cargill Dow has scaled-up the process of turning sugar into lactic acid and subsequently polymerises it into the polymer polylactic acid, NatureWorks™PLA. Lactic acid can be produced synthetically from hydrogen cyanide and acetaldehyde or naturally from fermentation of sugars, by *Lactobacillus*. Fermentation offers the best route to the optically pure isomers desired for polymerisation. Condensation polymerisation of lactic acid itself generally results in low molecular weight polymers. Higher molecular weights are obtained by condensation polymerisation of lactide, the intermediate monomer. When racemic lactides are used, the result is an amorphous polymer, with a glass transition temperature of about 60ºC, which is not suitable for packaging.[7]

24.2.3 Polylactic acid

Polylactic acid (PLA) is a polymer that behaves quite similarly to polyolefines and can be converted into plastic products by standard processing methods such as injection moulding and extrusion. It has potential for use in the packaging industry as well as hygiene applications. Currently a main obstacle is the high price of the raw material and the lack of a composting infrastructure in the European, Japanese and US markets. The current global market for lactic acid demand is 100,000 tons per annum, of which more than 75% is used in the food industry. Perhaps the biggest opportunities for PLA lie in fibres and films. For instance, worldwide demand for non-woven fabrics for hygiene application is 400,000 tons per annum. Other important market niches can be found in the agricultural industry such as crop covers and compostable bags.

The polymer of choice for most packaging applications may be 90% L-lactide and 10% racemic D,L-lactide. This material is reported to be readily polymerised, easily meltprocessable and easily oriented. Its Tg is 60ºC and its melting temperature is 155ºC. Tensile strength of oriented polymers is reported to be 80–110Mpa with elongation at break of up to 30%. Polylactide films are reported to be very similar in appearance and properties to oriented polystyrene films. Residual lactide is not a concern since it hydrolyses to lactic acid, which occurs naturally in food and in the body.[7] Therefore, PLA polymers are designed for food contact. Cargill Dow, the largest producer of PLA polymers, has confirmed that one of their grades is GRAS (Generally Recognised As Safe), permitting its use in direct food contact with aqueous, acidic and fatty foods under 60ºC and aqueous and acidic drinks served under 90ºC. In Europe, lactic acid is listed as an approved monomer for food contact applications in Amendment 4 of the Monomers Directive, 96/11/EC. All PLA polymer additives have appropriate EU national regulatory status.[8] However, PLA is not yet found in large applications of food packaging today.

24.2.4 Native starch

Starch is nature's primary means of storing energy and is found in granule form in seeds, roots and tubers as well as in stems, leaves and fruits of plants. Starch is totally biodegradable in a wide variety of environments and allows the development of totally degradable products for specific market needs. The two main components of starch are polymers of glucose: amylose (MW 10^5–10^6), an essentially linear molecule and amylopectin (MW 10^7–10^9), a highly branched molecule. Amylopectin is the major component of starch and may be considered as one of the largest naturally occurring macromolecules. Starch granules are semi-crystalline, with crystallinity varying from 15 to 45% depending on the source. The term 'native starch' is mostly used for industrially extracted starch. It is an inexpensive (< 0.5 Euro/kg) and abundant product, available from potato, maize, wheat and tapioca.[9]

24.2.5 Thermoplastic starch

Thermoplastic starch (TPS) or destructurised starch (DS) is a homogeneous thermoplastic substance made from native starch by swelling in a solvent (plasticiser) and a consecutive 'extrusion' treatment consisting of a combined kneading and heating process. Due to the destructurisation treatment, the starch undergoes a thermo-mechanical transformation from the semi-crystalline starch granules into a homogeneous amorphous polymeric material. Water and glycerol are mainly used as plasticisers, with glycerol having a less plasticising effect in TPS compared to water, which plays a dominant role with respect to the properties of thermoplastic starch.

24.2.6 Water resistance of starch-based products

Thermoplastic starch behaves as a common thermoplastic polymer and can be processed as a traditional plastic. TPS shows a very low permeability for oxygen (43cm^3/m^2/min/bar compared to 1880cm^3/m^2/min/bar of LDPE) which makes this material very suitable for many packaging applications. In contrast, the permeability of TPS for water vapour is very high (4708cm^3/m^2 compared to 0.7cm^3/m^2 of LDPE). This sensitivity to humidity (highly hydrophilic) and the quick ageing due to water evaporation from the matrix makes thermoplastic starch as such unsuitable for most applications. Due to this drawback there are no products available at the moment made from pure thermoplastic starch, which are form-stable (or even hydrophobic) in a wet atmosphere and mechanically stable over a sufficiently long period of time.

Producers of starch-based products overcome this problem by blending the thermoplastic starch with hydrophobic synthetic polymers (biodegradable polyesters) or by the production of more hydrophobic TPS derivatives (starch ester). Unfortunately, all theses production processes make the starch-based products rather expensive in comparison to the common plastic alternatives.

New concepts are required to solve the intrinsic problem of the hydrophilicity and mechanical instability of starch-based bioplastics without too much added cost.[9]

24.2.7 Polyhydroxyalkanoates

An industrial fermentation process in which microorganisms converted plant sugars into polyhydroxyalkanoates was developed by ICI, later Zeneca. Almost all living organisms may accumulate energy storage materials (e.g. glycogen in muscles and in livers, starch in plants and fatty compounds in all higher organisms) whereby polyhydroxyalkanoates (PHAs), as polyesters, represent the group of energy storage materials (e.g. carbon source that is exclusively found among bacteria). Generally PHAs are thermoplastic, water-insoluble biopolyesters of alkanoic acids, containing a hydroxyl group and at least one functional group to the carboxyl group. The FDA approved Biopol, the PHA produced by Monsanto who acquired the technology from Zeneca, as a food contact material. Important aspects were the biopolymer itself and the presence of breakdown products as crotonic acid. Also the incorporation of fermentation by-products – the microorganism *Ralstonia eutrophus* is not food grade – was of major concern. Other types of PHAs have not been approved for food contact applications yet.[10] Although its water-resistant properties give it a cutting edge in food packaging compared to other bioplastics, the plastic turned out to cost substantially more than its fossil fuel-based counterparts and offered no performance advantages other than biodegradability.[11]

29.2.8 Synthetic polyesters

These (aliphatic) polyesters are formed by polycondensation of glycols and dicarboxylic acids. They have tensile and tear strengths comparable to low density polyethylene and can be coextruded and readily heat-sealed. They can be processed into blown or extruded films, foams and injection moulded products and used in refuse and compost bags and cosmetic and beverage bottles. Due to their high price, aliphatic polyesters are used only in combination with starch. When tested, starch-polyester blends show in all cases an important decrease in water sensitivity whatever the thermoplastic starch and polyester type and content but for thermoforming applications such blends cannot provide sufficient stiffness due to the intrinsic softness of the polyester.[12, 13, 14]

24.2.9 Polycaprolactone and polyvinylalcohol

Polycaprolactone is made from synthetic (petroleum) sources, and has seen only limited use, apart from being used in starch-blends because of its low glass transition temperature of −60°C and melting temperature of 60°C.

Another polymer being used in packaging applications is polyvinylalcohol (PVOH), although its biodegradability is disputed. Some polymers like PVOH and starch are so water sensitive that they can in fact be water soluble. The most widely used water soluble polymer PVOH is prepared by hydrolysis of polyvinylacetate. Its water solubility can be adjusted to render it soluble in both hot and cold water or in hot water only. Control of the degree of hydrolysis can give control over the water solubility of the resulting resin. PVOH is not used as food packaging but in unit doses for agricultural chemicals, dyes and pigments, as well as water-soluble laundry bags for hospitals and detergent pouches.[7]

24.3 Developing novel biodegradable materials

24.3.1 Introduction

One of the major problems connected with the use of most of the natural polymers, especially of carbohydrates, is their high water permeability and associated swelling behaviour in contact with water. All this contributes to a considerable loss of mechanical properties, which prohibits straightforward use in most applications. Because of the hydrophilic and low mechanical properties of starch the property profile of these materials is insufficient for advanced applications like food packaging. The few applications for just thermoplastic starch, which do not involve the use of polymeric substances to form blends, are packaging chips, packaging for capsules and as packaging for food products (e.g. separate layers in boxes of chocolates) but never in direct contact with food. Their hydrophilic character, their reduced processability (with respect to polyolefines), and their insufficient mechanical properties represent particular drawbacks in this respect. Special processing or after-treatment procedures are necessary to sustain an acceptable product quality. As indicated before, presently applied methods for decreasing the hydrophility and increasing and stabilising the mechanical properties are blending with different, hydrophobic, biodegradable synthetic polymers (polyesters) and the application of hydrophobic coating(s). One recent new technology involves the application of the nano-composite concept that has proven to be a promising option.[9]

24.3.2 Barrier effect of nano clay particles in a biopolymer matrix

The incorporation of nano-clay sheets into biopolymers has a large positive effect on the water sensitivity and related stability problems of bioplastic products. The nature of this positive effect lies in the fact that clay particles act as barrier elements since the highly crystalline silicate sheets are essentially non-permeable even for small gas molecules like oxygen or water. This has a large effect on the migration speed of both incoming molecules (water or gases) as well as for molecules that tend to migrate out of the biopolymer, like the water used as a plasticiser in TPS. In other words, nano-composite materials with well-dispersed nano-scaled barrier elements will not only show increased mechanical properties but also an increased long-time stability of these properties and a related reduction of ageing effects.

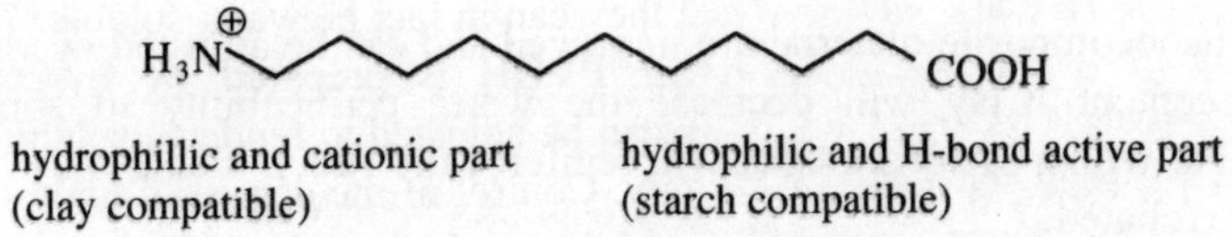

$H_3N^{\oplus}$ ~~~~~~~~~ COOH

hydrophillic and cationic part (clay compatible) — hydrophilic and H-bond active part (starch compatible)

Fig. 24.1 Example of possible modifiers for starch-clay nano-composites and requirements for clay modification.

In order to achieve the final clay-starch nano-composite material, a 'clay modification' and an 'extrusion' processing step can be distinguished, which are described below. For the preparation of nano-composite materials consisting of starch and clay, the use of special compatibilising agents (modifier) between the two basic materials is necessary as depicted in Fig. 24.1.

Layered silicates are characterised by a periodic stacking of mineral sheets with a weak interaction between the layers and a strong interaction within the layer. The space between the layers is occupied by cations. By cation exchange reactions between the clay and organic cations (such as alkyl ammonium salts) the layered silicate can be transformed into organically modified clay. The inter layer distance will increase by using voluminous modifiers. If this modifier is compatible with starch as well, a homogeneously and nanoscaled distribution (exfoliation) of the clay sheets can be effected in the polymer matrix. The modified clay can be analysed by X-ray investigation (XRD) to determine the inter-layer distance. The pure clay shows an interlayer distance of 1.26nm. It has been proven by XRD analysis that most of the layers are indeed 'swollen' after the modification reaction. The interlayer distance changes to 2.34nm – an increase of nearly 100% compared to the pure clay.

24.3.3 Extrusion

The starch and the modified clay are mixed at temperatures above the softening point of the polymer by polymer melt processing (extrusion). At these temperatures the polymer melt intercalates. The success of the polymer intercalation depends on the modification of the clay, on the degree of increased interlayer distance and on the interaction between the modifier and the matrix material. A full destructurisation is needed for a successful polymer melt process of starch. Therefore, it is very important to find the optimal starch/clay/plasticiser content, the most effective geometry of the screws and the right temperature profile within the extruder.

24.3.4 Properties of the starch-clay nanocomposites

A homogeneous incorporation of clay particles into a starch matrix on a true nano scale has proved to be possible. The addition of clay during processing supports and intensifies the destructuring process of starch, providing a means of easier processing. The obtained starch/clay nanocomposite films show a very strong decrease in hydrophilicity. The stiffness, the strength and the toughness of the nanocomposite material are improved and can be adjusted by varying the water content. Clay will decrease the water permeability to some extent (maximal with a factor 2). Clay will reinforce the starch blends only when it is fully exfoliated.

Hot pressed films made out from material indeed showed a great advantage compared to films made from pure thermoplastic starch. Ordinary TPS evaporates water very quickly upon ageing. Figure 24.2a shows a photograph of a hot pressed film of pure thermoplastic starch (after ageing the granulates for three hours at room temperature following the extrusion step). The apparent morphology indicates that it is not possible to form a true film any more. In contrast Fig. 24.2b shows a hot pressed film of a starch/clay nano-composite. Transparent and homogeneous films can be formed which show an increased mechanical stability and toughness as well.

24.4 Legislative issues

It is important to remark that biodegradability and compostability are different concepts.[2] While biodegradation may take place as a result of the disposal of a material in landfills, composting usually requires a pre-treatment of municipal solid waste; it is necessary in fact to remove all bulky non-compostable items before beginning the composting process, separating organic from inorganic waste. Moreover, before composting other steps are necessary: particle size reduction, magnetic removal of metals, moisture addition and mixing. Under ideal conditions the decomposition of organic material can take 30 to 60 days. International Standards Research (ISR) at the request of ASTM studied the performance of biodegradable plastics in full-sized composting facilities and under laboratory conditions. The ISR work determined that plastics needed to meet three criteria to be compostable. According to this standard ASTM D6400 they must be able to:

1. demonstrate inherent biodegradability at a rate and degree similar to natural biodegradable polymers
2. disintegrate during active composting, so that there are no visible, distinguishable pieces found on the screens
3. have no ecotoxicity – nor impact the ability of the resultant compost to support microbial and plant growth.[15]

A standard world-wide definition for biodegradable plastics has not been established, nevertheless all the definitions already in place (ASTM, CEN, ISO) correlate the degradability of a material to a specific disposal environment and to a specific standard test method which simulates this environment in a time period which determines its classification. The European Parliament on 20 December 1994 adopted a directive (94/62 EC) in order to harmonise national measures concerning the management of packaging and packaging waste, to provide a high level of environmental protection and to ensure the functioning of the internal

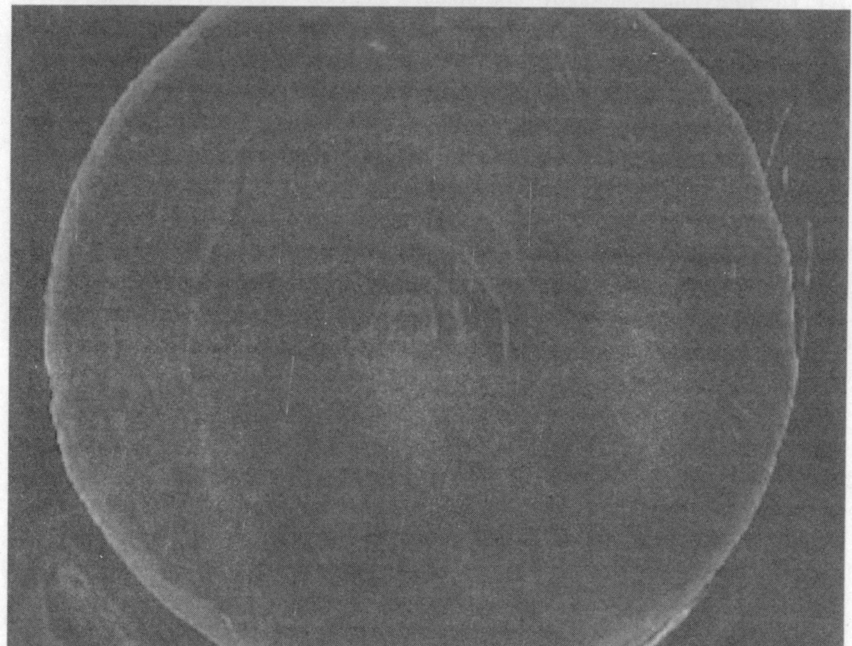

Fig. 24.2 Compression moulded films of a) pure TPS granulate and b) starch/clay nano-composite granulate

market. In the 94/62 EC Directive a very brief part is dedicated to compostable and biodegradable materials. In item three, 'compostability' is defined as organic recycling and it is pointed out that compostability can take place only under controlled conditions and not in landfills. Moreover, 'biodegradable packaging' is defined as a material that must be capable of physical, chemical, thermal and/or biological degradation such that this material used as compost ultimately decomposes completely into carbon dioxide and water.[2]

According to European directive No. 94/62 the producer or importer of packaging is responsible for the recovery of a substantial fraction of the annual amount of packaging it produces in the market. It states that at least 65% must be recovered, at least 45% must be recovered by material recycling and at least 15% of each packaging material must be recycled. The term recovery denotes the sum of recycling (material recovery), incineration (energy recovery) and composting (organic recovery). Furthermore, the directive prohibits packaging that does not fulfil the essential requirements. For products to be designed to be compostable the requirement is that 'they should be of such a biodegradable nature that it does not hinder the source-separated collection of biowaste, nor the composting activities in which it will be treated'. A draft standard, prEN 13432, has been made with requirements for compostable products. According to this standard, the following criteria are relevant for a compostable product.[16]

1. The individual packaging components shall be completely biodegradable.
2. The total product shall disintegrate completely during a composting process.
3. The addition of the product to the biowaste shall not have negative effects on the composting process.
4. The addition of the product to the biowaste shall not have negative effects on the quality of the final compost.

To demonstrate biodegradability, it is possible to use several internationally accepted standard methods for determining the biodegradability of organic compounds. Both aquatic tests and tests with high solids environments are allowed, although tests under controlled composting conditions are preferred. Evaluation criteria follow.

1. For a packaging material or the constituents of a packaging material which consists of only one polymer (homo-polymer or random copolymer) without any additives, the degree of biodegradation based on carbon dioxide release or oxygen consumption shall be more than 60% of the theoretical value.
2. For a packaging material or the constituents of a packaging material comprised of different components (polymer blends), or block copolymers and after addition of low molecular additives, the degree of biodegradation based on carbon dioxide release or oxygen consumption shall be more than 90% of the theoretical value.
3. The period of application of the test methods shall be a maximum of six months

Unless technically impossible, the packaging, packaging materials or packaging component shall be tested for disintegration in the form in which it will ultimately be used. In practice, packaging materials are tested and from this it is concluded that a complete packaging will be disintegrated if all its materials are capable of disintegration. A complete packaging should, however, be tested in cases where a direct conclusion is not possible, e.g., if two or more packaging materials are firmly joined together forming a fixed multi-layer structure.

Due to the nature and analytical condition of the disintegration test, the test results cannot differentiate between biodegradation and biotic disintegration but they are required to demonstrate that a sufficient disintegration of the test materials is achieved within the specified treatment time of biowaste. By combining these observations with the information obtained from the laboratory tests it can be concluded whether a test material is sufficiently biodegradable under the known conditions of biological waste treatment.

Food Contact Materials are all materials and articles intended to come into contact with foodstuffs, including not only packaging materials but also cutlery, dishes, processing machines, containers, etc.[17] The term also includes materials and articles that are in contact with water intended for human consumption, but it does not cover fixed public or private water supply equipment.

The harmonisation at EU level of the legislation on Food Contact Materials fulfils two essential goals: the protection of the health of the consumer and the removal of technical barriers to trade. Food contact materials shall be safe and shall not transfer their components into the foodstuff in unacceptable quantities. The transfer of constituents of the food contact materials into the food is called migration. To ensure the protection of the health of the consumer and to avoid adulteration of the foodstuff two types of migration limits have been established in the area of plastic materials:[18] an overall migration limit of 60mg (of substances)/kg (of foodstuff or food simulants) that applies to all substances that can migrate from the food contact material to the foodstuff and a specific migration limit (SML) which applies to individual authorised substances and is fixed on the basis of the toxicological evaluation of the substance. The SML is generally established according to the acceptable daily intake or the tolerable daily intake set by the Scientific Committee on Food. To set the limit, it is assumed that, every day throughout his/her lifetime, a person of 60kg eats 1kg of food packed in plastics containing the relevant substance at the maximum permitted quantity.

24.5 Current applications

24.5.1 Introduction

Despite the problems still encountered in properties and production of biopolymers, biodegradable food packaging products enter the market because of supermarkets being increasingly tricked by the marketing effect of a green image. Novamont has begun to supply Tesco with nets for fruits and bags to Swiss and German supermarkets. Albert Heijn in The Netherlands uses biodegradable packages from Natura Verpackung while Sainsbury in the UK is doing composting trials for food waste at 75 of its stores, using Mater-Bi bags.

Table 24.1 Properties of polymers used as packaging materials

Polymer	Tensile strength MPa	Tensile modulus GPa	Max. use temperature °C
LDPE	6.2–17.2	0.14–0.19	65
HDPE	20–37.2		121
PET	68.9	2.8–4.1	204
PS	41.3–51.7	3.1	78
PA 6	62–82.7	1.2–2.8	–
PP	33–37.9	1.1–1.5	121
PLA	40–60	3–4	50–60

Table 24.1 gives some properties of polymers used in packaging materials.[19] Among the biodegradable polymers PLA seems to be the polymer that can compete in terms of mechanical properties with conventional polymers. The drawback here is the low Tg of PLA.[20] The main players in the biodegradable polymer market are shown in Table 4.2. Despite the slow entry into of use of biodegradables in food packaging, some trends are visibly shifting towards sustainable chemistry and green plastics being applied in niche markets. Most of the efforts these days seem to be focused on foamed products for food packaging.

24.5.2 Starch based foams in food packaging

The use of foamed polymer packaging, for example, polystyrene (PS) clamshells, by prominent users such as McDonald's Restaurants has recently decreased significantly because of perceived environmental disadvantages.

Table 24.2 Main players in biodegradable polymers and their trade names

Material	Supplier	Trade name
Starch based	Novamont	MaterBi
Starch based	Biotec	Bioplast
Thermoplastic starch	Avebe	Paragon
Thermoplastic starch	National Starch (Novamont licensee)	Ecofoam Envirofil
Polylactide/PLA	Cargill Dow	Nature Works PLA
Polylactide/PLA	Mitsui	Lacea
Polylactide/PLA	Hycail	
Polylactide/PLA	Galactic	Galactic
(Co)polyester	BASF	Ecoflex
(Co)polyester	Eastman Chemical	Eastar Bio
(Co)polyester	Du Pont	Biomax
(Co)polyester	Showa Highpolymer	Bionolle
Polycaprolactone	Union Carbide	Tone polymer
Polycaprolactone	Solvay	CAPA

Polystyrene is derived from non-renewable resources, is non-degradable and for its processing blowing agents were used in the past that contributed to the depletion of the ozone layer. Paper-based products have a more favourable environmental perception but do not share the mechanical properties of polystyrene foams. It is well known that starch, containing sufficient moisture, can provide stable foams.

A step forward has been the introduction recently by Novamont of a new foamed tray based on starch particularly for the 'McDonalds' type of applications.[21] Apack has introduced another tray made from a baked starch formulation that has a coating of EastarBio aliphatic-aromatic copolyester. Sainsbury, a leading retailer in the UK, has been packaging its organic fruit and vegetables in these starch-based materials by Apack.[22]

Paperfoam has patented Paperfoam that is produced from a viscous suspension containing starch, cellulose fibres and water. The suspension is injected into the mould. Due to the mould's temperature (c. 200ºC) the starch granulate gelatinises and the water evaporates. The manufacturing cycle bakes from five seconds to two minutes, depending on wall thickness, from starch, natural fibres and water using an energy-efficient, one-step production technology. It can be recycled with paper and is biodegradable. Paperfoam combines a foamed inner structure with a smooth outer face and is applicable for a wide variety of uses. At present Paperfoam is used in the packaging of hand-held electronic consumer goods, such as telephones, but not yet in food packaging.[23]

Other types of foamed products at different stages of development are blends of starch with poly(vinyl alcohol-co-40%-ethylene), PVOH-40, a degradable, water resistant polymer that can be processed into viable alternatives to PS foam packages via wafer baking technology, extrusion, or expanded-bead moulding[24] and starch-based dough made by a baking process for various food containers.[25] Although one of the most versatile technologies for the production of starch-based foam is via this type of baking process where a starch dough is heated under pressure to form a moulded foam product, these starch products are moisture sensitive and have poor mechanical properties. Both of these attributes can be improved by the inclusion of fibres and/or fillers in the dough formulation. The resulting products are starch-based foam composites with mechanical and thermal properties rivalling those of polystyrene.[26]

24.5.3 PHA in food packaging

PHA properties show that it might be a very good alternative for conventional polymers in food contact packaging. However, when Monsanto bought the Biopol process in 1995 profitability still remained elusive.[11] The approach with the most potential was to grow PHA in plants, modifying the genetic make-up of the crop so that it could synthesise plastic as it grew and eliminate the fermentation process. However, it was found that producing one kilogram of PHA from genetically modified corn plants would require about 300% more energy than the 29 megajoules needed to manufacture 1kg of fossil-fuel-based polyethylene. This finding prompted Monsanto to terminate this method of producing PHA. The Biopol assets were obtained by Metabolix but food packaging is not a product line they intend to focus on in future. Proctor and Gamble together with Kaneka Corporation are working on a new development in PHA production but this product might not become available in large quantities until 2005.

24.5.4 PLA in food packaging

With PLA it is a similar story in terms of energy balance to PHA when one looks at the production site of Cargill Dow with its Nature Works™ PLA in Blair, Nebraska. The Blair plant with a capacity of 140 000 metric tons per year produces 1kg PLA with 56 megajoules of energy. However, in principle PLA processes can require between 20–50% fewer fossil resources than making plastics from oil but it is still significantly more energy intensive than most petrochemical processes.

Packaging solutions from Nature Works can be extruded, thermoformed and blow moulded, unlike other traditional products, such as paper. For packaging they have two film grades and one grade for thermoforming. The film grades are designed for applications like candy twistwrap and for laminations for packaging such as flavoured crewels, coffee packs and pet foods because of the additional advantageous properties such as the barrier to flavour and grease and superior oil resistance. The potential applications of thermoformed products within packaging is multifold, dairy containers, food service ware, transparent food containers, blister packs and cold drink cups. PLA polymer has been shown to biodegrade similarly to paper under simulated composting conditions (ASTM D5338, 58°C). Degradation of PLA packaging depends both on exposure conditions and on amount and type of plasticiser. Sainsbury would like to use PLA but will not do this because of the lactic acid coming from GM crops.[21] The consumer might not accept this at the moment although the GM label is destroyed at the fermenting stage. Much of the PLA of Cargill Dow is for fibre applications but the company is already working with many leading European packaging converters, including Trespaphan on oriented PLA films, Klöcknes pentaplast on thermoformed trays and lids, and Autobar on thermoformed dairy pots. Food retailers are also increasingly involved.

24.5.5 Proteins in food packaging

Proteins have long and empirically been used to make biodegradable, renewable, and/or edible packaging materials. Numerous vegetable proteins (corn zein, wheat gluten, soy proteins) and animal proteins (milk proteins, collagen, gelatine, keratin, myofibrillar proteins) are commonly used. Although protein materials have been studied extensively[27] a breakthrough is not yet imminent although some of the properties have increased extensively recently. Protein materials can be processed into transparent and water-resistant films by casting or thermo-forming. In packaging, collagen sausage casings are the best known of the commercial applications.[28, 29]

24.6 Future trends

When one looks at the present market for biodegradable food packaging materials then that market is still virtually non-existent compared to any conventional plastics used in food packaging. The reason for this is the high price, the sometimes inferior cost/performance relation and the fact that still only a few materials have received FDA approval.

Since 2001 the market for biodegradable/compostable products has definitely been growing after remaining at the same level of 20,000 tons worldwide for the last five years and although few in number, new products for food packaging have been introduced since.

24.7 References

1. JOGDAND S N, *Welcome to the world of eco-friendly plastics*, www.members.rediff.com, 1999 .
2. AVELLA M, BONADIES E, MARTUSSCELLI E, RIMEDIO R, 'European current standardization for plastic packaging recoverable through composting and biodegradation', *Polymer Testing,* 2001, **20**, 517–521.
3. www.rivm.nl/mieucompendium: C6.6 Hoeveelheid kunststoffen en PVC in huishoudelijk restafval.
4. GRUBER P 'Sustainable design of polymers and materials', Abstracts, *6th International Scientific Workshop on Biodegradable Polymers and Plastics*, Honolulu, Hawaii, 2000.
5. BASTIOLI C, 'Global status of the production of biobased packaging materials', Conference proceedings *The Food Biopack Conference*, Copenhagen, August 27–29, edited by C J Weber, 2–7, 2000.
6. BILBY G D, *'Degradable polymers'*, www.angelfire.com, 2000.
7. SELKE S, *'Biodegradation and packaging'* (2nd edn), Pira International Reviews, 2000.
8. CARGILL DOW LLC, Product information, published June, 2000.
9. FISCHER S, VLIEGER J J, DE KOCK T, BATENBURG L-, FISCHER H, 'Green' nano-composite materials – new possibilities for bioplastics' Materialen, 2000, **16,** 3–12.
10. WEUSTHUIS R A, WALLE G A M, VAN DER EGGINK G, 'Potential of PHA based packaging materials for the food industry', Conference proceedings *The Food Biopack Conference*, Copenhagen, August 27–29, edited by C J Weber, 24–7, 2000.
11. GERNGROSS T U, SLATER S C, 'How green are green plastics', *Scientific*

American, feature article of August isssue, 2000.

12. AVÉROUS L, FAUCONNIER N, MORO L, FRINGANT C, Blends of thermoplastic starch and polyesteramide: Processing and properties, *J. Appl. Polym. Sci.*, 1999, **76**, 1117.
13. AVÉROUS L, MORO L, DOLE P, FRINGANT C, 'Properties of thermoplastic blends: Starch-Polycaprolactone', *Polymer*, 1999, **41**, 4157.
14. AVÉROUS L, FRINGANT C, to be published.
15. NARAYAN R, 'The scientific rationale behind biodegradable/compostable standards', Abstracts *6th International Scientific Workshop on Biodegradable Polymers and Plastics*, Honolulu, Hawaii, 2000.
16. ZEE M VAN DER, '*Compostable products-Legislation standards and future policy*', Green-Tech Newsletter April 1999, **2**, 2, I.
17. LEBARON P C, WANG Z, PINNAVIA T J, 'Polymer-layered silicate nanocomposites: an overview', *Applied Clay Science*, 1999, **15**, 11–29.
18. Final report Bionanopack project 1999-2001, FAIR CT98-4416, Co-ordinator J J de Vlieger, 2002.
19. BRODY A L, 'Packaging materials', *Encyclopedia of Polymer Science and Engineering*, 1987, **10**, 684–710.
20. SÖDERGÅRD A, 'Lactic acid polymers for packaging materials for the food industry', Conference proceedings *The Food Biopack Conference*, Copenhagen, August 27–29, edited by C J Weber, 19–23, 2000.
21. *Modern Plastics International*, December 2001, 44.
22. *European Plastic News*, January 2002, 21.
23. Green-Tech Newsletter 2000, vol. 3, no 1, 4 Paperfoam, www.paper foam.nl.
24. ORTS W, NOBES G A R, GLENN G M, GRAY G M, HANSEN L U, HARPER M V, 'Blends of starch with poly(vinyl alcohol)/ethylene copolymers for use in foam containers', Abstracts *6th International Scientific Workshop on Biodegradable Polymers and Plastics*, Honolulu, Hawaii, 2000.
25. GLENN G M, NOBES G A R, GRAY G M, 'Insitu laminating process for baked starch based foams', Abstracts *6th International Scientific Workshop on Biodegradable Polymers and Plastics*, Honolulu, Hawaii, 2000.
26. NOBES G A R, ORTS W J, GLENN G M, 'Use and effeects of agricultural fibers and fillers in baked starch-based foam composites', Abstracts, *6th International Scientific Workshop on Biodegradable Polymers and Plastics*, Honolulu, Hawaii, 2000.
27. GUILBERT S, 'Potential of the protein based biomaterials for the food industry', Conference proceedings *The Food Biopack Conference*, Copenhagen, August 27–29, edited by C.J. Weber, 13–18, 2000.
28. GUILBERT S, CUQ B, GONTARD N, 'Recent innovations in edible and/or biodegradable packaging', *Food Additives and Contaminants* 14, 2000, **6**, 741–751.
29. KROCHTA J M, MUKDER-JOHNSTON C DE, 'Edible and biodegradable polymer films: challenges and opportunities', *Food Technol.*, 1997, **51**, 61–73

25

Integrating intelligent packaging, storage and distribution

T. Järvi-Kääriäinen, Association of Packaging Technology and Research, Finland

'All engineering, manufacturing, quality and sales efforts are wasted if your transport packaging fails and your customer receives a damaged product' (ISTA).[1]

25.1 Introduction: the supply chain for perishable foods

Food is a perishable product. It is temperature-, moisture-, and time-sensitive, compared to books, automobile parts, and clothes, however they are shipped globally. The present systems for improving logistics, ordering and networks may cause the special nature of food to be ignored. The new IT systems are first applied to expensive, valuable products, not to commodity products, such as food. Then the commodity products must adapt to the systems, which exist even if the producers have not taken part in the development work.

25.1.1 Growth seasons and specialities in different areas of the world

In many parts of the world there is only one yearly growing season yielding only one or two crops in a year. However, consumers prefer to eat both fresh produce and the specialities of specific areas the year round. This means that foodstuffs may be transported to the other side of the world.

Some foodstuffs (canned food, aseptic packages, dried food) can be stored at room temperature. They are not sensitive to small temperature changes if they have been correctly packed. The quality and safety of frozen, chilled, and fresh food tolerates only a very narrow temperature range. Also chilled and fresh food has a limited shelf-life. An average employee in the food industry knows these facts very well, but there are several steps during transportation where one has no or very limited knowledge of the special requirements or needs of perishable products (harbours, airports, transport terminals). Also, a consumer may purchase products and expose them to too high temperatures or otherwise wrong storage conditions before they are eaten.

As an example, let us look at the shipment of bananas to Finland where they are everyday commodities throughout the year. What follows are the stages on the route of bananas from tree to table.

- A half-ripe banana cluster is cut from a banana tree.
- It is lifted to a hook of a cableway.
- The cableway transports the banana cluster to the packaging area.
- The banana cluster is rinsed.
- The washed cluster is cut into smaller bunches, usually of about five bananas each.
- Cut bananas are washed.
- Bananas are lifted into a plastic tray, each tray containing about 18kg of bananas.
- The plastic tray is transported to the weighing station.
- Weighed bananas are sprayed with a biocide.
- Banana groups are brand labelled.
- Bananas are moved from the trays into transportation boxes.
- Boxes are stamped with packing date and location.
- Boxes are lifted on a pallet.
- Pallet loads are transferred to containers.
- Containers are transported to the harbour.
- Containers are lifted to a ship.
- Banana containers are monitored for temperature during the sea trip, which takes about two weeks.
- Containers are unloaded from the ship in Sweden.
- Banana pallet loads are transported to the ship terminal.
- The temperature of the bananas is checked.
- Banana pallet loads are loaded onto a lorry.
- The lorry drives onboard a ship sailing to Finland.
- Second sea voyage.
- The lorry drives the bananas to a building where the bananas are allowed to ripen.
- Bananas and their temperature are checked on arrival.
- Bananas are moved into a maturing room in order to ripen.
- During the ripening process, which takes about six days, the bananas are checked daily for their temperature and degree of ripeness.
- After ripening the bananas are transferred to a collection point.
- There the banana boxes are lifted either onto a trolley or onto a pallet.
- The full trolley or the pallet load is transported to the right gate on the shipping area.

- The driver brings the trolley or the pallet to his truck.
- Bananas are transported to the shop.
- The driver moves the trolley to the inspection area of the shop.
- The shopkeeper checks and accepts the product.
- Bananas are moved to the storage area of the shop.
- Depending on sales the products are put on sale.
- The boxes are opened and the bananas are put on display.
- A consumer chooses some bananas and packs them into a bag
- The bananas are weighed.
- A price label is attached to the bag.
- The customer goes to the till.
- Customer puts the product on the cashier line.
- The scanner reads the price and gives the storekeeper information on the amount of bananas sold.
- The customer packs the products into a carrier bag and takes them home (Leppäaho, 2002).[2]

The above list shows that bananas were exposed to several different temperature and moisture conditions. They were handled frequently and moved from place to place using a wide variety of transporting media.

25.1.2 Effect of distance, time, shock, vibration, air pressure, temperature and moisture to the products

Transportation can be a long and time consuming process involving several handling steps, as the banana case illustrates. Transportation of goods exposures them to shock, vibration, air pressure, and moisture variations in addition to time and temperature. There are several studies on the vulnerability of foods, and how different packaging can improve or destroy the quality of a specific product (such as Chonhenchob *et al.* on mangoes, 2002).[3]

Distribution packaging is generally tested by integrity and general simulation tests before shipment. The first step in the focused simulation test is to quantify, by actual field measurement, the distribution hazards on the packaged products in terms of their intensities and other conditions. For example, drops and impacts are measured, and the data is analysed according to the height or velocity, package orientation at impact, and frequency of occurrence. Vehicle vibrations are measured, with the data typically analysed as power spectral density plots according to the vehicle type and lading conditions, and time durations (or with a given relationship of time to trip length). Compression is measured in vehicles and warehouses, and data analysed to time and superimposed conditions. Atmospheric profiles are measured, and data analysed in terms of extremes, rates of change, and combinations. The measurements have become possible with the help of the currently available small, self-contained electronic field data recorders. (Kipp, 2002).[4] These instruments can record both static and dynamic information in order to get the required analyses. They are often smaller in size than a brick.

Unique systems must be designed, if temperature and humidity sensitive products are ordered by e-mail and shipped to other places (Singh, 2000).[5] Since fresh produce continues to respire after being harvested, this causes an intake of oxygen and release of carbon dioxide. The respiration rate of fruits, vegetables and flowers is dependent on temperature. An increase of package storage temperature results in an exponential increase in respiration rate that shortens the shelf-life of the produce, resulting in eventual decay. The United States Department of Agriculture has documented this information on recommended storage temperature, relative humidity and approximate shelf-life for various fresh produce (Welby *et al.*, 1997).[6] Most fresh produce has a high moisture content, so it is important to maintain a high humidity environment during transportation in order to prevent moisture loss that would result in drying of the produce. USDA recommends a high humidity environment (80–95% relative humidity) for most fresh produce. Possible solutions include cooling aids and specially insulated packaging materials (Singh, 2002a).[7]

A company in the USA selling expensive meat parts by e-mail has found an interesting method for chilling their goods. Instead of shipping meat pieces with cooling aids and fillers, frozen hamburgers are packed into the boxes serving as coolers and fillers. The customer gets a usable give-away and the extras increase the incentive for a further purchase (Singh, 2002b).[8]

25.2 The role of packaging in the supply chain

The main duties of packaging are to protect, contain, inform, and sell. The right packaging also preserves, as the products are received in a good and usable condition. The package needed for protection is a combination of product characteristics and logistical hazards. Well chosen packaging can reduce the cost of every logistical activity: transport, storage, handling, inventory control, and customer service. It can reduce the cost of damage, safety, and disposal. Integrated management of packaging and logistics is required, if a firm is to realise such opportunities to reduce costs (Twede, 1997).[9]

25.2.1 Interviewing the food industry and trade

Tekes, the National Agency of Finland, finances the ‘Safety and Information in Packaging’ program in Finland. There have been studies on the needs and wishes of industry and trade in order to gather information for directing the research program. During the summer of 2001 Pakkausteknologia – PTR (Association of Packaging Technology and Research) asked for the opinions of the Finnish food industry with a questionnaire of 93 questions (Pikkarainen *et al.*, 2002).[10] Answers were received from producers of dry foods (sugar, flour), beverages, and ready-to eat foods, dairy products, and so on, covering the different sectors of food industry. Shocks, compression, changes in temperature and packaging closing methods all cause packaging problems for the food industry (Fig. 25.1).

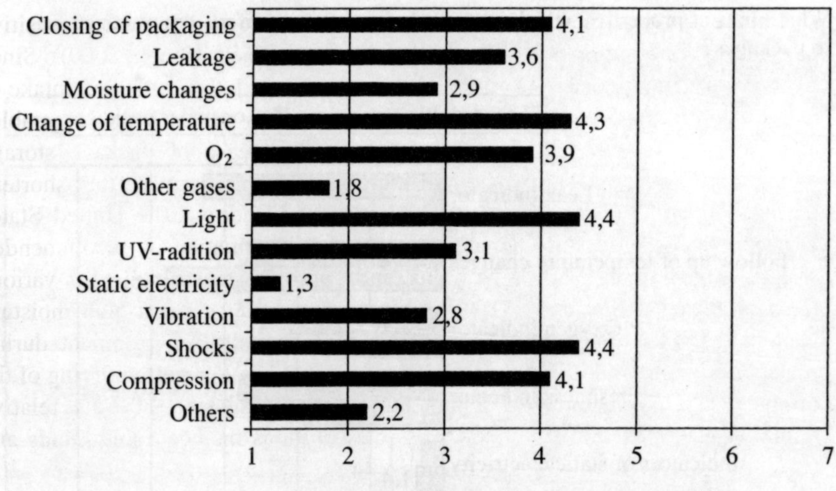

Fig. 25.1 Possible causes of trouble in packaging in food industry. Presented as an average of all answers. No problems at all is indicated by number 1, and 7 is given if major problems (Pikkarainen *et al.*, 2002).

The interviewees wished to gather more information on the following aspects: circumstances during the storage and transport, temperature changes, breakage of cold chain, and leak detection. They preferred additional properties to be within the packaging material, and also that the new properties would be in transport packages instead of consumer packages. They particularly wanted an active tag or a smart card containing memory to be developed for the distribution package rather than for the consumer package. The main reason is probably the price of the tags. If it were only a few cents then the tags would be accepted also for consumer packages. The profit margin in the food sector is low and therefore all additional costs must be carefully considered.

Different information is required in consumer packages than in transport and distribution packages. The information in a consumer package is aimed at the consumers; retailers and other parties in the trade need information on distribution and storage conditions. All who replied wished to know about the cold chain, especially if it had been interrupted. Monitoring the temperature during distribution was an important aspect (Fig. 25.2). Only those whose products were not sensitive to temperature did not consider it very important. The majority (70%) also preferred the indicator to record information during distribution.

In the 'Safety and Information in Packaging' program the storekeepers in Finland were also interviewed by Pakkausteknologia – PTR (Association of Packaging Technology and Research). All who were interviewed, mentioned that the most important aspects were the retail packages (the size should be right for the size of the store), environmental aspects, economics, alarm systems, and

What kinds of properties, which increase information or safety, are worth adding to packages?

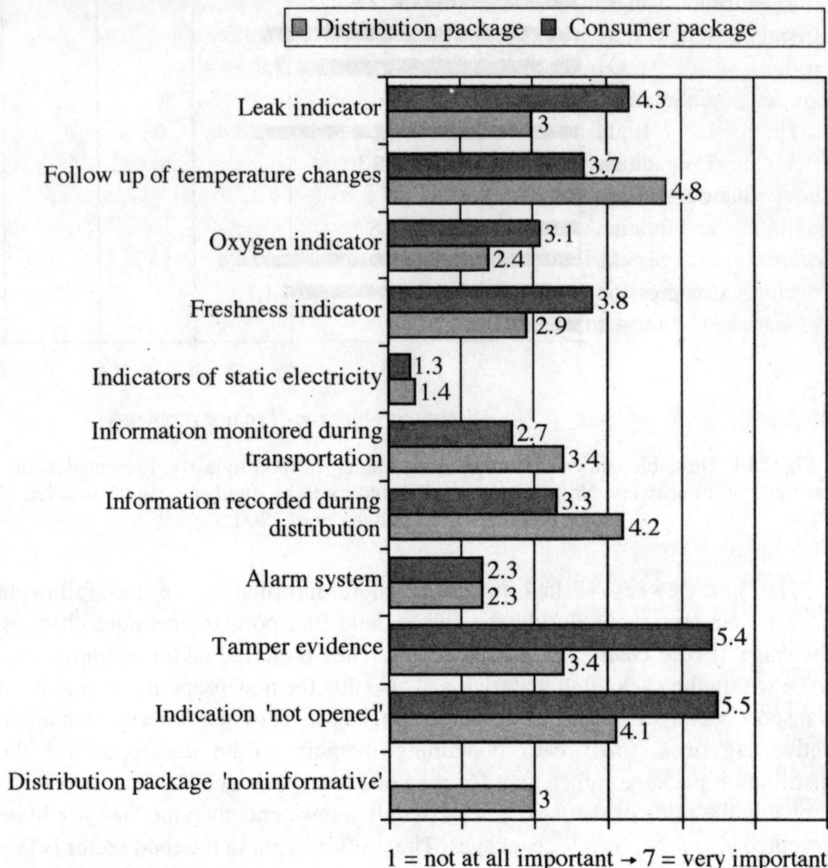

Fig. 25.2 Properties that increase information or safety on consumer or distribution packages. Presented as an average of all answers (Pikkarainen *et al.*, 2002).

consumer packages. The consumer packages should be appealing to the consumer and contain information that the customer needs (Leppänen-Turkula, 2002).[11] Ergonomics and ease of opening the packages at the retail level are important aspects of packaging design and are valued by the retailers.

25.3 Creating integrated packaging, storage and distribution: alarm systems and TTIs

The current logistic systems use EAN-codes, shipment labels and codes, alarm systems, separate in-house control systems, and manual check-ups. Sometimes the distribution conditions are also checked with shock and vibration devices or continuous expensive devices, which monitor the time and temperature and moisture conditions during distribution. There is a need to develop an economic integrated system that would serve the different aspects of the whole chain. A producer wants traceability and an easy way to recall a product from the market. Low inventories and feed-back systems are also desirable.

There is also a desire to check every actual shipment for environmental effects, and for shocks and vibrations. The main reason is to get information about where and when possible mishandlings occur so that responsibilities, liabilities, and improvements can be determined. Presently the price of the recorders and the difficulty of returning used recorders exert a pressure to develop cheaper systems that could even be added to each shipment. However, the achieved savings must be bigger than the costs.

25.3.1 Alarm systems

It is estimated that shoplifters account for a $10 billion annual loss in the American retail trade. Retailers have struggled to reduce these losses by various means. Early electronic devices were cumbersome, and their cost meant that they had to be removed at the checkout counter and used again. A modern EAS (electronic article surveillance) device is paper-thin and the size of a postage stamp. The EAS is attached to the package or product. As more and more retailers are asking their suppliers to include EAS tags on their products, the problem of tagging merchandise is shifting from the retailer to the package supplier. The two systems, which are currently much in use, are acoustomagnetic and radio frequency technology. EAS will set off an alarm when the active device is passed though an EAS detection system. Acoustomagnetic uses 58 kilohertz. Demagnetising the strip or altering its magnetic properties so that it resonates at different frequency inactivates the alarm (Soroka, 1999).[12]

25.3.2 In-house control

At the moment all different stages of the food chain do their own in-house control tests. The storage facilities usually have temperature recorders, but how well the right temperature is maintained near doors and walkways is open to question. Is the capacity enough to cool a warmed product fast enough? The most common tests are plain visual checks, but the lorries usually have temperature recorders if they transport frozen or chilled foods. The store checks product temperatures at arrival. These sporadic tests, however, do not give continuous information. By using TTI indicators the time and temperature combined effects could be better monitored. When used in distribution packages the cost of the indicator is divided between several consumer packages.

25.4 Traceability: radio frequency identification

25.4.1 Automatic identification

Automatic identification is a generic term describing various methods of data collection and entry. Automatic data collection reduces human error and speeds up the work. There are varying types of Auto ID operating in the world, many encountered on a daily basis without the users truly being aware of them. Examples of these are: card technologies, scanning devices, machine vision, optical character recognition, speech/voice recognition, radio frequency identification. Associated with an increase in entry accuracy is an overall reduction of costs as well as time savings. There are additional benefits that can be derived as well including, where appropriate, increased product or service quality, increased productivity and a reduction in inventory.

25.4.2 Radio Frequency Identification (RFID)

RFID is a non-contact, wireless, data communication form where tags of some material, usually embedded with an IC chip, are programmed with unique information and attached to objects for identification and tracking. The information can be location, product name or code, expiration/product date, etc., depending on what is required. As tagged items pass by readers, the data from the tags is decoded and transferred to a host computer for processing. RFID offers practical benefits over other automatic data capture systems: a line-of-sight is not needed, it has the ability to read multiple tags simultaneously even on the move, and it is possible to write information to the tag, and so on (Clarke, 2002).[13] The main difference with EAS is that EAS gives only an alarm, but a RFID tag also identifies the article as a unique product. This increases the possibility of using the technique for recall and tracing.

There are several recent articles on RFID. Articles portray a world where the items give detailed information and can communicate with each other without human intervention (Chips, 2002, Covell, 2002).[14, 15] There are also several names used: RFID tag, smart card, smart label. One of the questions is when should the RFID tag replace EAN-code in fast-moving consumer goods (FMCG)? Will it be in the next ten years or will there be only a limited use of RDIF tags? The present guess is that RFID will come first to returnable systems, such as pallets, crates and so on and later to distribution packages and to more expensive consumer goods. The main incentive to pursuing the system will be that some big retail chain such as Wal-Mart, or a producer, demands it in order to get real-time visibility in their supply chains. Actually Wal-Mart and Procter & Gamble are carrying out preliminary tests with transponders on pallets. Wal-Mart has fitted out two distribution centres as well as one Sam's Club and one Wal-Mart store with the technology in the supply chain. The software provider is SAP (LebensmittelZeitung.com, 2002).[16]

Actually, simple RFID tags are already are in use, library systems, bus and ski tickets, garment tracking, product handshaking (embedded systems). In warehouses and distribution centres some applications are in development. To read a pallet load manually with bar codes can take up to 30 minutes, but smart labels can be read up to 1000 + tags per second automatically. The ultimate goal is real-time 'people-free' visibility throughout the supply chain. Trends and factors that affect the success of RFID tags, are the need for end-to-end visibility, the kind of investment that will be necessary, how open the infrastructure will be and the price level of systems.

RFID technology may need some explanation. It generally uses passive tags that contain a chip and an antenna. When a host system sends power to the tag, it responds to the reader giving the information inside the chip. The tag is passive so that it needs no battery with it. There are also active tags, which have a battery. Active tags can be used to record information during the shipment or can be used to give information from a longer distance. Even the passive tags (transponders) and integrated host systems and readers make wireless data carrier technology available to manufacturer, supplier, retailer, and consumer. Transponders can be added to labels, packages, and products.

RFID and standardisation

There are several RFID frequencies in use: 125–134KHz, 13.56MHz, UHF 862–928MHz, 2.45GHz, and 5.8GHz. In order to get a system that can work globally, there is a need for RFID standards. The following standards are available:

- ISO/IEC 15693-1:2000 Identification cards – Contactless integrated circuit(s) cards – Vicinity cards – Part 1: Physical characteristics
- ISO/IEC 15693-2:2000 Identification cards – Contactless integrated circuit(s) cards – Vicinity cards – Part 2: Air interface and initialisation (available in English only)
- ISO/IEC 15693-2:2000/Cor 1:2001 (available in English only)
- ISO/IEC 15693-3:2001 Identification cards – Contactless integrated circuit(s) cards – Vicinity cards – Part 3: Anticollision and transmission protocol (available in English only)

AIM (Association for Automatic Identification and Data Capture Technologies) has listed in its web pages all different standards that may affect the usage of RFID (www.aimglobal.org, 2001):[17]

JTC 1/SC 31 Automatic identification and data capture techniques
JTC 1/SC 17 Identification cards and related devices
ISO *TC 104 / SC 4* Identification and communication
ISO TC 23 / SC 19 Agricultural electronics
CEN/TC 278 Road transport and traffic telematics
CEN/TC 23/SC 3/WG 3 Transportable gas cylinders – operational requirements – identification of cylinders and contents
ISO/TC204 Transport information and control systems
ETSI European Telecommunications Standards Institute
ERO European Radiocommunications Office

ANSI American National Standards Institute
Universal Postal Union
ASTM

RFID now

There was a major decision made in June 2002. EAN · UCC (EAN International and Uniform Code Council) has taken some steps to launch GTAG (EAN 2002).[18] The technical foundation for GTAG-compliant products is aligned with standards being developed under the auspices of the international Organization for Standardization (ISO), and specifically the work of JTC1/SC31/WG4, and SG3. One of the deliverables from this Working Group and Sub-Group is a standard for the air interface (RF communications link) between reader and tags, which, through technical due diligence, has been chosen by the GTAG Project Team as the interface for GTAG-compliant systems. This specification is ISO 18000-6. The present state of this specification is a Committee Draft (CD).

Auto-ID Center is a strong global promoter of wireless identification technologies. It is founded to develop network solutions and software for RFID technology and create an infrastructure throughout the entire retail supply chain. Its message is to get the next wave of the Internet revolution going; 'all sorts of objects communicating with each other without people being involved' (www.autoidcenter.com).

Finnish Rafsec supplies the intelligent tags, which support the Auto-ID Center's specifications for the low-cost tags (mass production from roll to roll). They cost now about 50 cents each. Rafsec produces the RFID transponders, the antenna and the fixing of the chip to the tag and antenna. The RFID chip generally contains memory from 512 bits to 2kb. The transponders are thin and flexible in form, they have read and write capabilities, and each chip is unique! So every product is unique after it is tagged! The information is coded in, the transponder is robust and it survives the whole life cycle and operates at temperatures from −25°C to +70°C. It can for a short time stand even a temperature of 150°C, which means that it can survive in a car tyre or in an injection moulding. When using 13.56MHz, the tag can be read from up to a one metre range at a relatively high data rate. A number of pilot projects are under way, but the threshold to use the systems is high (Savolainen, 2002a and b).[19,20]

The possibility of reading simultaneously through material at a distance provides interesting features. One can get exact information of a pallet load even though some products are removed. Of course there are also weaknesses in RFID technology; metal affects the signal so it will not read products in a metal cage. Also liquid materials can affect readability (Cole, 2002).[21]

EAN · UCC has been active with RFID. The use of a RFID tag in the supply chain offers a solution to multiple labelling during logistics. When a product is produced the only known information is the product identification, the batch number, and the production date. The logistic, transport and customer information will be known later on. The unique RFID tag would not replace the entire printed label, but it will be the unique support on a logistic unit for capturing or writing computerised data. The new support could provide significant improvement in the supply chain, such as

- reduction in the number of labels printed by each party
- reduction in the number of barcode printers and label applicators in plants, warehouses and plate-form
- simplification of the reading and writing process.

The EAN · UCC report specially mention the possibility of tagging the support pallet. This makes the tag reusable, simplifies the replenishment of tags and allows the costs to be divided by the number of times it is reused. It also provides the opportunity to use this tag for the tracking of the pallet support itself. According to the general EAN · UCC specifications, the only mandatory data that is required on a logistics unit is the serial shipping container code (SSCC). Additional common information is the batch number, best before date, net weight, the consignment number and ship to or deliver to postal code. In addition, one of the advantages of the RFID tag is that it enables the repetition of a particular piece of information, so that on a mixed pallet all required information of all components can be found (EAN · UCC, 2002).

EU and RFID research
The EU has financed ten projects relating to RFID technology (www.cordis.lu, 2002):[22]

1. Laundry application using RFID tags for enhanced logistics.
2. Trial of intelligent tag on industrial environment.
3. Passive long distance multiple access high radio frequency identification system.
4. Grocery store commerce electronic resource.
5. ParcelCall – an open architecture for intelligent tracing solutions in transport and logistics.
6. Development of an integration method of low-cost RFID tags in injection moulded and blow moulded packaging systems.
7. Inductively powered universal telemetry systems.
8. Intuitive physical interfaces to the www.
9. Hardtag – development of an industrial RF tagging technology to enable automated manufacture of 'one-off' and small batch products.
10. Radio frequency identification system I.CODE.

25.5 Future trends

25.5.1 The next wave of the Internet

The next wave of the Internet revolution provides that all sorts of objects communicate with each other without people being involved. The products come to the warehouse along a line, all codes of pallet loads are read to the computer and the automatic transport vehicles shift the products to the right shelves, from where they are then shipped further, on customer demand, without anyone touching them. With RFID, Bluetooth and Internet technologies it is already possible to detect when an expensive delivery is coming to a building site. Systems, of course, shall be first applied to more expensive products and to products that must be delivered to a site at an exact time. As computers have changed in 30 years, so these technologies will become cheaper and more common in the near future.

The RFID technology is already in use in some 'handshaking' cases, the main usage areas now being in assembly lines. There are also some printers, which accept only the right colour carriage, as there is a microchip and antenna in the carriage and a reader in the printer to detect this. There are several businesses and industries that can benefit from the use of RFID tags; package delivery, consumer product manufacturers third-party logistics, retail, airlines, legal, medical, insurance, electronics, automotive, government. There are several places where tags can be read; in yards, on trucks, at sea, rooms, zones, dock doors, shelves, doors, conveyors, en route. The list of tagged assets is even longer; items, boxes, totes, pallets, trays, tubs, sacks, cartons, books, documents, files, handling equipment, containers, rail cars, trucks, forklifts, trailers, capital equipment (d'Hont, 2001).[23]

How can the food industry use tags and their capabilities effectively for its own special needs? It must actively take part in research in this field or the systems and standards will be developed by others. The present development of RFID tags solves the problems of traceability, and information banks can be of great help. Maybe in the future the consumers also will get the required information easily, entertainingly and accurately. What intelligent and active features can be added to the new technology? Can techniques such as TTI or product freshness indicators be added to this technology? Can the RFID tag tell that a product is getting close to its best before date? Can it record the time and temperature? Can it be used also in-house control systems?

25.5.2 Traceability

There are more and more demands for traceablity. The BSC and other incidents, such as food poisonings, have been the main cause of plans to increase traceability. We should be able to trace and recall if necessary not only the product but also all its ingredients and packaging materials. The idea is that product recall could be made precisely within an in-house control system in the whole food chain. There would be a pallet with a barcode lot number. The pallet contains products whose batch identity is known, and all of the ingredients, premixes and packaging materials can be traced by their own ID numbers. Of course we would then also know where each pallet is delivered and where the products are put on sale.

25.5.3 Information banks

EAN is collecting large information banks, from where typical information on fast moving consumer goods (FMCG) can be retrieved for the wholesalers, retailers and in the future maybe also for the customers. Currently the EAN-number provides information on the product and its package. If the RFID tags in the future become common as EAN-code is today, all parties will get more information on the product and its properties in an easy way. The communicating information banks would be at our doorsteps.

25.5.4 Consumer behaviour

A number of very different segments of consumers already exist. Here are only a few examples served by the food sector:

Not willing to spend time in mega stores

Some of us are not keen on everyday shopping. Shall we go to the shops or order only via Internet and expect the products to be delivered to our home? There are already automats available that heat up the chosen frozen food according to the instructions on the package and tell the customer when the food is ready. Automats are replacing daily meal-to-home delivery systems with fortnightly refills.

My diet is this, find suitable new products

There are more and more people who must follow a certain diet. If your mobile phone communicates with the RFID tags on the products in the store, it could in future tell you that this or that product is good for your diet. 'You need more vegetables in your diet, try those. Get the new instructions at the store computer.'

I am bored, give me pleasure, enjoyment ...

Enjoyment, excitement and other psychological aspects are becoming more and more important. Active systems in packages may give the consumer a new interest in buying a certain product.

25.6 Sources of further information and advice

In addition to general packaging books and magazines, the following Internet pages are worth visiting as certain aspects such as the RFID are in rapid developmental stages.

www.aimglobal.org
www.autoidcenter.com
www.cordis.lu
www.ean-ucc.org
www.pakkausteknologia-ptr.fi
www.rafsec.com

25.7 References

1. ISTA, The Association for Transport Packaging (2002), 'INSTA and package performance testing: how they contribute to the quality of life' *Worldpak 2002, Improving the quality of life through packaging innovation, Proceedings of the 13th IAPRI conference on packaging*, June 23–28, 2002 East Lansing, Michigan, USA, CRC press, ISBN 1-158716-154-0 p. 179.
2. LEPPÄAHO K (2002) Personal information in Finnish, Kesko 30.10.2002.
3. CHONHENCHOP V and SILAYOI P (2002) 'Comparison of Various Packages for Mango Distribution in Thailand', *Worldpak 2002, Improving the quality of life through packaging innovation, Proceedings of the 13th IAPRI conference on packaging*, June 23–28, 2002 East Lansing, Michigan, USA, CRC press, ISBN 1-158716-154-0 pp. 3–11.
4. KIPP W (2002) 'The technology of transport packaging evaluation', *Worldpak 2002, Improving the quality of life through packaging innovation, Proceedings of the 13th IAPRI conference on packaging*, June 23-28, 2002 East Lansing, Michigan, USA, CRC press, ISBN1-158716-154-0 pp. 180–9.
5. SINGH S P (2000) 'Packaging Requirements for Shipping Fresh Produce Using E-commerce', Marcondes J, *Advance in Packaging Development and Research, Proceedings from the 20th International Association of Packaging Research Institutes Symposium,* San Jose State University pp. 500—6.
6. WELBY E and MCGREGOS B (1997) *Agricultural Export Transportation Handbook*, USDA.
7. SINGH S P and SINGH J (2002a) 'E-commerce distribution of fresh produce and flowers', *Worldpak 2002, Improving the quality of life through packaging innovation, Proceedings of the 13th IAPRI conference on packaging*, June 23–28, 2002 East Lansing, Michigan, USA, CRC press, ISBN1-158716-154-0 pp. 446–54.
8. SINGH, S P (2002b) Personal communication during the presentation 24.6.2002.
9. TWEDE D (1997) 'Logistical/distribution packaging' in Brody A and Marsh K, *The Wiley Encyclopedia of Packaging Technology,* second edition, John Wiley & Sons, Inc ISBN 0-471-06397-5 pp. 572–9.
10. PIKKARAINEN K and JÄRVI-KÄÄRIÄINEN T (2002) 'SIP -pakkaustutkimus käyntiin tarvekartoituksella', *Kehittyvä elintarvike* 2/2002 pp. 38–9.
11. LEPPÄNEN-TURKULA A (2002) 'Mitä kuluttajat odottavat pakkauksilta päivittäistavarakaupoissa, kauppiaiden näkemyksiä', Oral presentation at SIP vuosisemiaari Vantaa, Finland on 8.10.2002.
12. SOROKA W (1999) *Fundamentals of Packaging Technology,* Institute of Packaging Professionals, ISBN 1-939268-06-8, pp. 499.
13. CLARKE R H (2002) 'Radio Frequency identification: will it work in your supply chain?' *Worldpak 2002, Improving the quality of life through*

packaging innovation, Proceedings of the 13th IAPRI conference on packaging, June 23–28, 2002 East Lansing, Michigan, USA, CRC press, ISBN1-158716-154-0 pp. 655–62.

14. 'Chips with everything' (2002) *New Scientist*, 19.10.2002 pp. 44–7.
15. COVELL P (2002) 'Smart money on RFID', *Packaging Today International* October 2002, pp. 28–31.
16. LebensmittelZeitnung.com June 13.2002 (http://english.lz-net.de/news/webtechnews/pages/showmsg.prl?id=1217) (available via Internet on 1.11.2002).
17. http://www.aimglobal.org/standards/rfidstds/RFIDStandard.htm 2001 (available via Internet on 8.11.2002).
18. EAN·UCC (2002) *White Paper on Radio Frequency Identification*, June 2002, EAN International and Uniform Code Council, Inc. http/www.ean-ucc.org/gtag.htm
19. SAVOLAINEN S (2002a), RFID:llä älykäs ja turvallinen pakkauskonsepti, *Älykkäät ja aktiiviset materiaalit pakkauksissa 21–22.5.2002* Suomen Kemian Seura.
20. SAVOLAINEN S (2002b), Personal communications on 5.11.2002.
21. COLE P H (2002) *White paper: A study of factors affecting the design of EPC antennas & readers for supermarket shelves*. Auto-id centre. June 1, 2002.
22. www.cordis.lu. search about RFID on 5.11.2002.
23. D'HONT, SUSY, Smart labels – hype or hope? *The RFID Summit,* Nov 12, 2001 available on the web at www.matrics.com.

26

Testing consumer responses to new packaging concepts

L. Lähteenmäki and A. Arvola, VTT Biotechnology, Finland

26.1 Introduction: new packaging techniques and the consumer

New packaging techniques promise consumers safe food products that keep their high quality throughout shelf-life. The improved quality has been achieved by applying tailored technological solutions that require highly specialised knowledge. From consumers' point of view these new techniques require explanations if food can keep fresh for an unexpected and thereby unnaturally long time. Consumers in general tend to be suspicious towards novelty in food products as any new element can be potentially harmful (Rozin and Royzman, 2001). Furthermore, applying technology to achieve benefits can add to distrust as technology by itself can have negative connotations. Understanding how the benefits have been achieved requires advanced consumer education on the principles of food spoilage.

The basic functions of the package have been described as containing the foodstuff, protecting and maintaining its quality, providing information for the consumer, convenience in use, being environmentally friendly, and selling the product (Hurme *et al.*, 2002). For consumers, the favourable packaging attributes include convenience in opening, resealing, storing and disposing (Eastlack *et al.*, 1993; Mikkola *et al.*, 1997). These positive attributes are almost all related to the practical properties of packages and how easy they are to use, but include no safety issues. Similarly, most negative attributes referred to lack of convenience, the only safety related attributes listed were 'product spoils easily' and 'can spill or leak' (Eastlack *et al.*, 1993).

Most active and intelligent packaging methods aim at improving the quality and safety of food products. The improvement of safety by producing longer safe shelf-life may be a hard concept to sell to consumers. Safety is likely to be for consumers a self-evident feature and therefore regarded as a basic requirement in packed food products. Therefore, consumers do not assess the package based on its safety merits, rather they assess the convenience of using the pack when taking the presumably safe foodstuff from the package. This implies that consumers need to be educated about the possible benefits that active and intelligent packaging can provide them and treating the different types of packaging solutions as integral parts of the product rather than the foodstuff and packaging as separate issues. Although active and intelligent packaging methods have been studied widely and innovations have been developed very few of them have been developed into commercially available products (Hurme *et al.*, 2002). One reason for the slow progress may have been the anticipated consumer concerns of these new applications. Surprisingly, however, very few consumer studies have been published on this topic.

This chapter describes how different approaches can be used to study consumers' attitudes towards active and intelligent packaging technology. The first section calls attention to the special problems that are encountered when novel technologies are studied. Then the principles of most frequently used qualitative and quantitative methods are introduced and their strengths and weaknesses are discussed. A short overview of our current knowledge on consumer attitudes towards active and intelligent packaging will follow the methodological section. The few studies carried out have mostly dealt with consumers' attitudes towards oxygen absorbers and time-temperature indicators. The last section in this chapter will discuss the future prospects of active and intelligent packaging from a consumer standpoint; what the issues are that need to be taken into account and how to approach possible consumer concerns.

26.2 Special problems in testing responses to new packaging

The novelty aspect and the fact that food products are regarded as entities including both package and foodstuff create challenges for studying consumer responses to new packaging technologies. When asked about familiar issues consumers tend to have either positive or negative attitudes that are activated by asking questions related to them. This process depends on the importance and topicality of the subject. Information on important or on relevant matters are given more attention and the belief structures tend to be more complex for relevant than for non-relevant issues. Recent exposure to the topic, on the other hand, makes the beliefs more accessible. When required to give answers about new food products or technologies these responses can be very arbitrary. People give responses although they are not sure what the question actually involves since this is the socially most appropriate and easiest way of handling questions. The issues that come out are highly dependent on the associations these new technologies create in consumers' minds and what other matters are relevant for the consumer at the time.

In order to gain meaningful responses, consumers need to be made more familiar with intelligent packaging. This can be done by explaining what a concept, whether active packaging or special indicator, means or by showing concrete examples of these active or intelligent package solutions. A simple way to explain to interviewees what the applications are and how they function is a set of photographs that are easy to take to different places. Furthermore, they are the same for all interviewees regardless of the time and location. If real food packages or indicators are used, they have to be replaced at each demonstration. This will raise the expenses of the study, not to mention the amount of products that need to be carried to different locations and stored at accurate temperatures. Modern technology makes it possible to carry out research by using the internet or computer-aided data collection systems. With these applications it is possible to demonstrate how the indicators work with no need to use actual food packages as samples.

The most feasible way of demonstrating these package solutions is to show food products with and without the indicators, absorbers or emitters. The responses are then related both to the example food and the packaging technology. This raises the question whether packaging technologies can be studied separately from their applications in consumer studies, as they provide improvements for the quality of food, not improvements for packaging. For consumer acceptance the perceived benefits are important. Consumers will assess the benefits they gain, but they also have concerns about how these benefits have been achieved. Furthermore, any technology that solely provides advantages for the other actors in the food chain are not easily accepted by consumers especially if they raise prices.

26.3 Methods for testing consumer responses

The central objective in consumer research is to find out whether consumers are willing to accept new packaging technologies, whether there are concerns that may obstruct or delay acceptance and how the benefits provided by the new technologies are perceived. The methods used can be broadly divided into two categories; qualitative and quantitative approaches. With qualitative methods we can get systematic information about how consumers think and formulate their opinions about food and packaging related issues. These techniques are valuable when we want to gather information about the different possible concerns consumers attach to novel technologies or we want to define what the reasoning is behind these concerns. The advantage of qualitative techniques is that consumers can use their own language and expressions to describe their opinions. Often qualitative techniques are used as pilot studies for quantitative approaches, but they are gaining value as independent tools. The most frequently used qualitative methods are focus group discussion and individual interviews. Both these types of methods can be applied with different techniques depending on the question on the hand.

Qualitative methods describe how consumers think about certain issues but they do not give the frequency of these ideas or how important the ideas are to different people. Quantitative methods are used when we want to find out how many people have a certain opinion or estimate the strength of an opinion. The quantitative surveys finding out people's opinions can be carried out as interviews or questionnaires or a combination of these. Experimental designs are a special type of quantitative study in which respondents are given different treatments, e.g., samples to try, and their responses are measured and compared in different experimental groups or with a control group. Below are short descriptions of typical features of most typically used methods and implications of their use in studying novel packaging materials. Detailed descriptions of the methods can be found in textbooks.

26.3.1 Focus group discussions

Focus group discussions provide information on how consumers talk about particular issues (Casey and Krueger, 1994). Moderating focus groups require careful preparation and the questions need to be outlined beforehand. The moderator needs to be well-trained for the task and possess appropriate social skills on diplomacy and bringing all participants into discussion as equal members of the group. The basic principle is that the moderator does not lead the discussion in any specific direction, as long as the conversation remains topical. The participants in the discussion group respond with comments and opinions from each other and thus the discussion deals with aspects coming from several individuals. This social interaction enables the pondering of the importance of matters that have been raised during the discussion. Analysing the focus group data is a relatively difficult task because the material produced during interactive discussions tends to be vast and branch in various directions. Due to this heterogeneity of material Casey and Krueger (1994) recommend that at least three groups with the same questions and similar participants should be run to cover the variation.

Where packaging issues are concerned focus group discussion works well with consumers because new technologies can be demonstrated as part of the group session and there is no pressure to be an expert on the topic. Experts working for retailers, food industry, authority or consumer associations may find group discussion less relaxing than consumers, since these individuals should be knowledgeable about the novel packaging developments. This may cause tension in a group discussion. If the aim in discussion is a free exchange of ideas and views about the future, tension may exclude some participants from the discussion or ideas and opinions are carefully controlled. Therefore respondents with vested interests in the topic are easier to handle in a one-to-one interview situation.

26.3.2 Qualitative and quantitative interviews

Interviews allow direct interaction between respondent and interviewer. Individual interviews can be carried out using several techniques. Some techniques follow very structured procedures with a predefined order and form of questions; others allow an interviewee's responses to delineate how to continue as long as the relevant topics are discussed. The type of interview is typically selected on the basis of research questions. Packaging issues are rarely sensitive issues and are therefore easy to talk about. Often in this type of study either semi-structured or structured interviews have been used.

Qualitative interviews are used when we want know how respondents think about packaging and we do not have enough previous knowledge about what the possible responses can be. The approach is suitable for examining more complex issues as participants are not restricted in predefined response alternatives. Data analysis with a qualitative approach tends to be time consuming and the researcher has to be very skilful in analysing transcripts of focus group discussions.

If we want to quantify responses the interviews are typically carried out with structured outlines and sometime the possible response alternatives are preselected. The advantage of carrying out an interview survey rather than a questionnaire is that interviewees can ask for explanations if they do not understand questions and also interviewers can ask for elaboration if the responses contain ambiguous expressions. With novel packaging solutions, using interviews enables a demonstration of what these absorbers and indicators are like when they are attached to the food package.

26.3.3 Questionnaires

Questionnaires offer a relatively inexpensive method to study what people think about an issue on average. A questionnaire approach can be selected if we know well enough what the possible response alternatives are that consumers are likely to give or we have an explicit predefined question. With appropriate sampling techniques the respondents can be selected to fulfil certain predefined criteria. Typically respondents are selected based on their socio-demographic background (sex, age, education, profession) or based on their consumption or buying habits. Often food-related studies are targeted on those who typically use the product or questions are asked of those who have the main responsibility for food choice in their own household. Due to the latter criterion, the majority of the food or packaging related studies have had mostly female respondents (Anon., 1991; Korhonen *et al.*, 1999; Mikkola *et al.*, 1997).

The limitation of questionnaires in packaging related research is that items in a questionnaire should refer to familiar things. If consumers are asked opinions about themes they are not familiar with, the reliability and validity of these responses may not be very good. There are several textbooks describing how to construct a questionnaire and ask factual and attitudinal questions, but the basic rule is that the questions should be easily comprehensible and provide alternatives that consumers can relate to.

26.3.4 Experimental designs

Experimental designs are useful when novel applications in the food domain are studied as they provide a chance to familiarise the respondents with the new technologies and thus reduce fears that rise from uncertainty. The designs also enable controlled comparisons of consumer responses to different types of packaging solutions. Consumers can experience concretely how indicators or absorbers look and function and what their advantages are. In most experimental set-ups there is a need for a control product, which is often the same product packed without the indicator or other active component. This enables a direct comparison of how acceptable the new applications are in relation to the existing packaging methods. As most consumer responses tend to be relative, the experimental design can produce more reliable information in this sense, although the drawback is that instructions tend to make the assessments rather artificial.

26.4 Consumer attitudes towards active and intelligent packaging

26.4.1 General attitudes

The idea of active and intelligent packaging has received a generally positive response from consumers and their representatives. The reason may be that they seem to provide solutions to consumer concerns. According to Korhonen *et al.* (1999) about half of Finnish respondents ($n = 460$) did not trust that all food products would still be edible on their expiry date. Half of the consumers also reported that they would choose packages from the bottom of a chilled counter to ensure the freshness of the product. The active and intelligent packaging methods were more familiar to those who were involved with packaging issues. A small number of individuals ($n = 21$) who are responsible for delivering information to consumers about the packaging issues in Finland were interviewed in 1995 (Mikkola *et al.*, 1997). The group consisted of retailers, journalists and government officials. When asked whether they were familiar with modified atmosphere and vacuum packs, four out of five interviewees could recognise both of these. Furthermore, about half of the respondents could recognise moisture absorbers (57%), oxygen absorbers (52%) and time-temperature indicators (42%). The interviewees had a positive attitude towards these examples of active and intelligent packaging, especially if applied to foods that are easily perishable, such as chilled foods, vegetables, some bakery products, meat or fish products.

People have different requirements for food packaging. In a study carried out in the UK (Anon., 1991) consumers could be divided into three groups according to their attitudes towards the safety of chilled foods. The 'ultra-cautious' are likely to throw away all foods that have passed the use by date, the 'cautious' use their own judgement and believe in some safety margins around the given dates, whereas the 'non-cautious' care very little about dates. A considerable number of consumers fell into the 'ultra-cautious' category whereas the 'non-cautious' were in a minority. The study itself targeted consumer acceptance of time-temperature indicators and reflected the acceptance of these new devices to the needs these three respondent groups had.

26.4.2 Acceptance of oxygen absorbers

When asked about the possible benefits of absorbers or emitters, the interviewees ($n = 21$) mentioned that food products retain their good quality longer, which may be especially helpful for small households and those who shop once a week (Mikkola *et al.*, 1997). The absorbers were believed to improve safety by reducing microbial risks and thereby contributing to a decrease in the use of additives in food products. On the negative side, the added components can increase price and produce more waste. People also may eat older food if it keeps a longer time in good condition. Furthermore, the possibility that these absorbers or emitters could contain harmful substances that may be ingested by vulnerable consumer groups, such as older people and children, caused concern.

Acceptance of oxygen absorbers among Finnish consumers was examined with an experimental design. Mikkola *et al.* (1997) carried out a study where consumers ($n = 346$) were given two types of food products to take home. Sliced rye bread and pizza filled with ham were packed with or without oxygen absorbers. The products were stored at the research institute so that their delivery date was close to the best by date. A trained laboratory panel assessed the samples and gave higher quality points on appearance, flavour and freshness for pizza when it was packed with the absorber than when it contained no absorber, but there was no difference in the assessed quality of sliced rye bread. Consumers, however, assessed both products with oxygen absorbers as having higher quality, although the difference between oxygen absorber product and conventional product was small for rye bread. In the trained panel evaluation the samples were blind coded and the panel did not know what the samples were when they tasted them. Consumers, on the other hand received the samples clearly labelled and based their assessment on both sensory quality and on information they received. In addition to overall quality, respondents were asked to evaluate whether they were willing to accept the absorbers and buy these products if they were available on the market.

The oxygen absorber used in the study was a loose sachet enclosed in the package and half of the respondents also received an information leaflet that described what the oxygen absorber was, how it functioned and how it could be disposed of (Mikkola *et al.*, 1997). After the demonstration with real food products 72% on average were ready to accept these additional sachets, 23% were unsure and 5% were clearly negative. From those who received the additional leaflet 76% accepted the oxygen absorber vs. 67% in the no-information group. Information decreased the number of unsure people among the respondents but had no effect on the size of the negative group.

Respondents' attitudes towards oxygen absorbers were positive (3.8/max 5), respondents would rather favour than avoid them (3.6/max 5) and evaluated them more necessary than unnecessary (3.4/max 5) (Mikkola *et al.*, 1997). Those who were most positive about oxygen absorbers were also positive about pre-packed food, use of additives and long shelf-life. When asked about the acceptance of oxygen absorbers in different types of meat products, use in pizza (62%), meatballs (48%), sausages (37%) were accepted best, while in fresh meat only 29% would accept them. The high acceptance rate in pizza illustrates the usefulness of the demonstration material in the study. Consumers could experience with their own senses what the benefits in pizza were and thus the acceptance rate is high. The low acceptance rate in fresh meat indicates that an idea of prolonged shelf-life is not considered as acceptable in fresh products. In bakery products the highest acceptance rate was again in the product used in the demonstration, namely rye bread (57%), but all other examples were also accepted by half of the respondents (50–55%). Furthermore, when asked about willingness to pay more if the products contained an oxygen absorber, 40% of the respondents were willing to pay 0.15€ more.

26.4.3 Acceptance of time-temperature indicators

The concept of time-temperature indicators (TTIs) has been well received in consumer studies (Anon., 1991; Korhonen *et al.*, 1999; Sherlock and Labuza, 1992). In a UK study (Anon., 1991) the majority of respondents (95%; $n = 511$) considered TTIs as being a good idea because they show whether food is safe (28%), whether it is kept at the right temperature (21%) and whether food is fresh (16%). In an American questionnaire study ($n = 104$) 90% considered TTI tags as a desirable addition and 97% believed that they would increase confidence in the freshness of the product (Sherlock and Labuza, 1992). The study was carried out to find out how consumers react to the use of TTIs in refrigerated dairy products. In a small interview study ($n = 21$) carried out in Finland by Mikkola *et al.* (1997) a time-temperature indicator (TTI) created less uncertainty than an oxygen absorber. Increasing safety was perceived to be an obvious benefit because consumers do not have to trust merely their own senses. The suspicion that these indicators may give inaccurate information and thereby cause a safety hazard was mentioned as a drawback together with adding price and waste (Anon., 1991; Mikkola *et al.*, 1997). In three focus groups (Sherlock and Labuza, 1992) run in Nebraska, the TTI tags were considered to be clever devices that could be used to differentiate products on the market, but they were not perceived to replace date markers. Furthermore, the discussion brought to light a need for a consumer campaign before these tags could be used as a marketing tool, since consumers need to be informed about their benefits.

The TTIs were perceived to be most suitable for frozen food and freshly prepared refrigerated entrées, but not dairy products (Sherlock and Labuza, 1992). In a Finnish study (Korhonen *et al.*, 1999) TTIs were regarded as necessary to most products but the most necessary targets were packaged fresh meat or fish, smoked fish, meat products, foods for children or ready prepared foods. Over 80% regarded TTIs as necessary in these applications although they were told that TTIs would increase the price of the product by 8.5 cents. This study was carried out as a survey in which participants ($n = 460$) were asked to fill in a questionnaire. While responding to the questionnaire consumers could observe models of TTIs used in packages. Similarly, 59% of the respondents in the UK expressed their willingness to pay more for chilled products that contained a TTI tag (Anon., 1991).

The result that TTIs are more suitable for fresh meat (Korhonen *et al.*, 1999), whereas oxygen absorbers were considered acceptable in fresh meat by only by a minority of respondents (Mikkola *et al.*, 1997) elevates the importance of perceived consumer benefit and understanding the reasons for food choices. The apparently contradictory result may be easily explained by the different functional principles of these two packaging devices, which may have a different appeal to consumers. The oxygen absorber could prolong the shelf-life of fresh meat, whereas the time temperature indicator shows only how it has been operated through the chill chain. The idea of extending the shelf-life of fresh meat is not attractive, but it is important to know if the fresh product is still in prime condition. This highlights the fact that all these different applications have to be studied as separate concepts in consumer studies. Measuring an overall attitude towards active and intelligent packaging is not feasible, as the benefits and possible concerns are specific to each application.

Some worries about possible tampering with TTIs in the shop were brought forward (Anon., 1991; Korhonen *et al.*, 1999). One worry was that the shopkeeper could possibly change the indicator and thus mislead consumers. In the UK (Anon., 1991) the non-cautious respondents perceived the TTIs to the unnecessary and some reported that they would deliberately sabotage them if they appeared on the market. The technical reliability of the indicators was also questioned; other markings should be clear so that consumers would not have to trust solely the indicator.

In general, people seemed to trust the TTI indicators. When respondents had to make assessments on the quality of a food product they seemed to place more trust on the TTI tag than on the date mark (Anon., 1991; Sherlock and Labuza, 1992). A vast majority in a study carried out in the UK (Anon., 1991) said that they would not buy a product even though the product was not past the best before mark, if the indicator had changed. If the situation was the other way around and after the best before date but the indicator showed that the product was good, about half of the respondents thought it was safe to eat. Over half of the respondents would use their own judgement to decide whether the food was edible, a third would adjust the temperature in the fridge and one in five would throw the food away. In an American study (Sherlock and Labuza, 1992) 80% would not purchase a product if the date stamp indicated freshness but the TTI tag had changed. If the situation was the other way around 49% said that they would not be likely to buy the product. Although respondents seemed to trust the indicators more, having both date marks and indicators were perceived to be the best solution in these studies. In the UK 88% thought both should be on the package and only about 11% would have been happy with either date mark or TTI (Anon., 1991). In the USA 75% thought that both should be attached to the package, but acceptance for the date mark (23%) and TTI tag (24%) were equal (Sherlock and Labuza, 1992). This may be due partly to the way the question was asked. In the UK the study respondents had to make choices between the alternatives, whereas in the American study the questions were asked on separate rating scales. Therefore the same people could support the self-sufficiency both of date stamps and TTIs. Having TTIs in the package increased respondents' willingness to buy the product by 72% (Sherlock and Labuza, 1992). The date marks and TTIs were regarded as tools that can complement each other and thus give a better guarantee of product quality (Anon., 1991; Sherlock and Labuza, 1992).

In the study carried out in the UK (Anon., 1991) the time-temperature indicators were also regarded as tools to educate consumers on how to keep food at home. If the product is in prime condition when bought and then the indicator changes rapidly at home, this may tell the consumer that the product has been stored in too warm an environment. The indicator would clearly demonstrate to consumers the need for appropriate practices in handling foods that should be kept refrigerated.

26.5 Consumers and the future of active and intelligent packaging

Active and intelligent packaging technology offers several benefits to consumers. The different absorbers and indicators can be used for various purposes. The basic purpose is to guarantee that the food products are safe and keep their quality better. The performance of distinct applications of active and intelligent packaging is based on several mechanisms: some measure time and temperature sum, others absorb certain compounds that promote spoilage, and others, excrete beneficial compounds (Hurme *et al.*, 2002). The technological possibilities are well ahead of commercial applications, which may be due to suspicion about consumer attitudes towards these new devices. Consumers tend to be sensitive about novelty in the food domain, as food ingested and incorporated in the body could be an unknown substance and a potential source of risk.

As with all innovations, innovators themselves and early adapters are the first to adopt them, then acceptance spreads to the majority of the population. According to Eastlack *et al.* (1993) adoption happens relatively rapidly for new packaging solutions. This may be due to the high exposure consumers have to new packaging solutions during their weekly visits to supermarkets or grocery stores and the low risk of these products. Nevertheless, to gain success in the market the new packaging solutions need to provide consumers with benefits or solutions to their current problems.

The challenge for new packaging solutions is how they and their benefits are made familiar to the consumers. Experts and consumers in the few studies that have been carried out have emphasised the need for information (Anon., 1991; Mikkola *et al.*, 1997; Sherlock and Labuza, 1992). The tools mentioned were both product-related information in the stores and packages and wider campaigns in the media, which is the main source of information for many (Anon., 1991). A public campaign can explain what the indicators and absorbers are, what they are used for and what their limitations are. Providing this information, such as a description of the operating principles, is a basic requirement but it may often not be sufficient to gain public acceptance. In written texts the information tends to be on an abstract level and it does not remove the unfamiliarity of the new applications effectively. Making it possible to observe what the absorbers and indicators look like, and how they work and change in different conditions, makes these devices realistic options for consumers. As the benefits tend to be on the product rather than the package, consumers need demonstrations with those products that are the target applications of active and intelligent packaging.

As both information and demonstration are required, the promotion of new packaging devices needs to done carefully. Although information as such is a weak motivator for choices (Mikkola *et al.*, 1997), consumers need to know how the different indicators work, what they tell about the product and also what they do not tell. The open information policy enables consumers to make their own decisions whether to buy the products with indicators and assess how trustworthy they are in different situations. The familiarising process was described in focus group discussions carried out in the UK (Anon., 1991). Participants did not know very much about the TTIs before the principles behind the indicators were explained. The attitude towards indicators turned from scepticism to something more positive during the group discussion when different possible benefits and disadvantages were debated.

The few example studies on oxygen absorbers and time-temperature indicators show that the improved freshness and safety of products are regarded as real benefits by consumers and the responses to these new packaging tools have been positive in general. Monitoring the freshness of the product is an obvious and definite advantage for consumers as it provides better tasting products for consumers. The improved safety may be a more complex benefit for consumers, as it is avoidance of a negative effect. Safety in food products is an attribute that is assumed to be in order if food is sold in the store. Everyone agrees that safety is a crucial quality factor, but when consumers are asked for the reasons behind their food choices safety is not typically mentioned (Lappalainen *et al.*, 1998). Also, emphasising improved safety raises a question in consumers' minds about whether the food products have not been safe before. As was expressed by consumers in the few studies carried out on active and intelligent packaging, these new techniques may be more beneficial for the food industry and retailers than consumers, but consumers still have to pay the price (Anon., 1991). The worries included the fact that the shelf-life of products will extend and thus consumers will receive food less fresh than formerly (Mikkola *et al.*, 1997).

In addition to oxygen absorbers and TTIs, a wide range of absorbers, emitters and indicators have been developed or are under development (Hurme *et al.*, 2002). Some of them offer benefits for all actors in the food chain, others to only some. Leak indicators are developed to detect if modified atmosphere packages leak and thus the quality and safety of the product is in jeopardy. If damaged packages can be removed from the shelf before the consumer buys them this will guarantee better quality for the consumer and improve safety. The drawbacks are additional cost and waste. The crucial question is how these indicators will affect the price and who is going to pay. If the price rises, the consumer will be the final payer but if the indicators are financed through decreased spoilage and losses the consumer benefit is clear. Ethylene absorbers can keep fruit and vegetables fresh for longer and reduce waste but some of the compounds used can be toxic if ingested. Flavour-scalping materials can modify the flavour of the product, maintain it fresh by absorbing unwanted compounds and by emitting desired compounds to the product. Some materials are used to mask bitter flavours in citrus fruit (Hurme *et al.*, 2002). The success of these packaging solutions will depend on how consumers perceive their benefits and whether they are willing to pay extra for these.

The existing studies illustrate that asking consumers their attitudes towards active and intelligent packaging in general bears little relevance to the acceptance of distinct packaging solutions, since most consumers have only a vague idea about what different terms mean. Nonetheless, when consumers are presented with different applications that belong to this category, they can accurately evaluate the possible benefits these applications can provide for them. Therefore the acceptance of active and intelligent packaging has to be studied separately for each application. The general attitude studies and focus group discussions give an idea of the factors that cause concern among consumers in packaging issues but the product related responses can reflect these worries to a varying extent and often differ from general concerns. The realistic examples of products presented to consumers may help them to evaluate their responses in relation to other motivations present in food choice situation. Clear demonstrations also provide information about how the indicators work and increase trust in them. When something is presented as an abstract idea the application may sound more technical, distant and also scary than when the real application can be observeded.

Further development in intelligent and active packaging will provide completely new benefits to consumers. So far the intelligent packaging concepts have dealt with the safety and quality aspects of foods. In the future, it is likely that intelligent and smart tags can contain abundant information about the product characteristics, the amount of information being now limited by the available space on the package. Each product can be labelled to provide targeted information about the origin and composition of the product. The information may include the nutrient content and possible allergens in the products. Also the environmental load of the product and packaging material can be included together with instructions on how to dispose of the package.

26.6 References

ANON. (1991), *Time-temperature indicators: Research into consumer attitudes and behaviour*, MAFF, Food Safety Directorate, UK.

CASEY MA and KRUEGER RA (1994), 'Focus group interviewing', in MacFie HJH and Thomson DMH , *Measurement of Food Preferences*, London, pp. 77–96.

EASTLACK JO, DI BENEDETTO CA and CHANDRAN R (1993), 'Consumer Goods packaging Innovation and Its Role in the Product Adoption Process', *J Food Prod Market*, 1, 117–33.

HURME E, SIPILÄINEN-MALM T and AHVENAINEN R (2002), 'Active and intelligent packaging', in Ohlsson T and Bengtsson N, *Minimal processing technologies in the Food Industry*, Woodhead, 87–123.

KORHONEN V, JÄRVI-KÄÄRIÄINEN IT and LEPPÄNEN-TURKULA A (1999), 'Consumers' attitudes towards active and intelligent packaging technologies' Conference presentation in *Challenges of the Packaging in the 21st Century*, IAPRI World Conference on Packaging, Singapore.

LAPPALAINEN R, KEARNEY J and GIBNEY M (1998), 'A Pan EU survey of consumer attitudes to food, nutrition and health: an overview', *Food Qual Pref*, 9, 467–478.

MIKKOLA V, LÄHTEENMÄKI L, HURME E, HEINIÖ RL, JÄRVI-KÄÄRIÄINEN T and AHVENAINEN R (1997), *Consumer attitudes towards oxygen absorbers in food packages*, VTT Research notes 1858, VTT.

ROZIN P and ROYZMAN EB (2001), 'Negativity bias, negativity dominance, and contagion', *Personality and Social Psychology Bulletin*, 5, 296–320.

SHERLOCK M and LABUZA TB (1992) 'Consumer perceptions of Consumer Type Time-Temperature Indicators for Use on Refrigerated Dairy Foods', *Dairy, Food & Env Sanit*, 12, 559–65.

27

MAP performance under dynamic temperature conditions

M.L.A.T.M. Hertog, Katholieke Universiteit Leuven, Belgium

27.1 Introduction

Modified atmosphere (MA) techniques for horticultural products are based on the principle that manipulating or controlling the composition of the surrounding atmosphere affects the metabolism of the packaged product. By creating favourable conditions, quality decay of the product can be inhibited. The different MA techniques come with different levels of control to realise and/or maintain the composition of the atmosphere around the product. Passive MA packaging (MAP), as an extreme, relies solely on the metabolic activity of the packaged product to modify and subsequently maintain the gas composition surrounding the product.

Temperature has a major effect on the rates of all processes involved in establishing the gas conditions in MAP (rates of gas exchange by the product and rates of diffusion through the packaging materials) and also on the rates of all metabolic processes that will inevitably lead to deterioration of the product and finally death. Ideally, steady state gas conditions should be obtained that, from the point of retaining quality, are optimal for the product packed. The time needed for a package to reach a steady state is extremely important as only from that moment in maximum benefit from MA being realised. Depending on conditions, the time to reach a steady state could theoretically outlast the shelf life of the packaged product. Given the ubiquitous role of temperature in MAP, success or failure of the ultimate MA package for a certain product largely depends on the level of integral temperature control from the moment of packing up to the moment of opening the package by the consumer. In logistic chains without integral temperature control, the application of MAP is often a waste of time, money and produce.

In spite of the important role of temperature in MAP, most MAP research trials are performed at constant temperatures, at temperatures often close to what is known as the optimum storage temperature for the product under study. No extensive literature data is available on monitoring MAP in terms of temperature, gas conditions and product quality throughout a logistic chain. Without such a complete set of data it is difficult, if not impossible, to know why a certain MAP design failed. This could, for instance, be due to a direct temperature effect on the product's metabolism, or due to an indirect effect through a failure to establish the intended steady state gas conditions (too high or too low), or an unfortunate combination of other factors like leakage or issues related to product quality (maturity, microbial load, etc.).

This chapter will focus on the effects of dynamic temperature conditions on the performance of MAP. First of all it will discuss how to define MAP performance; when MAP can be regarded as being successful and how this can be measured. Subsequently it will discuss what risks are involved in MAP and how these risks are affected by a lack of integral temperature control in a logistic chain. This chapter will conclude with a discussion of several simple strategies to maximise MAP performance, making the best of MAP given the limited resources available. The different aspects discussed in this chapter are illustrated using simulation results from a fully dynamic MA model[12] using realistic settings for both film and product characteristics.

27.2 MAP performance

The first question to answer when discussing MAP performance is how MAP performance should be defined. The aim of MAP is to inhibit retardation of product quality, the means employed to reach this aim is the application of certain optimal MA temperature and gas conditions. To grade the performance of MAP one can test whether the aim was reached (in terms of product quality) or whether the means were employed correctly (in terms of temperature or gas conditions). If life were simple these two measures would be interchangeable, as they would be strongly correlated to each other.

From a technical point of view, tracing and tracking gas conditions and temperature in the logistic chain is much easier than tracing and tracking those product properties responsible for the overall product quality. However, assessing the benefits and losses in terms of product quality gives much more insight than just the observation of MA conditions getting below or above their target levels. The question that should always be asked is how possible deviations in temperature or gas condition affect the quality and keeping quality. Product quality gives static information on the status of the product at a certain moment, for instance at the point of sale. Keeping quality provides dynamic information on how long a product can be stored, kept for sale, transported to distant markets or remain acceptable to the consumer.

A wide range of equipment is available to monitor temperature throughout a logistic chain. Given that most MA packages are relatively small consumer packs and given the potentially large spatial and temporal variation in temperature within cold stores and truckloads, there is a need to measure temperature at the level of the individual packs. Cheap versatile time temperature indicators (TTI) have been developed to give an indication of the temperature history to which individual packs have been exposed (See chapter 6). Even though these TTIs can give an indication of temperature abuse somewhere in the chain, they are not intended to reconstruct a complete temperature history and, therefore, cannot be expected perfectly to explain the resulting product quality.

To give an example, a TTI will not discriminate between one week's storage at 4ºC disrupted by either 12 hours of continuous 12ºC or six two-hour periods at 12ºC. However, for the packed product this might make a difference, especially as the product needs time to heat up. With 12 hours of continuous 12ºC the product will actually be at 12ºC for part of that time. Exposed to the six two-hour periods of 12ºC it depends on the time in between the warm periods how warm the product eventually will get. As a consequence, the two identical TTI readings from this example, can relate to two completely different qualities in the final product. Also the order of imposed temperatures will not make a difference to a TTI reading. However, for product quality, the order of the subsequent temperatures the product was exposed to might make a difference. For instance, pre-climacteric fruit generally responds less vigorously to temperature than the same fruit in its climacteric stage. With the effect of temperature on fruit physiology depending on the physiological stage of the fruit, two comparable temperature profiles (in terms of the total temperature sum) can have different effects in terms of product quality as this depends on the timing of the temperature relative to the physiological development of the fruit.

The other important aspects of the established MA conditions are the gas conditions, which are inextricably related to temperature. As for temperature, several indicators have been developed to monitor oxygen (O_2) and carbon dioxide (CO_2) in individual packages.[16] As with TTIs, these gas indicators give only an indicative value. The potential strength of the different types of indicators arises from their combined application where information on temperature and gas conditions together can give a better indication of the realised MA conditions in individual MA packs. However, defining MAP performance by the realised MA conditions in terms of temperature and gas conditions is only an indirect measure.

The ultimate unambiguous measure of the success of MAP is the final quality of the product. Some aspects of product quality can be related to volatiles produced by the product (ethylene as a measure of ripening stage, specific volatiles produced during spoilage or anaerobic conditions, etc.). This opens the door to adding product specific indicators to the range of indicators already available, resulting in the type of integrated freshness indicators as described in Chapter 7. Such freshness indicators might come close to giving a good evaluation of MAP performance incorporating several aspects of product quality into the equation. However, other aspects of product quality might never lend themselves to measurement in this way.

In spite of the importance of product quality as the ultimate determinant of MAP performance, this chapter will mainly focus on the effect of dynamic temperature conditions on the gas conditions developing inside MAP. Most of this is ruled by relatively simple physics. The link to product quality will be made when possible, but given the vast range of products and their different ways of responding to the applied MA conditions,[2, 11] no simple rules can be laid down on how dynamic temperature conditions will affect the quality of an MA packed product. For this, product specific knowledge is required on how product physiology responds to surrounding gas and temperature conditions in relation to the product at its own developmental stage. For now, one should be made aware that MAP performance is determined by more than just temperature and/ or the established gas conditions.

27.3 Temperature control and risks of MAP

Like most techniques, MAP comes with a number of potential risks that largely depend on the level of integral temperature control in a logistic chain.

27.3.1 Low oxygen

Generally, MAP is designed to create low levels of O_2 that give maximum benefit by suppressing the metabolism without getting into the range of O_2 levels that might induce fermentation. The critical O_2 level at which fermentation starts to occur is defined as the fermentation threshold.[18] The O_2 level in the package is the resultant of the influx through the package and consumption by the product. Both processes depend on temperature. O_2 consumption by the product generally increases much faster with increasing temperature (3- to 10-fold from 0–15°C[11]) rather than the permeance of the packaging material (2- to 3-fold from 0–15°C[9]). As a result, the steady state O_2 levels in the pack will decrease with increasing temperature. The O_2 level in a MA package designed to operate just above the fermentation threshold will, as a result of an increase in temperature, drop below this fermentation threshold; the product will start to ferment resulting in the development of off-odours and off-flavours. To make life more complicated, the fermentation threshold is not a constant but can vary with temperature.[1, 3, 18] When MA packed blueberries are exposed to a temperature increase, the drop in O_2 level is combined with an increase in fermentation threshold resulting in very little scope before anaerobic conditions are reached.

Polymeric packaging materials that have the same responsiveness to temperature as the packed product can prevent induction of anaerobic conditions following increased temperature. In such cases an increase in O_2 consumption rate is counteracted by exactly the same increase in O_2 influx through the packaging material with the steady state gas conditions becoming independent of temperature. One such example was described for capsicums packed using

LDPE film.[7] One can argue whether a temperature-independent atmosphere inside the package is important in its own right. The aim of MAP is to retain quality. With constant gas conditions at increasing temperatures, respiration rate and the rate of quality decay will still increase due to the increased temperature.

The O_2 levels in MA packages that make use of perforated films are even more sensitive to changes in temperature, as diffusion through the holes (i.e. diffusion through a barrier of standing air) is almost independent of temperature. An increase in temperature will induce increased O_2 consumption by the product without inducing a substantial increased influx through the packaging material, resulting in a fast drop of the steady state O_2 levels.

27.3.2 High carbon dioxide

Besides reducing O_2 levels in MAP, CO_2 levels are increased to further inhibit the product's metabolism.[2] High CO_2 levels also inhibit decay by suppressing the growth of microbes, although sometimes the CO_2 levels needed to suppress microbial growth exceed the tolerance levels of the vegetable produce packaged.[4,6] This identifies another dilemma in controlling the gas conditions in MAP.

For most polymeric packaging films the permeance for CO_2 is 2- to 10-fold higher than for O_2,[9] under aerobic conditions O_2 depletes much faster than CO_2 will accumulate. Assuming a respiratory quotient of 1 and a steady state O_2 level of 2kPa, the maximum achievable steady state CO_2 level varies between 2 and 9kPa depending on the film material. To achieve higher steady state CO_2 levels without inducing fermentation, microperforated films should be used that have comparable permeances for O_2 and CO_2. When using microperforated films, O_2 will deplete about as fast as CO_2 accumulates, such that the sum of O_2 and CO_2 partial pressure remains around 20kPa. A microperforated MA package designed for 2kPa O_2 can therefore generate CO_2 levels of around 18kPa. For soft fruit like strawberries, these high CO_2 levels are needed to prolong shelf-life.[8,13] However, after prolonged storage at high CO_2 (>15 kPa) CO_2 injury becomes visible from tissue defects and fermentation off-flavours.[10,14]

When exposing MA packages to dynamic temperature conditions there is a direct risk of inducing fermentation and an added secondary risk of inducing CO_2 damage due to the accumulating fermentative CO_2. Especially for microperforated packs where the permeance does not increase with temperature, the risk of inducing fermentation and consequently the accumulation of high CO_2 levels is much larger. Scavengers to constrain the accumulation of CO_2 (Chapter 3) might limit the secondary risk of CO_2 damage but cannot prevent the direct risk of inducing fermentation.

27.3.3 High humidity

With horticultural products generally consisting of up to 90% water and with their economic value often determined by the saleable weight of the crop, moisture loss needs to be limited under all conditions. Depending on how and for how long horticultural products are stored, they can easily lose up to 5% or more of their harvested weight before they reach the consumer. Generally, MAP films, either perforated or not, are relatively impermeable to water vapour and therefore quickly generate high humidity levels in the package atmosphere close to saturation.

With dropping temperatures the saturating vapour pressure drops as well and the colder air cannot continue to hold as much water vapour. Due to the extremely low water vapour permeance of most films, water vapour cannot leave the package fast enough, resulting in condensation in the package. This will happen with temperature fluctuations as small as 0.5ºC. In the heat of harvest activities, there is often not enough capacity properly to cool the product before packing. Packing warm product in plastic film, either for MA purposes or as liners in carton boxes, followed by cool storage, also results in extreme condensation inside the package thus wetting the product. The high humidity levels generated inside MAP prevent excessive water loss from the product retaining product quality, but the presence of free water following temperature fluctuations creates favourable conditions for microbes to flourish and break down this same product quality.

27.4 The impact of dynamic temperature conditions on MAP performance

As outlined in Chapter 16, different sources of variation interact with the performance of MA packages. In this chapter we discuss the effect of temperature variation over time, and how that can affect MA conditions and final product quality. To allow for some temperature flexibility, MAP should be designed to prevent those risks outlined in the previous sections (too low O_2, too high CO_2, too high humidity). The closer package atmospheres are targeted to what is feasible, the more likely temperature variation can induce these risks. How closely the theoretical ideal gas conditions can be approximated depends not only on the amount of temperature variation one wants to allow for but also on the amount of variation in other relevant aspects and on how temperature interacts with these. For instance, when aimed for O_2 levels are close to the fermentation threshold, depending on the variation in gas exchange rate, there is a risk that some of the packages result in O_2 levels dropping below the fermentation threshold.[5] Depending on the variation in the fermentation threshold itself and the variation in film permeability, tightness of seal, number of layers wrapped around the product, etc., the targeted safe gas conditions might need to be far removed from the theoretical ideal gas conditions.

With the number of variables encountered in MA packaging it is difficult to give full coverage of all aspects of the impact of dynamic temperature profiles on MAP, as this strongly depends on the specifications of the package of interest. Some of the important aspects are now discussed using simulations of MA packaging of shredded lettuce.

The quality of shredded lettuce is often limited by browning of the cut edges. This can be controlled by packaging in $< 1\%$ O_2 and 10% CO_2 atmospheres.[15,17] Shredded lettuce is a product with a relative high respiration rate and a high responsiveness to temperature as expressed by the energy of activation of respiration (see Chapter 16). As a reference we simulated MAP of pre-cooled lettuce stored at a constant 4°C and packed in a polymeric bag with an energy of activation of about one-third of the lettuce itself (Fig. 27.1b and c). Steady state gas conditions (10kPa CO_2 and 1kPa O_2) are reached after about two and a half days of storage with O_2 levels reaching 2 kPa after one-day storage. The realised steady state gas conditions correspond to the targeted optimum values for shredded lettuce. When one realises that minimally processed products generally have a limited shelf-life, the two and a half days needed to establish steady state conditions is relatively long.

For subsequent simulations an artificial dynamic temperature profile was created (Fig. 27.1a) consisting of one day at a constant 4°C followed by a two-day period of slow fluctuating temperature around 4°C and a subsequent one-day period of fast fluctuating temperature. After this, temperature was rapidly increased to a constant 12°C. Instead of assuming the lettuce to be pre-cooled, lettuce was assumed to be at room temperature when packed. As a result of packing warm lettuce, depletion of O_2 and accumulation of CO_2 was accelerated in comparison to the reference situation (Fig. 27.1b and c), the O_2 level of 2 kPa was reached only half a day after packing. Both O_2 and CO_2 show fluctuating levels in response to the fluctuating temperature of the environment.

The fluctuations in O_2 and CO_2 follow the fluctuations in temperature after a short delay, as the product needs time to warm up and cool down. The larger the thermal mass and heat capacity of the product, the slower the product will respond to fluctuations in temperature. This explains why gas levels follow slow temperature fluctuations more clearly than they follow the fast temperature changes. Another reason why gas levels do not follow fast temperature changes is because of the void volume in the package, which buffers the change in gas conditions.

The direction of the fluctuation in CO_2 level is the same as for temperature while the direction of the fluctuation in O_2 level is the opposite. As temperature increases, film permeance increases. However, the rate of O_2 consumption increases faster than the increase in film permeance resulting in dropping O_2 levels. With dropping O_2 levels fermentative CO_2 production increases resulting in increasing levels of CO_2. During the period of fluctuating temperature the same average gas levels are reached as seen before. When temperature is increased to 12°C, the O_2 level drops to 0.5kPa while CO_2 accumulates up to 18kPa due to the fermentation induced. It will be clear that such an increase in temperature to 12°C when a package is designed to operate around 4°C is fatal for the packed product. Depending on the product such temperature increase might irreversibly affect product quality.

Packing warm product has the advantage of rapidly establishing the targeted gas conditions. The downside is the induction of condensation as the warm

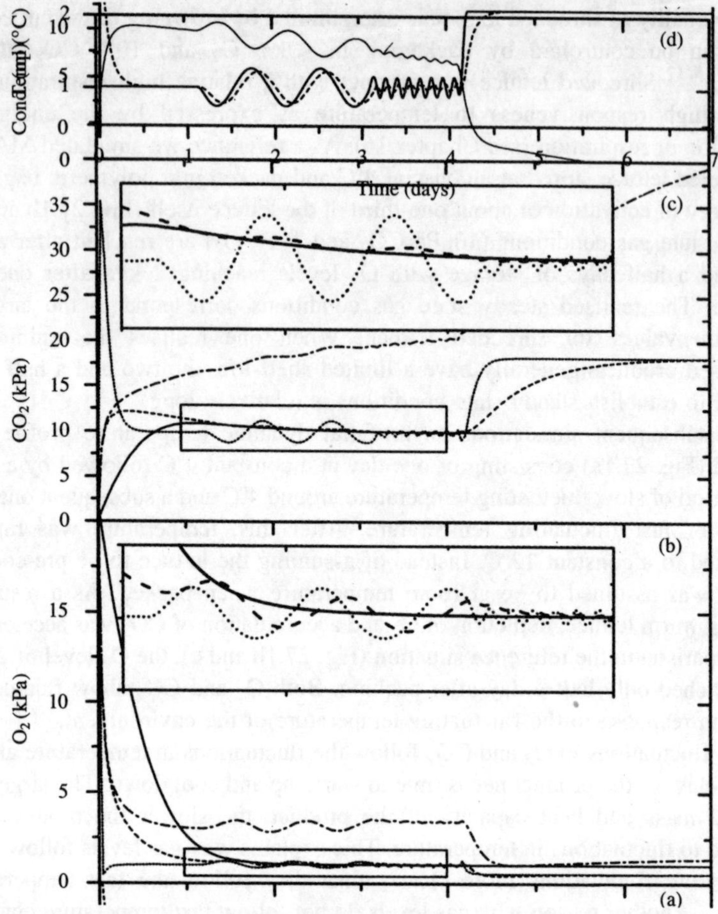

Fig. 27.1 Simulation results of MA packed shredded lettuce stored at a constant 4°C or at dynamic temperature conditions. (a) Temperature profile used for the dynamic temperature conditions; air temperature (——) and product temperature (– – –). (b) O_2 levels observed in the package during different simulation runs. (c) CO_2 levels observed in the package during different simulation runs. (d) Condensation formed during dynamic temperature conditions. The following simulations are depicted in (b) and (c): reference simulation of pre-cooled lettuce packed in polymeric film and stored at a constant 4°C (——), lettuce packed warm using polymeric film and stored at dynamic temperature conditions (– – –), lettuce packed warm and stored at dynamic temperature conditions but with a reduced void volume (·····), lettuce packed warm using microperforated polymeric film and stored at dynamic temperature conditions (·–·–·–). The boxes in (b) and (c) contain an enlargement of what is happening during the period with fluctuating temperatures.

product evaporates more water than the cold air can contain, quickly oversaturating the air with an excess water condensating on the inside of the cold packaging material (Fig. 27.1d). During the subsequent period, condensation slowly disappears again by evaporation and diffusion through the film. With fluctuating temperatures the amount of condensate fluctuates as well. Once temperature is increased to 12°C there is a fast drop in the amount of condensate. These relative fast changes are due to changes in the air saturation levels for water vapour as a function of temperature. This example shows that condensation can be rapidly induced but once present is hard to remove without increasing temperature again.

When the void volume in the package is eliminated (Fig. 27.1b and c) steady state gas conditions are rapidly realised within half a day. Because of the warm lettuce, the CO_2 level peaks to initially extremely high levels, rapidly disappearing when the product cools down. By reducing the void volume we have removed the buffering capacity of the system as a consequence of which the gas levels respond much more vigorously to the fluctuating temperature and also become more sensitive to fast fluctuations. When temperature is increased to 12°C, the increase in CO_2 is much faster than before.

When the film is replaced by a microperforated material, permeance of the packaging film has become almost independent of temperature. The resulting gas conditions are now different (Fig. 27.1b and c) with O_2 going towards 3kPa and CO_2 continuing to increase with time. The reason for not reaching steady state conditions is the relatively much lower permeance for CO_2 as compared to the permeance for O_2. Therefore the steady state conditions for CO_2 are at much higher CO_2 levels than before, which takes more time and the MA package never reaches this situation. Because of the temperature independency of film permeance the fluctuations in O_2 levels respond vigorously to changes in temperature. The final temperature increase to 12°C results in a drop of O_2 to 1kPa and an increase of CO_2 towards 40–50kPa. This increase is clearly the result of fermentative CO_2 production that, due to the low permeance for CO_2 is trapped inside the package. As the accumulating CO_2 has an inhibitive effect on the respiration of lettuce, O_2 consumption is inhibited, resulting in a subsequent slight increase of the O_2 level.

The outlined simulations were focused on a single average MA pack. When the dynamic temperature condition is applied to a batch of MA packages, each prepared package will differ slightly from another. Given that biological variance is the most variable parameter, we simulated a batch of 500 packages assuming 25% variation on product respiration rates, and 10% variation on packed product weight and film thickness (Fig. 27.2). The simulation result clearly shows the effect of variation in MA design parameters on the resulting MA gas conditions. At the same time it shows that variation in MA gas conditions depends on time and temperature. As, depending on the respiration rates, some packages establish MA conditions faster than others, initially a large variation in MA gas conditions is observed. Some packages reached a level of 2kPa O_2 within three hours after packing while others took two days to reach

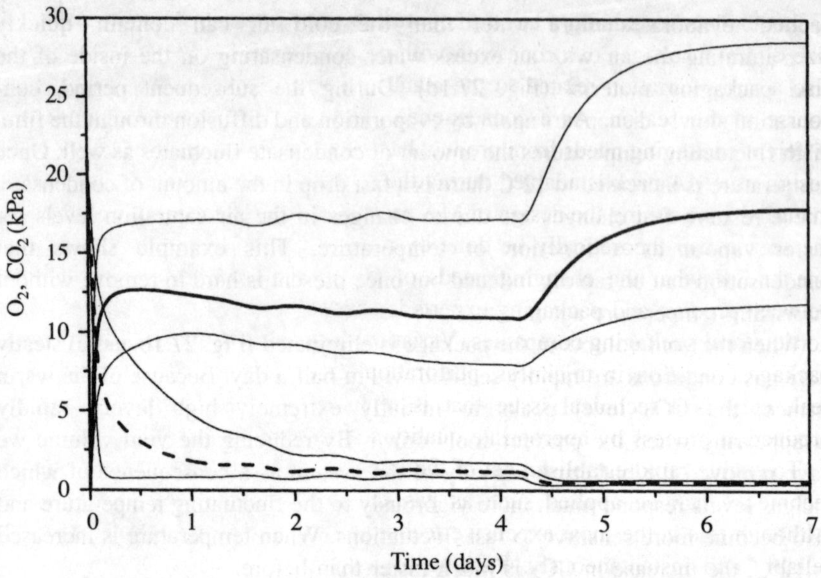

Fig. 27.2 Simulation results of 500 MA packages of shredded lettuce packed using polymeric film stored at dynamic temperature conditions (Fig. 27.1a). The average CO_2 (——) and O_2 levels (– – –) are plotted together with their 95% confidence intervals (·····).

this stage. By reducing the void volume, packing warm product, or flushing the package with nitrogen, the process of establishing MA conditions can be facilitated reducing the initial large variation in MA gas conditions.

The variation in O_2 levels is generally much smaller than the variation in CO_2 levels, especially when the temperature increase to 12°C induces fermentation. Under these conditions the high CO_2 levels in some of the packs will induce CO_2 injury. Controlling temperature in such a way that none of the packs develop fermentation would keep the variation in CO_2 levels within limits.

27.5 Maximising MAP performance

From the simulations in the previous section it became clear that it is of the utmost importance to prevent all sources of variation, whether that is temperature variation (time but also spatial variation), variation in the product (maturity differences causing variation in respiration rate or variation in the amount of product packed), or variation in the homogeneity of the package itself (variation in thickness, perforations, layers of wrapping, tightness of seal, etc.).

Biological variation tends to average itself out when large enough batches of product are packed. The variation between consumer MA packs containing a limited amount of product will be much larger than variation between MA packed pallets containing a large amount of product per pallet. So increasing the size of MA packages can cope with within-batch variation. Potentially, there is also a large variation related to the maturity of the packed product during the course of the season. As a consequence, early harvested product might have different packaging needs from product harvested later in the season. Ideally, the design of a MA pack is adapted during the season to cope with these changes in maturity. Fine-tuning the design of MA packages to these changing needs during the season can theoretically be done by relatively simple measures as long as one knows what the changing needs of the product are. Close co-operation between product and packaging experts is needed to develop guidelines for the horticultural packaging companies. Variation in the homogeneity of the physical package (variation in thickness, perforations, layers of wrapping, tightness of seal, etc.) is a technical issue that is relatively easy to control during the production process by appropriate quality control.

To enable rapid establishment of the intended MA conditions several simple techniques can be applied such as gas flushing the package before sealing. Although this is the most expensive technique, it can establish steady state gas reliably and instantaneously. Packing of warm fruit is the simplest way but comes with the risk of inducing lots of free water in the pack. Depending on how vulnerable the product is to microbial breakdown this might not be an option. Reducing the void volume is the third way of speeding up the process of establishing steady state gas conditions. However, this is not only speeding up the initial process of establishing steady state gas conditions but is increasing the overall responsiveness of the package allowing it rapidly to follow any temperature fluctuations in the logistic chain.

Temperature variation can be minimised only by an integral temperature control throughout the whole logistic chain from field to table. It is of the utmost importance to involve all partners in the chain in this integral temperature control as any temperature abuse might nullify the efforts of all other partners. In the end, the success of a chain is determined by the weakest link in the chain. When designing MAP for a certain product one should consider whether the potential benefits are worth the possible risks of a lack of temperature control. If this is questionable, one might consider designing a safe MA system by designing it for the highest temperature likely to be encountered. Although this approach does not utilise the maximum benefits it rules out all associated risks. In the end, MAP can only be successful when good temperature control can be guaranteed.

27.6 Future trends

The eventual success of MA depends on temperature control between the moment of packing and the moment of opening of the package by the consumer. Instead of relying solely on one's gut feeling when optimising MAP, a MAP model to simulate a package going through a logistic chain will give insight into the strong and weak parts of that chain in terms of temperature control.[12] It will make clear which parts of the chain are responsible for the largest quality losses of the packaged product and need improvement. It enables the optimisation of a whole chain considering the related costs and benefits. To operate such a model, information is needed on temperature, O_2 and CO_2 dependencies of gas exchange and on temperature dependency of film permeance.

With regard to the temperature effect on the oxidative respiration of different fruits and vegetables there is some data available.[11] Information on fermentation and on the effects of O_2 and CO_2 on gas exchange is much more fragmentary. This makes it almost impossible to identify at what temperature anaerobic conditions are going to be induced. Also a good database on permeance of packaging films that includes their temperature dependency is lacking. Before a new film can be used for MAP its temperature characteristics need to be identified at temperatures relevant to MAP (0–25°C). To be able to bring MAP to the next level and to predict what the effect of certain dynamic temperature conditions is on a particular MAP design it is vital to establish such databases on product and film characteristics. Without this elementary knowledge, MAP will remain at the level of trial and error. Ultimately, any temperature variation in the logistic chain should be ruled out. Meanwhile, technical solutions like temperature sensitive films are emerging to cope with some of the existing dynamic temperature conditions.

27.7 References

1. BEAUDRY R M, CAMERON A C, SHIRAZI A and DOSTAL-LANGE D L, 'Modified-atmosphere packaging of blueberry fruit: effect of temperature on package O_2 and CO_2', *J. Amer. Soc. Hort. Sci.*, 1992 **117** 436–41.
2. BEAUDRY R M, 'Effect of O_2 and CO_2 partial pressure on selected phenomena affecting fruit and vegetable quality'. *Postharvest Biology & Technology*, 1999 **15** 293–303.
3. BEAUDRY R M and GRAN C D, 'Using a modified-atmosphere packaging approach to answer some postharvest questions: Factors affecting the lower oxygen limit.' *Acta Hort.* 1993 **362** 203–12.
4. BENNIK M H J, *Biopreservation in modified atmosphere packaged vegetables*, Thesis Wageningen Agricultural University. ISBN 90-5485-808-7, 1997.
5. CAMERON A C, PATTERSON B D, TALASILA P C and JOLES D W, 'Modeling the risk in modified-atmosphere packaging: a case for sense-and-respond packaging', in: Blanpied. G D, Bartsch, J A and Hicks, J R (eds), *Proc. 6th Intl. Controlled Atmosphere Research Conference*, Ithaca NY, 1993.
6. CHAMBROY Y, GUINEBRETIERE M H, JACQUEMIN G, REICH M, BREUILS L and SOUTY M, 'Effects of carbon dioxide on shelf-life and post harvest decay of strawberries fruit', *Sciences des Aliments*, 1993 **13** 409–23.
7. CHEN X Y, HERTOG M L A T M and BANKS N H, 'The effect of temperature on

gas relations in MA packages for capsicums (Capsicum annuum L., cv. Tasty): an integrated approach'. *Postharvest Biology & Technology*, 2000 **20** 71–80.

8. COLELLI G and MARTELLI S, 'Beneficial effects on the application of CO_2-enriched atmospheres on fresh strawberries (Fragaria ananassa Duch.).' *Adv. Hortic. Sci.*, 1995 **9** 55–60.
9. EXAMA A, ARUL J, LENCKI RW, LEE L Z and TOUPIN C, 'Suitability of plastic films for modified atmosphere packaging of fruit and vegetables.' *J. Food Sci.*, 1993 **58** 1365–70.
10. GIL M I, HOLCROFT D M and KADER A A, 'Changes in strawberry anthocyanins and other polyphenols in response to carbon dioxide treatments.' *J. Agric. Food Chem.* 1997 **45** 1662–7.
11. GROSS, K C, WANG C Y and SALTVEIT M E, 'The Commercial Storage of Fruits, Vegetables, and Florist and Nursery Crops.' An Adobe Acrobat pdf of a draft version of the forthcoming revision to US Department of Agriculture, *Agriculture Handbook 66* on the website of the USDA, Agricultural Research Service, Beltsville Area <http://www.ba.ars.usda.gov/hb66.html> (November 8, 2002).
12. HERTOG M L A T M, PEPPELENBOS H W, TIJSKENS L M M and EVELO R G, 'Modified atmosphere packaging: optimisation through simulation', in: Gorny, J R (ed.), *Proc. 7th Intl. Controlled Atmosphere Research Conference*, Davis CA, 1997.
13. HERTOG M L A T M, BOERRIGTER H A M, VAN DEN BOOGAARD G J P M, TIJSKENS L M M and VAN SCHAIK A C R, 'Predicting keeping quality of strawberries (cv. 'Elsanta') packed under modified atmospheres: an integrated model approach.' *Postharvest Biology & Technology*, 1999 **15** 1–12.
14. KE D, GOLDSTEIN L, O'MAHONY M and KADER A A, 'Effects of short-term exposure to low O_2 and high CO_2 atmospheres on quality attributes of strawberries.' *J. Food Sci.* 1991 **56** 50–4.
15. LOPEZ-GALVEZ G, SALTVEIT M E and CANTWELL M, 'The visual quality of minimally processed lettuce stored in air or controlled atmospheres with emphasis on romaine and iceberg types.' *Postharvest Biology & Technology*, 1996 **8** 179–90.
16. SMOLANDER M, HURME E and AHVENAINEN R, 'Leak indicators for modified-atmosphere packages.' *Trends in Food Science & Technology*, 1997 **8** 101–6.
17. SMYTH A B, SONG J and CAMERON A C, 'Modified atmosphere packaged cut iceberg lettuce: effect of temperature and O_2 partial pressure on respiration and quality.' *J. Agric. Food Chem.* 1998 **46** 4556–62.
18. YEARSLEY C W, BANKS N H and GANESH S, 'Temperature effects on the internal lower oxygen limits of apple fruit.' *Postharvest Biology & Technology*, 1997 **11** 73–83.

Index

absorbers see scavengers
absorption 22, 144–5
  see also packaging-flavour interactions
absorption-type antimicrobial packaging 51
acetaldehyde scavengers 514
acetate buffer 392
acetic acid 129, 298–9
acid tolerance response (ATR) 250–1
Actipak study 6, 13, 459, 460, 461–8
  classification of active and intelligent systems 461–5
  inventory 461
  microbiological safety and shelf-life extending capacity 465–6
  recommendations for legislative amendments 467–8
  toxicological, economic and environmental evaluation 466–7
activated carbon scavengers 38, 39, 41
activation energy 347
active packaging 2, 6, 337–8, 459–60
  antimicrobial packaging *see* antimicrobial packaging
  classification 461–5
  and colour control *see* colour control
  consumers and *see* consumers
  current research 13
  current use 12
  fish *see* fish
  future trends 16–17
  integrated active packaging, storage and distribution 5, 18, 204, 535–49
  legislation *see* legislation
  meat *see* meat
  NMBP *see* non-migratory bioactive polymers
  scavengers *see* scavengers
  scope 469
  techniques 6–11
activity
  locus of 74–5, 77–8
  reduced 78
  specific activity 60
adsorption 22, 144–5
  clay 175, 176
  irreversible 179–80
  molecular sieves 177–8
  silica gel 173, 174, 175
adsorption capacity 184–5
adsorption rate 185
aerobic bacteria 23–5
Aeromonas
  hydrophila 212–14, 237–8, 250, 290
  species 212–14, 235
aflatoxin 24
Ageless absorbers 29, 42, 280, 388, 389–90
agricultural production practices 252, 255
Air Liquide 191–2
air pressure 537–8

alarm systems 540–1
alcohol dehydrogenase 93
alcohol oxidase 30
aldehydes 165
  absorbers 8, 43, 393–4
algae 320
alginates 393
alpha-tocopherol (vitamin E) 44
ambient moisture 181
Ambitemp indicator 108
amines 165
  absorbers 8, 43, 393–4
  biogenic 130–1, 137
amino-cyclopropane carboxylic acid (ACC) 35
ammonia 130
Amosorb oxygen scavenger 32
amylopectin 522
anaerobiosis 221
ANICO bags 43, 165, 394
anthocyanins 220, 417, 426–7, 429
  degradation 422–3
antibiotics 56, 57
anti-browning agents 432
antimicrobial agents 52–8
  carbon dioxide 208–10, 387
  naturally occurring 244
antimicrobial dipping 245–7, 253
antimicrobial packaging 8–11, 16, 50–70
  factors affecting effectiveness 60–4
  fish 391–2, 395–6
  NMBP 92–4
  system construction 58–60
antimicrobial preservative releasers 8, 9
antioxidant releasers 9
antioxidants 57, 392
Apack trays 531
APNEP system 87
apolar polymers 164–5
appearance 366–8
  see also colour control
applicability of TTIs 111
argon MAP 191–2, 431, 480
  testing 193–6
Arrhenius equations 163, 347, 348
Arun Foods Ltd 202–3
ascorbate oxidase 216, 218
ascorbic acid (AA) 30–1, 216–18, 432
Aspergillus parasiticus 23–4
ATP degradation products 131
Australia 12, 14–15
Auto-ID Center 544
automatic identification 542
availability of technology 79

Bacillus
  cereus 241, 289–90
  subtilis 333, 334
bacteria 320
  aerobic 23–5
  LAB 231, 249–50, 255, 300–1, 370–1, 386
  spore-forming 293–4
  see also microorganisms
bacteriocins 54, 57, 292, 300–1
bananas 536–7
barium oxide 180
barrier layers 26
barrier properties 153–6
barriers, functional 509–12
beef 404–8, 411
Bentonite 174
benzoic acid 297, 393
benzophenone 506–7
'best before' date 558–9
beta-carotene 218
betalains 417
bioactive materials 71
  see also non-migratory bioactive polymers
biocides 483–4
biodegradable packaging materials 519–34
  current applications 529–33
  developing novel materials 524–6
  future trends 533
  legislative issues 526–9
  problem of plastic packaging waste 519–20
  range of biopolymers 520–4
biogenic amines 130–1, 137
Bioka 30
biological variation 345–6, 350–1, 356, 572–3
biomimetic antimicrobial polymers 83–4
Biopol 523, 531–2
biopolymers *see* biodegradable packaging materials
biosensors 95, 137
biphenyl 484
bone 367
bottle closures 33
bound moisture 182
Brassica vegetables 221–2
Brochothrix thermosphacta 386
bromothymol blue 134
browning 220, 221, 423–4, 425, 428, 430
  inhibitors 432
bulk MAP 374–5

bulk modification 85–6
bulk transport 483
butylated hydroxytoluene (BHT) 392

cadaverine 130–1
calcium hydroxide (slaked lime) 42, 179
calcium oxide (quicklime) 179–80
calcium sulphate 179
calicivirus 239, 247
Campylobacter jejuni 235, 238–9, 241
capillary condensation 184
carbodiimide coupling method 89–90
carbohydrates 151–2
carbon dioxide 94, 190, 279, 480
  absorption in meat 403
  CAP for meat products 376–7
  generation inside package 300, 387
  high carbon dioxide
    and colour stability of fresh produce 426–8, 429–31
    and dynamic temperature conditions 567, 569–72
  high oxygen MAP 197–8
  indicators 11, 282
  MAP and food preservation 312, 313
  MAP for meat products 372, 373–4
  and microbial growth 131, 208–10, 248, 372, 387
  pH change indicator 134
  pH and solubility of 298
  SGS 300, 387
carbon dioxide emitters 9, 387, 388–90
carbon dioxide scavengers 7, 41–2
carbon monoxide 367–8, 372–3, 374, 375–6, 379, 431
carboxymyoglobin 366, 368, 375–6
carotenogenic fruits 420
carotenoids 218–19, 417
casting methods 63
catechol oxygen scavenging sachets 30
catecholases 423
cellular penetration 209
cellulose acetate (CA) 43
certification procedure 474, 475
challenge test 504–5
cheese slices 449–50, 451, 452
chicken balls 454–6
chicken legs, raw 450–4
chitosan 55, 57, 80–2, 393
chlorine 245–7
chlorine dioxide 58
chlorophyll 417, 427–8
  degradation 419–22
chloroplasts 417, 420
cholesterol reductase 91–2
cholesterol remover 8
chroma 418, 424
chromoplasts 417, 420
CIELAB colour measurement system 418
citric acid 298–9
clay
  moisture regulation 172, 174–6
  starch-clay nano-composites 524–6, 527
climacteric fruits 420
Clostridium botulinum 240, 250
  fish 395–6
  measuring pathogen risks 235, 236
  microbial safety of MAP 210–11
  oxygen scavengers and 24–5
  preservation technologies 289–90
coffee 41
collagen sausage casings 533
colorimeters 417–18
colorimetric nose 138
colour control
  fruit and vegetables 416–38
    colour changes and stability 349, 417–18
    colour measurement 418–19
    colour stability and MAP 424–6
    combining low oxygen, high carbon dioxide and other gases 429–32
  future trends 423
    high carbon dioxide effects 426–8
    process of colour change 419–24
  meat 401–15
    cured meat 409–10, 410, 411–12, 414
    factors affecting colour stability 371, 402–3
    fresh meat 404–8, 411, 412, 413
    future trends 412–14
    modelling impact of MAP 403–10
    pre- and post-slaughter factors 410–12
colours 480–1
combined preservation techniques 287–311
  combining MAP with other techniques 288–93
  consumer attitudes 301–2
  future trends 302–3
  heat treatment and irradiation 292, 293–6
  hygienic conditions 291–3
  $Na_2CaEDTA$ 299–300

preservatives 292, 297–9
protective microbes and bacteriocins 300–1
soluble gas stabilisation 292, 300
specific groups of microorganisms 289–91
combined UV/ozone systems 317–18, 327–32, 335–6
compostability 526–8
computer simulations *see* modelling
concentration 147–8
condensation 181, 568, 569–72
consumers 15–16, 550–62
attitudes towards combining MAP with other technologies 301–2
attitudes towards novel packaging 555–9
convenience and optimization 443, 447, 448–9, 449–58
and future of novel packaging 559–62
general attitudes 555–6
methods for testing responses 552–5
new packaging technologies and 550–1
problems in testing responses 551–2
segments and behaviour 547
contamination levels, and recycling 500–3
controlled atmosphere packaging (CAP) 365, 378–9
meat products 376–7
controlled release 60–1, 63, 64
corona discharge (CD) 87, 321–2
costs
indirect packaging costs 448, 448–9, 449–58
NMBP 79
package optimization 443, 447–8, 448–9, 449–58
TTIs 111
coupling chemistries 88–90
critical temperature indicators (CTIs) 105, 107, 108
critical temperature/time integrators (CTTIs) 105, 107, 108
Cryovac OS1000 32
Cryptosporidium parvum 241, 242, 247
crystallinity 146–7
cured meat 410, 411–12, 414
ham 409–10
Cyclospora cayetanensis 241, 242
cysts, parasitic 241, 242, 247, 320

dairy products 92–3, 449–50, 451, 452, 510–11
dangerous substances 490
Darex oxygen scavenging technology 33
date marks 485, 486, 487, 558–9
dedicated MAP models 351–2
dedicated tests 477
degreening 420, 421
dehydroascorbic acid (DHA) 216–18
demonstrations 560
deoxymyoglobin 366, 367
desiccants *see* moisture regulation
destructive leak test methods 277
destructurised (thermoplastic) starch 522, 526, 527, 530
dew point 181
diacetyl 135
diffusion 77–8, 145, 163
meso-level modelling 340–1, 342, 347–8
micro-level modelling 342–4, 349
diffusion-based TTIs 108–9
dimensioning MAP 352–3
dimethylamine 130
dinitrogen oxide *see* nitrous oxide MAP
dipping, antimicrobial 245–7
discolouration 371
disinfectants 245–7
disintegration test 528–9
distance 537–8
distribution 63
and butchering of meat 377–9
integrating active packaging, storage and 5, 18, 204, 535–49
monitoring shelf-life during 112–16
optimization using TTIs 116–21
package leak indicators during 279–82
reducing pathogen risks 254, 255–6
'do not eat' symbol 489–90
drying 172
recycling processes 505–6
see also moisture regulation
dynamic temperature conditions 563–75
future trends 573–4
impact on MAP performance 568–72
maximising MAP performance 572–3
temperature control and risks of MAP 566–8

EAN UCC 544–5
economic evaluation 467
edible coatings 255
antimicrobial packaging and 59–60, 61
chitosan 81–2
fish 392–3
elasticity, modulus of 146, 147
electrochemical sensors 323, 324
electronic article surveillance (EAS) tags 541, 542

electronic field date recorders 537
electronic labelling 17, 138, 541, 542
electronic nose 137–8, 513
Enterobacteriaceae 249
environment
  evaluation in Actipak study 467
  green plastics *see* biodegradable packaging materials
  packaging waste legislation 491–2
  stresses and package optimization 443, 444, 446–7, 448–9, 449–58
enzymatic TTIs 109, 110
enzyme-based oxygen scavengers 30
enzymes 53–4, 71, 209
  see also non-migratory bioactive polymers
equilibrium modified atmosphere (EMA) 190
Escherichia coli (*E. coli*) 80, 240
  factors affecting survival 242–3, 246, 250
  indicator 136
  measuring pathogen risks 235, 237
  new germicidal techniques 324–5, 332, 333, 334
essential oils 292, 299, 392
ethanol
  antimicrobial agent 56, 57, 58, 94
  indicator of freshness 129–30, 135
ethanol emitters 9, 12, 391
equilibrium relative humidity (ERH) 182
ethylene 344–5, 421–2
ethylene scavengers 7, 12, 34–41, 561
  economic aspects 41
  measuring ethylene sorption 40
  principle of ethylene sorption 37–40
  role of 34–7
European Union 12
  ILSI document 515
  legislation *see* legislation
  recycling projects 514
  RFID research 545
EVOH (ethylene vinyl alcohol) 25
experimental designs 555
external indicators 11
extrusion 63, 525
exudate losses 369

facultative anaerobes 247–8, 370–1
fat 150–1, 152–3
fermentation threshold 566
Fick's law 347
films
  with antimicrobial properties 8–11
  choice for MAP 327, 373
  developing new films 353
  gas impermeable 376
  microperforated 190, 567, 571
  multilayer PET 511–12
  oxygen scavenging 31–3
  perforated 347, 567
  permeability 347, 353, 356, 402–3
Finland 538–40
First In First Out (FIFO) system 116, 118–20, 121
fish 384–400
  active packaging
    antimicrobial and antioxidant applications 391–2
    atmosphere modifiers 387–90
    edible coatings and films 392–3
    taint removal 393–4
    water control 390–1
  future trends 395–6
  intelligent packaging applications 394–5
  microbiology of fish products 385–7
  monitoring shelf-life 113–15
flame treatment 87
flavonoids 219, 220
flavour
  absorbers of off flavours 8
  control 368–9
  packaging-flavour interactions *see* packaging-flavour interactions
flavour absorption
  factors affecting 145–9
  modelling 160–4
flavour modification 156–8
flavour-releasing materials 9, 44, 482, 561
flavouring substances 482–3
Flory-Huggins theory 160
flow pack machines 198, 325–6
fluorescent dye 281
foamed polymer packaging 530–1
focus group discussions 553
food additives
  European legislation 478–81, 487
  US regulations 75–6
food contact materials (FCM) 460, 470–8, 529
  basic rules for migration tests 474–8
  Framework Directive 470–1
  plastics directives 472–4
  symbol for 471–2
Food and Drug Administration (FDA) 504, 514–15

Food Hygiene Directive 485–7
Food Imitation Directive 490–1
food matrix 149–53
food poisoning hazards 314
food safety *see* safety
Food Sentinel System 136
food spoilage *see* spoilage
foodborne infections 232–3, 240–1
form-fill-seal machines 198, 325–7
free volume 146
freezing 294–5
fresh meat 411, 412, 413
  beef 404–8
fresh produce 1
  colour control *see* colour control
  effect of MAP on nutritional quality 215–22
  novel MAP applications 189–207
    future trends 202–4
    high oxygen MAP 196–202
    novel MAP gases 191–2
    testing 193–6
  reducing pathogen risks 231–75
    factors affecting pathogen survival 242–51
    future trends 254–6
    improving MAP 251–4
    measuring pathogen risks 232–42
  supply chain 535–8
Fresh-Check 109–10
Freshilizer 29
FreshLock 42
FreshMax 33
Freshness Check 108
freshness indicators 11, 17–18, 127–43, 565
  biosensors 137
  compounds indicating quality 128–32
  electronic nose 137–8
  future trends 138
  pathogen indicators 11, 17–18, 134, 136, 395, 396
  types of 132–6
Freshness Monitor 109–10
FreshPad 10
FreshPax 29–30
FreshTag 135, 395
Frisspack 38
frozen foods 115–16
fruit juices 43, 156–8
fruit and vegetables *see* fresh produce
functional barriers 509–12
functionalisation of polymers 79, 85–8
fungi 194, 195
fungicides 54, 57
gas composition
  colour stability of fresh produce 429–32
  colour stability of meat 402–3
  high oxygen MAP 197–8
  pathogen survival 247–8
gas diffusion
  meso-level modelling 340–1, 342, 347–8
  micro-level modelling 342–4, 349
gas exchange 356, 402–3
  meso-level modelling 341, 342
  micro-level modelling 341, 348, 349
gas flushing 573
gas indicators 565
  carbon dioxide 11, 282
  oxygen 11, 280–2
gases
  absorption in meat 403
  antimicrobial agents 56, 58
  EU legislation and packaging gases 480
  MAP for fish 387–90
  novel MAP gases 191–2
  SGS 292, 300, 387–8, 388–90
  see also under individual gases
GEMANOVA model 405–8
general attitude studies 555–6
Germany 515
germicidal techniques 312–36
  future trends 332–6
  installation of UV/ozone systems 327–32
  integration with MAP 325–32
  new 314–15
  ozone 315, 321–5
  ultraviolet radiation 314–15, 315–20
Giardia lamblia 241, 242
glass transition temperature 146, 147
glucalase 30
glucose 129, 136
glucose oxidase 30, 93, 393
glucosinolates 221–2
glutaraldehyde coupling 89
good agricultural practices (GAP) 252, 255
good manufacturing practices (GMP) 252, 252–3
grapefruit juice 43
green plastics *see* biodegradable packaging materials
growth seasons 535–7
GTAG 544
gypsum 179

Hafnia alvei 250
ham, cured 409–10
ham pizza 457–8
Hazard Analysis and Critical Control Point (HACCP) system 103, 252, 253, 254, 485
headspace gas composition *see* gas composition
heat stress response 250–1
heat transfer 339, 346–7
heat treatment 292, 293–4, 303
helium 279, 480
hepatitis A viruses (HAV) 239, 241, 291
hexamethylenetetramine 56
high density polyethylene (HDPE) 498–9, 502–3, 530
  monolayer bottles 512–13
high impact polystyrene (HIPS) 512
high oxygen MAP 189–90, 191, 192
  applying 196–202
  fresh produce applications 201–2
  future trends 202–4
  optimal gas levels 197–8
  packaging materials 199–201
  produce volume/gas volume ratio 198–9
  safety 197
  temperature control 201
  testing 193–6
histamine 130–1
homeostasis 289
horizontal form-fill-seal (HFFS) machines 198, 325–6
hot-fill conditions 508, 568, 569–72, 573
hue angle 418, 424
humectant salts 178–9
humidity 538
  modelling MAP 341, 344, 349, 356
  relative 149, 181, 184
  temperature control and risks of MAP 567–8
  see also moisture regulation; water
humidity absorbers *see* moisture absorbers
hurdle technology 51, 52, 289
hydration, water of 179
hydrogen 279
hydrogen sulphide 131–2, 135
hydroxypropyl cellulose 393
hygiene 254
  combining MAP with other preservation technologies 291–3
  European legislation 485–7

I-Point indicator 108, 109
imitation, food 490–1
immobilisation-type antimicrobial packaging 51–2
immobilised bioactive-type polymers 84–95
  antimicrobial packaging/shelf-life extension 92–4
  applications 90–5
  developing NMBP by immobilisation 85–90
  in-package processing 90–2
  intelligent packaging 94–5
in-house control 541
in-package processing 90–2
incineration, suitability for 447, 448–9, 449–58
indigenous microflora 244, 249–50
indirect packaging costs 448, 448–9, 449–58
information 560
information banks 547
inherently bioactive synthetic polymers 79–84
  chitosan 80–2
  UV irradiated nylon 10, 82–3
insect damage 23
insulating materials 10
integrated active packaging, storage and distribution 5, 18, 204, 535–49
  alarm systems and TTIs 540–1
  future trends 545–7
  role of packaging in supply chain 538–40
  supply chain for perishable foods 535–8
  traceability 18, 542–5, 546
intelligent packaging 2, 6, 459–60
  classification 461–5
  consumers and *see* consumers
  current research 13
  current use 12
  fish 394–5
  freshness indicators *see* freshness indicators
  future trends 17–18
  integrating active packaging, storage and distribution 5, 18, 204, 535–49
  leak detectors *see* leak detectors
  legislation *see* legislation
  NMBP 94–5
  pathogen indicators 11, 17–18, 134, 136, 395, 396

scope 469
technologies 11–12
TTIs *see* time-temperature indicators
internal indicators 11
Internet 545–6
interviews 553–4
ionising radiation (irradiation) 255, 292, 295–6, 302
invisible oxygen indicators 281
iron-based oxygen scavengers 27–30, 388
irradiation (ionising radiation) 255, 292, 295–6, 302
ISO 9000:2000 quality management standard 103

Japan 12, 14

kimchi 41, 42
kinetic modelling 113–16

labelling 302
European legislation 471, 481, 485, 487–9, 489–90
labels 8
oxygen scavenging 33
lactase-active packaging 8, 77, 90–1
lactic acid 129, 298–9, 521
lactic acid bacteria (LAB) 231, 249–50, 255, 300–1, 370–1, 386
lactoperoxidase system of milk 92–3
lactose-free milk 77, 90–1
lactose remover 8
leak detectors 276–86
detection during processing 277–9
future trends 282–3
indicators during distribution 279–82, 561
leakage, product safety and quality 276–7
Least Shelf-Life First Out (LSFO) system 116–20, 121
legislation 13–15, 200
Australia 14–15
European 15, 75, 459–96
Actipak study 6, 13, 459, 460, 461–8
biocides and pesticides 483–4
biodegradable plastics 526–9
current legislation and recommendations for change 468–70
food additives 478–81, 487
food contact materials 460, 470–8
food flavouring 482–3
food hygiene 485–7
food labelling 471, 481, 485, 487–9, 489–90
initiatives to amend 461–8
Nordic report 460, 468
product safety and waste 489–92
Japan 14
USA 14, 75–6
lettuce, shredded 568–72
lidded trays 375
light
bacteria deactivation 332–3, 334
exposure and colour stability of meat 405–7
light-activated oxygen scavenging films 31–2
limonene 157–8, 501–2, 503, 510–11
limonin 43, 165
Lifelines Freshness Monitor 109–10
lipase 424
lipid oxidation 368–9, 392
modelling 162–4
packaging-flavour interactions 159, 162–4
lipoxygenase 424
Listeria monocytogenes 240, 289–90, 395
factors affecting survival 242, 245–6, 248, 249, 251
measuring pathogen risks 233–6
microbial safety of MAP 211–12
LLDPE film 159
locus of activity 74–5, 77–8
logistics
optimizing logistic chains 353–4
performance and packaging optimization 442–3
low density polyethylene (LDPE) 154, 155, 157, 158, 498, 530
low-pressure mercury vapour lamp 317
low temperature preservation 292, 294–5, 396
luminescent dyes 281
lycopene 218–19
lysozyme 92

macro-level modelling 339, 346–7
magnesium chloride 178
magnesium oxide 180
malic acid 298–9
marketing 76
properties and packaging optimization 443, 447, 448–9, 449–58
marking/tagging 18

mass transfer 144–5, 339, 346–7
meat 1
  active packaging 365–83
    CAP 376–7
    control of appearance 366–8
    delaying microbial spoilage 369–71
    flavour and texture 368–9
    future trends 377–9
    MAP technology 372–6
    temperature and storage life 371–2
  colour control *see* colour control
  consumer attitudes towards novel packaging 557, 558
mechanical integrity 63–4
mechanical strength 446, 448–9, 449–58
mechanisms of action 78
mechanistic models 355
membranes, cell 210
melanins 220
mercury lamp, low-pressure 317
meso-level modelling 340–1, 342, 347–8
methyl cellulose 393
metmyoglobin 366, 367, 368
metmyoglobin reduction activity 366–7
Michaelis Menten approach 348
microaerophilic microorganisms 247–8
micro-level modelling 341–6, 348–51
microorganisms 36
  Actipak study and safety 465–6
  combining MAP with other preservation technologies 287–301
  fish 385–7, 395–6
  growth kinetics 60–1
  integrating MAP with new germicidal techniques *see* germicidal techniques
  meat
    delaying microbial spoilage 369–71
    temperature and storage life 371–2
  mechanisms of carbon dioxide inhibition 208–10, 387
  microbial growth indicators *see* freshness indicators
  modelling MAP 345, 350
  novel MAP applications for fresh produce 191, 192, 193–5
  pathogen risks *see* pathogen risks
  safety of MAP 210–14
microperforated films 190, 567, 571
microwave heating modifiers 10
microwave susceptors 10
microwave UV lamps 317–18, 327–32, 335–6
migration 144–5
  evaluation and recycling 507–8
  from oxygen scavengers and moisture absorbers 461–2, 463
  see also packaging-flavour interactions
migration limits 460, 472, 529
migration tests, basic rules for 474–8
milk 92–3
milk bottles 512–13
milk products 510–11
minerals, finely dispersed 38–40, 41
minimal processing 252–3
  and pathogen survival 244–7
  and stress responses 250–1
minimum durability dates 485, 486, 487, 588–9
MINIPAX sachet 44, 394
misuse of plastics 500–3
mixing 77–8
mixtures of flavour compounds 147–8
modelling 337–61
  advantages and disadvantages of models 354–5
  applying models to improve MAP 352–4
  current MAP-related models 346–51
  dedicated models 351–2
  dynamic changes in headspace gas composition 402–3
  flavour absorption 160–4
  future trends 356
  impact of MAP and colour stability in meat 403–10, 412–13
  macro-level 339, 346–7
  meso-level 340–1, 342, 347–8
  micro-level 341–6, 348–51
  migration 507–8
  principles and methods 338–46
modified atmosphere packaging (MAP) 12, 22–3, 365
  colour stability in fresh produce 424–32
  combining with other preservation techniques 287–311
  defining MAP performance 564–6
  dimensioning 352–3
  and food preservation, food spoilage and shelf-life 312–14
  gases *see* gas composition; gases
  improving through modelling 337–61
  integrating with new germicidal techniques 312–36
  leak detection 276–86
  novel applications for fresh produce 189–207
  packaging optimization 449–58

performance under dynamic
temperature conditions 563–75
product safety and quality 208–30
reducing pathogen risks 231–75
technologies for meat products 372–6
modulus of elasticity 146, 147
moisture 537–8
ingress 182–3
sources in packaging 181–3
see also humidity; moisture regulation; water
moisture absorbers 7, 477, 489
fish 390–1
migration from 461–2, 463
moisture regulation 172–85
clay 172, 174–6
fish 390–1
future trends 185
humectant salts 178–9
irreversible adsorption 179–80
molecular sieves 172, 176–8
planning a moisture defence 180–5
selection of desiccant 183–5
silica gel 173–4, 175
moisture vapour transmission rate (MVTR) 182–3
molecular sieves 172, 176–8
molecular size 148
molecular structure 148
Monitor Mark TTI 108–9
monolayer recycled plastics 512–13
monolithic system 62–3, 64
Monte Carlo simulation 118–20, 350–1
Montmorillonite 174, 176
moulds 231, 320
oxygen scavengers and 23–5
multilayer adsorption 184
multilayer PET 511–12
Multivac Rollstock machine 326–7
mycotoxins 23–4
myoglobin 135, 366–8, 404, 411

$Na_2CaEDTA$ 299–300
nano-composites 524–6, 527
naringin 90, 165
native starch 522
natural extracts 55, 57–8
naturally occurring antimicrobials 244
Nature Works PLA 521, 532
nisin 301
nitric oxide 431–2
nitrite 411–12, 414
nitrogen 190, 313, 373, 374, 376, 480
nitrogen compounds, volatile 130, 135
nitrous oxide MAP 191–2, 480
testing 193–6
noble gases 431
non-climacteric fruits 420
non-destructive package leak testing 276, 277–9
non-migratory bioactive polymers (NMBP) 71–102
benefits 72–7
food processor's perspective 77
marketing aspects 76
regulatory advantages 75–6
technical 72–5
current limitations 77–9
future trends 95
inherently bioactive polymers 79–84
polymers with immobilised bioactive compounds 84–90
applications 90–5
non-respiring food products 215
Nordic report 460, 468
Norwalk virus 239, 241
nutritional quality *see* quality
nylon, UV irradiated 10, 82–3

odour scavengers 8, 42–4, 393–4
off flavours, absorbers of 8
oil 150–1, 152–3
one-sided contact 476–7
optimization 2, 5, 441–58
examples of 449–58
improving decision-making 458
issues in 442–4
VTT Precision Packaging Concept 444–9
Orega bag 40
oregano essential oil 299
organic acids 53, 57, 129, 298–9
orientated polypropylene (OPP) 199–200
overall migration limit 460, 472, 529
Oxbar 33
oxidation
lipid oxidation *see* lipid oxidation
prevention by oxygen scavengers 23
oxygen 312, 313, 480
barrier films 199–200
and growth of *Listeria monocytogenes* 212
high oxygen MAP *see* high oxygen MAP
indicators 11, 280–2
low oxygen MAP
colour stability of fresh produce 424–6, 429–31

oxygen (*cont.*)
  dynamic temperature conditions 566–7, 569–72
  meat
    appearance 366–7
    colour stability 404–8, 409–10
    MAP technology 372, 374
  permeability 26, 153–6
oxygen emitter/carbon dioxide scavenger dual action sachet 203
oxygen scavengers 7, 12, 22–34, 56, 481
  consumer acceptance 556–7
  economic aspects 33–4
  fish 387, 388–90
  meat 377, 378
  migration from 461–2, 463
  role of 23–5
  selection of right type of 25–34
Oxyguard 32
oxymyoglobin 366, 367
Oxysorb 30–1
ozone 315, 321–5
  combined UV/ozone systems 317–18, 327–32, 335–6
  future trends 332–6
  integration with MAP 327–32

PA 530
package integrity 276
  see also leak detectors
package weight/product weight ratio 446, 448–9, 449–58
packaging: role in supply chain 5–6, 538–40
packaging-flavour interactions 144–71
  and active packaging 164–6
  factors affecting flavour absorption 145–9
  flavour modification and quality 156–8
  lipid oxidation 159, 162–4
  modelling flavour absorption 160–4
  role of differing packaging materials 153–6
  role of food matrix 149–53
packaging gases *see* gas composition; gases
packaging materials
  green plastics *see* biodegradable packaging materials
  high oxygen MAP 199–201
  improving MAP to reduce pathogen risks 253–4
  recycling *see* recycling
role in packaging-flavour interactions 153–6
  waste *see* waste
packaging optimization *see* optimization
palletised packs 339, 346–7
palmitates 393
Paperfoam 531
parasites 241, 242, 247, 320
partial freezing 294–5
partition coefficients 161–2
passive MAP 338
pathogen indicators 11, 17–18, 134, 136, 395, 396
pathogen risks 231–75
  factors affecting pathogen survival 242–51
  future trends 254–6
  improving MAP to reduce 251–4
  measuring 232–42
  see also microorganisms
peptides 71
  see also non-migratory bioactive polymers
perforated films 347, 567
permeability 204
  films 347, 353, 356, 402–3
  modelling dynamic changes in headspace gas composition 402–3
  oxygen 26, 153–6
permeability coefficient 183
permeability rate 183
permeation 144–5
  see also packaging-flavour interactions
peroxidases 424
pesticide emitters 9
pesticides 484
pH 209, 244, 292, 298
  freshness indicators sensitive to pH change 132–4
phenolic compounds 219–21
pheophytin 420–1
Photobacterium phosphoreum 386–7
photodiode 318–19
physical integrity 63–4
pichit film 391
pigment composition 418–19
pizza 457–8, 557
plasma processing 87–8
plaster of paris 179
plastics
  directives 472–4, 475
  packaging-flavour interactions *see* packaging-flavour interactions
  recyclability of 498–500
    see also recycling

polarity 148, 161–2
polyalkylene imine (PAI) 43
polycaprolactone 523, 530
polycarbonate (PC) 153, 154, 155, 158
polyesters 148, 165
  synthetic 523, 530
polyethylene (PE) 153
polyethylene glycol (PEG) 84, 88
polyethylene naphthalate (PEN) 498–9
polyethylene terephthalate (PET) 530
  packaging-flavour interactions 153, 154, 155, 156, 158
  recycling 492, 497, 498–9, 500–3, 513
    monolayer bottle 512
    multilayer PET bottles 511
    multilayer PET films 511–12
    testing safety and quality 505–8
polyhydroxyalkanoates (PHAs) 523, 531–2
polylactic acid (PLA) 84, 520, 521, 530, 532
poly-1-lysine 84
polymer-based TTIs 109–10
polymeric spacers 88
polymers
  antimicrobial 55, 57
  apolar 164–5
  NMBP *see* non-migratory bioactive polymers
  swelling 86
polyolefins 148
polyphenol oxidase (PPO) 191, 192, 196, 220, 422–3, 423
polypropylene (PP) 153, 154, 155, 498, 530
  three-layered PP cups 510–11
polystyrene (PS) 498–9, 530, 530–1
polyunsaturated fatty acids (PUFAs) 31
polyvinyl chloride (PVC) 498–9
polyvinylalcohol (PVOH) 523–4, 531
polyvinylidene chloride (PVDC) 26, 199–200
positive lists 460, 472
potassium permanganate scavengers 37–8, 39, 41
potassium sorbate 297, 392
poultry products 1, 368, 450–6
preformed tray and lidding film (PTLF) machines 198
preservation 1, 5–6
  combining MAP with other preservation techniques 287–311
  MAP and 312–14
  traditional methods 51
preservatives 292, 297–9
pressure differential leak testing 278
processing 77
  minimal 244–7, 250–1, 252–3
    package leak detection during 277–9
    reducing pathogen risks 255–6
product/headspace volume ratio 198–9, 409–10
Profresh sachet 394
protective microbes 244, 300–1
  see also lactic acid bacteria
proteins 71, 532–3
  see also non-migratory bioactive polymers
protozoan parasites 241, 242, 247, 320
provitamin A carotenoids 218–19
Pseudomonas
  aeruginosa 80, 332
  species 249–50, 370
psychotrophic microorganisms 247–8, 450
pulsed UV light 333–5
putrescine 130–1

qualitative consumer testing 552–3, 553–4
quality 208–30, 303
  compounds indicating 128–32
  effect of MAP on nutritional quality 215–22
    fresh produce 215–22
    non-respiring food products 215
  extending quality of fish 396
  freshness indicators *see* freshness indicators
  improving MAP through modelling 345, 355, 356
  intelligent packaging techniques 11–12
  leakage, product safety and 276–7
  packaging-flavour interactions and 156–8
quantitative consumer testing 553, 553–5
questionnaires 554
quinones 220

radio frequency identification (RFID) 542–5, 546
  standardisation 543–4
ready meals 1–2, 293–4, 396
recovery 491, 492, 528
recycling 491, 492, 497–518, 528
  conventional recycling processes 505–6
  direct contact applications 512–13
  future trends 513–14

recycling (*cont.*)
indirect contact applications 509–12
recyclability of packaging plastics 498–500
recyclability of plastic packaging 500–4
super-clean processes 503–4, 506–7
technology 503–4
testing safety and quality of recycled material 504–9
using recycled plastics in packaging 509–13
redox indicators 26, 280
reducing compounds 280
regeneration of silica gel 173–4
regulation 64, 75–6
see also legislation
relative humidity 149, 181, 184
release-type active packaging systems 6–8, 9
release-type antimicrobial packaging 51
reliability 111
remelting 506
research 13, 256
respiration 35–6, 190
respiration rate 204
ripening 35–6, 349–50
roasted chicken balls 454–6

sachets 8
oxygen scavenging 27–31
safety 76, 208–30, 288, 303, 481
consumers and 560–1
European legislation 489–92
high oxygen MAP 197
leakage and 276–7
microbial safety of MAP 210–14
'Safety and Information in Packaging' programme 538–40
Safety Monitoring and Assurance System (SMAS) 122
Salmonella
species 235, 238, 240, 242–3, 246, 250, 290
typhimurium 80
salt (sodium chloride) 172, 178, 292, 297–8
scavengers 6, 7–8, 22–49
carbon dioxide 7, 41–2
ethylene *see* ethylene scavengers
future trends 44–5
moisture *see* moisture absorbers
odour 8, 42–4, 393–4
oxygen *see* oxygen scavengers
seasons, growth 535–7
security tags 17, 138
self-cooling cans/containers 10
self-heating cans/containers 10
sensitivity studies 354
sensory quality *see* quality
sensory testing 508–9
shape 490
shelf-life 303, 485
Actipak study and extending capacity 465–6
antimicrobial packaging and extension of 92–4
high oxygen MAP and extending 203
MAP and food preservation 312–14
novel MAP applications and fresh produce 193, 194
packaging optimization 441–2, 444–5
TTIs and monitoring during distribution 112–16
Shelf Life Decision System (SLDS) 120–1
Shewanella putrefaciens 386
Shigella
sonnei 240
species 239
shock 537–8
shredded lettuce 568–72
silica gel 173–4, 175
silicates 172, 176–8
silver nitrate 56
silver zeolite 56, 391
simulants 474, 476–7
size 490
slaughtering facilities 378
Smellrite/Abscents 44
sniffing device 513
see also electronic nose
snorkel type (ST) machines 198, 374–5, 377
sodium chloride (salt) 172, 178, 292, 297–8
sodium lactate (Na-lactate) 292, 299
softdrink bottles 511, 512
solubility 62–3
soluble gas stabilisation (SGS) 292, 300, 387–8, 388–90
solvent casting 63
sorbic acid 297
source control 499, 500
spacers, polymeric 88
specific activity 60
specific migration limit (SML) 460, 529
specific spoilage organisms (SSOs) 386
spoilage

fish 385–7, 395–6
MAP, food preservation and 312–14
meat 369–72
delaying microbial spoilage 369–71
mechanisms 314
micro-level modelling 345
see also microorganisms
spore-forming bacteria 293–4
stability 72–4
standards, suitability with respect to 446, 448–9, 449–58
Staphylococcus aureus 324–5, 326, 328, 332, 333, 334
starch 522
native 522
thermoplastic 522, 526, 527, 530
starch-based polymers 520, 522–4, 530, 531
starch-clay nano-composites 524–6, 527
Stefan-Maxwell equations 348
stock rotation 116–21
storage 63, 445
integrating active packaging, distribution and 5, 18, 204, 535–49
meat
colour stability 402–3, 404–8
effects of temperature 371–2
reducing pathogen risks 254
strain variation 251
Strecker degradation reaction 41
stress responses 250–1, 255–6
STRIPPAX sachet 44, 394
succinimidyl succinate (SS) esters 89–90
sulphur dioxide emitters 9
sulphuric compounds 131–2
super-chilling 294–5, 396
super-clean processes 503–4
cleaning efficiency 506–7
supermarkets 377–8
supply chain 535–49
intelligent supply chain 18
for perishable foods 535–8
role of packaging in 5–6, 538–40
see also integrated active packaging, storage and distribution
surface functionalisation 79, 85–8
surface-treated packaging materials 10
sweeteners 480–1
swelling, polymer 86
symbols
'do not eat' 489–90
for food contact materials 471
synthetic polyesters 523, 530
tagging 18
Temchron indicator 108
temperature 181, 313, 537–8
control and high oxygen MAP 201
glass transition temperature 146, 147
low temperature preservation 292, 294–5, 396
MAP performance under dynamic temperature conditions 563–75
meat
colour stability 404–8
effects on storage life 371–2
packaging-flavour interaction 149, 158
reducing pathogen risks 254, 256
storage temperature and pathogen survival 242–3
surface temperature after irradiation 335
temperature-sensitive films 10
Tempil indicator 108
tenderisation 369
texture 368–9
thermoform-fill-seal (TFFS) machines 192, 326–7
thermoplastic starch (TPS) 522, 526, 527, 530
three-layered PP cups 510–11
3M Monitor Mark 108, 108–9
time 537–8
time-temperature indicators (or integrators) (TTIs) 11, 94–5, 103–26
consumer acceptance 557–9
current TTI systems 108–10
defining and classifying 104–6
development of 106–8
fish 394, 396
future trends 121–2
integrated packaging, storage and distribution 540–1
legislation and 473–4, 486
MAP performance 565
maximising effectiveness of 111
monitoring shelf-life during distribution 112–16
optimization of distribution and stock rotation 116–21
requirements for 106
total immersion 476–7
toxicity 490
toxicological evaluation 466–7
Toxin Alert indicator 395
Toxin Guard 136
traceability 18, 542–5, 546
tracer gas leak detection 279

traditional preservation methods 51
transportation 483, 536–7, 537–8
trimethylamine (TMA) 130, 386
trimethylamine-oxide (TMAO) 385–6
tristimulus colorimetric measurements 418–19
2–in-1 44, 394
tyramine 130–1

ultraviolet (UV) radiation 255, 314–15, 315–20
  combined UV/ozone lamp 317–18, 327–32, 335–6
  future trends 332–6
  UV irradiated nylon 10, 82–3
  UV-light absorbers 8
UNIFAC group contribution model 160
'unintended' atmospheres 247–8
United States (USA) 12, 14, 75–6
  FDA 504, 514–15
'use by' date 485, 486, 487

vacuum chamber (VC) machines 192
vacuum packaging 12, 22–3
variation 345–6, 350–1, 356, 572–3
Verifrais package 390
vertical form-fill-seal (VFFS) machines 198, 326
vibration 537–8
viruses 239–42, 243, 247, 320
visual oxygen indicators 280, 281–2
vitamin C 216–18
vitamin E 44
VITSAB TTIs 109, 110, 394
volatile nitrogen compounds 130, 135
volume control 488–9, 573
volume ratio 198–9, 409–10
VTT Precision Packaging Concept 441–58
  basic requirements 444–5
  calculating the coefficients 448–9
  examples of use 449–58
  importance of package characteristics 448
  improving decision-making 458
  optimization results 449
  package and storage combinations 445
  scoring selected package types 445–8

warm product, packing 508, 568, 569–72, 573
washing
  fresh produce 245–7
  recycling technology 503–4, 505–6
  see also super-clean processes
waste
  legislation 491–2
  problem of plastic packaging waste 519–20
  volume of package waste 446, 448–9, 449–58
water
  modelling diffusion and loss 344, 349
  packaging-flavour interaction 149, 154–5
  resistance of starch-based products 522–3
  see also humidity; moisture regulation
water activity 182
water of hydration 179
weight control 488–9
wet chemical oxidation procedures 86–7
Wexler's equation 178
white light 332–3, 334
wholesale 491
wine barrels 482

xenon flashlamp 332–3, 334

yeasts 231, 320
Yersinia
  enterocolitica 212–14, 238, 241, 246, 290
  species 235

$Zero_2$ oxygen scavenger 32